TRANSGENIC PLANTS

Volume 1

Engineering and Utilization

TRANSGENIC PLANTS

Volume 1

Engineering and Utilization

EDITED BY

Shain-dow Kung

The Hong Kong University of Science and Technology
Kowloon, Hong Kong
and
Center for Agricultural Biotechnology
Maryland Biotechnology Institute
Department of Botany
The University of Maryland
College Park, Maryland

Ray Wu

Section of Biochemistry
Molecular and Cell Biology
Cornell University
Ithaca, New York

Academic Press, Inc.
Harcourt Brace Jovanovich, Publishers
San Diego New York Boston
London Sydney Tokyo Toronto

This book is printed on acid-free paper. ♾

Academic Press, Inc.
1250 Sixth Avenue, San Diego, California 92101-4311

United Kingdom Edition published by
Academic Press Limited
24–28 Oval Road, London NW1 7DX

Library of Congress Cataloging-in-Publication Data

Transgenic plants / edited by Shain-dow Kung, Ray Wu.
p. cm.
Includes indexes.
Contents: v. 1. Engineering and utilization – v. 2. Present status and social and economic impacts.
ISBN 0-12-428781-6 (v. 1). – ISBN 0-12-428782-4 (v. 2)
1. Transgenic plants. 2. Crops–Genetic engineering. 3. Plant genetic engineering. I. Kung, Shain-dow. II. Wu, Ray.
SB123.57.T73 1992
631.5'23–dc20 92-11016
CIP

PRINTED IN THE UNITED STATES OF AMERICA
92 93 94 95 96 97 QW 9 8 7 6 5 4 3 2 1

Contents

Part II

Transgenic Plants and Basic Research

Contributors

Kenneth A. Barton, Agracetus, Inc., Middleton, Wisconsin 53562
Donald Boulter, Department of Biological Sciences, University of Durham, Durham, DH1 3LI, United Kingdom
Benjamin A. Bowen, Department of Biotechnology Research, Pioneer Hi-Bred International, Johnston, Iowa 50131
Karen Broglie, Agricultural Products Department, E. I. Du Pont de Nemours and Company, Wilmington, Delaware 19880
Richard Broglie, Agricultural Products Department, E. I. Du Pont de Nemours and Company, Wilmington, Delaware 19880
Jun Cao, Plant Science Center, Biotechnology Building, Cornell University, Ithaca, New York 14853
Ilan Chet,[1] Agricultural Products Department, E. I. Du Pont de Nemours and Company, Wilmington, Delaware 19880
Shirley A. Coomber,[2] E. I. Du Pont de Nemours and Company, Wilmington, Delaware 19880
Gloria M. Coruzzi, Department of Biology, New York University, New York, New York 10003
Xavier Delannay, Monsanto Company, St. Louis, Missouri 63198
Kenneth A. Feldmann,[3] E. I. Du Pont de Nemours and Company, Wilmington, Delaware 19880
Robert T. Fraley, Monsanto Company, St. Louis, Missouri 63198
Angharad M. R. Gatehouse, Department of Biological Sciences, University of Durham, Durham, DH1 3LI, United Kingdom

[1] Current affiliation: Department of Pathology and Plant Microbiology, Faculty of Agriculture, The Hebrew University, Rehovot, 76100 Israel
[2] Current affiliation: Division of Biosphere Sciences, King's College, London, London W8 7AH, United Kingdom
[3] Current affiliation: Department of Plant Sciences, University of Arizona, Tucson, Arizona 85721

Stanton B. Gelvin, Department of Biological Sciences, Purdue University, West Lafayette, Indiana 47907

Vaughan A. Hilder, Department of Biological Sciences, University of Durham, Durham DH1 3LI, United Kingdom

Maud A. W. Hinchee, Monsanto Company, St. Louis, Missouri 63198

Thomas K. Hodges, Department of Botany and Plant Pathology, Purdue University, West Lafayette, Indiana 47907

Stephen H. Howell, Boyce Thompson Institute for Plant Research, Ithaca, New York 14853

Barnabás Jenes, Institute for Plant Sciences, Agricultural Biotechnology Center, Gödöllö, 2101 Hungary

Jacobus M. Kaper, Microbiology and Plant Pathology Laboratory, Plant Sciences Institute, ARS, United States Department of Agriculture, BARC-W, Beltsville, Maryland 20705

Ganesh M. Kishore, Monsanto Company, St. Louis, Missouri 63198

Shain-dow Kung, The Hong Kong University of Science and Technology, Kowloon, Hong Kong, and Center for Agricultural Biotechnology, Maryland Biotechnology Institute, and Department of Botany, The University of Maryland, College Park, Maryland 20742

Brian A. Larkins, Department of Plant Sciences, University of Arizona, Tucson, Arizona 85721

Peter B. McGarvey, Microbiology and Plant Pathology Laboratory, Plant Sciences Institute, ARS, United States Department of Agriculture, BARC-W, Beltsville, Maryland 20705

Michael J. Miller, Agracetus, Inc., Middleton, Wisconsin 53562

Helen Moore, Section of Biochemistry, Molecular and Cell Biology, Biotechnology Building, Cornell University, Ithaca, New York 14853

Stephen R. Padgette, Monsanto Company, St. Louis, Missouri 63198

Steven W. Ritchie, Department of Botany and Plant Pathology, Purdue University, West Lafayette, Indiana 47907

Dominique Roby,[4] Agricultural Products Department, E. I. Du Pont de Nemours and Company, Wilmington, Delaware 19880

Samuel S. M. Sun, Department of Plant Molecular Physiology, University of Hawaii, Honolulu, Hawaii 96822

Fong-Ying Tsai, Department of Biology, New York University, New York, New York 10003

Frank F. White, Department of Plant Pathology, Kansas State University, Manhattan, Kansas 66506

Ray Wu, Section of Biochemistry, Molecular and Cell Biology, Biotechnology Building, Cornell University, Ithaca, New York 14853

Wanggen Zhang, Section of Biochemistry, Molecular and Cell Biology, Biotechnology Building, Cornell University, Ithaca, New York 14853

[4] Current affiliation: Laboratoire de Biologie Moleculaire des Relations Plantes-Microorganismes, CNRS/INRA, 31326 Castanet Tolosan Cedex, France

Preface

Biotechnology offers new ideas and techniques applicable to agriculture. It uses the conceptual framework and the technical approaches of molecular biology and plant cell culture systems to develop commercial processes and products. With the rapid development of biotechnology, agriculture has moved from a resource-based to a science-based industry. Indeed, plant breeding has been dramatically broadened by the introduction of genetic engineering techniques based on knowledge of gene structure and function. This has ushered agriculture into a new era where it joins ranks with the most sophisticated of biological sciences in using molecular and cell culture approaches that promise to yield a new generation of plants of almost any desirable trait. Biotechnology has taken us from the era of hybrid plants to the era of transgenic plants. We have advanced far beyond the "Green Revolution" to the "Gene Revolution."

Since 1985, more than fifty species of transgenic plants have been engineered. However, reports on this work have been published in different journals and in various formats. We have attempted to harvest the first group of transgenic crops by assembling the important information in a two-volume work on "Transgenic Plants." These volumes represent a significant milestone in plant/agricultural biology, promote the practical application of recombinant DNA technology, and assist in transforming the agricultural industry.

Reports on transgenic plants that have been engineered to date include the development of new techniques for transformation, basic study on a specific gene, and crop improvement for a specific cause. A number of publications on transgenic plants focus mainly on the development of efficient and reproducible methods for the transformation and regeneration of major crops. To date, the plants transformed have been predominantly dicot, particularly from the family Solanaceae, with only a few monocots and woody plants. Thus, the utilization of transgenic plants to improve agriculture is still in its infancy.

Volume 1, Engineering and Utilization, has been organized into three parts. Part I provides an overview of basic techniques for engineering transgenic plants. Part II presents selected examples of the value of transgenic plants to basic research. Part III includes some of the most significant contributions transgenic plants have made thus far to crop improvements, ranging from herbicide tolerance to the improvement of seed storage proteins. In Volume 2, Present Status and Social and Economic Impacts, organized into three major parts, Part I covers transgenic monocotyledonous plants: rice, maize, and asparagus. Part II is directed to the transgenic dicotyledonous plants, including the most popular family of Solanaceae, legumes, oilseeds and vegetables, woody plants, cotton, and *Arabidopsis*. Part III focuses on the social and economic impacts of transgenic plants. These volumes were designed so that they will be suitable for use in classrooms as well as in research laboratories.

We thank the authors for their valuable contributions. Their generosity, efforts, and patience are deeply appreciated. We also thank the staff of Academic Press for their help in making these volumes possible.

Shain-dow Kung
Ray Wu

Contents of Transgenic Plants, Volume 2

PRESENT STATUS AND SOCIAL AND ECONOMIC IMPACTS

Introduction: From Hybrid Plants to Transgenic Plants

Shain-dow Kung

The Hong Kong University of Science and Technology
Kowloon, Hong Kong
and
Center for Agricultural Biotechnology
Maryland Biotechnology Institute
Department of Botany
The University of Maryland
College Park, Maryland 20742

Food is essential to sustain life. An ancient Chinese motto expresses this truth: "food is the haven for the people, agriculture is the foundation of the country." The emergence of civilization was tied to agriculture as our ancestors moved from hunting and gathering to cultivating and improving crops for their primary sustenance. Since all food is derived directly or indirectly from plants, selection and upgrading of plant products constituted the first step toward modern plant breeding. Plant breeding became an important branch of science immediately after the rediscovery of Mendel's laws of genetics (Borlaug, 1983).

Conventional plant breeding techniques have resulted in dramatic crop improvements and will continue to provide future benefit. Among the 3000 species of plants used for food, only 29 species are major crops. These species include eight cereals, seven legumes, seven oil seeds, three root crops, two sugar crops, and two tree crops (Borlaug, 1983). In addition, these 29 basic food crops are supplemented by about 15 major species of vegetables and 15 major species of fruits (Borlaug, 1983). Collectively, these provide the proteins, calories, vitamins, and minerals necessary for the human diet. However, there are strong pressures for further improvements in crop quality and quantity exerted from population growth, social demands, health requirements, environmental stresses, and ecological considerations. World population is predicted to reach eight billion by the year 2010. To feed three billion additional people in the next twenty years will require dramatic increases in crop production, a formidable task by any standard. Conventional plant breeders and related scientists have worked diligently and skillfully to upgrade quality and raise yields by employing various crop improvement techniques with commendable results. For example, the combined production of 17 major crops in the United States increased over 242% between 1940 and 1980, while acreage under cultivation increased approximately 3% (Borlaug, 1983). Had crop yields remained at 1940 levels until 1980, 177 million additional acres of good United States cropland would have been required to produce the 1980 harvest. Notwithstanding these impressive gains in productivity, there are limitations to conventional plant breeding technology, which have stimulated the development of more advanced technologies.

It is the intention of the author to introduce this book by describing the road we have traveled from the era of hybrid production to a new era of transgenic plant engineering. This transition was a response to the limitations of conventional hybridization techniques. In order to produce hybrid plants with much wider ranges of genetic variabilities, somatic hybrid plants were generated (Carlson, 1973). The failure of somatic hybrid plants to be of immediate and economic value, in turn, stimulated the development of transgenic plant engineering. These remarks focus on the leap from hybrid to transgenic plants viewed from the perspective of a plant molecular biologist.

I.1 Plant Breeding

Since 1900, Mendel's laws of genetics provided the scientific basis for plant breeding. Plant breeding principles consist of identifying and selecting desirable traits and combining these into one individual plant. Since all traits are controlled by genes located on chromosomes, plant breeding can be considered as the manipulation of chromosomes. In general, there are four major ways to manipulate plant chromosomes. First, similar chromosomes can be sorted out and retained in one individual plant to reach a homozygous state, a method termed pure-line selection. Second, different chromosomes can be combined together to obtain a heterozygous state. This method is termed hybridization. Third, new genetic variability can be introduced through spontaneous or artificially induced mutations. Finally, polyploidy also contributes to crop improvement.

I.1.1 PURE-LINE SELECTION

Pure-line selection involves selection from the extremes for the desired phenotype. In genetic terms, a selected and inbred population is more homozygous than its wild relatives. Pure-line selection generally involves three distinct steps. First, a large number of selections are made from the genetically variable original population. The number of initial selections should be as great as consideration of time, expense, space, and competitive plant breeding projects will permit. Second, progeny rows are grown from the individual plant selections for observational purposes. After obvious elimination, the selections are grown over a shorter or longer period of years to permit observations of performance under different environmental conditions for making further eliminations. Finally, when the breeder can no longer decide between lines solely on the basis of observation, he/she must turn to replicated traits, comparing among remaining selections with established commercial varieties in relative yielding ability and other aspects of performance. This stage of evaluation lasts at least three years. If a large number instead of a single pure-line are likely to be retained, this procedure is usually referred to as mass selection.

One of the most important features of pure-lines is the great precision with which they produce themselves. Associated with this advantage are some potential problems in unmasking some harmful genes as well as eliminating some desirable genes.

I.1.2 HYBRIDIZATION

In contrast to pure-line selection, the most frequently employed plant breeding technique is hybridization. Hybridization demonstrated that crops such as maize could be inbred for six or seven generations until there was no further reduction in vigor or size. When these highly inbred plants were hybridized with other inbred varieties, very vigorous, large-sized, large-fruited plants were produced. This led to the origin of hybrid maize in 1919 (Chrispeels and Sadava, 1977), then the most significant improvement in American agriculture at that time. The term "heterosis" was used to describe this phenomenon of hybrid vigor.

The first step in hybrid production is to generate homozygous inbred lines. This is normally done using self-pollinating plants where pollen from male flowers fertilizes female flowers on the same plants. Once the pure-lines are generated, these are outcrossed. A problem with such hybridization is that the farmer must buy new hybrid seed each year.

Another useful breeding technique is backcrossing. Backcrossing makes it possible to transfer specific genes from one plant variety to another. In this way, desirable characteristics of one variety can be combined with those of another variety. This trick circumvents the problem of trying to select simultaneously for many traits in the same variety.

I.1.3 MUTATIONS

Instead of relying only on the introduction of genetic variability from the wild species gene pool, another alternative is the introduction of spontaneous mutations or those induced by either chemicals or radiation. The mutant obtained is tested and further selected to meet standards of an established cultivar or used as a donor in a crossing program. This method is applicable for both modes of pollination. It has not been widely used in breeding programs because the great majority of mutations carry undesirable traits.

1.1.4 POLYPLOIDY

Normal plants are diploid. Plants with three or more complete sets of chromosomes are common, and are referred to as polyploids. The increase of chromosome sets can be induced by applying the chemical colchicine, which disrupts spindle formation during cell division so that daughter chromosome sets remain in the same cell. Thus, the chromosome number is doubled. Generally, the main effect of chromosome doubling is to increase size. Morphologically, polyploids tend to be larger than diploids, with thicker leaves, and they may respond differently to environmental conditions. Genetically, the plants may also differ. For example, an autotetraploid with two allelic forms of a gene at a locus (A and a) can have five genotypes in the population (AAAA, AAAa, AAaa, Aaaa, aaaa). This increases its genetic variability. Autopolyploids contain genes similar to their diploid progenitors, whereas allopolyploids combine the gene content of two different species. The latter possesses higher potential capacity for variation.

Success in using plant breeding principles and agricultural practices to improve crops reached its peak when high-yielding wheat and rice strains were cultivated in the 1960s (Borlaug, 1983). The doubling or tripling of productivity of these important crops in Asia signaled a new agricultural revolution in the developing countries. This breakthrough in food production was termed the "Green Revolution" to describe the social, economic, and nutritional impact of the new high-yielding wheat and rice strains (Chrispeels and Sadava, 1977). Norman Borlaug was awarded the Noble Peace Prize in 1970 for his contribution in breeding new high-yielding strains of cereals. Since that time the Green Revolution has been both praised and damned. In the 1960s, as use of the new strains spread rapidly in Asia, claims were made that many underdeveloped countries would soon be self-sufficient in cereal grains. However, these new high-yielding strains were heavily dependent on fertilizer and irrigation and required energy-intensive investments. In the early 1970s poor weather, high energy costs, and global economic constraints slowed the progress of the revolution considerably. Thus, the Green Revolution illustrates both the potential and limitations of conventional plant breeding technology.

1.2 Hybrid Plants

The first drastically improved crop species was probably a hybrid plant. Plant breeding was recently described as the selection of plants with desired traits after sexual exchange of genes by cross-fertilization between two parents (Goodman *et al.*, 1987). This clearly indicates that hybridization is an essential technique in plant breeding and the resulting plants are hybrid plants. There are, in general, two types of hybrid plants: interspecific and intergeneric hybrids. Beyond this biological boundary hybridization cannot be accomplished due to sexual incompatibility. The successful breeding of wheat (McFadden, 1930), tomato (Bohn and Tucker, 1939), and soybean (Newell and Hymowitz, 1982) are noted examples of interspecific hybridizations. The origin of many tobacco species was believed to arise from the interspecific hybridizations followed by the doubling of the chromosomes (Gray *et al.*, 1974). On the other hand, some modern crop species, such as rapeseed and certain wheat strains originated in nature by hybridization between different species or genera (Simmonds, 1976).

The most notable and successful hybrid plant ever produced is the hybrid maize more than half a century ago (Borlaug, 1983). Many improved elite hybrids with continually higher yields, increased disease and insect resistance, and shorter and stronger stalks suitable for mechanical harvesting have since been developed.

The first maize hybrid plant was developed after recognition that inbreeding in maize leads to reduced vigor in the following generation and that vigor can be restored by crossing. In 1908 Shull reported that the hybrid plants produced by a cross between two different pure-lines quadrupled the yield, from 20 to 80 bushels per acre (Chrispeels and Sadava, 1977). By 1919, the first commercial hybrid maize was available in the United States, and two decades later, nearly all the maize was hybrid, as it is to this day (Chrispeels and Sadava, 1977).

There are two technical steps in the production of hybrid maize plants: the production of desirable homozygous lines and the crossing of these lines. Hybrid maize plants produced are the result of double crossing. One parental plant possessing the unique property of cytoplasmic male sterility was used to avoid inbreeding. Currently, hybrid maize plants produced in the United States can have yields as high as 130 bushels per acre.

Although interspecific as well as intergeneric hybrid plants can be produced in certain species, there are still severe limitations in obtaining hybrid plants between many genera and even species because of sexual incompatibility. Sexual incompatibility, in most cases, is genetically controlled and regulated by the pollen–stigma recognition processes (Heslop-Harrison, 1978). This recognition is vitally important in preserving species identity by preventing cross-hybridization between genetically unrelated species. Otherwise, distinctions among plant families, genera, and species would disappear. However, this incompatibility limits our ability to widen genetic variability within a given plant species because it precludes the natural genetic exchange between distant or unrelated species. Even within similar species, there may be self- or cross-incompatibility. To overcome this problem, techniques for hybridization which bypassed the sexual cross were developed, the subject of the next section.

1.3 Somatic Hybrid Plants

Somatic hybrid plants are hybrids derived from the fusion of somatic cells. Since all fusions are possible in principle, regardless of the extent of genetic relatedness, somatic hybrid plants theoretically offer unlimited possibilities for genetic exchanges. The development of somatic hybrids was a new approach to plant breeding based on the unique property of plant cells. All plant cells can be isolated, cultured, and induced to regenerate into complete plants identical to the parent from which these cells were obtained, a property known as totipotency.

Cell fusion was developed after the successful culture of a large number of plant cells stripped of their walls. The resulting naked plant cells have been referred to as protoplasts (Shepard *et al.*, 1983). In the 1960s, Takebe and coworkers prepared tobacco protoplasts for efficient viral infection experiments. Protoplasts were soon adapted for regeneration (Nagata and Takebe, 1971). Success in regenerating complete plants from protoplasts eventually led to attempts to combine cells with different genetic background. In 1972, Carlson *et al.* were the first group to succeed in fusing tobacco protoplasts from two genetically compatible *Nicotiana* species, *N. glauca* and *N. langsdorffii*. Initially the fused products were termed parasexual hybrids (Carlson *et al.*, 1973). Later the term somatic hybrid was adopted (Shepard *et al.*, 1983).

The dramatic success in producing somatic hybrids generated great excitement and great expectations in the 1970s. Hopes were high that this technology could permit unlimited generic exchange. Plant biologists speculated about the possibility of producing a somatic hybrid between soybean and maize that could fix nitrogen, contained high protein content, and produced high yield. Furthermore, a somatic hybrid combining tomato, potato, and tobacco could produce an efficient plant whose leaves, fruits, and tubers could be harvested and economically utilized. To move toward this goal, many fusions between protoplasts of phylogenetically unrelated species were attempted. Notable examples as reviewed and compiled by Ouellette and Cheremisinoff (1985) included fusion between soybean and tobacco, soybean and barley, soybean and maize, soybean and clover, soybean and alfalfa, soybean and rapeseed, carrot and barley, potato and tomato, carrot and parsley, sorghum and maize, and even plant and human cells (Lima-De-Faria *et al.*, 1977). In most cases, not surprisingly, these attempts failed. No progress could made beyond the simple fusion stage since the initial fusion between any two protoplasts of different background is achievable but subsequent division and growth of the fused product is almost impossible.

The success in fusing two sexually compatible *Nicotiana* species led to the fusion between two sexually incompatible petunia species, *Petunia parodii* and *P. parviflora* (Power *et al.*, 1980). Interspecific protoplast fusion can be considered as a conceptual extension of interspecific sexual crosses of incompatible species, a substantial contribution. Beyond this plant breeding technology, success of intergeneric protoplast fusion was very limited. Many fertile intergeneric somatic hybrid plants were generated in the Brassicaceae family (Glimelius *et al.*, 1991). These have been used for backcrossing to the cultivated species. Nevertheless, the well-publicized somatic hybrids between potato and tomato deserve special mention.

Potato and tomato are members of the same family (Solanaceae) but different genera. Potato belongs to the genus *Solanum* and tomato to *Lycopersicon* and their somatic chromosome numbers are 48 and 24, respectively (Shepard *et al.*, 1983). They are, of course, not sexually compatible. In 1978, Melchers *et al.* were the first to produce an intergeneric somatic hybrid between potato and tomato. They fused protoplasts between cultured diploid potato line and tomato leaf cells. Hybrid plants displayed morphological features of both parents like most hybrids. The hybrid plants were named "pomatoes" (Melchers *et al.*, 1978). Some hybrid plants formed "tuber-like stolons"; however, none set fertile flowers or fruits or produced true tubers (Shepard *et al.*, 1983). Somatic hybrid plants produced to date, such as the pomato are not of immediate economic value but they serve as the starting point of a genetic introgression scheme. However, it should be mentioned that the contributions made from these fusion studies on the interaction between the nucleus and organelles (Kung *et al.*, 1975) and on the regeneration and development of cell wall and others are invaluable.

The evidence from the era of hybrid plants to the era of somatic hybrid plants permitted expansion of the scope of plant breeding to both the organismic and cellular levels. Further advances in plant breeding—to the molecular level—are the subject of the next section. This demonstrates that we are entering the new era of biotechnology or genetic engineering.

I.4 Transgenic Plants

As described in this volume, there are four essential components, in addition to desirable genes, for a successful production of transgenic plants. They are suitable vectors (Chapter 2, this volume), selectable markers (Chapter 3, this volume), efficient techniques for transformation (Chapter 3, this volume), and regeneration (Chapter 4, this volume). It is quite obvious that the dramatic progress made recently in the development of the gene transfer system (Gasser and Fraley, 1989) makes it possible to produce the numerous transgenic plants (dicot and monocot). Our current ability to insert foreign genes into plant cells and tissues and to regenerate viable, stable, and fertile novel plants has provided unique opportunities to plant breeders unmatched in history. This allows for explosive expansion of our understanding in the field of plant biology and provides us with the technology to modify and improve crop plants in ways unthinkable only 10 years ago. This progress is remarkable. This is clearly reflected by the steady

TABLE I.1
Transgenic Plants Produced[a]

Type of plant	Examples
Monocotyledons	Asparagus, maize, millet, wheat, orchard grass, rice, rye
Dicotyledons	
Legumes	Alfalfa, clover, peas, soybean, moth bean
Solanaceae	Eggplant, petunia, potato, tobacco, tomato
Vegetable and fruit crops	Cabbage, carrot, cauliflower, celery, cucumber, horseradish, lettuce, rape, grape, muskmelon, strawberry
Special crop	Cotton, flax, lotus, sugar beet, sunflower
Woody plants	Apple, pear, poplar, walnut
Special plant	Arabidopsis
Other groups	*Medicago varia, Vigna aconitifolia, Stylesanithes humilis,* morning glory, kalanchoe

[a]From Gasser and Fraley (1989); Fraley (1989); Ratner (1989); Rogers (1988); and Chapter 4, this volume.

1990 = 40 T.G.P.

increase in numbers of transgenic plants produced every year. Up to 1988, there were only a total of 23 transgenic plants produced (Rogers, 1988). Their number increased to 30 in 1989 (Gasser and Fraley, 1989; Ratner, 1989) and over 40 a year later (Table I.1).

A brief review of the literature on transgenic plants indicates that transgenic plants have been engineered with several purposes: developing new techniques for transformation (Chapter 3, this volume), basic study on a specific gene (Chapters 5–8, this volume), and crop improvement for a specific cause (Chapters 9–14, this volume). Among the 50 publications on transgenic plants reviewed by the author, most focus still on the development of suitable methods for the transformation and regeneration of major crops. The plants transformed were predominantly dicot (Volume 2, Chapters 4–9), particularly from the family Solanaceae (Volume 2, Chapter 4), with only a few monocots (Volume 2, Chapters 1–3) and woody plants (Volume 2, Chapter 7). This is reflected in the organization and content of this volume. Since social, economic, regulatory, and commercial considerations of transgenic plants play important roles, these form an integral part of this treatise (Volume 2, Chapters 10–12).

The first transgenic plants were engineered in the 1980s. Hall, Kemp and coworkers transferred the β-phaseolin gene from bean to sunflower and tobacco plants (Murai *et al.*, 1983). Independently, several transgenic tobacco plants were produced to express foreign genes engineered by the *Agrobacterium tumefaciens* vectors (Horsch *et al.*, 1984; De Block *et al.*, 1984). Now, a variety of free DNA delivery systems are available (Chapter 3, this volume). This was followed by the highly publicized transformation of luciferase gene from firefly to tobacco (Ow *et al.*, 1986). The early transformation experiments often utilized plant protoplasts as the recipient cells; the subsequent development of transformation methods based on regenerable explant (Horsch *et al.*, 1985) such as leaves, stems, and roots simplified significantly the transformation techniques that are widely used today. The recent breakthrough in the regeneration of transformed monocot plants rice and maize (Volume 2, Chapters 1 and 2) has removed a major obstacle in improving cereal crops. Production of transgenic plants in most laboratories is now a routine operation. Many papers on transgenic plants are published each year. This activity leads to publication of this treatise on transgenic plants.

1.5 Biotechnology Revolutionizes Plant Breeding

Biotechnology offers new ideas and techniques applicable to agriculture. It uses the conceptual framework and the technical approaches of molecular biology to develop commercial processes and products. With the rapid development of biotechnology, agriculture has moved from a resource-based to a science-based industry. Indeed, plant breeding has been dramatically broadened by the introduction of genetic engineering techniques based on knowledge about gene structure and function. This has ushered agriculture into a new era where it joins ranks with the most sophisticated of biological sciences in using molecular approaches whose promise to yield new generations of plants of any desirable trait is matched by their widely demonstrated potential in medicine. The technology required for engineering transgenic plants is considerably more sophisticated than the one for producing somatic hybrid plants by protoplast fusion and much more so than that required for production of hybrid plants by cross-fertilization. In 1983, direct gene transfers became possible using recombinant DNA technologies (Fraley, 1989). The newly acquired ability to transfer genes among organisms without sexual crossing provides breeders with new opportunities to improve the efficiency of production and to increase the utility of agricultural crops. Plants with new traits such as herbicide resistance have been genetically engineered using genes from unrelated organisms (Chapter 9, this volume). Scientists are also trying to improve the quality of our agricultural products by altering the nutritional value of proteins and oils (Chapter 14, this volume). However, it should be emphasized that the biotechnology is not a substitute or replacement for the conventional breeding methods. Rather, it can improve on past, conventional methods. The major differences between conventional breeding and biotechnology lie neither in goals nor processes, but rather in speed, precision, reliability, and scope.

By using recombinant DNA technology many transgenic organisms have been engineered since 1985. Table I.1 lists over 40 transgenic plants produced to date. The list is growing rapidly because the technology for inserting and expressing genes in plants is in hand, and more and more genes are being identified, together with the traits they control.

The majority of plant biotechnology research thus far has focused on developing plants that resist insects, diseases, and frost damage and tolerate certain herbicides. Insect resistance research has centered on inserting a gene obtained from the bacteria *Bacillus thuringiensis* (BT) into plants or bacteria (Chapter 12, this volume). The gene produces a toxin that is fatal to insects. Efforts have been concentrated on the use of BT because it has been used commercially in agriculture for 25 years and dissipates rapidly in the environment.

Herbicide resistance research has focused on developing tomato and tobacco cultivars that are resistant to herbicides (Fraley, 1989) such as glyphosate (Roundup, Monsanto, St. Louis, Missouri). Environmental groups have expressed concern about the benefits from this research because it is not clear whether it will lead to increased or decreased use of herbicides. Commercialization of these plant technologies may be a few years away as a result of economic incentives, environmental regulation, and public acceptance. For example, a major economic consideration will be whether or not we could produce the fruits that meet the standards of processors.

Plant technologies that involve multigenic traits such as nitrogen fixation and stress tolerance are even further away. Researchers are concerned because plants that are engineered to fix nitrogen may use so much energy in doing it that yields may fall dramatically. Much more needs to be learned about genetic structures, gene function, and gene regulation before these applications are achievable and marketable.

I.6 Future Prospects

Crop improvement represents continuous efforts to seek new ways and to solve problems associated with conventional methods of improving groups. The application of genetic principles in breeding programs has been very successful and will continue to be. It formulated the basis for the "Green Revolution." This was further complemented by the development of tissue culture and protoplast fusion techniques. The combination of conventional breeding approaches and tissue culture techniques has opened many avenues for crop improvements which were not available previously. The current advancement of biotechnology has ushered agricultural research into a new era of transgenic organisms. It is notable that as the range of technologies available for crop improvement expands, the role of technological progress is rapidly increasing. The era of the gene revolution rests on a solid foundation of agricultural research. It requires infinite knowledge of plant biochemistry and physiology, and development at the molecular level. This is a striking departure from the traditional view of agricultural research. Thus, we have advanced from the hybrid plants to the transgenic plants, or from the green revolution to the gene revolution.

Currently, genes used for plant transformation are not of plant origin. It is anticipated that this may change in a few years. Major efforts are under way to identify the genes of disease or pest resistance. Our next giant leap in crop improvement will be the transfer of plant genes into plants for fortifying their natural defense capability. The ultimate goal is to construct a stable agro-ecosystem through biotechnology that supports high productivity without constant environmental or chemical interventions.

Acknowledgment

Greg Silsbee's help in preparing this manuscript and Shyam Dube's review are greatly appreciated.

References

Bohn, G. W., and Tucker, C. M. (1939). *Science* 89, 603–605.

Borlaug, N. E. (1983). *Science* 219, 689–693.

Carlson, P. S. (1973). *Proc. Natl. Acad. Sci. U.S.A.* 70, 598-602.

Carlson, P. S., Smith, H. H., and Dearing, R. D. (1972). *Proc. Natl. Acad. Sci. U.S.A.*69, 2292–2294.

Chrispeels, M. J., and Sadava, D. (1977). *Plants, Food and People,* p. 192. San Francisco: W. H. Freeman.

De Block, M., Herrera-Estrella, L., Van Montagu, M., Schell, J., and Zambryski, P. (1984). *EMBO J.* 31, 681–686.

Fraley, R. (1989). *Plant Biotechnology.* (Ed. by S. D. Kung and C. J. Arntzen), pp. 395–407. Boston: Butterworths.

Gasser, C. S., and Fraley, R. T. (1989). *Science* 244, 1293–1299.

Glimelius, K., Fahlesson, J., Landgren, M., Sjödin, C., and Sundberg, E. (1991). *Trend in Biotech.* 9, 24–30.

Goodman, R. M., Hauptli, H., Crossway, A., and Knaup, V. C. (1987). *Science* 236, 48–54.

Gray, J. C., Kung, S. D., and Wildman, S. G. (1974). *Nature (London)* 525, 226–227.

Heslop-Harrison, J. (1978). *Cellular Recognition Systems in Plants.* Studies in Biology. London: Edward Arnold.

Horsch, R. B., Fraley, R. T., Rogers, S. G., Sanders, P. R., Lloyd, A., and Hoffman, N. (1984). *Science* 223, 496–498.

Horsch, R. B., Frey, J. E., Hoffman, N. L., Eichholtz, D., Rogers, S. G., and Fraley, R. T. (1985). *Science* 227, 1229–1231.

Kung, S. D., Gray, J. C., Wildman, S. G., and Carlson, P. S. (1975). *Science* 187, 353–355.

Lima-De-Faria, A., Eriksson, T., and Kjellen, L. (1977). *Hereditas* 87, 57–61.

McFadden, E. S. (1930). *J. Am. Soc. Agron.* 22, 1050–1050.

Melchers, G., Sacristan, M. D., and Halder, A. A. (1978). *Carlsberg Res. Commen.* 43, 203–209.

Murai, N., Sutton, D. W., Murray, M. E., Slighton, J. L., Merlo, D. J., Reichert, N. A., Sengupta-Gopalan, C., Stock, C. A., Barber, R. F., Kemp, J. D., and Hall, T. C. (1983). *Science* 22, 476–482.

Nagata, T., and Takebe, I. (1971). *Planta* 99, 12–16.

Newell, C. A., and Hymowitz, R. (1982). *Crop Sci.* **22,** 1062–1066.

Ouellette, R. R., and Cheremisinoff, P. N. (1985). *Application of Biotechnology.* Lancaster, Pennsylvania: Technormic Pub. Co.

Ow, D. W., Wood, K. U., DeLuca, M., De Wet, J. R., Helinski, D. R., and Howell, S. H. (1986). *Science* 234, 856–859.

Power, J. B., Berry, S. F., Champman, J. V., and Cocking, E. C. (1980). *Theor. Appl. Genet.* 57, 1–6.

Ratner, M. (1989). *Biotechnology* 7, 337–341.

Rogers, S. (1988). *Proceedings of Transgenic Plant Conference,* pp. 31–40. Annapolis, Maryland.

Shepard, J. F., Bidney, D. F., Basby, T., and Kemble, R. (1983). *Science* 219, 683–688.

Simmonds, N. W. (1976). *Evaluation of Crop Plants* (Ed. by N. W. Simmonds). New York: Longman.

Part I

Basic Techniques for Engineering Transgenic Plants

1

Vectors for Gene Transfer in Higher Plants

Frank F. White

Department of Plant Pathology
Kansas State University
Manhattan, Kansas 66506

1.1 Introduction

Progress in the development of plant gene transfer vectors can be attributed to developments of genetic and recombinant DNA techniques, plant tissue culture techniques, and the design of selection methods for transgenic plant cells or tissue. In terms of use and apparent progression to field trials, the majority of gene transfer experiments have involved the *Agrobacterium* system. However, while this chapter emphasizes the *Agrobacterium* system, methods of direct DNA transfer are well developed and, for many plant species, represent the clearest choice for gene transfer. Viral vectors also present interesting possibilities for plant transformation and may have advantages in certain applications. However, viral vectors or schemes involving viral replicons have been slower in development.

This chapter will not delve into the development of plant cell manipulation and regeneration methods. However, progress in plant cell transformation, particularly with direct DNA transfer methods, has been largely dictated by the cell culture techniques. Thus, one of the critical factors in the transformation of a particular plant species is the ability to manipulate and regenerate whole plants from the explants or protoplasts. Fortunately, many important crop species are amenable to plant regeneration. On the other hand, other important crop species, notably soybeans, corn, and wheat, have been more recalcitrant in regeneration schemes. In general, the ability to introduce DNA into the genomes of these

TRANSGENIC PLANTS, VOLUME 1
Engineering and Utilization

species is not the limiting factor. Stable gene transfer has been demonstrated in most cases, and progress in genetic engineering depends on developments in techniques for tissue differentiation. These impediments are continually being overcome in species such as rice, corn, and soybeans and are likely to be overcome in others in the near future (Hinchee *et al.*, 1988; Shimamoto *et al.*, 1989; Gordon-Kamm *et al.*, 1990).

1.2 *Agrobacterium*-Mediated Transformation

The most developed system for higher plant transformation is derived from the tumor-inducing mechanism of the soil bacterium *Agrobacterium tumefaciens* (for reviews see Binns and Thomashow, 1988; Zambryski, 1988). *Agrobacterium tumefaciens* is the causal agent of crown gall disease. A closely related disease called hairy root is caused by *A. rhizogenes* (for review see White and Sinkar, 1987). The utility of these bacteria has developed from an understanding of the molecular basis of the disease symptoms of tissue hyperplasia, namely, the transfer of DNA from the bacterium to the plant nuclear genome. *Agrobacterium* was recognized early to be the etiological agent of crown gall disease and was, in fact, one of the first bacterial diseases of plants to be characterized. Hairy root was recognized as a variant of crown gall disease, but due to cultural and pathological differences between the *A. tumefaciens* and hairy root strains which were studied at that time separate species were defined. The distinction is now largely for convenience since relatively few genes are known to control the differences between the ability to induce grown gall or hairy root tumors. From a practical perspective, two strains of *A. rhizogenes* commonly used in gene transfer experiments, namely A4 and 15834, are now known as biotype 2 strains (Keane *et al.*, 1970). These strains generally require biotin (2 μg/liter) for growth on minimal media. Other strains of *A. tumefaciens* and *A. rhizogenes* that have been used for transfer experiments have been biotype 1 strains (Keane *et al.*, 1970). These strains do not have any special nutritional requirements.

The aberrant plant growths caused by either type of *Agrobacterium* strain are due largely to genes present on large plasmids within the bacteria (Van Larabeke *et al.*, 1974; Watson *et al.*, 1975; White and Nester, 1980). Only a relatively small discrete portion of the plasmid is transferred to the plant cells during the tumor formation process, and this region is now familiarly known as the T-DNA (transferred DNA). The types of morphological effects observed in the transformed plant tissues are controlled by the complement of genes that are transferred. Plasmids that confer the tumor-inducing ability on *Agrobacterium* strains have become known as Ti (tumor-inducing) plasmids. The plasmids controlling the hairy root morphology are commonly called the Ri (root-inducing) plasmids. Three T-DNA genes of the Ti plasmids are now known to control the biosynthesis of two phytohormones, indole-3-acetic acid (an auxin) and isopentyladenosine (a cytokinin precursor). The genes encode the enzymes for phytohormone synthesis (Akiyoshi *et al.*, 1984; Barry *et al.*, 1984; Schröder *et al.*, 1984; Thomashow *et al.*, 1984; Thomashow *et al.*, 1986), and incorporation of the genes into the plant genome leads to the constitutive synthesis of the phytohormones. The constitutive synthesis of phytohormones in turn alters the developmental program of the transformed plant tissue. In addition to the genes of known biochemical function, a number of genes from both Ti and Ri plasmids affect growth and development when transferred to plants and as yet have no assigned biochemical functions (Messens *et al.*, 1985; White *et al.*, 1985; Spena and Schell, 1987; Sinkar *et al.*, 1988; Spanier *et al.*, 1989; Tinland *et al.*, 1989).

A second characteristic trait of *Agrobacterium*-induced tumor tissue was later identified by Morel and colleagues (Menagé and Morel, 1964) when they discovered that tumor cells synthesized unusual amino acid compounds which are given the general name opines. Opines are derived from a variety of compounds occurring in the plant cell (for review see Tempé and Goldmann, 1982; Smith and Neilands, 1987). The first two discovered were called octopine and nopaline, and are derived from arginine and pyruvate, and arginine and ketoglutarate, respectively. The type of opine that was produced by the plant cells is dependent on the type of inciting bacterial strain. The discovery that *Agrobacterium* infection results in the production of specific opines has led to the so-called opine hypothesis which states that the bacteria attempt to create a ecological niche in the plant tumor and surrounding environs by genetically diverting plant compounds into substances which are preferentially utilized by the bacterium. Many organisms including the plant cannot effectively utilize opines, while genes for the utilization of a particular opine also reside on the Ti plasmid. Opine production is controlled by T-DNA genes which encode the enzymes for opine synthesis (Murai and Kemp, 1982). The genes function in the plant genome, and the presence of opine synthesis can be used as a genetic marker. The gene for nopaline synthase (nopaline dehydrogenase) has been the most widely used opine gene and was derived from the nopaline plasmid pTiT37 (Depicker *et al.*, 1982; Bevan *et al.*, 1983a).

Ti and Ri plasmids from several strains of *Agrobacterium* have been well characterized by a variety of physical and genetic techniques. The findings indicate that the plasmids share the same basic architecture and contain many genes which are of marginal interest in regards to vector development, including genes involved in replication, conjugal transfer between bacteria, and utilization of the various opines that are synthesized in the tumors. Two critical regions for gene transfer are also found on the plasmids, although the exact organization and content of these regions may differ. The first region of the plasmid is known as the virulence or *vir* region. This region contains genes which are essential for gene transfer, but are not themselves generally transferred to the plant genome.

The virulence region from strain A6 covers approximately 35 kilobases (kb) and includes seven genetic loci which can affect virulence as indicated by symptoms on test plants (Klee *et al.*, 1983; Stachel and Nester, 1986; Kanemoto *et al.*, 1989). The second important region is the T-DNA. The length and number of T-DNA regions vary in Ti and Ri plasmids. The nopaline-type Ti plasmid has a single 23-kb T-DNA region (Engler *et al.*, 1981), and the octopine-type Ti plasmid has two closely adjoining regions of 13 kb and 8 kb (Barker *et al.*, 1983), respectively. The phytohormone and octopine synthesis genes are located in the left (T_L-DNA) region of the octopine-type Ti plasmid. More recently, several plasmids in addition to the octopine-type plasmid have been demonstrated to possess multiple T-DNA regions. The agropine-type Ri plasmids have two T-DNA regions which are separated by approximately 16 kb (Huffman *et al.*, 1984; Jouanin, 1984). Two T-DNA regions were also found in the so-called limited host range and supervirulent Ti plasmids (Hood *et al.*, 1984; Yanofsky *et al.*, 1985b; Komari *et al.*, 1986). A single T-DNA region has been identified in the mannopine-type Ri plasmids (Koplow *et al.*, 1984). Multiple T-DNA regions may turn out to be the rule as more wild-type Ti and Ri plasmids are characterized, but all have similar essential features. Introduction of an individual T-DNA region from one plasmid into a strain containing another type of plasmid results in the transfer of the newly acquired T-DNA segment (Hoekema *et al.*, 1984; Hooykaas-Van Slogteren *et al.*, 1984; White *et al.*, 1985). Therefore, the T-DNA regions from various Ti and Ri plasmids appear to be mobilized to plant cells by a conserved mechanism.

The T-DNA regions of octopine-type Ti plasmid provide a good model for the basic features of all T-DNA regions (Figure 1.1). The genes which are located in the T-DNA region contain the necessary structural features for expression by the plant. The messenger RNA is polyadenylated and contains a 7-methylguanosine cap at the 5′ end. A total of eight mRNAs have been detected in plant tissue containing the T_L-DNA of the octopine-type Ti plasmid (Willmitzer *et al.*, 1983). Five mRNA species have been mapped to the T_R-DNA (Salomon *et al.*, 1984). None of the transcriptional units contain introns. Two T_L-DNA genes control the synthesis of indole-3-acetic acid from tryptophan. Transcript 1 (using the nomenclature of Willmitzer *et al.*, 1983) encodes a tryptophan monooxygenase which converts tryptophan to indoleacetamide (Schröder *et al.*, 1984; Thomashow *et al.*, 1984), while the product of the transcript 2 encodes a hydrolase that converts the indoleacetamide to indole-3-acetic acid (Thomashow *et al.*, 1986). The transcript 4 product has been identified as an isopentyltransferase which catalyzes the formation of isopentyladenosine from adenosine monophosphate and isopentyl pyrophosphate (Akiyoshi *et al.*, 1984; Barry *et al.*, 1984). With the exception of the octopine synthase (octopine dehydrogenase) gene the functions of other T_L-DNA loci are not understood. In the case of the octopine-type T-DNA, mutations at the *tml* locus (which corresponds to transcript 6b)

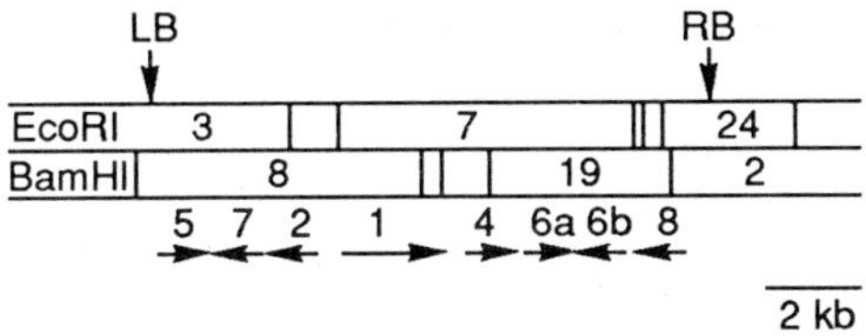

Transcript	Enzymatic activity
1	Tryptophan Monooxygenase
2	Indoleacetamide Hydrolase
4	Dimethylallypyrophosphate Transferase
8	Ocotopine Synthase

Figure 1.1 Left T-DNA region of pTIA6. Arrows beneath the map represent recognized transcripts from the T-DNA. The functions of the encoded proteins for several of the genes are listed. Vertical arrows designate the positions of the left (LB) and right (RB) borders.

result in larger tumors on some plants (Garfinkel *et al.*, 1981). The biochemical basis for this response is not known nor is the phenotype observed when the homologous region is mutated in the nopaline-type T-DNA (Joos *et al.*, 1983). More recent results indicate that the 6b gene product in some unknown way modulates the response of the plant cell to cytokinin and/or auxin (Spanier *et al.*, 1989; Tinland *et al.*, 1989). Evidence regarding transcript 6a of the nopaline plasmid suggests that the gene product allows leakage of opines from the plant cells, and therefore may encode a permease activity (Messens *et al.*, 1985). The molecular basis of leakage is unknown. The genes of the T_R-DNA appear to be predominantly involved with the synthesis of agropine-type opines (Salomon *et al.*, 1984). While no vectors have been designed from the T_R-DNA, several of the agropine synthesis genes have been the source of promoters for chimeric genes (Velten *et al.*, 1984).

One fortunate aspect of the T-DNA structure in regards to gene transfer is that none of the T-DNA genes are involved in the transfer process (Garfinkel *et al.*, 1981; Leemans *et al.*, 1982a; Ream *et al.*, 1983). Any or all of the genes can be removed or mutated, and the T-DNA region can still be transferred to the plant genome. The key elements of the T-DNA are the 25-bp imperfect direct repeats present at the boundaries of the T-DNA (Yadav *et al.*, 1982; Caplan *et al.*, 1983). Most plant/T-DNA junction sequences which were isolated from transformed cells occur within or near the border sequences (Simpson *et al.*, 1982; Yadav *et al.*, 1982; Zambryski *et al.*, 1982). As will be discussed later, recent experiments demonstrate that the interaction of *vir* genes and the border sequences initiates the transfer events. As represented by standard convention, the right border is critical for T-DNA transfer and tumorigenesis, while removal of the left border appears not to affect tumor formation (Ooms *et al.*, 1982; Shaw *et al.*, 1984; Wang *et al.*, 1984; Peralta and Ream, 1985). A sequence, which has been termed *overdrive,* has been identified immediately adjacent to the octopine T-DNA right border that is required for maximal transfer efficiency (Peralta *et al.*, 1986). All Ti and Ri plasmids that have been sufficiently analyzed appear to contain related border sequences (Slightom *et al.*, 1986). The regions of the virulence genes also for the most part appear to be closely related among the different plasmids as indicated by DNA hybridization experiments (Huffman *et al.*, 1984).

T-DNA transfer begins with the introduction of the bacteria into a plant wound. Wounding is a necessary event in the process and, at least in part, may be required for the synthesis of compounds by the plant which induce the expression of the *vir* genes (Bolton *et al.*, 1986; Janssens *et al.*, 1986; Stachel *et al.*, 1986a). A group of compounds have been identified in tobacco (*Nicotiana tabacum*) wound exudate that when added to *Agrobacterium* cultures cause *vir* gene induction (Stachel *et al.*, 1985). One of the most active substances yet identified is acetosyringone. Three additional conditions have also been found to be conducive to T-DNA transfer: a low pH medium, low phosphate, and the proper sugar source (Winans *et al.*, 1988; Ankenbauer and Nester, 1990; Cangelosi *et al.*, 1990). Two *vir* genes, *virA* and *VirG,* are constitutively expressed and are involved in mediating the induction of the other genes (Stachel and Zambryski, 1986a). On the basis of amino acid sequence homology, *virG* belongs to a class of positive regulatory genes in bacteria which link gene induction to different external stimuli (Nixon *et al.*, 1986; Winans *et al.*, 1986). In turn, *virA* belongs to a second group of genes which lead to the activation of the positive regulatory gene proteins (Ronson *et al.*, 1987; Winans *et al.*, 1986). VirA is a membrane-associated protein which monitors the presence of the induction compounds (Winans *et al.*, 1989). The presence of inducers is then signaled to the positive regulatory protein and induction of the other *vir* genes occurs (Stachel and Zambryski, 1986b). As with several other members of the response genes, *virA* has been shown to be autophosphorylated at one specific histidine residue under appropriate induction conditions (Jin *et al.*, 1990b). The phosphate group is then transferred to *virG* product increasing the transcriptional activating activity of VirG (Jin *et al.*, 1990a). VirG binds specifically to upstream regions of, at least, some of the *vir* genes (Pazour and Das, 1990). The dynamics of VirG phosphorylation, binding, and *vir* gene induction remained to be described in more detail.

A particularly intriguing area of research involves deciphering the functions of the inducible *vir* genes in the transfer process. Circular T-DNA molecules can be isolated from the bacteria after induction with acetosyringone or plant wound extracts (Koukolíková-Nicola *et al.*, 1985; Machida *et al.*, 1986). Originally thought to represent transformation intermediates, the circles appear to result from an excision and subsequent recombination or ligation events at the right and left border sequences. The covalently closed circular T-DNA molecules may in fact be artifacts or inadvertent products of the transfer process. More recently, induction has been shown to result in the appearance of single-stranded nicks within the bottom strand of the 25 base pair (bp) border (Albright *et al.*, 1987; Wang *et al.*, 1987). The nick site occurs predominately after the third nucleotide from the left end of the right border (T-DNA side). The DNA nicking is followed by the generation of a single-stranded copy (called the T-strand) or the T-DNA region which is thought to be the intermediate in plant cell transformation (Stachel *et al.*, 1986b; Albright *et al.*, 1987). The *virD* codes for at least two polypeptides called VirDI and VirDII (Wang *et al.*, 1990). VirDII has been shown to be required for the nicking activity and single-strand synthesis and encodes an endonuclease activity (Yanofsky *et al.*, 1986; Yamamoto *et al.*, 1987). VirDI has been shown to possess helicase activity and is also required for T-strand formation (Ghai and Das, 1989).

The recent findings of T-DNA processing and *vir* gene function suggest a model (Figure 1.2) with striking similarity to bacterial conjugal transfer mechanisms (Willetts and Wilkins, 1984; Stachel and Zambryski, 1986b). In fact, one report indicates that bacterial origins of transfer (*oriT*) and the corresponding gene for DNA cleavage (*mob*) can mediate plant cell transformation (Buchanan-Wollaston *et al.*, 1987). Another report indicated that transfer can occur between other bacterial/eukaryotic cell combinations, in this case *Escherichia coli* and *Saccharomyces cerevisiae* (Heinemann and Sprague, 1989). DNA transfer during conjugation occurs by the generation of a single-strand copy of the plasmid. The *oriT* contains the site where strand cleavage occurs and DNA synthesis begins. Both *oriT* and the T-DNA border direct transfer in a polar manner. Presumably the left T-DNA border provides a termination point for single-strand synthesis but is not required. If the T-DNA border is placed on a relatively small plasmid the orientation of the border sequence is less critical, because in this situation genes are transferred in either orientation of the border since there is high probability that the entire plasmid will be transferred to the plant cells (Horsch and Klee, 1986; Jen and Chilton, 1986).

VirE has also been shown to behave as a single-stranded DNA binding protein (Gietl *et al.*, 1987; Christie *et al.*, 1988; Citovsky *et al.*, 1988; Das, 1988), although VirE is not absolutely required for T-DNA transfer (Gardner and Knauf, 1986; Horsch *et al.*, 1986). The VirE and VirDII proteins may provide protection to the T-strand intermediate during the transfer process. VirE appears to be a nonspecific single-stranded binding protein that coats the T-strand, presumably affording protection from endonucleolytic catalysis. The VirDII protein can be found covalently attached to the 3′ end of the T-strand (Herrera-Estrella *et al.*, 1988; Ward and Barnes, 1988; Young and Nester, 1988; Howard *et al.*, 1989) and has been shown to protect the T-DNA from exonucleolytic degradation (Dürrenberger *et al.*, 1989). VirDII may have a third function in that the N-

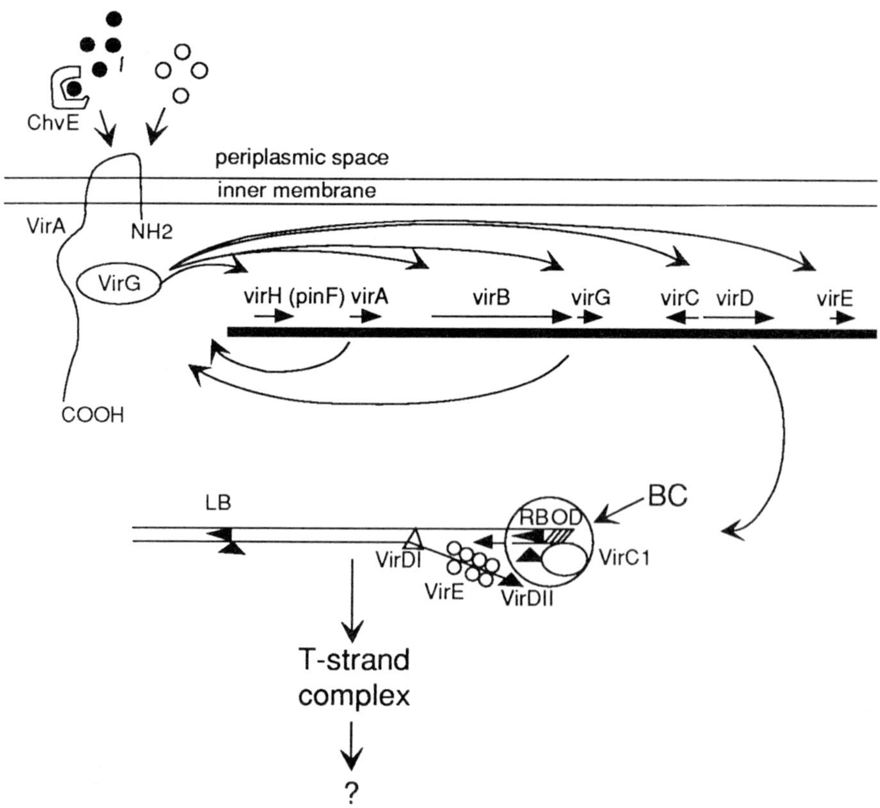

Figure 1.2 Schematic model of possible *vir* gene function and T-strand formation. Inducers (I) such as glucose and acetosyringone are recognized by *virA* gene product. Sugar induction requires the product of the *chvE* gene, a periplasmic binding protein. VirA and VirG protein interaction results in transcription of other *vir* genes. VirDII protein causes nick at right and left borders (RB and LB, respectively). VirC protein may be involved in a border complex (BC) at the overdrive sequence (OD) that directs nicking by VirDII at the right border. VirE protein is involved as a single-strand binding protein. Other possible *vir* functions include bacterial membrane or membrane protein modifications for plant cell interaction, and delivery mechanism of T-strand to plant cell.

terminal portion of the protein has been shown to direct a fusion protein to the plant cell nucleus (Herrera-Estrella *et al.*, 1990).

VirC1 protein is another *vir* gene product that has been shown to have DNA binding activity. VirC1 binds specifically to the *overdrive* sequence of the octopine-type A6 T-DNA (Toro *et al.*, 1989). VirC like VirE is not absolutely required for T-DNA transfer and tumorigenesis on some plants (Yanofsky *et al.*, 1985a; Horsch *et al.*, 1986). VirC1 appears to mediate the enhancer effect of the *overdrive* sequence. One model proposed for VirC1 suggests that VirC1 and *overdrive* facilitate the interaction of VirDII with the border nick site (Toro *et al.*, 1989). In support of this model, VirDII was found to associate with the *overdrive* and T-DNA border sequence (Toro *et al.*, 1988). Whether the VirDII interaction with *overdrive* is mediated by VirC1 is unknown.

The functions that the *virB* gene products play in the tumorigenesis process are unknown. The analogy with the conjugation system suggests involvement in functions thought to be involved in conjugal transfer and function in DNA binding, replication, covalent modification of termini or other form of packaging the transfer intermediate, and/or transport across the cell envelope. The *virB* locus encodes 11 potential polypeptides, some of which appear to be membrane-spanning proteins (Ward *et al.*, 1988). Further insight into the transfer process will undoubtedly come from experiments involving *virB* products (Ward *et al.*, 1990). It should be emphasized that little is known regarding the actual form and mechanism of the transfer intermediate.

1.3 *Agrobacterium* Vectors

The *Agrobacterium*/Ti plasmid system exploits the elements of the *Agrobacterium* transformation mechanism. Although a considerable number of plant transformation vectors now exist, all consist of modifications to two elements. The first element consists of a plasmid vector containing a border sequence such that the vector functions as a T-DNA region. The vector also provides a shuttle that allows the recombinant DNA manipulations to be performed in *Escherichia coli* hosts before subsequent transfer to *Agrobacterium*. Second, transfer of vector DNA to the plant is mediated by a strain of *Agrobacterium* that contains a full complement of *vir* genes. Basically, two types of *Agrobacterium* vectors are presently in use. Vectors which recombine via DNA homology into a resident Ti plasmid are often referred to as integrative or intermediate vectors (Leemans *et al.*, 1981; Matzke and Chilton, 1981; Fraley *et al.*, 1983). The second type of vector need not recombine into the resident Ti plasmid and is referred to as a binary or autonomous vector (de Framond *et al.*, 1983; Hoekema *et al.*, 1983; Bevan, 1984; An *et al.*, 1985; Klee *et al.*, 1985). Both types provide three essential features: convenient (unique) endonuclease restriction sites for cloning a target gene or DNA fragment within the T-DNA region; a plant-expressed gene marker for selection and/or screening of transgenic plant cells; and a bacterially expressed gene marker for selection of the vector in *Agrobacterium* and *E. coli*. The development of the binary vectors came after the discovery that a T-DNA region would function *in trans* to the *vir* genes when present on a separate replicon (de Framond *et al.*, 1983; Hoekema *et al.*, 1983). Binary vectors are the most commonly used for plant transformations and exist in the most varieties.

However, until wide host range vectors are trimmed into smaller plasmids, integrative vectors have a greater cloning capacity and make good T-DNA cosmid vectors for transfers of genomic clones (Yanofsky *et al.*, 1990).

The strategy for the construction of integrative vectors was based on earlier techniques for site-directed mutagenesis (Ruvkun and Ausubel, 1981). In simplest form an integrative vector only requires a plasmid which shares homology with the T-DNA. Homologous recombination between the plasmid and the T-DNA will produce a hybrid structure which can then be mobilized to the plant cell. However, in practice a number of important modifications have been incorporated into the design of integrative and binary vectors, and some variations in different vectors exist. Because the wild-type T-DNA contains oncogenic genes, these genes can be removed if normal and stable plant development is desired (Zambryski *et al.*, 1983; Fraley *et al.*, 1985). Without the selection for tumorous growth, alternative selection schemes have been devised. Kanamycin resistance remains the most widely used selection for transformed plant cells (or the related aminoglycoside antibiotic G418) (Bevan *et al.*, 1983b; Fraley *et al.*, 1983; Herrera-Estrella *et al.*, 1983). A variety of other selection techniques have been devised or envisioned (Table 1.1). In addition, some vectors include a second marker for screening putative transformants after the initial selection (Table 1.1). A second marker provides a convenient check for cells that have escaped the selection regimen. The second marker may be a selectable type gene or, more commonly, an opine synthesis gene such as the nopaline synthesis gene.

The integrative vector is well illustrated by the pMON200 vector (Figure 1.3; Rogers *et al.*, 1986). In this particular vector the right T-DNA border from pTi37 has been utilized. The critical feature is that the vector by itself will not replicate in *Agrobacterium* without first integrating into the resident Ti plasmid. In this case the origin of replication is derived from pBR322 which has a ColE1-type origin. Replication of the plasmid is limited to *Escherichia coli* and several related enteric bacteria. The plasmid also contains the *bom* site which allows mobilization to another bacterium if other transfer functions are supplied *in trans*. Integrative vectors are mobilized into *Agrobacterium* by a triparental conjugation system (Ditta *et al.*, 1980). In triparental conjugations the functions necessary for plasmid transfer are supplied on another helper plasmid which is often in a separate bacterial strain. A mating, therefore, consists of three members: the recipient *Agrobacterium,* the helper strain, and the strain harboring the plant transformation vector. Additional *E. coli* helper strains have been designed with the RK2 transfer genes positioned on the chromosome (Simon *et al.*, 1983), and the T-DNA vector is first introduced into the helper strain. The vector is then transferred to *Agrobacterium* in a two-member conjugation. Vectors can also be introduced by electroporation (Miller, 1988) or by the freeze–thaw transformation procedure (Holsters *et al.*, 1978). The *Agrobacterium* transconjugant is then isolated on the basis of selection for stable antibiotic resistance conferred by the

TABLE 1.1
Plant Transformation Markers

Trait	Refs.
I. Selection[a]	
Kanamycin or G418 resistance	Bevan *et al.* (1983b), Fraley *et al.* (1983), Herrera-Estrella *et al.* (1983)
Methotrexate resistance	Herrera-Estrella *et al.* (1983), Brisson *et al.* (1984)
Hygromycin resistance	van den Elzen *et al.* (1985b), Waldron *et al.* (1985)
Bleomycin resistance	Hille *et al.* (1986)
Aminoethylcysteine resistance	Dahl and Tempé (1983)
Glyphosate resistance[b]	Comai *et al.* (1985), Shah *et al.* (1986)
Streptomycin resistance	Jones *et al.* (1987)
Sulfonylurea resistance[b]	Haughn and Somerville (1986)
Phosphinothricin resistance[b]	De Block *et al.* (1987)
Gentamicin resistance	Hayford *et al.* (1988)
Indoleacetamide utilization or sensitivity[c]	Inzé *et al.* (1984), Depicker *et al.* (1988)
Hairy root formation	Stougaard *et al.* (1986)
II. Screens and assays	
Opine synthesis	Otten and Schilperoort (1978)
Neomycin phosphotransferase	Reiss *et al.* (1984)
Chloramphenicol acetyltransferase	Gorman *et al.* (1982)
β-Galactosidase	Helmer *et al.* (1984)
β-Glucuronidase[d]	Jefferson *et al.* (1986)
Firefly luciferase	Ow *et al.* (1986)
Bacterial luciferase	Koncz *et al.* (1987)

[a]These traits provide selectable phenotypes for plants or plant cells that are transformed. The list does not include the tumorous phenotype of phytohormone autonomy.

[b]Herbicide resistance traits have not been reported in use as selection traits but obvious potential for their use exists.

[c]Cells can be selected on the basis of use of indoleacetamide as an auxin, either at nontoxic or toxic levels.

[d]Toxic, or "suicide," substrates may be developed for use of β-glucuronidase as a selective agent.

bacterially expressed resistance gene on the plant transformation vector, which in the case of pMON200, is the gene for spectinomycin resistance. A minimal media is routinely used and is generally adequate to counterselect against the auxotrophic *E. coli* cells present in the mating mixture. Stabilization of the vector in *Agrobacterium* requires integration of the vector into a resident Ti plasmid via a region of homology shared between the resident plasmid and vector. In the case of pMON200, the integration site is a short segment of the Ti

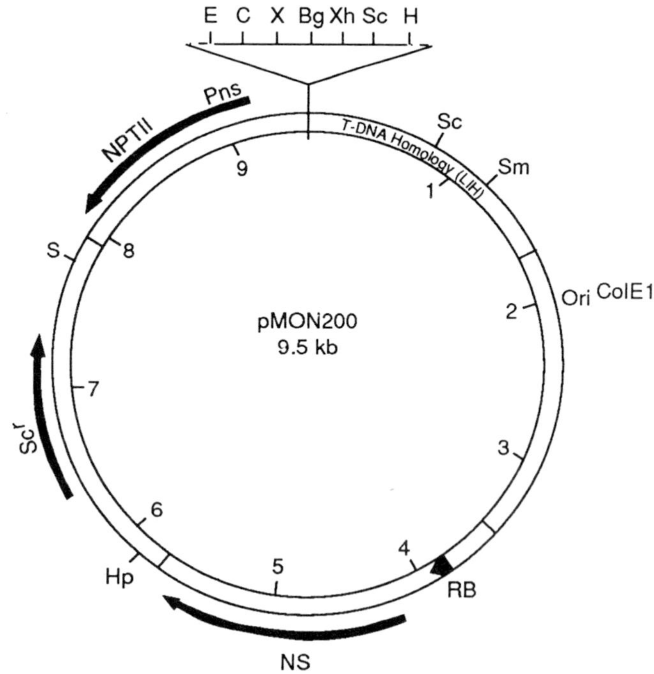

Figure 1.3 Integrative or intermediate-type T-DNA vector. The vector pMON200 contains two plant-expressed genes: the chimeric kanamycin resistance gene (neomycin phosphotransferase or NPTII) driven by the nopaline synthase promoter (Pns), and the nopaline synthase (NS) gene. The T-DNA homology region is a sequence that allows homologous recombination between the vector and the resistant Ti plasmid (B6 or A6 octopine Ti T-DNA). Sc^r, bacterially expressed gene for spectinomycin resistance; Ori^{ColE1}, origin or replication from ColE1 (pBR322) type plasmid; RB, right T-DNA border; Bg, *Bgl*II; C, *Cla*I; E, *Eco*RI; H, *Hind*III; Hp, *Hpa*I; S, *Stu*I; Sc, *Sac*I; Sm, *Sma*I; X, *Xba*I; Xh, *Xho*I.

plasmid (limited internal homology or LIH site; Fraley *et al.*, 1985). The exact requirements for efficient recombination have not been extensively studied. However, evidence from site-directed mutagenesis suggests that a DNA sequence of 500 bp or greater is adequate (Ruvkun and Ausubel, 1981). In theory, a large homologous region creates a large duplication in the T-DNA which could result in instability due to reversal of the integration process. However, given continual antibiotic resistance selection, the rate of loss has not prevented the use of this approach. An alternative approach is to design a vector that permits selection of a double recombination event which would eliminate sequence duplication provided that flanking sites of homology are present in the vector. A more detailed description of T-DNA vector manipulations can be found in Lichtenstein and Draper (1985).

Integrative vectors can be designed in several fashions. The T-DNA borders and the plant-selectable marker gene can reside on the disarmed Ti plasmid (Zambryski *et al.*, 1983) or these elements can be introduced with the shuttle vector (Fraley *et al.*, 1985). The Ti plasmid pGV3850, one of the earliest disarmed resident Ti plasmids for use with integrative vectors, was constructed such that the oncogenic genes of the T-DNA region were replaced with pBR322 sequences (Zambryski *et al.*, 1983). The T-DNA borders and the nopaline synthase (nopaline dehydrogenase) genes remained intact. The shuttle vector portion contains a plant selectable marker (in later uses; Herrera-Estrella *et al.*, 1983), homologous recombination site, and bacterially expressed antibiotic resistance gene for selection and maintenance in bacteria. The pMON200 vector involves a slightly different approach. The shuttle vector contains the plant-expressed kanamycin resistance gene, nopaline synthase gene, spectinomycin resistance gene, and the T-DNA border. The pMON200 plasmid has been referred to as a split and vector (SEV) since one border is supplied by the shuttle vector and the other border resides on the Ti plasmid (Fraley *et al.*, 1985).

Binary vectors are designed very similar to the integrative vectors with the exception that the components are assembled on a replicon that is stably maintained in *Agrobacterium*. The vector pGA482 is illustrative of the binary vector (Figure 1.4; An, 1986). The vector contains a plant-selectable marker [NPTII coding sequence fused to the nopaline synthase (nopaline dehydrogenase) promoter] and a multiple restriction endonuclease cloning site which are situated between right and left borders. In addition, pGA482 contains a gene for resistance to tetracycline for selection of bacteria containing the vector and a *cos* site which allows utilization of the λ bacteriophage packaging system. The replication functions of pGA482 are derived from the RK2 broad host range plasmid (Ditta *et al.*, 1980). Theoretically, other bacterial plasmids that can be used for the basis of a binary vector include the RSF1010-related plasmids (Bagdasarian *et al.*, 1981), pSa (incompatibility group W) plasmids (Leemans *et al.*, 1982b; Tait *et al.*, 1983), and endogenous *Agrobacterium* plasmids which are compatible with resident Ti or Ri plasmids (Jouanin *et al.*, 1985). The discovery that *mob* and *oriT* (*bom*) functions of other plasmids can replace the T-DNA border suggests that an even wider variety of plasmids can serve not only as transformation vector replicons but that they can supply the T-DNA border sequence as well (Buchanan-Wollaston, 1987).

Binary vectors like integrative vectors can be used with either disarmed or wild-type Ti plasmid containing strains. For most gene transfer studies involving regenerated plants, disarmed strains are preferred. LBA4404 is representative of a disarmed strain (Hoekema *et al.*, 1983). The strain is derived from an octopine-type Ti plasmid strain. The *vir* genes are intact. However, the complete T_R-DNA and T_L-DNA segments have been removed by *in vivo* deletion techniques (Ooms *et al.*, 1982). This strain can mediate transfer if a T-DNA region is supplied *in trans*.

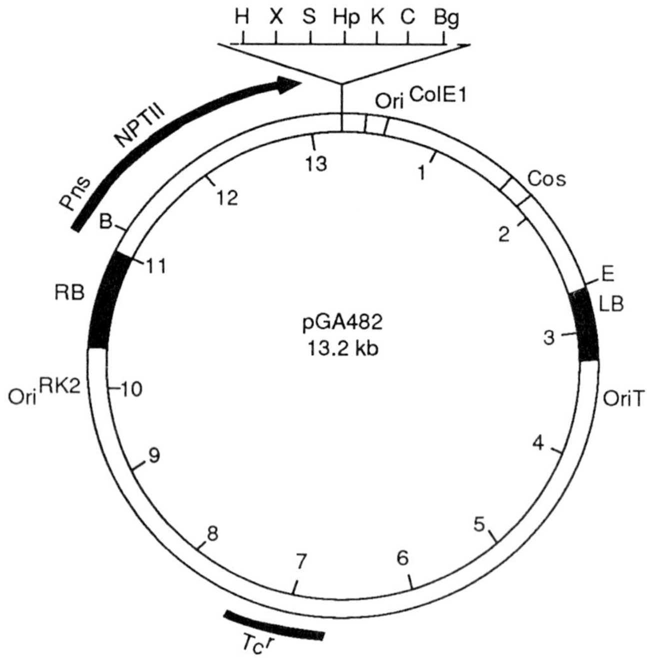

Figure 1.4 Autonomous or binary type T-DNA vector. The vector pGA482 contains a single plant-expressed chimeric kanamycin resistance gene (NPTII) driven by the nopaline synthase promoter (Pns). The plasmid has two origins of replication; the pBR322 origin (Ori^{ColE1}) and the pRK2 origin (Ori^{RK2}) which allow replication in *Escherichia coli* and *Agrobacterium*. RB, right T-DNA border; LB, left T-DNA border; Cos, λ phage cos site; OriT, pRK2 origin of transfer; Tc^r, bacterially expressed tetracycline resistance gene; C, *Cla*I; B, *Bam*HI; B, *Bgl*II; H, *Hin*dIII; Hp, *Hpa*I; K, *Kpn*I; S, *Sst*I; X, *Xba*I.

If the undesirable effects of the T-DNA oncogenes are not important, the cotransformation approach in which a binary vector is introduced with a resident wild-type Ti or Ri plasmid can be useful. Independent integration of T-DNA from a vector and a coresident Ti or Ri plasmid occurs frequently enough that cells can generally be selected and screened for inheritance of the vector T-DNA alone (Depicker *et al.*, 1985). This approach has been exploited most often with the Ri plasmid strains. Shahin *et al.* (1986b) introduced pARC8 (a standard binary vector conferring plant resistance to kanamycin; Simpson *et al.*, 1986) into *Agrobacterium rhizogenes* strain A4. Upon infection of punctured tomato cotyledons, no Ri T-DNA was found in plants when transformants were initially selected for kanamycin resistance via callus induction. However, the outcome of regenerated plants can vary with the method of selection. If roots were selected from inoculated inverted cotyledons virtually all the transformants contained vector and Ri T-DNA. Instead of selecting for cells which have not received the Ri T-DNA, Stougaard *et al.* (1986) exploited the propensity of Ri-transformed cells to differentiate into roots in order to transform the leguminous plant *Lotus corniculatus*. Transformation and regeneration of *L. corniculatus* is apparently facilitated by the effects of the Ri plasmid T_L-DNA. Whether the Ri T_L-DNA improves regeneration is not clear. However, roots containing the Ri T_L-DNA do appear to form more lateral branches. A Ri T_L-DNA integrative vector was designed to not disturb the T-DNA genes affecting root formation (Stougaard *et al.*, 1987). A drawback to this system is that the T_L-DNA genes also alter the developmental behavior of other plant organs, particularly the leaves (Tepfer, 1984; Taylor *et al.*, 1985; Spena and Schell, 1987). This aspect precludes the system for general gene transfer with plants other than species that are recalcitrant to regeneration by other names. On the other hand, Ri-transformed roots may offer potential for the production of plant metabolites and may be desirable in certain circumstances (for review see Hamill *et al.*, 1987).

Methods for inoculation of plant tissues vary depending on the plant species and the *Agrobacterium* delivery system. Initial *in vitro* infections were performed with regenerating protoplasts (Wullems *et al.*, 1981). This approach has been largely circumvented by the leaf disk procedure (Horsch *et al.*, 1985). In practice, the procedure can be performed with any tissue explant that provides a good source for initiation of whole plant differentiation. In this regard, newly emerged cotyledons have provided useful material which can be prepared fairly rapidly (Shahin *et al.*, 1986b). In the disk or explant method, a tissue segment is excised and inoculated with bacteria. The infected tissue is allowed to incubate for 1 to 3 days depending on the extent of bacterial growth. Application of a vacuum and subsequent release during bacterial infection has been suggested to improve transformation rate (Burow *et al.*, 1990). Some plants appear to promote more rapid bacterial growth and, therefore, the explants should be rescued before overgrowth by the bacteria occurs. Selection against bacterial growth is usually accomplished by the inclusion of carbenicillin and/or cefotaxione in the plant media. Nurse tissue may be included during this step to help promote division and growth of the infected cells (Horsch *et al.*, 1985; Burow *et al.*, 1990). Nurse tissue may also help in *vir* gene induction. Tissue explants are then moved onto a selective media that eliminates the bacteria. The media is also designed for the selection of transformed plant cells and, therefore, contains the appropriate antibiotic such as kanamycin or hygromycin. The appropriate levels of phytohormones are also included in the media for the type of tissue development desired. Generally, callus and shoot formation have been induced. However, in the case of Ri plasmid-mediated transformation in one *Arabidopsis* transformation scheme (Valvekens *et al.*, 1988) root development is desired. A radical new method of inoculation is the treatment of imbibing *Arabidopsis thaliana* seeds with bacteria (Feldmann and Marks, 1987). Subsequent germination of the progeny of the treated seeds results in a low percentage of transformed plants. Examination of the T-DNA structure of these plants indicates that transformation occurred through the normal T-DNA delivery process. Over 8000 transformed lines have been generated by the seed treatment procedure (Feldmann, 1991).

Perhaps the final issue regarding T-DNA transformation is the nature of the integration events and the stability of integrated sequences. T-DNA copy number can vary considerably in transformed plants (Thomashow *et al.*, 1980; Zambryski *et al.*, 1982; Spielmann and Simpson, 1986; Jones *et al.*, 1987), and two unlinked T-DNA elements can integrate into the same plant cell (Depicker *et al.*, 1985; de Framond *et al.*, 1986; Petit *et al.*, 1986). T-DNA junctures can vary from tandem inverted repeats, direct repeats, or single copy insertions (Zambryski *et al.*, 1982; Spielmann and Simpson, 1986; Jorgensen *et al.*, 1987). Recently, the rearrangement of target plant sequences at the site of insertion has also been documented (Gheysen *et al.*, 1987). The type of insertion event which predominates may vary depending on different factors such as the particular vector employed. Considerable differences (approximately 10% vs. 70%, respectively) in the percentage of inverted tandem insertions between LBA4404/pKLBG2 (pARC8 derivative) and the C58/pGV3850KanR systems have been observed (Jorgensen *et al.*, 1987). The presence of both multiple insertion sites and inverted tandem structures suggests some differences in the mode of DNA uptake from direct DNA transformation of mammalian and plant cells (De Jonge and Bootsma, 1984; Potrykus *et al.*, 1985a; Roth *et al.*, 1985). Direct DNA transfer and incorporation in cells often result in large direct repeated arrays of introduced DNA. Presumably these differences reflect the efficiency and mode of transfer of the T-DNA.

Under the appropriate conditions, homologous recombination between T-DNA and genomic DNA has been reported (Lee *et al.*, 1990). Transformed cells were selected on the basis of *in vivo* production of a chlorsulfuron-resistant allele of the acetolactate synthase gene. Homologous recombination has also been reported to occur between two DNA sequences that were introduced by direct DNA uptake methods (Baur *et al.*, 1990). Without selection for the homologous recombination event, the far greater number of nonhomologous integration events makes screening impractical. While initial experiments have been performed in cell cultures, improvements may eventually permit targeted mutagenesis in plants.

The type of insertion probably has some bearing on the stability of the insertion, and large tandem arrays are likely to be less stable than single insertions (Potrykus *et al.*, 1985b; Müller *et al.*, 1987). Several plants with multiple insertions were noted to produce untransformed progeny at a higher frequency than plants containing simple insertions (Chyi *et al.*, 1986; Feldmann and Marks, 1987). However, with the possible exception of plants containing large tandem inserts, single and low copy T-DNA insertions appear to be very stable (Müller *et al.*, 1987). Frequencies of T-DNA loss in transgenic tobacco plants with low copy number inserts were measured at 1.3 and 5.6 per 10^4 gametes. One should note, however, that loss of T-DNA gene expression, probably by DNA methylation, has also been observed (Heberle-Bors *et al.*, 1988). Factors affecting the probability of a gene being suppressed in plants have not been determined (Budar *et al.*, 1986; Feldmann and Marks, 1987). Therefore, studies suggest that the majority of T-DNA insertions by *Agrobacterium*-mediated transfer are stably inherited in normal Mendelian fashion (Budar *et al.*, 1986; Chyi *et al.*, 1986; Wallroth *et al.*, 1986; Feldmann and Marks, 1987; Müller *et al.*, 1987). On the other hand, not all T-DNA copies in a given line are expressed, and increased copy number does not always correlate with higher levels of gene expression (Feldmann and Marks, 1987; Jones *et al.*, 1987). Several studies have indicated that suppression of T-DNA gene expression is associated with DNA methylation (Hepburn *et al.*, 1983; Amasino *et al.*, 1984). Factors affecting the probability of gene expression being suppressed in plants have not been determined.

1.4 Uses of *Agrobacterium* Vectors

The greatest use of T-DNA gene transfer has been in the analysis of plant gene expression. Although an in-depth review of plant gene expression is not intended nor possible here, a number of basic observations can be discussed. In general, gene transfer as a tool to study plant gene expression has exceeded most expectations. In most cases, transferred genes function similar to the endogenous copy. Thus, regulatory elements from plants that are very difficult to work with in culture can be studied by isolation, modification, and reintroduction to a manageable species. For example, light-regulated genes from wheat also respond to light induction in tobacco (Lamppa *et al.*, 1985). In one case, the maize alcohol dehydrogenase (ADH) gene was not functional after transfer to tobacco. However, the expression elements were not completely defective in tobacco, and upon introduction of an octopine synthesis promoter in the promoter region of ADH, anaerobic regulated expression was observed (Ellis *et al.*, 1987). On the other hand, genes from more widely divergent organisms such as mammalian cells and yeast have not been found to express efficiently in plant cells (Barton *et al.*, 1983; Caplan *et al.*, 1983). Expression can be analyzed after stable incorporation and callus growth or plant regeneration, or expression can be measured transiently by assaying soon after bacterial infection (Horsch *et al.*, 1985; An, 1986). Low-level expression of otherwise developmentally regulated genes has been noted for callus tissue and primary regenerated tissue that was assayed after gene transfer. Expression can revert to the normal pattern after passage through seed (An, 1987). Many studies involving examples of gene expression analysis

TABLE 1.2
Transgenic Plant Species Produced by *Agrobacterium* Transformation

Species[a]	Refs.
Nicotiana tabacum	De Block *et al.* (1984)
Nicotiana plumbaginifolia	André *et al.* (1986)
Nicotiana glauca	Taylor *et al.* (1985)
Petunia hybrida	Horsch *et al.* (1985)
Solanum tuberosum	Ooms *et al.* (1985b)
Lycopersicon esculentum	Sukhapinda *et al.* (1987), McCormick *et al.* (1986)
Helianthus annuus	Everett *et al.* (1987)
Daucus carota	Chilton *et al.* (1982)
Brassica oleracea L. var *Botrytis*	David and Tempé (1988)
Brassica napus	Ooms *et al.* (1985a), Guerche *et al.* (1987), Pua *et al.* (1987)
Apium graveolens	Catlin *et al.* (1988)
Cucumis melo	Fang and Grumet (1990)
Cucumis sativus	Trulson *et al.* (1986)
Malus pumila	James *et al.* (1989)
Populus species	Fillatti *et al.* (1987b)
Arabidopsis thaliana	Lloyd *et al.* (1986), Feldmann and Marks (1987)
Pisum sativum	De Kathen and Jacobsen (1990)
Convolvulus arvensis	Tepfer (1984)
Medicago sativa	Shahin *et al.* (1986a)
Lotus corniculatus	Stougaard *et al.* (1986)
Fragaria × *ananassa*	Nehra *et al.* (1990)
Stylosanthes humilis	Manners (1988)

[a]The list includes only species that have been regenerated to whole plants and is not all inclusive, but is intended to convey the variety of species that are amenable to genetic engineering procedures using *Agrobacterium*. Root cultures of numerous species have been established (for review see Hamill *et al.*, 1987).

are now in progress, and one can expect a sizable increase in information regarding promotor control of plant gene expression.

T-DNA transfer has numerous other applications in addition to studies designed to characterize gene regulatory elements. The system affords an ever-increasing array of interspecific and modified gene transfers (Table 1.2). Perhaps the greatest attention has been given to the prospects of improved plant protection by the engineering of herbicide (Comai *et al.*, 1985; Haughn and Somerville, 1986; Shah *et al.*, 1986; Fillatti *et al.*, 1987a), insect (Fischhoff *et al.*, 1987; Vaeck *et al.*, 1987), and plant viral resistance (Abel *et al.*, 1986; Loesch-Fries *et al.*, 1987; Tumer *et al.*, 1987; Golemboski *et al.*, 1990). Although built on the same principle as the integrative and binary vectors, a number of vectors are available that have multiple cloning sites situated between a constitutive promoter and a RNA polyadenylation site in order to facilitate cloning and expression of heterologous genes (Velten and Schell, 1985; Rothstein *et al.*, 1987b). In one case, the chloramphenicol acetyltransferase gene was also incorporated between the cloning site and the polyadenylation site to facilitate expression assays (An, 1986).

In addition to the great potential for new crop protection strategies, T-DNA gene transfer mechanisms have already provided important tools for the study of virus biology in general. By making infectious clones for viruses and viroids as has been done for gemini viruses maize streak virus (Grimsley *et al.*, 1987), tomato golden mosaic virus (Rogers *et al.*, 1986), and potato spindle tuber viroid (Gardner and Knauf, 1986), specific mutations can be engineered in the viral genome and tested for the effects on viral replication and disease development. The approach is equally applicable to single viral genes such as the coat protein (Bevan *et al.*, 1985; Abel *et al.*, 1986; Loesch-Fries *et al.*, 1987; Tumer *et al.*, 1987), replicase (Golemboski *et al.*, 1990), and cell-to-cell (Deom *et al.*, 1987) movement genes or different components of viral genomes such as the satellite RNA of cucumber mosaic virus (Baulcombe *et al.*, 1986).

The study of organellar genes is also assisted by the T-DNA transfer system. Chloroplast and mitochondrial protein synthesis is directed by the chloroplast genome and synthetic machinery or by the nuclear and cytoplasmic machinery and imported into the chloroplast. One early report indicated that some T-DNA insertions occur in the chloroplast genome (De Block *et al.*, 1984). This finding opened the way for the study of organellar encoded proteins. However, no confirmation of the chloroplast genome transformation has ensued and no other organelle transformants have been obtained, suggesting they are very rare. Progress has been made regarding cytoplasmic proteins that are transported into the chloroplast or mitochondria. Nuclear genes whose products are destined for organelles function normally when reintroduced to plasmid cells via the T-DNA. When reporter genes (NPTII and CAT) were properly fused to the coding sequences of nuclear chloroplast and mitochondrial genes, these proteins were also transported (Van den Broeck *et al.*, 1985; Kuntz *et al.*, 1986; Boutry *et al.*, 1987). Therefore, T-DNA-mediated transformation can at least be used to target proteins to organelles. One can envision in the future that some chloroplast genes might be functional when properly engineered and expressed as nuclear genes. More recently, evidence for transformation of chloroplasts has been obtained using microprojectile technology (Boynton *et al.*, 1988; Svab *et al.*, 1990).

T-DNA transfer has also been used to introduce transposable elements from heterologous species such as the transfer of the *Ac* element from maize to tobacco (Baker *et al.*, 1986). The transposon can then in turn be used for insertional mutations (transposon tagging; Freeling, 1984). The T-DNA itself can be used as a mutagen, and an interesting variation of this type of approach has been to fuse a promoterless NPTII gene near the T-DNA border (André *et al.*, 1986). The absence of a promoter allows selection for plant cells where the T-DNA insertion has occurred behind an endogenous promoter element by virtue of the cells ability to grow in the presence of kanamycin. A hypothetical application of T-DNA-mediated transfer involves the possible shotgun approach whereby an entire genomic or cDNA library is transferred to plant cells *en masse* (van den Elzen *et al.*, 1985a; Simoens *et al.*, 1986). The shotgun approach requires a good selection scheme for the relatively rare transformants containing the desired gene. An *adh* mutant in *Arabidopsis thaliana* was the first mutation to be complemented by T-DNA transfer of the *adh* gene (Chang *et al.*, in press). However, complementation was not accomplished by shotgun transformation because a probe was available for the gene. A model ten-member library was transferred to tobacco *en masse* (Prosen and Simpson, 1987). At least two of 66 plants were obtained which contained each of the library members. One plant contained at least four members, while the average was 1.3 different T-DNA regions per plant. Without further improvements, shotgun experiments would require the manipulation of extremely large numbers of plants to insure a 95% probability of transferring a given gene or fragment. The transfer of traits that can be selected in tissue culture appears to offer the best candidates for shotgun transfer at present.

The seed treatment procedure for *Arabidopsis* transformation has proven to be a powerful tool in analyzing plant genes (Feldmann and Marks, 1987). Several genes have been isolated using the T-DNA as a mutagen and a tag. The GL1 and AG genes, which affect trichome and floral development, respectively, were both isolated by cloning the mutated copy containing the T-DNA and subsequent cloning and reintroduction of the wild-type copy (Herman and Marks, 1989; Yanofsky *et al.*, 1990). A large number of additional mutant lines have been isolated and await further analysis (Feldmann, 1991).

1.5 Host Range of *Agrobacterium*

The *Agrobacterium* system is often considered to be limited to the transformation of susceptible plants, and, while the list of susceptible plants was extensive (De Cleene and De Ley, 1976) these plants were largely dicotyledonous. However, susceptibility was always judged on the basis of tumorous growths or swellings, not DNA transfer. From the earlier discussions, it is apparent that susceptibility of or lack of tumor formation can be the consequence of dysfunction at different steps in the transformation process and not simply the lack of T-DNA transfer. However, the limiting factor is not necessarily DNA transfer, and, for genetic engineering purposes, a distinction between T-DNA transfer and disease symptoms (galls) must be made. The evidence clearly demonstrates that T-DNA transfer does occur with some monocot species. Opine assays have first suggested that T-DNA enters monocot cells since opines have been detected first in maize embryos, *Chlorophytum capense* (Hooykaas-Van Slogteren *et al.*, 1984), *Narcissus* cv. Paperwhite (Hooykaas-Van Slogteren *et al.*, 1984), and *Asparagus officinalis* (Hernalsteens *et al.*, 1984) after exposure to *Agrobacterium*. Opine synthesis is always accepted with caution in the absence of T-DNA analysis, particularly in light of a report that some plants can synthesize opines without the T-DNA genes (Christou *et al.*, 1986). However, in the reports showing opine synthesis in infected tissue, appropriate controls would seem to indicate that opine synthesis is a consequence of T-DNA transfer. More recently, infectious viral genomes have been introduced into monocots (Grimsley *et al.*, 1987). Disease symptoms are a sensitive assay for the introduction of DNA into the monocot cells, and disease symptoms were observed when the virus was transmitted by *Agrobacterium* to maize plants. Recently, actual T-DNA transfer data demonstrated DNA integration in *Dioscorea bulbifera* (Schafer *et al.*, 1987), *A. officinalis* (Bytebier *et al.*, 1987), and rice (Raineri *et al.*, 1990). Thus, monocot transformation may prove less intransigent than initially perceived, and the utility of T-DNA transfer may expand even further.

1.6 Alternative DNA Transfer Methods

Agrobacterium delivery of genetic material has no precedence in other eukarotic organisms. However, other gene transformation methods for mammalian and yeast cells were developed concomitant to the development of *Agrobacterium*-mediated transfer in plants, and these strategies have also been adapted to plant cells. The methods have been overshadowed by the alternative *Agrobacterium* method, but some techniques will undoubtedly gain more prominence as methods improve. Alternative methods can be divided into two types of approaches, direct physical introduction of DNA and transmission of genetic material by modified plant viruses. Viral gene transfer can also involve physical transmission to the plant (e.g., rub inoculation). The difference lies in that the approach attempts to exploit the replicative and, in some cases, the expression properties of the viral genome. Alternative methods have been suggested as ways to circumvent the apparent inability of *Agrobacterium* to deliver DNA efficiently to some species. Several of the techniques also offer the potential of greatly simplifying gene transfer procedures and eliminating the need for extensive tissue culture facilities. Some alternatives are confined to single reports and have not been developed to the point of routine use in gene transfer.

An interesting variety of approaches have been explored for the direct incorporation of DNA, or free DNA delivery as these protocols are sometimes generally termed. The most prominent method for the direct introduction of DNA into plant cells is the transformation of plant cell protoplasts (reviewed in Potrykus *et al.*, 1985b). The approach necessarily requires the ability to isolate viable protoplasts. Whole plant studies require further the ability to regenerate plants from the protoplasts. Initial experiments involved addition of the DNA to protoplasts in the presence of poly(L-ornithine) (Davey *et al.*, 1980). This treatment has now largely been replaced with $Ca^{2+}PO_4$/polyethylene glycol (Krens *et al.*, 1982) or Mg^{2+} and polyethylene glycol (Negrutiu *et al.*, 1987). Recently, electroporation has emerged as an alternative method for protoplast transformation (Neumann *et al.*, 1982; Fromm *et al.*, 1986; Morikawa *et al.*, 1986; Okada *et al.*, 1986). The protoplasts are exposed to a strong electric field and briefly shocked. This treatment is believed to create small depolarized pores in the protoplast membrane which allow passage of small molecules including DNA or RNA. Originally developed for mammalian cell transformation, the electroporation method has been adapted to protoplasts from both monocot and dicot species. Where regeneration techniques exist, stable transformed calli have been obtained. The method is very useful for transient expression assays (Ou-Lee *et al.*, 1986; Ow *et al.*, 1986), and has been shown to be useful for intact tissues (Dekeyser *et al.*, 1990).

A number of transformation protocols have been developed with the goal of avoiding the isolation of protoplasts and the inherent difficulties of callus and plant regeneration. Microinjection is one such method (Crossway *et al.*, 1986). However, so far only protoplasts have proven to be manageable for injection due to the difficulty of monitoring cell penetration and injection. Toward the same end, DNA has been absorbed to magnesium tungsten or gold particles and literally blasted into the cells by means of detonating a gunpowder charge behind a load of particles (Klein *et al.*, 1987). Again, the idea is to introduce DNA directly into cells with or without protoplast isolation. The power of the technique may lie in the ability to transform tissue without first making protoplasts (Gordon-Kamm *et al.*, 1990). Pollen transformation has been explored as a means to introduce genes but the technique has not been convincingly demonstrated (Ohta, 1986).

One technique, referred to here as macroinjection, has been used in successfully transforming rye plants (de la Pená *et al.*, 1987). DNA was injected into the immature tillers of the plant and was incorporated into germplasm (archsporial) cells. Previous work had established that the archsporial cells were sensitive to macromolecules of caffeine and colchicine 14 days before the cells' first meiotic division. Two transformed plants were obtained from 4023 seedlings. A total of 135 plants were injected. The transformed nature of the plants was confirmed by Southern blot analysis of the transgenic plants. This technique has great potential for wide practical application. However, no independent reports using the technique have appeared. The frequency may be too low for mass transformation experiments such as shotgun-type cloning or screening experiments.

The techniques of free DNA transformation have not involved the complex vectors as in the *Agrobacterium* system since DNA uptake and incorporation rely on the plant physiology. Most of the DNA used for transformation involves an *Escherichia coli* replicon (often pBR322 or pUC derivative) containing a plant-expressed marker. Transformation of monocot tissue culture cell lines has been selected for kanamycin resistance using either the nopaline synthase or CAMV 35S promoter (Lörz *et al.*, 1985; Fromm *et al.*, 1986). In the case of the rye macroinjection procedure the nopaline synthase promoter fused to the NPTII gene was used (de la Pená *et al.*, 1987). The actual process of DNA incorporation is not well understood and refinement of vectors may be possible once the process is better understood. One possible improvement of the technique might involve designing a vector which utilizes the activity of a transposon to improve the efficiency of DNA integration. Such use would be similar to the P element transformation vectors in *Drosophila* (Rubin and Spradling, 1982).

Plant DNA and RNA viruses offer possibilities for plant transformation vectors. Spurred by efforts both to further understand viral biology and toward the development of vectors, DNA and RNA viruses have been demonstrated to mediate expression of foreign genes in plant cells (Brisson *et al.*, 1984; French *et al.*, 1986; Takamatsu *et al.*, 1987). Theoretically, these vectors could have some advantages in certain circumstances beyond studies of virus function. Advantages of such vectors might involve improvements in the ease of gene introduction. Once the proper construct is obtained, the plant could be simply rub-inoculated. If high levels of the transgenic product are desired, the high titer that is often achieved by plant viruses could facilitate high levels of product synthesis. The timing of gene expression could also be controlled by the timing of inoculation.

On the other hand, viral vectors present unique problems. The presence of undesirable disease symptoms may interfere with the intended use. Viral vectors may also involve the release of infectious agents into the environment and therefore may be difficult to control in terms of other plants becoming infected. Finally, the problem of stable transformation, if desired, must be overcome. Only the DNA viruses appear to offer the possibility of integration into the host genome. While integration has not been observed, techniques enabling integration could expand the use of viral vectors for transformation of species which are difficult to regenerate in tissue culture.

DNA viruses would seem to offer the best system in terms of a classical vector for inserting foreign genes and hence DNA viruses have received considerable attention as possible gene transfer agents. Two types of plant DNA viruses are known to exist, the cauliflower mosaic (caulimo) virus group and the gemini virus group. DNA clones of both types of virus are infectious (Howell *et al.*, 1980; Gronenborn *et al.*, 1981; Hamilton *et al.*, 1983). Cauliflower mosaic virus (CaMV) is a circular double-stranded virus which replicates via an RNA intermediate. Several open reading frames (ORF II and ORF VII) on the virus genome are dispensable for infectious DNA, and these regions can be substituted with alternate DNA sequences. A bacterially derived gene for methotrexate resistance has been inserted into CaMV, and the engineered virus is able to infect and to confer resistance on turnip plants (Brisson *et al.*, 1984). A problem with CaMV is the present limit to the size of gene which can be delivered, and the problem of stable integration if so desired (Brisson and Hohn, 1986). One could possibly engineer a specific plant to be a complementation host for defective CaMV vectors which are capable of encoding a much larger insert. Whether copies of CaMV become inserted into the plant genome at a relatively low frequency has been difficult to assess. With marker genes inserted in the genome it should be possible to determine if viral DNA does become inserted into germline cells by screening large numbers of seed from infected plants for drug resistance.

RNA viruses present special problems when considering the construction of gene vectors. The vectors embody a different concept than *Agrobacterium* T-DNA vectors or DNA transformation. The primary consideration is designing a vector which produces infectious transcripts and, therefore, has to preserve the terminal structure of the RNA in order for the transcript to be a good substrate for the viral replicase (Ahlquist, 1986). The bacterial phage promoters have proven ideal for directing the synthesis of RNA from DNA clones (Melton *et al.*, 1984). Thus, the basic plant virus vector consists of a DNA plasmid with a bacteriophage promoter. The promoter is designed to initiate a transcript with the proper 5′ and 3′ termini (and other portions of the plant virus which are desired). The gene of interest is cloned into the vector in a manner which does not disrupt the terminal structure. Transcripts are synthesized *in vitro,* processed by adding a 7-methylguanosine cap, and rub-inoculated onto a plant. The first RNA virus expression vector, pPM1, utilized a modified λ P_r promoter (Ahlquist and Janda, 1981). This vector was used to synthesize infectious transcripts of each brome mosaic virus (BMV) genomic RNA from BMV cDNA clones. More recent vectors have been designed around the more highly efficient T7 RNA polymerase promoter (Langereis *et al.*, 1986; Janda *et al.*, 1987). Gene chimeras of BMV-CAT and TMV-CAT have demonstrated that this approach can mediate the expression of foreign genes that are properly engineered into the viral genome (French *et al.*, 1986; Takamatsu *et al.*, 1987).

1.7 Conclusions

Great strides have been made in our ability to transfer genes to higher plants, and presently a variety of plant transformation vectors are available for gene transfer. One might even argue that our ability to manipulate plants genetically has improved faster than our understanding of basic plant biochemical and developmental processes. The question that is confronted in many situations is not whether a particular gene can be transferred but which gene or genes need to be manipulated for a desired plant phenotype. Single genetic traits such as herbicide resistance and possibly insect resistance as conferred by *Bacillus thuringiensis* toxin will likely first reach the field as practical applications of the new technology. Other important traits of agronomic interest such as cold hardiness, drought and disease resistance, or salt tolerance await further characterization of the genes controlling the phenotypes. In addition to the immediate application of gene transfer technology to crop improvement, the techniques have already been demonstrated to provide important tools in the further understanding of plant molecular biology.

Given the array of transformation schemes, new and ingenious methods to modify plant content and growth can be expected in the future. The development of antisense expression and ribozyme technology already shows promise in extending our ability to modify the expression of traits in plants (Rothstein *et al.*, 1987a; Haseloff and Gerlach, 1988). The more recent development regarding the production of antibodies in plants is another example with great potential for the modulation of plant traits (Hiatt *et al.*, 1989). The *Agrobacterium* system has steadily expanded to include both dicotyledonous and monocotyledonous species. This progress may be even further accelerated as a better understanding of the T-DNA transfer process is attained. Current work on the factors controlling the host range of *Agrobacterium* infections may also contribute to a wider application of T-DNA-mediated gene transfer. In addition, inroads are continually being made in regards to plant regeneration. The continuous development of new plant transformation technology is providing an exciting period in plant science.

Acknowledgments

The author would like to thank Marie Ward for preparation of the manuscript and acknowledge the support from the Kansas Agricultural Experiment Station, Kansas State University, Manhattan.

References

Abel, P. P., Nelson, R. S., De, B., Hoffmann, N., Rogers, S. G., Fraley, R. T., and Beachy, R. N. (1986). *Science* 232, 738–743.

Ahlquist, P. (1986). *Methods Enzymol.* 118, 704–716.

Ahlquist, P., and Janda, M. (1981). *Mol. Cell. Biol.* 4, 2876–2882.

Akiyoshi, D. E., Klee, H., Amasino, R. M., Nester, E. W., and Gordon, M. P. (1984). *Proc. Natl. Acad. Sci. U.S.A.* 81, 5994–5998.

Albright, L. M., Yanofsky, M. F., Leroux, B., Ma, D., and Nester, E. W. (1987). *J. Bacteriol.* 169, 1046–1055.

Amasino, R. M., Powell, A. L. T., and Gordon, M. P. (1984). *Mol. Gen. Genet.* 197, 437–146.

An, G. (1987). *Mol. Gen. Genet.* 207, 210–216.

An, G., Watson, B. D., Stachel, S., Gordon, M. P., and Nester, E. W. (1985). *EMBO J.* 4, 277–284.

An, G., Watson, B. D., and Chiang, C. C. (1986). *Plant Physiol.* 81, 301–305.

André, D., Colau, D., Schell, J., Van Montagu, M., and Hernalsteens, J.-P. (1986). *Mol. Gen. Genet.* 204, 512–518.

Ankenbauer, R. G., and Nester, E. W. (1990). *J. Bacteriol.* 172, 6442–6446.

Baba, A., Hasezawa, S., and Syono, K. (1986). *Plant Cell Physiol.* 27, 463–471.

Bagdasarian, M., Lurz, R., Ruckert, B., Franklin, F. C. H., Bagdasarian, M. M., Frey, J., and Timmis, K. N. (1981). *Gene* 16, 237–247.

Baker, B., Schell, J., Lörz, H., and Fedoroff, N. (1986). *Proc. Natl. Acad. Sci. U.S.A.* 83, 4844–4848.

Barker, R. F., Idler, K. B., Thompson, D. V., and Kemp, J. D. (1983). *Plant Mol. Biol.* 2, 335–350.

Barry, G. F., Rogers, S. G., Fraley, R. T., and Brand, L. (1984). *Proc. Natl. Acad. Sci. U.S.A.* 81, 4776–4780.

Barton, K. A., Binns, A. N., Matzke, A. J. M., and Chilton, M.-D. (1983). *Cell* 32, 1033–1043.

Baulcombe, D. C., Saunders, G. R., Bevan, M. W., Mayo, M. A., and Harrison, B. D. (1986). *Nature (London)* 321, 446–449.

Baur, M., Potrykus, I., and Paszkowski, J. (1990). *Mol. Cell. Biol.* 10, 492–500.

Bevan, M. W. (1984). *Nucl. Acids Res.* 12, 8711–8721.

Bevan, M., Barnes, W. M., and Chilton, M.-D. (1983a). *Nucl. Acids Res.* 11, 369–379.

Bevan, M., Flavell, R., and Chilton, M.-D. (1983b). *Nature (London)* 304, 184–187.

Bevan, M. W., Mason, S. E., and Goelett, P. (1985). *EMBO J.* 4, 1921–1926.

Binns, A. N., and Thomashow, M. F. (1988). *Ann. Rev. Microbiol.* 42, 575–606.

Bolton, G. W., Nester, E. W., and Gordon, M. P. (1986). *Science* 232, 983–985.

Boutry, M., Nagy, F., Poulsen, C., Aoyagi, K., and Chua, N.-H. (1987). *Nature (London)* 328, 340–342.

Boynton, J. E., Gillham, N. W., Harris, E. H., Hosler, J. P., Johnson, A. M., Jones, A. R., Randolph-Anderson, B. L., Robertson, D., Klein, T. M., Shark, K. B., and Sanford, J. C. (1988). *Science* 240, 1534–1538.

Brisson, N., and Hohn, T. (1986). *Methods Enzymol.* 118, 659–684.

Brisson, N., Paskowsky, J., Penswick, J. R., Gronenborn, B., Potrykus, I., and Hohn, T. (1984). *Nature (London)* 310, 511–514.

Buchanan-Wollaston, V., Passiatore, J. E., and Cannon, F. (1987). *Nature (London)* 328, 172–175.

Budar, F., Thia-Toong, L., Van Montagu, M., and Hernalsteens, J.-P. (1986). *Genetics* 114, 303–313.

Burow, M. D., Chlan, C. A., Sen, P., Lisca, A., and Murai, N. (1990). *Plant Mol. Biol. Rep.* 8, 124–139.

Bytebier, B., Deboeck, F., DeGreve, H., Van Montagu, M., and Hernalsteens, J. P. (1987). *Proc. Natl. Acad. Sci. U.S.A.* **84,** 5345–5349.

Cangelosi, G. A., Ankenbauer, A. R., and Nester, E. W. (1990). *Proc. Natl. Acad. Sci. U.S.A.* 87, 6708–6712.

Caplan, A., Herrera-Estrella, L., Inzé, D., Van Haute, E., Van Montagu, M., Schell, J., and Zambryski, P. (1983). *Science* 222, 815–821.

Caplan, A. B., Van Montagu, M., and Schell, J. (1985). *J. Bacteriol.* 161, 655–664.

Catlin, D., Ochoa, O., McCormick, S., and Quiros, C. F. (1988). *Plant Cell Rep.* 7, 100–103.

Chang *et al.*, in press.

Chilton, M.-D., Tepfer, D. A., Petit, A., David, C., Casse-Delbart, F., and Tempé, J. (1982). *Nature (London)* 295, 432–434.

Christie, P. J., Ward, J. E., Winans, S. C., and Nester, E. W. (1988). *J. Bacteriol.* 170, 2659–2667.

Christou, P., Platt, S. G., and Ackerman, M. C. (1986). *Plant Physiol.* 82, 218–221.

Christou, P., Murphy, J. E., and Swain, W. F. (1987). *Proc. Natl. Acad. Sci. U.S.A.* 84, 3962–3966.

Chyi, Y. S., Jorgensen, R. A., Goldstein, D., Tanksley, S. D., Loaiza-Figueroa, F. (1986). *Mol. Gen. Genet.* 204, 64–69.

Citovsky, V., De Vos, G., and Zambryski, P. (1988). *Science* 240, 501–504.

Comai, L., Facciotti, D., Hiatt, W. R., Thompson, G., Rose, R., and Stalker, D. (1985). *Nature (London)* 317, 741–744.

Crossway, A., Oakes, J., Irvine, J., Ward, B., Knauf, V., and Shewmaker, C. (1986). *Mol. Gen. Genet.* 202, 179–185.

Dahl, G. A., and Tempé, J. (1983). *Theor. Appl. Genet.* 66, 233.

Das, A. (1988). *Proc. Natl. Acad. Sci. U.S.A.* 85, 2909–2913.

Davey, M. R., Cocking, E. C., Freeman, J., Pearce, N., and Tudor, I. (1980). *Plant Sci. Lett.* 18, 307–313.

David, C., and Tempé, J. (1988). *Plant Cell Rep.* 7, 88–91.

De Block, M., Herrera-Estrella, L., Van Montagu, M., Schell, J., and Zambryski, P. (1984). *EMBO J.* 3, 1681–1689.

De Block, M., Botterman, J., Vandewiele, M., Dockx, J., Thoen, C., Gossele, V., Movva, N. R., Thompson, C., Van Montagu, M., and Leemans, J. (1987). *EMBO J.* 6, 2513–2518.

De Cleene, M. D., and De Ley, J. (1976). *Bot. Rev.* 42, 389–466.

de Framond, A., Barton, K. A., and Chilton, M.-D. (1983). *Bio/Technology* 1, 262–269.

de Framond, A. J., Back, E. W., Chilton, W. S., Kayes, L., Chilton, M.-D. (1986). *Mol. Gen. Genet.* 202, 125–131.

De Jonge, A. J. R., and Bootsma, D. (1984). *Int. Rev. Cytol.* 92, 133–158.

De Kathen, A., and Jacobsen, H.-J. (1990). *Plant Cell Rep.* 9, 276–279.

Dekeyser, R. A., Claes, B., De Rycke, R. M. U., Habets, M. E., Van Montagu, M. C., and Caplan, A. B. (1990). *The Plant Cell* 2, 591–602.

de la Pená, A., Lörz, H., and Schell, J. (1987). *Nature (London)* 325, 274-276.

Deom, C. M., Oliver, M. J., and Beachy, R. N. (1987). *Science* 237, 389–394.

Depicker, A., Stachel, S., Dhaese, P., Zambryski, P., and Goodman, H. (1982). *J. Mol. Appl. Genet.* 1, 561–573.

Depicker, A., Herman, L., Jacobs, A., Schell, J., and Van Montagu, M. (1985). *Mol. Gen. Genet.* 201, 477–484.

Depicker, A., Jacobs, A. M., and Van Montagu, M. C. (1988). *Plant Cell Rep.* 7, 63–66.

Ditta, G., Stanfield, S., Corbin, D., and Helinski, D. R. (1980). *Proc. Natl. Acad. Sci. U.S.A.* 77, 7347–7351.

Dürrenberger, F., Crameri, A., Hohn, B., and Koukolíková-Nicola, Z. (1989). *Proc. Natl. Acad. Sci. U.S.A.* 86, 9154–9158.

Ellis, J. G., Llewellyn, D. J., Dennis, E. S., and Peacock, W. J. (1987). *EMBO J.* 6, 11–16.

Engler, G., Depicker, A., Maenhaut, R., Villarroel-Mandiola, R., Van Montagu, M., and Schell, J. (1981). *J. Mol. Biol.* 152, 183–208.

Everett, N. P., Robinson, K. E. P., and Mascarenhas, D. (1987). *Bio/Technology* 5, 1201–1204.

Fang, G., and Grumet, R. (1990). *Plant Cell Rep.* 9, 160–164.

Feldmann, K. A. (1991). *J. Cell. Biochem. Suppl.* 15A, 26.

Feldmann, K. A., and Marks, M. D. (1987). *Mol. Gen. Genet.* 208, 1–9.

Fillatti, J. J., Kiser, J., Rose, R., and Comai, L. (1987a). *Bio/Technology* 5, 726–730.

Fillatti, J. J., Sellmer, J., McCown, B., Haissig, B., and Comai, L. (1987b). *Mol. Gen. Genet.* 206, 192–199.

Fischhoff, D. A., Bowdish, K. S., Perlak, F. J., Marrone, P. G., McCormick, S. M., Niedermeyer, J. G., Dean, D. A., Kusano-Kretzmer, K., Mayer, E. J., Rochester, D. E., Rogers, S. G., and Fraley, R. T. (1987). *Bio/Technology* 5, 807–813.

Fraley, R. T., Rogers, S. G., Horsch, R. B., Sanders, P. R., Flick, J. S., Adams, S. P., Bittner, M. L., Brand, L. A., Fink, C. L., Fry, J. S., Galluppi, G. R., Goldberg, S. B., Hoffmann, N. L., and Woo, S. C. (1983). *Proc. Natl. Acad. Sci. U.S.A.* 80, 4803–4807.
Fraley, R. T., Rogers, S. G., Horsch, R. B., Eicholtz, D. A., Flick, J. S., Fink, C. L., Hoffmann, N. L., and Sanders, P. R. (1985). *Bio/Technology* 3, 629–635.
Freeling, M. (1984). *Annu. Rev. Plant Physiol.* 35, 277–298.
French, R., Janda, M., and Ahlquist, P. (1986). *Science* 231, 1294–1297.
Fromm, M. E., Taylor, L. P., and Walbot, V. (1986). *Nature (London)* 319, 791–793.
Gardner, R. C., and Knauf, V. C. (1986). *Science* 23, 725–727.
Garfinkel, D. J., Simpson, R. B., Ream, L. W., White, F. F., Gordon, M. P., and Nester, E. W. (1981). *Cell* 27, 143–153.
Ghai, H., and Das, A. (1989). *Proc. Natl. Acad. Sci. U.S.A.* 86, 3109–3113.
Gheysen, G., Van Montagu, M., and Zambryski, P. (1987). *Proc. Natl. Acad. Sci. U.S.A.* 84, 6169–6173.
Gietl, C., Koukolíková-Nicola, Z., and Hohn, B. (1987). *Proc. Natl. Acad. Sci. U.S.A.* 84, 9006–9010.
Golemboski, D. B., Lomonosoff, G. P., and Zaitlin, M. (1990). *Proc. Natl. Acad. Sci. U.S.A.* 87, 6311–6315.
Gordon-Kamm, W. J., Spencer, T. M., Mangano, M. L., Adams, T. R., Daines, R. J., Start, W. G., O'Brien, J. V., Chambers, S. A., Adams, W. R., Jr., Willetts, N. G., Rice, T. B., Mackey, C. J., Krueger, R. W., Kausch, A. P., and Lemaux, P. G. (1990). *The Plant Cell* 2, 603–618.
Gorman, C. M., Moffat, L. F., and Howard, B. H. (1982). *Mol. Cell Biol.* 2, 1044–1051.
Graves, A. C. F., and Goldman, S. L. (1986). *Plant Mol. Biol.* 7, 43–50.
Graves, A. C. F., and Goldman, S. L. (1987). *J. Bacteriol.* 169, 1745–1746.
Grimsley, N., Hohn, T., Davis, J. W., and Hohn, B. (1987). *Nature (London)* 325, 177–179.
Gronenborn, B., Gardner, R. C., Schaefer, S., and Shepherd, R. J. (1981). *Nature (London)* 294, 773.
Guerche, P., Jouanin, L., Tepfer, D., and Pelletier, G. (1987). *Mol. Gen. Genet.* 206, 382–386.
Gusew, N., Nepveu, A., and Chartrand, P. (1987). *Mol. Gen. Genet.* 206, 121–125.
Hamill, J. D., Parr, A. J., Rhodes, M. J. C., Robins, R. J., and Walton, N. J. (1987). *Bio/Technology* 5, 800–804.
Hamilton, W. D. O., Bisaro, D. M., Coutts, R. H. A., and Buck, K. W. (1983). *Nucl. Acids Res.* 11, 7387–7396.
Haseloff, J., and Gerlach, W. L. (1988). *Nature (London)* 334, 585–591.
Haughn, G. W., and Somerville, C. (1986). *Mol. Gen. Genet.* 204, 430–434.
Hayford, M. B., Medford, J. I., Hoffmann, N. L., Rogers, S., and Klee, H. (1988). *Plant Physiol.* 86, 1216–1222.
Heberle-Bors, E., Charvat, B., Thompson, D., Schernthaner, J. P., Barta, A., Matzke, A. J. M., and Matzke, M. A. (1988). *Plant Cell Rep.* 7, 571–574.
Heinemann, J., and Sprague, G. (1989). *Nature (London)* 340, 205–209.
Helmer, G., Casadaban, M., Bevan, M., Kayes, L., and Chilton, M.-D. (1984). *Bio/Technology* 2, 520–527.
Hepburn, A. G., Clarke, L. E., Blundy, K. S., and White, J. (1983). *J. Mol. Appl. Genet.* 2, 211–224.
Herman, P. L., and Marks, M. D. (1989). *The Plant Cell* 1, 1051–1055.
Hernalsteens, J.-P., Thia-Toong, L., Schell, J., and Montagu, M. V. (1984). *EMBO J.* 3, 3039–3042.
Herrera-Estrella, A., Chen, Z.-M., Van Montagu, M., and Wang, K. (1988). *EMBO J.* 7, 4055–4062.
Herrera-Estrella, A., Van Montagu, M., and Wang, K. (1990). *Proc. Natl. Acad. Sci. U.S.A.* 87, 9534–9537.

Herrera-Estrella, L., De Block, M., Messens, E., Hernalsteens, J.-P., Van Montagu, M., and Schell, J. (1983). *EMBO J.* 2, 987–995.

Hiatt, A., Cafferkey, R., and Bowdish, K. (1989). *Nature (London)* 342, 76–78.

Hille, J., Verheggen, F., Roelvink, P., Franssen, H., van Kammen, A., and Zabel, P. (1986). *Plant Mol. Biol.* 7, 171–176.

Hinchee, M. A. W., Connor-Ward, D. V., Newell, C. A., McDonnell, R. E., Sato, S. J., Gasser, C. S., Fischhoff, D. A., Re, D. B., Fraley, R. T., and Horsch, R. B. (1988). *Bio/Technology* 6, 915–922.

Hoekema, A., Hirsch, P. R., Hooykaas, P. J., and Schilperoort, R. A. (1983). *Nature (London)* 303, 179–181.

Hoekema, A., Hooykaas, P. J., and Schilperoort, R. A. (1984). *J. Bacteriol.* 158, 383–385.

Holsters, M., de Waele, D., Depicker, A., Messens, E., Van Montagu, M., and Schell, J. (1978). *Mol. Gen. Genet.* 163, 181–187.

Hood, E. E., Jen, G., Kayes, L., Kramer, J., Fraley, R. T., and Chilton, M.-D. (1984). *Bio/Technology* 2, 702–708.

Hooykaas-Van Slogteren, G. M. S., Hooykaas, P. J. J., and Schilperoort, R. A. (1984). *Nature (London)* 311, 763–764.

Horsch, R. B., and Klee, H. J. (1986). *Proc. Natl. Acad. Sci. U.S.A.* 83, 4428–4432.

Horsch, R. B., Fry, J. E., Hoffmann, N. L., Eichholtz, D., Rogers, S. G., and Fraley, R. T. (1985). *Science* 233, 1229–1231.

Horsch, R. B., Klee, H. J., Stachel, S., Winans, S. C., Nester, E. W., Rogers, S. G., and Fraley, R. T. (1986). *Proc. Natl. Acad. Sci. U.S.A.* 83, 2571–2575.

Howard, E. A., Winsor, B. A., De Vos, G., and Zambryski, P. (1989). *Proc. Natl. Acad. Sci. U.S.A.* 86, 4017–4021.

Howell, S. H., Walker, L. L., and Dudley, R. K. (1980). *Science* 208, 1265.

Huffman, G. A., White, F. F., Gordon, M. P., and Nester, E. W. (1984). *J. Bacteriol.* 157, 269–276.

Inzé, D., Follin, A., Van Lijsebettins, M., Simoens, C., Genetello, C., Van Montagu, M., and Schell, J. (1984). *Mol. Gen. Genet.* 194, 265–274.

James, D. J., Passey, A. J., Barbara, D. J., and Bevan, M. (1989). *Plant Cell Rep.* 7, 658–661.

Janda, M., French, R., and Ahlquist, P. (1987). *Virology* 158, 259–262.

Janssens, A., Genetello, C., Van Montagu, M., and Zambryski, P. (1986). *Plant Sci.* 47, 185–193.

Jefferson, R. A., Burgess, S. M., and Hirsh, D. (1986). *Proc. Natl. Acad. Sci. U.S.A.* 83, 8447–8451.

Jen, G. C., and Chilton, M.-D. (1986). *Proc. Natl. Acad. Sci. U.S.A.* 83, 3895–3899.

Jin, S., Prusti, R., Roitsch, T., Ankenbauer, R., and Nester, E. W. (1990a). *J. Bacteriol.* 172, 4945–4950.

Jin, S., Roitsch, T., Ankenbauer, R. G., Gordon, M. P., and Nester, E. W. (1990b). *J. Bacteriol.* 172, 525–530.

Jones, J. D. G., Gilbert, D. E., Grady, K. L., and Jorgensen, R. A. (1987). *Mol. Gen. Genet.* 207, 478–485.

Joos, H., Inzé, D., Caplan, A., Sormann, M., Van Montagu, M., and Schell, J. (1983). *Cell* 32, 1057–1067.

Jorgensen, R., Snyder, C., and Jones, J. D. G. (1987). *Mol. Gen. Genet.* 207, 471–477.

Jouanin, L. (1984). *Plasmid* 12, 91–102.

Jouanin, L., Vilaine, F., d'Eufert, C., and Casse-Delbart, F. (1985). *Mol. Gen. Genet.* 201, 370–374.

Kamada, H., Okamura, N., Satake, M., Harada, H., and Shimomura, K. (1986). *Plant Cell Rep.* 5, 239–242.

Kanemoto, R., Powell, A., Akiyoshi, D., Regier, D., Kerstetter, R., Nester, E., Hawes, M., and Gordon, M. (1989). *J. Bacteriol.* 171, 2506–2512.

Keane, P. J., Kerr, A., and New, P. B. (1970). *Aust. J. Biol. Sci.* 23, 585–595.
Klee, H. J., White, F. F., Iyer, V. N., Gordon, M. P., and Nester, E. W. (1983). *J. Bacteriol.* 153, 878–883.
Klee, H. J., Yanofsky, M. F., and Nester, E. W. (1985). *Bio/Technology* 3, 637–642.
Klein, T. M., Wolf, E. D., Wu, R., and Sanford, J. C. (1987). *Nature (London)* 327, 70–73.
Komari, T., Halperin, W., and Nester, E. W. (1986). *J. Bacteriol.* 166, 88–94.
Koncz, C., Olsson, O., Langridge, W. H. R., and Schell, J., and Szalay, A. A. (1987). *Proc. Natl. Acad. Sci. U.S.A.* 84, 131.
Koplow, J., Byrne, M., Jen, G., Tempé, J., and Chilton, M.-D. (1984). *Plasmid* 11, 17–27.
Koukolíková-Nicola, Z., Shillito, R., Hohn, B., Wang, K., Van Montagu, M., and Zambryski, P. (1985). *Nature (London)* 313, 191–196.
Krens, F. A., Molendijk, L., Wullems, G. J., and Schilperoort, R. A. (1982). *Nature (London)* 296, 72–74.
Kuntz, M., Simons, A., Schell, J., and Schreier, P. H. (1986). *Mol. Gen. Genet.* 205, 454–460.
Kyozuka, J., Hayashi, Y., and Shimamoto, K. (1987). *Mol. Gen. Genet.* 296, 408–413.
Lamppa, G., Nagy, F., and Chua, N.-H. (1985). *Nature (London)* 316, 750–752.
Langereis, K., Neeleman, L., and Bol, J. F. (1986). *Plant Mol. Biol.* 6, 281–288.
Lee, K. Y., Lund, P., Lowe, K., and Dunsmuir, P. (1990). *The Plant Cell* 2, 415–425.
Leemans, J., Shaw, C., Deblaere, R., De Greve, H., Hernalsteens, J.-P., Maes, M., Van Montagu, M., and Schell, J. (1981). *J. Mol. Appl. Genet.* 1, 149–164.
Leemans, J., Deblaere, R., Willmitzer, L., De Greve, H., Hernalsteens, J.-P., Van Montagu, M., and Schell, J. (1982a). *EMBO J.* 1, 147–152.
Leemans, J., Langenakens, J., DeGreve, H., Deblaere, R., Van Montagu, M., and Schell, J. (1982b). *Gene* 19, 361–364.
Lichtenstein, C., and Draper, J. (1985). In *DNA Cloning, A Practical Approach* (Ed. by D. M. Glover), pp. 67–118. Oxford: IRL Press.
Lloyd, A. M., Barnason, A. R., Rogers, S. G., Byrne, M. C., Fraley, R. T., and Horsch, R. B. (1986). *Science* 234, 464–466.
Loesch-Fries, L. S., Merlo, D., Zinnen, T., Burhop, L., Hill, K., Krahn, K., Jarvis, N., Nelson, S., and Halk, E. (1987). *EMBO J.* 6, 1845–1851.
Lörz, H., Baker, B., and Schell, J. (1985). *Mol. Gen. Genet.* 199, 178–182.
Machida, Y., Usami, S., Yamamoto, A., Niwa, Y., and Takebe, I. (1986). *Mol. Gen. Genet.* 204, 374–382.
Manners, J. M. (1988). *Plant Sci.* 55, 61–68.
Matzke, A., and Chilton, M.-D. (1981). *J. Mol. Appl. Genet.* 1, 39–49.
Matzke, A. J. M., and Matzke, M. A. (1986). *Plant Mol. Biol.* 7, 357–365.
McCormick, S., Niedermeyer, J., Fry, J., Barnason, A., Horsch, R., and Fraley, R. (1986). *Plant Cell Rep.* 5, 81–84.
Melton, D. A., Krieg, P. A., Rebagliati, M. R., Maniatis, T., Zinn, K., and Green, M. R. (1984). *Nucl. Acids Res.* 12, 7035–7056.
Menagé, A., and Morel, G. (1964). *C.R. Acad. Sci.* 259, 4795–3796.
Messens, E., Lenaerts, A., van Montagu, M., and Hedges, R. W. (1985). *Mol. Gen. Genet.* 199, 344–348.
Miller, J. F. (1988). *Mol. Biol. Rep.* 5, 1–4.
Morikawa, H., Iida, A., Matsui, C., Ikegami, M., and Yamada, Y. (1986). *Gene* 41, 121–124.
Müller, A. J., Mendel, R. R., Schiemann, J., Simoens, C., and Inzé, D. (1987). *Mol. Gen. Genet.* 207, 171-175.
Murai, N., and Kemp, J. D. (1982). *Proc. Natl. Acad. Sci. U.S.A.* 79, 86–90.
Negrutiu, I., Shillito, R., Potrykus, I., Biasini, G., and Sala, F. (1987). *Plant Mol. Biol.* 8, 363–373.

Nehra, N. S., Chibbar, R. N., Kartha, K. K., Datla, R. S. S., and Crosby, W. L. (1990). *Plant Cell Rep.* 9, 293–298.
Neumann, E., Schaeffer-Ridder, M., Wang, Y., and Hofschneider, P. H. (1982). *EMBO J.* 1, 841–845.
Nixon, B. T., Ronson, C. W., and Ausubel, F. M. (1986). *Proc. Natl. Acad. Sci. U.S.A.* 83, 7850–7854.
Ohta, Y. (1986). *Proc. Natl. Acad. Sci. U.S.A.* 83, 715–719.
Okada, K., Takebe, I., and Nagata, T. (1986). *Mol. Gen. Genet.* 205, 398–403.
Ooms, G., Hooykaas, P. J. J., Van Veen, R. J. M., Van Beelen, P., Regensburg-Tuink, T. J. G., and Schilperoort, R. A. (1982). *Plasmid* 7, 15–29.
Ooms, G., Bains, A., Burrell, M., Twell, D., and Wilcox, E. (1985a). *Theor. Appl. Genet.* 721, 325–329.
Ooms, G., Karp, A., Burrell, M. M., and Twell, D., and Roberts, J. (1985b). *Theor. Appl. Genet.* 70, 440–446.
Otten, L. A., and Schilperoort, R. A. (1978). *Biochim. Biophys. Acta* 527, 497–500.
Otten, L., Piotrowiak, G., Hooykaas, P., Dubois, M., Szegedi, E., and Schell, J. (1985). *Mol. Gen. Genet.* 199, 189–193.
Ou-Lee, T.-M., Turgeon, R., and Wu, R. (1986). *Proc. Natl. Acad. Sci. U.S.A.* 83, 6815–6819.
Ow, D. W., Wood, K. V., DeLuca, M., de Wet, J. R., Helinski, D. R., and Howell, S. H. (1986). *Science* 234, 856–859.
Pazour, G. J., and Das, A. (1990). *J. Bacteriol.* 172, 1241–1249.
Peralta, E. G., and Ream, L. W. (1985). *Proc. Natl. Acad. Sci. U.S.A.* 82, 5112–5116.
Peralta, E. G., Hellmiss, R., and Ream, W. (1986). *EMBO J.* 5, 1137–1142.
Petit, A., Berkaloff, A., and Tempé, J. (1986). *Mol. Gen. Genet.* 202, 388-383.
Potrykus, I., Saul, M. W., Petruska, J., Paszkowski, J., and Shillito, R. D. (1985a). *Mol. Gen. Genet.* 199, 183–188.
Potrykus, I., Shillito, R. D., Saul, M. W., and Paszkowski, J. (1985b). *Plant Mol. Biol. Reptr.* 3, 117–128.
Prosen, D. E., and Simpson, R. B. (1987). *Bio/Technology* 5, 966–971.
Pua, E.-C., Mehra-Palta, A., Nagy, F., and Chua, N.-H. (1987). *Bio/Technology* 5, 815–817.
Raineri, D. M., Bottino, P., Gordon, M. P., and Nester, E. W. (1990). *Bio/Technology* 8, 33–38.
Ream, L. W., Gordon, M. P., and Nester, E. W. (1983). *Proc. Natl. Acad. Sci. U.S.A.* 80, 1660–1664.
Reiss, B., Sprengel, R., Will, H., and Schaller, H. (1984). *Gene* 30, 211–218.
Rogers, S. G., Horsch, R. B., and Fraley, R. T. (1986). *Methods Enzymol.* 118, 627–641.
Ronson, C., Nixon, B., and Ausubel, F. (1987). *Cell* 49, 579–581.
Roth, D. B., Porter, T. N., and Wilson, J. H. (1985). *Mol. Cell. Biol.* 5, 2599–2607.
Rothstein, S. J., DiMaio, J., Strand, M., and Rice, D. (1987a). *Proc. Natl. Acad. Sci. U.S.A.* 84, 8439–8443.
Rothstein, S. J., Lahners, K. N., Lotstein, R. J., Carozzi, N. B., Jayne, S. M., and Rice, D. A. (1987b). *Gene* 53, 153–161.
Rubin, G., and Spradling, A. (1982). *Science* 218, 348–353.
Ruvkun, G. B., and Ausubel, F. M. (1981). *Nature (London)* 289, 85–88.
Salomon, F., Deblaere, R., Leemans, J., Hernalsteens, J.-P., Van Montagu, M., and Schell, J. (1984). *EMBO J.* 3, 141.
Schafer, W., Gorz, A., and Kahl, G. (1987). *Nature (London)* 327, 529–532.
Schröder, G., Waffenschmidt, S., Weiler, E. W., and Schröder, J. (1984). *Eur. J. Biochem.* 138, 387–391.
Sederoff, R., Stomp, A.-M., Chilton, W. S., and Moore, L. W. (1986). *Bio/Technology* 4, 647–649.
Shah, D. M., Horsch, R. B., Klee, H. J., Kishore, G. M., Winter, J. A., Tumer, N. E., Hironaka,

C. M., Sanders, P. R., Gasser, C. S., Aykent, S., Siegel, N. R., Rogers, S. G., and Fraley, R. T. (1986). *Science* 233, 478–481.

Shahin, E. A., Spielmann, A., Sukhapinda, K., Yashar, M., and Simpson, R. B. (1986a). *Crop Sci.* 26, 1235–1239.

Shahin, E. A., Sukhapinda, K., Simpson, R. B., and Spivey, R. (1986b). *Theor. Appl. Genet.* 72, 770–777.

Shaw, C., Watson, M., Carter, G., and Shaw, C. (1984). *Nucl. Acids Res.* 12, 6031–6041.

Shimamoto, K., Terada, R., Izawa, T., and Fujimoto, H. (1989). *Nature (London)* 338, 274–276.

Simoens, C., Alliotte, Th., Mendel, R., Müller, A., Schiemann, J., Van Lijsebettens, M., Schell, J., Van Montagu, M., and Inzé, D. (1986). *Nucl. Acids Res.* 14, 8073–8090.

Simon, R., Priefer, U., and Pühler, A. (1983). *Bio/Technology* 1, 784–791.

Simpson, R., O'Hara, P., Lichtenstein, C., Montoya, A. L., Kwok, W., Gordon, M. P., and Nester, E. W. (1982). *Cell* 29, 1005–1014.

Simpson, R. B., Spielmann, A., Margossian, L., and McKnight, T. D. (1986). *Plant Mol. Biol.* 6, 403–415.

Sinkar, V. P., Pythoud, F., White, F. F., Nester, E. W., and Gordon, M. P. (1988). *Genes & Dev.* 2, 688–697.

Slightom, J. L., Durand-Tardif, M., Jouanin, L., and Tepfer, D. (1986). *J. Biol. Chem.* 261, 108–121.

Smith, M. J., and Neilands, J. B. (1987). In *Molecular Strategies for Crop Protection* (Ed by C. J. Erntzen and C. Ryan), pp. 157–168. Alan R. Liss.

Spanier, K., Schell, J., and Schreier, P. H. (1989). *Mol. Gen. Genet.* 219, 209–216.

Spena, A., and Schell, J. (1987). *Mol. Gen. Genet.* 206, 436–440.

Spielmann, A., and Simpson, R. B. (1986). *Mol. Gen. Genet.* 205, 34–41.

Stachel, S. E., and Nester, E. W. (1986). *EMBO J.* 5, 1445–1454.

Stachel, S. E., and Zambryski, P. C. (1986a). *Cell* 46, 325–333.

Stachel, S. E., and Zambryski, P. C. (1986b). *Cell* 47, 155–157.

Stachel, S. E., Messens, E., Van Montagu, M., and Zambryski, P. (1985). *Nature (London)* 318, 624–629.

Stachel, S. E., Nester, E. W., and Zambryski, P. C. (1986a). *Proc. Natl. Acad. Sci. U.S.A.* 83, 379–383.

Stachel, S. E., Timmerman, B., and Zambryski, P. C. (1986b). *Nature (London)* 322, 706–712.

Stougaard, J., Marcker, K. A., Otten, L., and Schell, J. (1986). *Nature (London)* 321, 667–674.

Stougaard, J., Abildsten, D., and Marcker, K. A. (1987). *Mol. Gen. Genet.* 207, 251–255.

Sukhapinda, K., Spivey, R., Simpson, R. B., and Shahin, E. A. (1987). *Mol. Gen. Genet.* 206, 491–497.

Svab, Z., Hajdukiewicz, P., and Maliga, P. (1990). *Proc. Natl. Acad. Sci. U.S.A.* 87, 8526–8530.

Tait, R. C., Close, T. J., Lundquist, R. C., Hagiya, M., Rodriquez, R. L., and Kado, C. (1983). *Bio/Technology* 1, 269-274.

Takamatsu, N., Ishikawa, M., Meshi, T., and Okada, Y. (1987). *EMBO J.* **6,** 307–311.

Taylor, B. H., Amasino, R. M., White, F. F., Nester, E. W., and Gordon, M. P. (1985). *Mol. Gen. Genet.* 201, 554–557.

Teeri, T., Herrera-Estrella, L., Depicker, A., Van Montagu, M., and Palva, E. (1986). *EMBO J.* 5, 1755–1760.

Tempé, J., and Goldmann, A. (1982). In *Molecular Biology of Plant Tumors* (Ed. by G. Kahl and J. Schell), pp. 427–449. Orlando, Florida: Academic Press, Inc.

Tepfer, D. (1984). *Cell* 37, 959–967.

Thomashow, M. F., Nutter, R., Montoya, A. L., Gordon, M. P., and Nester, E. W. (1980). *Cell* 19, 729–739.

Thomashow, L. S., Reeves, S., and Thomashow, M. F. (1984). *Proc. Natl. Acad. Sci. U.S.A.* 81, 5071–5075.

Thomashow, M. F., Hugly, S., Buchholz, W., and Thomashow, L. S. (1986). *Science* 231, 616–618.

Tinland, B., Huss, B., Paulus, F., Bonard, G., and Otten, L. (1989). *Mol. Gen. Genet.* 219, 217–224.

Toro, N., Datta, A., Yanofsky, M., and Nester, E. W. (1988). *Proc. Natl. Acad. Sci. U.S.A.* 85, 8558–8562.

Toro, N., Datta, A., Carmi, O. A., Young, C., Prusti, R. K., and Nester, E. W. (1989). *J. Bacteriol.* 171, 6845–6849.

Trulson, A. J., Simpson, R. B., and Shahin, E. A. (1986). *Theor. Appl. Genet.* 73, 11–15.

Tumer, N. E., O'Connell, K. M., Nelson, R. S., Sanders, P. R., Beachy, R. N., Fraley, R. T., and Shah, D. M. (1987). *EMBO J.* 6, 1181–1188.

Uchimiya, H., Hirochika, H., Hashimoto, H., Hara, A., Masuda, T., Kasumimoto, T., Harada, H., and Ikeda, J.-E. (1986). *Mol. Gen. Genet.* 205, 1–8.

Usami, S., Morkawa, S., Takebe, I., and Machida, Y. (1987). *Mol. Gen. Genet.* 209, 221–226.

Vaeck, M., Reynaerts, A., Hofte, H., Jansens, S., De Beuckeleer, M., Dean, C., Zabeau, M., Van Montagu, M., and Leemans, J. (1987). *Nature (London)* 328, 33–37.

Valvekens, D., Van Montagu, M., and Van Lijsebettens, M. (1988). *Proc. Natl. Acad. Sci. U.S.A.* 85, 5536–5540.

Van den Broeck, G., Timko, M. P., Kausch, A. P., Cashmore, A. R., Van Montagu, M., and Herrera-Estrella, L. (1985). *Nature (London)* 313, 358–363.

van den Elzen, P., Lee, K. Y., Townsend, J., and Bedbrook, J. (1985a). *Plant Mol. Biol.* 5, 149–154.

van den Elzen, P., Townsend, J., Lee, K. Y., and Bedbrook, J. (1985b). *Plant Mol. Biol.* 5, 299–302.

Van Larebeke, N., Engler, G., Holsters, M., Van den Elsacker, S., Zaenen, I., Schilperoort, R. A., and Schell, J. (1974). *Nature (London)* 252, 169–170.

Velten, J., and Schell, J. (1985). *Nucl. Acids Res.* 13, 6981–6998.

Velten, J., Velten, L., Hain, R., and Schell, J. (1984). *EMBO J.* 3, 2723–2730.

Waldron, C., Murphy, E., Roberts, J., Gustafson, G., Armour, S., and Malcolm, S. (1985). *Plant Mol. Biol.* 5, 103.

Wallroth, M., Gerats, A. G. M., Rogers, S. G., Fraley, R. T., and Horsch, R. B. (1986). *Mol. Gen. Genet.* 202, 6–15.

Wang, K., Herrera-Estrella, L., Van Montagu, M., and Zambryski, P. (1984). *Cell* 38, 455–462.

Wang, K., Stachel, S. E., Timmerman, B., Van Montagu, M., and Zambryski, P. C. (1987). *Science* 235, 587–591.

Wang, K., Herrera-Estrella, A., and Van Montagu, M. (1990). *J. Bacteriol.* 172, 4432–4440.

Ward, E. R., and Barnes, W. M. (1988). *Science* 242, 927–930.

Ward, J. E., Akiyoshi, D. E., Regier, D., Datta, A., Gordon, M. P., and Nester, E. W. (1988). *J. Biol. Chem.* 263, 5804–5814.

Ward, J., Dale, E., Christie, P., Nester, E. W., and Binns, A. (1990). *J. Bacteriol.* 172, 5187–5199.

Watson, B., Currier, T. C., Gordon, M. P., Chilton, M.-D., and Nester, E. W. (1975). *J. Bacteriol.* 123, 255–254.

White, F. F., and Nester, E. W. (1980). *J. Bacteriol.* 141, 1134–1141.

White, F. F., and Sinkar, V. (1987). In *Advances in Plant Gene Research* (Ed. by T. Hohn), pp. 149–178. New York: Springer-Verlag.

White, F. F., Taylor, B. H., Huffman, G. A., Gordon, M. P., and Nester, E. W. (1985). *J. Bacteriol.* 164, 33–44.

Willetts, N., and Wilkins, B. (1984). Processing of plasmid DNA during bacterial conjugation. *Microbiol. Rev.* 48, 24–41.

Willmitzer, L., Simons, G., and Schell, J. (1982). *EMBO J.* 1, 139–146.

Willmitzer, L., Dhaese, P., Schreier, P., Schmalenbach, W., Van Montagu, M., and Schell, J. (1983). *Cell* 32, 1045–1056.

Winans, S. C., Ebert, P. R., Stachel, S. E., Gordon, M. P., and Nester, E. W. (1986). *Proc. Natl. Acad. Sci. U.S.A.* 83, 8278–8282.

Winans, S. C., Kerstetter, R. A., and Nester, E. W. (1988). *J. Bacteriol.* 170, 4047–4054.

Winans, S., Kerstetter, R., Ward, J., and Nester, E. W. J. (1989). *J. Bacteriol.* 171, 1616–1622.

Wullems, G. J., Molendijk, L., Ooms, G., and Schilperoort, R. A. (1981). *Cell* 24, 719–727.

Yadav, N. S., Vanderleyden, J., Bennett, D. R., Barnes, W. M., and Chilton, M.-D. (1982). *Proc. Natl. Acad. Sci. U.S.A.* 79, 6322–6326.

Yamada, Y., Zhi-Qi, Y., Ding-Tai, T. (1986). *Plant Cell Rep.* 5, 85–88.

Yamamoto, A., Iwahashi, M., Yanofsky, M. F., Nester, E. W., Takebe, I., and Machida, Y. (1987). *Mol. Gen. Genet.* 206, 174–177.

Yanofsky, M. F., and Nester, E. W. (1986). *J. Bacteriol.* 168, 244–250.

Yanofsky, M., Lowe, B., Montoya, A., Rubin, R., Krul, W., Gordon, M., and Nester, E. W. (1985a). *Mol. Gen. Genet.* 201, 237–246.

Yanofsky, M., Montoya, A., Knauf, V., Lowe, B., Gordon, M., and Nester, E. (1985b). *J. Bacteriol.* 163, 341–348.

Yanofsky, M. F., Porter, S. G., Young, C., Albright, L. M., Gordon, M. P., and Nester, E. W. (1986). *Cell* 47, 471–477.

Yanofsky, M. F., Ma, H., Bowman, J. L., Drews, G. N., Feldmann, K. A., and Meyerowitz, E. M. (1990). *Nature (London)* 346, 35–39.

Young, C., and Nester, E. W. (1988). *J. Bacteriol.* 170, 3367–3374.

Zambryski, P. (1988). *Annu. Rev. Genet.* 22, 1–30.

Zambryski, P., Depicker, A., Kruger, K., and Goodman, H. (1982). *J. Mol. Appl. Genet.* 1, 363–370.

Zambryski, P., Joos, H., Genetello, C., Leemans, J., Van Montagu, M., and Schell, J. (1983). *EMBO J.* 2, 2143–2150.

2

Molecular Genetics of T-DNA Transfer from Agrobacterium to Plants

Stanton B. Gelvin
Department of Biological Sciences
Purdue University
West Lafayette, Indiana 47907

Agrobacterium tumefaciens is a gram-negative phytopathogen that causes the disease crown gall on a wide variety of dicotyledonous plants. *Agrobacterium tumefaciens* can also induce tumors on certain monocots and gymnosperms. The infection cycle of *Agrobacterium* is complex, involving a number of chemical signals emitted by both the pathogen and its host. Infection commences when the bacteria are attracted to a number of chemicals secreted by wounded plant cells. Phenolic molecules synthesized by these wounded cells induce genes on the Ti (tumor-inducing) plasmid harbored by all virulent strains of *Agrobacterium.* The bacteria bind to the plant cells and transfer a portion of this Ti plasmid, called the T-(transferred)DNA, to the plant cell. The T-DNA covalently integrates into plant nuclear DNA, where genes encoded by the T-DNA direct the synthesis of the phytohormones auxin and cytokinin, as well as novel low molecular weight metabolites called opines. Opines are secreted from the tumor cells and can be used as carbon and sometimes nitrogen sources by the inciting strain of *Agrobacterium.* Certain of these opines can also induce the conjugal transfer of the Ti plasmid between *Agrobacterium* cells. In addition, some opines can also interact in complex ways with the *vir* (virulence) genes located on the Ti plasmid. These genes regulate T-DNA processing and transfer. Certain sugars, both neutral and pectic, can also affect gene activity in *Agrobacterium.*

Because of the ability of *Agrobacterium* to transfer genetic information to

TRANSGENIC PLANTS, VOLUME 1
Engineering and Utilization

plant cells, it had long been touted as an ideal system for creating transgenic plants, at least for plant species within the natural host range of the bacterium. This promise was fulfilled in the early to mid-1980s with the development of the first series of T-DNA vectors useful for plant genetic engineering. The development of these vectors relied heavily upon basic research in the areas of T-DNA gene function, T-DNA processing and transfer, *vir* gene induction and plant–microbe chemical signaling, T-DNA transcriptional regulation, and the microbial genetics of *Agrobacterium*. In this chapter, I shall review several aspects of the *Agrobacterium* infection cycle as it relates to T-DNA processing and transfer. In addition, I shall review a number of basic and applied scientific investigations that led to the development of *Agrobacterium* as a plant genetic engineering tool. A number of these important concepts have recently been discussed in several excellent reviews (Stachel and Zambryski, 1986a; An, 1987; Deblaere *et al.*, 1987; Klee *et al.*, 1987a; Rogers *et al.*, 1987, 1988; Schell, 1987; Weiler and Schroder, 1987; Binns and Thomashow, 1988; Gallie *et al.*, 1988; Hooykaas, 1988, 1989; Zambryski, 1988; Gasser and Fraley, 1989; Ream, 1989; White, 1989; Zambryski *et al.*, 1989; Gelvin, 1990).

2.1 *vir* Gene Inducers

The excision of the T-DNA from the Ti plasmid and its subsequent transfer from *Agrobacterium* to plant cells is mediated by a group of genes, called *vir* genes, found on the Ti plasmid (Garfinkel and Nester, 1980; Klee *et al.*, 1983; Hooykaas *et al.*, 1984; Lundquist *et al.*, 1984). Except for the genes *virA* and, to a limited extent, *virG*, the *vir* genes are transcriptionally inactive when *Agrobacterium* grows vegetatively in either rich broth or in minimal medium. All *vir* genes, including *virA* and *virG*, are induced when *Agrobacterium* is incubated in a suitable medium in the presence of extracts from wounded plant cells. This induction is easily monitored when *vir* genes are linked as transcriptional fusions to easily assayable reporter genes, such as *lacZ* (Stachel *et al.*, 1985a) or a gene encoding bacterial luciferase (*lux*) (Rogowsky *et al.*, 1987).

Stachel *et al.* (1986) first described such induction of the *vir* genes by incubating *Agrobacterium* harboring the *lacZ* fusion transposon Tn*3*-HoHo1 with NT1 tobacco cells or *Nicotiana tabacum* root cultures. The inducing factor(s) derived from these plants was small (it could pass through a dialysis membrane), stable to heat and cold treatments, and partially hydrophobic (as determined by its elution behavior from a Sep-Pak C_{18} column). Although the compound itself was stable to extremes of pH, optimal induction conditions required incubation of the bacteria at a pH less than 6.0. If the plant cells were first incubated in the translation inhibitor cycloheximide, the subsequent induction of *Agrobacterium vir* genes was markedly inhibited. This result suggested that the induction of the *vir* genes was not merely a response of *Agrobacterium* to necrotic plant cells, but required active plant cellular metabolism. In addition to inducing the *vir* genes, incubation of *Agrobacterium* cells with tobacco cells, or the phenolic *vir* gene inducer acetosyringone (see below), resulted in the accumulation in the bacterium of circular T-DNA molecules (T-circles) that could be rescued and amplified by *in vitro* transfer to *Escherichia coli* (Koukolikova-Nicola *et al.*, 1985; Alt-Moerbe *et al.*, 1986; Machida *et al.*, 1986).

Stachel *et al.* (1985b) identified two major *vir* gene inducers in tobacco exudates as the phenolic compounds acetosyringone [3,5-methoxy-4-hydroxyacetophenone (AS)] and hydroxyacetosyringone (OH-AS). They noted that such exudates contained 0.5–1.0 μ*M* AS, a concentration sufficient to induce the *vir* genes of *Agrobacterium*. Optimal induction of the *vir* genes, however, required higher concentrations of the inducing compound. A preliminary examination of related chemical compounds indicated that two methoxy groups at the ring carbon atoms 3 and 5 rendered the compound more potent in *vir* gene induction than compounds containing only one methoxy substituent. The finding that sinapinic acid, a lignin precursor, was an excellent *vir* gene inducer, and that wounded leaf sections served as better inducers than intact tobacco leaves, suggested that many potent *vir* gene inducers were derived from defense-related phenolic compounds synthesized by wounded plant cells to ward off microbial attack or to strengthen plant cell walls. It thus appeared that *Agrobacterium* had adapted itself to recognize such plant defense compounds as inducers of the genes necessary to infect plant cells.

Subsequent to the initial findings of Stachel *et al.* (1985b, 1986), a number of other plant-derived phenolic compounds were identified as *vir* gene inducers. Bolton *et al.* (1986) showed that a mixture of commonly found plant phenolics, including catechol, gallic acid, pyrogallic acid, *p*-hydroxybenzoic acid, protocatechuic acid, β-resorcylic acid, and vanillin, stimulated *vir* gene activity. Zerback *et al.* (1989) found that certain flavonol glycosides from *Petunia* pollen could induce *vir* genes. These compounds were only active at extremely high concentrations (5 m*M*), however, and may not play a role as inducers in nature. Spencer and Towers (1988) and Melchers *et al.* (1989b) investigated a large number of chemical compounds related to AS, and were able to establish a number of chemical "rules" governing the phenolic compounds that served as active *vir* gene inducers. In general, the most potent inducers contained a guaiacyl or syringyl substitution on a benzene ring, and an electron withdrawing group, such as a carbonyl or a carbon–carbon double-bond substituent, para to the hydroxyl group on the ring. A number of compounds more commonly found than AS in plants were potent *vir* gene inducers. These compounds included methyl sinapate, methyl ferulate and 5-hydroxy methyl ferulate, methyl syringate, syringaldehyde, and coniferyl alcohol. Two compounds in particular (2′,4′,4-trihydroxy-3-methoxychalcone and 2′,4′,4-trihydroxy-3,5-dimethoxychalcone) were 10 to 100-fold more potent inducers than was AS.

Although many *Agrobacterium* strains are virulent on a wide variety of dicotyledonous (and some monocotyledonous) plants, most monocots, many gymnosperms, and even some dicots are relatively recalcitrant to infection. Soon after the discovery of phenolic *vir* gene inducers in extracts of wounded dicots, the question was raised as to whether the lack of infectivity of *Agrobacterium* on many monocots resulted from a lack of such inducing compounds in the extracts or exudates of these plants. Although an initial report suggested that seedling exudates of many monocots did not contain *vir* gene inducers (Usami *et al.*, 1987), subsequent reports indicated that extracts of certain monocots did. Usami *et al.* (1988) found that extracts from certain tissues of wheat and oats contained substances that could induce *vir* genes. That these substances were perhaps different from those previously characterized from dicots was indicated by their relatively high molecular weight (exclusion from a Sephadex G-25 column) and hydrophilic nature. The authors suggested that the inducers may be a phenolic molecule conjugated to a hydrophilic molecule. Messens *et al.* (1990) identified a potent *vir* gene inducer in *Triticum monococcum* suspension cultures as ethyl ferulate.

The findings of *vir* gene inducers in extracts of monocotyledonous plants refractory to *Agrobacterium* infection suggested that this block to infection may occur at a step different from or subsequent to *vir* gene induction. That this may in part be true was suggested by an observation of Sahi *et al.* (1990). They discovered that homogenates of maize seedlings contained a substance that both inhibited the growth of *Agrobacterium* and inhibited the induction of *vir* genes by AS. This substance was identified as 2,4-dihydroxy-7-methoxy-2*H*-1,4-benzoxazin-3(4*H*)-one (DIMBOA). DIMBOA is a potent antimicrobial compound produced by many cultivars of maize, and is frequently present at concentrations at least 10-fold greater than that necessary to inhibit *vir* gene induction by AS. In addition to its effects on *Agrobacterium* growth and *vir* gene induction, the authors found that low concentrations (0.5 m*M*) of DIMBOA could inhibit crown gall tumorigenesis on *Kalanchoe,* a dicotyledonous plant normally susceptible to *Agrobacterium* infection.

Different strains of *Agrobacterium tumefaciens* display a broad range of virulence characteristics on many gymnosperms. Morris and Morris (1990) found that some strains, such as B3/73, were highly virulent on *Pseudotsuga menziesii* (Douglas fir), whereas other strains, such as MFM83.4, were only weakly tumorigenic. The *vir* genes were induced by acetosyringone in both strains, but only the strongly tumorigenic strain responded to extracts from Douglas fir. The *vir* gene-inducing compound found in these extracts was identified as the phenylpropanoid glucoside coniferin. Because *Agrobacterium* strains that contained high β-glucosidase activity were the most virulent on Douglas fir and displayed the highest degree of *vir* gene induction by extracts from these plants, the authors suggested that the virulence of *Agrobacterium* on this host was in part determined by the ability of the bacterium to convert an invert *vir* gene-inducing precursor to an active molecule.

There have been a few reports of increasing the virulence of *Agrobacterium* by the application of plant extracts or known *vir* gene inducers at the time of infection. Bouckaert-Urban and Vendrig (1982) reported enhanced virulence of *A. tumefaciens* B6 on *Helianthus giganteus* by the application of extracts from 6-day-old cotyledons, and Schafer *et al.* (1987) were able to induce crown gall tumors on the monocot *Dioscorea bulbifera* only when the inciting *Agrobacterium* strain C58 was pretreated with wound extracts from potatoes. Similarly, Owens and Smigocki (1988) increased the number of tumors formed on soybean cotyledons by the weakly virulent *A. tumefaciens* strain A348 by treatment of the bacteria with the potent *vir* gene inducer syringaldehyde. Veluthambi *et al.* (1989) reported an increased transformation frequency of cotton when the wounded apical meristems were incubated with acetosyringone or acetosyringone plus nopaline (see below) at the time of infection. Such reports of increasing the virulence of *Agrobacterium* by incubation with *vir* gene inducers are few in number, however, and it appears unlikely that such treatments will be a general solution to the problem of limited infectivity of *Agrobacterium* on many recalcitrant plants.

2.2 T-DNA Transfer and Expression

Although not usually regarded as such, the T-(transferred) DNA region of the Ti (tumor-inducing) or Ri (root-inducing) plasmid can be considered a chemical signal that travels from *Agrobacterium* to plant cells. The T-DNA, generally 10–20 kbp, is the region of these large (200–800 kbp) plasmids destined to be transferred to plant cells, and represents approximately 5–10% of the plasmid DNA. Some Ti plasmids contain only one T-DNA region, whereas other Ti and Ri plasmids contain two such regions that can be transferred to plants independently (Lemmers *et al.*, 1980; Merlo *et al.*, 1980; Thomashow *et al.*, 1980a,b; Zambryski *et al.*, 1980; DeBeuckeleer et al., 1981; Ooms *et al.*, 1982a; Zambryski *et al.*, 1982; Byrne *et al.*, 1983; Hepburn *et al.*, 1983; Holsters *et al.*, 1983; Knauf *et al.*, 1984; Taylor *et al.*, 1985; Yanofsky *et al.*, 1985; Vilaine and Casse-Delbart, 1987a). DNA sequence analyses of the T-region of several Ti and Ri plasmids and the junctions of T-DNA and plant DNA in crown gall and hairy root tumors suggested that specific DNA sequences delimited the T-region. These sequences, termed T-DNA border repeat sequences (or T-DNA borders), have been found in all Ti and Ri plasmids investigated to date. These T-DNA borders are made up of 25 bp DNA sequences that are highly conserved, and the borders are in a directly repeated orientation (Yadav *et al.*, 1982; Zambryski *et al.*, 1982; Shaw *et al.*, 1984; Wang *et al.*, 1984; Hepburn and White, 1985; Peralta and Ream, 1985; Gardner and Knauf, 1986; Horsch and Klee, 1986; Jen and Chilton, 1986a,b; Caplan *et al.*, 1985; Rubin, 1986; Slightom *et al.*, 1986; Van Haaren *et al.*, 1987, 1988, 1989; Wang *et al.*, 1987).

The importance of these border repeat sequences in tumorigenesis was shown by mutational analysis. The alteration of some nucleotides within the border repeat sequence resulted in an "inactive" (i.e., unable to be processed or to transfer the T-DNA) border (Wang *et al.*, 1987; Van Haaren *et al.*, 1989). Deletion of both borders from the T-DNA region of the Ti plasmid resulted in an avirulent bacterial strain (Jen and Chilton, 1986a; Van Haaren *et al.*, 1987). In addition, the inversion of one of the T-DNA borders, resulting in a T-DNA delimited by inverted repeats, also rendered the *Agrobacterium* strain harboring such a plasmid at best weakly virulent (Wang *et al.*, 1984; Peralta and Ream, 1985; Van Haaren *et al.*, 1987a). Experiments that removed only one T-DNA border suggested that right borders may play a more essential role in T-DNA transfer than do left borders. The deletion of the right border rendered the resulting *Agrobacterium* strain virtually avirulent, especially on tobacco plants (Shaw *et al.*, 1984; Wang *et al.*, 1984; Caplan *et al.*, 1985; Hepburn and White, 1985; Peralta and Ream, 1985; Horsch and Klee, 1986; Van Haaren *et al.*, 1987), whereas deletion of just the left border had little effect upon virulence (Joos *et al.*, 1983; Jen and Chilton, 1986a,b). In addition, Ti plasmid derivatives or "binary" plasmids harboring a replicon and a marker selectable in plant tissues only required a right T-DNA border for transfer of the marker to plant cells (Caplan *et al.*, 1985; Horsch and Klee, 1986; Jen and Chilton, 1986a; Rubin, 1986; Van Haaren *et al.*, 1987). These experiments suggested a possible polarity of these borders, perhaps indicating that certain molecular events of T-DNA processing initiated at the right border and terminated at the left border (Zambryski *et al.*, 1982; Wang *et al.*, 1984; Peralta and Ream, 1985; Jen and Chilton, 1986a,b; Rubin, 1986). Ream and co-workers (Peralta and Ream, 1985; Peralta *et al.*, 1986) and others (Van Haaren *et al.*, 1987) identified a DNA sequence, called "overdrive," near all right borders investigated but not near left borders. This sequence was important for efficient T-DNA transfer because deletion of the overdrive sequences resulted in an *Agrobacterium* strain that was only weakly virulent. Restoration of the natural or a synthetic overdrive sequence resulted in enhanced virulence. Overdrive, as a T-DNA transfer enhancer, had properties similar to known eukaryotic transcriptional enhancers. It could function in both an orientation and a position-independent manner (Peralta *et al.*, 1986; Culianez-Macia and Hepburn, 1988a).

The processing of the T-DNA from the Ti plasmid and its transfer to the plant cell is a complex process dependent upon genes encoded both by the Ti plasmid and the *Agrobacterium* chromosome. Garfinkel and Nester (1980) originally defined, by Tn*5* transposon mutagenesis, a region of the Ti plasmid that was essential for virulence. This region, termed the *vir* region, is composed of 6–10 genetic complementation groups, depending upon the Ti plasmid investigated (Iyer *et al.*, 1982; Klee *et al.*, 1983; Hooykaas *et al.*, 1984; Lundquist *et al.*, 1984). More highly refined mutagenesis studies, using the *lacZ* transcriptional fusion transposon Tn*3*-HoHo1 (Stachel *et al.*, 1985a), mapped both the extent and the direction of transcription of each of these complementation groups in the octopine-type Ti plasmid pTiA6 (Stachel and Nester, 1986; Stachel and Zambryski, 1986b). Similar experiments, using as a reporter of gene activity a gene encoding bacterial luciferase, identified *vir* operons in the nopaline-type Ti plasmid pTiC58 (Rogowsky *et al.*, 1987). The gene *virA* was active to a high level, and *virG* to a low level, in *Agrobacterium* cells grown in rich medium (Stachel and Nester, 1986; Stachel and Zambryski, 1986b; Rogowsky *et al.*, 1987). The genes *virB, virC, virD, virE, virH* (formerly called *pinF*) (Kanemoto *et al.*, 1989), and *virF* (Melchers *et al.*, 1990) found on octopine-type Ti plasmids, and similar genes harbored by nopaline-type Ti plasmids [although these plasmids contain a *tzs* gene (John and Amasino, 1988) and lack a *virF* and a *virH* gene], are relatively inactive in bacteria grown in broth (Stachel and Nester, 1986; Stachel and Zambryski, 1986b; Rogowsky *et al.*, 1987). Upon incubation of *Agrobacterium* cells in medium of low pH (5.0–5.8) containing such phenolic compounds as acetosyringone, these genes are induced to high levels (Stachel and Nester, 1986; Stachel and Zambryski, 1986b; Rogowsky *et al.*, 1987). In addition, *virG* and *virA* can also be induced several fold (Winans *et al.*, 1988). Because of their common mode of regulation by acetosyringone and related molecules, these genes have been termed the *vir* regulon. A map of the *vir* region of pTiA6 is shown in Figure 2.1.

Genetic analysis of the various *vir* genes suggested that *virA* and *virG* comprised a two-component system necessary for the expression of the *vir* regulon. The VirA protein was postulated to be a sensor of the phenolic-inducing

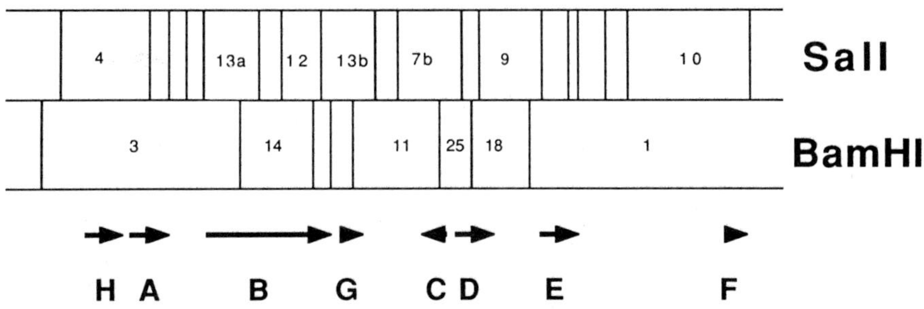

Figure 2.1 Map of the *vir* region of a typical wide host range octopine-type Ti plasmid. The direction of transcription of each of the *vir* operons is indicated by the arrows. This map is based upon data from Stachel and Nester (1986), Stachel and Zambryski (1986b), Kanemoto *et al.* (1989), and Melchers *et al.* (1990).

molecules, and the VirG protein a transcriptional activator of the *vir* genes (Melchers *et al.*, 1986; Stachel and Zambryski, 1986b; Leroux *et al.*, 1987; Melchers *et al.*, 1987; Powell *et al.*, 1987; Melchers *et al.*, 1989a; Morel *et al.*, 1989; Powell *et al.*, 1989; Huang *et al.*, 1990b; Jin *et al.*, 1990a,b; Pazour and Das, 1990; Roitsch *et al.*, 1990). Similar two-component, environmental-sensing, transcriptional-activating systems had previously been described in other bacteria. Such systems regulated the response of bacteria to environmental cues, resulting in the activation of genes involved in, among other cellular processes, phosphate and nitrogen utilization, osmotic regulation, chemotaxis, and sporulation (Nixon *et al.*, 1986; Winans *et al.*, 1986; Ronson *et al.*, 1987; Albright *et al.*, 1989).

Subsequent molecular analyses indicated that the VirA protein was localized to the inner (periplasmic) membrane of *Agrobacterium* cells (Leroux *et al.*, 1987). DNA sequence analyses suggested that VirA contained an amino-terminal region localized within the cell cytoplasm, two membrane-spanning regions delimiting a periplasmic domain, and a carboxy-terminal region within the cytoplasm (Melchers *et al.*, 1989a). Protein engineering experiments indicated that the periplasmic domain was responsive to pH and temperature, and that the second membrane-spanning domain was responsive to acetosyringone. Following induction by acetosyringone, a specific histidine residue within the carboxy-terminal region of the protein was autophosphorylated (Huang *et al.*, 1990b; Jin *et al.*, 1990b; Morel *et al.*, 1990). This phosphate group could subsequently be transferred to an aspartic acid residue in the amino-terminal region of the VirG protein (Jin *et al.*, 1990a; Roitsch *et al.*, 1990). It has been suggested that such phosphorylation of VirG may activate the protein to serve as a transcriptional activator. VirG protein can bind to *vir* boxes [DNA sequence motifs upstream of all the *vir* genes (Winans *et al.*, 1987; Steck *et al.*, 1988) and necessary for induced expression by acetosyringone] *in vitro* (Jin *et al.*, 1990c; Pazour and Das, 1990). However, such *in vitro* binding is independent of phosphorylation of the VirG protein.

Although the *vir* regulon is controlled by the genes *virA* and *virG*, the processing of the T-DNA is carried out by enzymes encoded by the *virD* operon (Yanofsky *et al.*, 1986; Albright *et al.*, 1987; Jayaswal *et al.*, 1987; Stachel *et al.*, 1987; Veluthambi *et al.*, 1987; Yamamoto *et al.*, 1987; De Vos and Zambryski, 1989; Wang *et al.*, 1990). Both of the first two open reading frames, and only these ORFs, are essential for T-DNA processing (Jayaswal *et al.*, 1987). The endonuclease activity encoded by *virD* cleaves the T-DNA between nucleotides 3 and 4 of the border repeat sequences (Albright *et al.*, 1987; Wang *et al.*, 1987). Such cleavage can result either in the single-stranded nicking of the "bottom" T-DNA strand (Albright *et al.*, 1987; Wang *et al.*, 1987; Culianez-Macia and Hepburn, 1988a,b), or in the double-stranded cleavage of the border repeat sequence (Veluthambi *et al.*, 1987, 1988; Jayaswal *et al.*, 1987; Steck *et al.*, 1989). VirD ORF1 has topoisomerase activity (Ghai and Das, 1989), and following cleavage the VirD ORF2 polypeptide covalently associates with the 5′ ends of the processed T-DNA molecules (Herrera-Estrella *et al.*, 1988; Ward and Barnes, 1988; Young and Nester, 1988; Durrenberger *et al.*, 1989; Howard *et al.*, 1989). The VirD2 protein contains within both its amino- and carboxy-terminal regions amino acid sequences homologous to motifs that, in animal and fungal cells, target proteins to the nucleus (Yanofsky *et al.*, 1986; Jayaswal *et al.*, 1987). Indeed, Herrera-Estrella *et al.* (1990) have shown that a translational fusion of the amino-terminal VirD2 region, containing this putative nuclear targeting domain, fused to *E. coli* β-galactosidase, directs β-galactosidase activity to the nucleus of transgenic tobacco plants. Recently, Howard *et al.* (1992) have also shown that sequences within the carboxy-terminal portion of VirD2 can serve as nuclear targeting signals.

Following induction of the *Agrobacterium vir* genes by acetosyringone or by incubation of the bacteria with plant cells, several different forms of "processed" T-DNA accumulate within the bacteria. Early reports suggested that circular T-DNA molecules (T-circles), that could be recovered by plasmid rescue in *E. coli,* played a prominent role in T-DNA transfer (Koukolikova-Nicola *et al.*, 1985). It is now generally accepted, however, that such molecules are rare and may result from recombination events between the similar T-DNA borders (Timmerman *et al.*, 1988). Two other forms of the T-DNA can accumulate to significant extents in induced *Agrobacterium* cells, however. These are full-length double-stranded and full-length single-stranded linear T-DNA molecules.

The predominant form of processed T-DNA found by most investigators is the full-length linear single-stranded T-DNA molecule, termed the T-strand (Stachel *et al.*, 1986; Albright *et al.*, 1987; Jayaswal *et al.*, 1987; Stachel *et al.*, 1987; Culianez-Macia and Hepburn, 1988a,b; Veluthambi *et al.*, 1988; 1989). These molecules most likely result from the single-stranded nicking by the VirD endonuclease of the "bottom" T-DNA strand between nucleotides 3 and 4 of the border repeat sequences, followed by the replicative displacement of the bottom T-DNA strand. In Ti plasmids that contain multiple T-DNA regions, all possible combinations of borders could serve to delineate the T-strand molecules (Stachel *et al.*, 1987; Veluthambi *et al.*, 1988). Upon phenol extraction of DNA from induced *Agrobacterium* cells, the T-strands tended to accumulate at the phenol–aqueous interphase, suggesting that they were complexed with protein. VirD2 protein is tightly (probably covalently) associated with the 5′ ends of T-strands (Herrera-Estrella *et al.*, 1988; Ward and Barnes, 1988; Young and Nester, 1988; Howard *et al.*, 1989), and it was demonstrated that this VirD2 protein "cap" can protect the 5 ′ end of T-strands against exonucleolytic degradation (Durrenberger *et al.*, 1989). As mentioned above, the VirD2 protein may also direct the T-strand to the nucleus of the plant cell. In addition, T-strands are noncovalently coated by a single-stranded DNA binding protein encoded by the *virE2* gene (Christie *et al.*, 1988). This protein cooperatively binds to any single-stranded DNA molecule (Gietl *et al.*, 1987; Citovsky *et al.*, 1988, 1989; Das, 1988; Sen *et al.*, 1989), and may protect the T-strand from endonucleolytic cleavage during its passage from *Agrobacterium* to the plant cell nucleus. This complex of T-strands capped at their 5′ ends by VirD2 protein and coated by the VirE2 protein has been termed the T-complex (Howard and Citovsky, 1990), and it has been postulated that this is the form of the T-DNA that is transferred to the plant cell. These authors have suggested that the transfer of the single-stranded T-DNA molecule from *Agrobacterium* to plant cells is an adaptation of the mechanism by which *E. coli* conjugally transfers F plasmids. However, such a "conjugation" now uses a eukaryotic cell as the transfer recipient (Stachel and Zambryski, 1986a). Although this model has been supported by a large amount of circumstantial evidence, there is no direct evidence that T-strands actually are the transferred T-DNA intermediate.

The presence of other forms of processed T-DNA in acetosyringone-induced *Agrobacterium* cells has been noted by a number of laboratories. That double-stranded T-DNA molecules may accumulate was first suggested by experiments conducted by Veluthambi *et al.* (1987, 1988), in which they found the relatively efficient double-stranded cleavage of the T-DNA borders following incubation of *Agrobacterium* with tobacco protoplasts or acetosyringone. Such double-stranded cleavage was also directed by the VirD endonuclease (Veluthambi *et al.*, 1987; Jayaswal *et al.*, 1987) and, in some Ti plasmids, was reported to be the major processing activity (Steck *et al.*, 1989). Durrenberger *et al.* (1989) demonstrated that full-length double-stranded T-DNA molecules could accumulate in acetosyringone-induced *Agrobacterium* cells, and that these molecules also were capped at the 5′ end by VirD2 protein. The relative abundance of single- and double-stranded T-DNA molecules in such induced bacteria depended upon a number of experimental parameters, such as the method of DNA extraction from the cells and the processes of electrophoresis and DNA blotting used. These authors thus suggested that one could not necessarily infer the form of the transferred T-DNA molecule merely from abundance arguments in that the relative levels of single- or double-stranded T-DNA molecules could reflect artifacts of the experimental manipulations.

Experiments in this author's laboratory have confirmed the presence of full-length double-stranded T-DNA molecules from octopine-type Ti plasmids in acetosyringone-induced *Agrobacterium* cells (Steck and Gelvin, unpublished). Figure 2.2 shows an analysis of DNA extracted from such cells. Both single- and double-stranded T-DNA molecules could be detected in approximately equal molar ratios (Figure 2.2C). These molecules were derived from the T-DNA flanked by borders A and C as well as by borders A and B. In an attempt to determine which form of the T-DNA (single- or double-stranded) was the true transferred intermediate, these investigators followed the kinetics of disappearance of the various forms of processed T-DNA from preinduced *Agrobacterium* cells following transfer of the bacteria to medium lacking acetosyringone

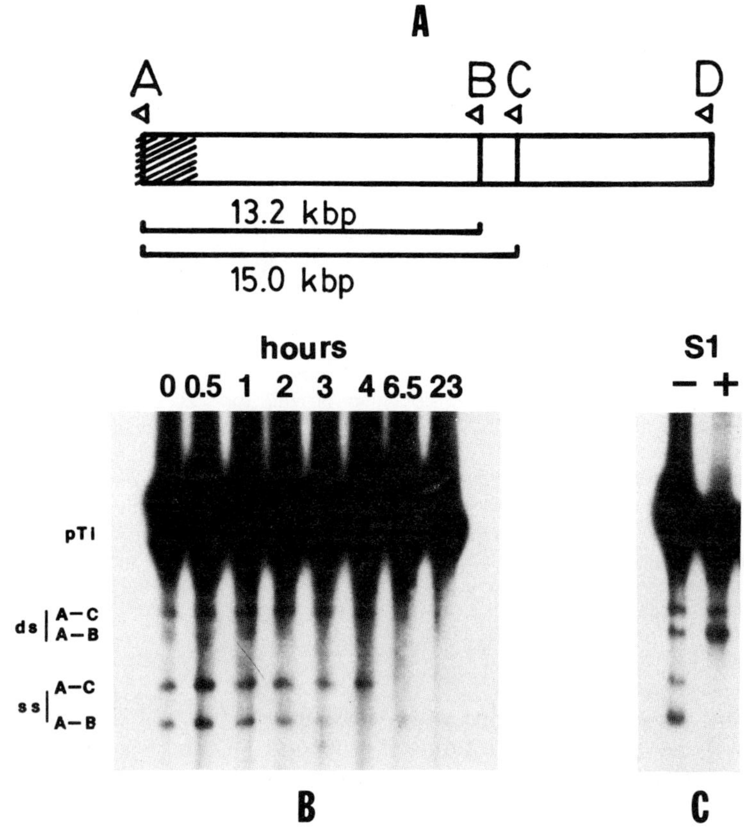

Figure 2.2 Characterization and decay kinetics of processed T-DNA molecules in *A. tumefaciens* harboring a mutation in the gene *pinF*. (A) Schematic map of the T-DNA region of a typical wide-host range octopine-type Ti plasmid. The triangles represent the four T-DNA border repeat sequences (A–D), and the shaded area represents the hybridization probe (*Hin*dIII fragment 18C) used for these analyses. The brackets beneath the map indicate the sizes of the expected processed T-DNA molecules. (B) DNA blot of total cellular DNA extracted from *Agrobacterium* cells. The cells were preinduced overnight with acetosyringone and octopine, then washed and resuspended in K3 tissue culture medium in the absence of the inducers. Hybridization signals from the unprocessed Ti plasmid (pTi) and the processed T-DNA [single-stranded (ss) and double-stranded (ds) T-DNA molecules from borders A to B and from borders A to C] indicate that the decay of all forms of the processed T-DNA is approximately equal, and is not significant until approximately 4 hours following removal of the inducer. (C) Following incubation with S1 nuclease (+), the two fastest migrating processed T-DNA molecules disappear, indicating that they are single-stranded.

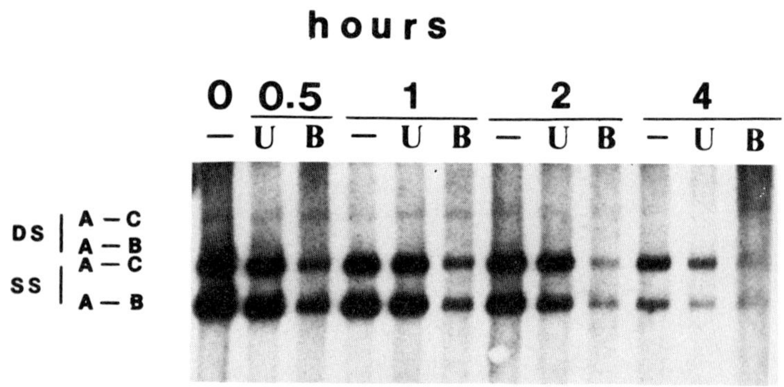

Figure 2.3 Decay kinetics of processed T-DNA molecules in *A. tumefaciens* harboring a mutation in the gene *pinF* following cocultivation with tobacco protoplasts. Following preinduction of the bacteria, DNA was extracted from bacteria incubated for various times in the absence of protoplasts (−), or from bacteria either bound to plant cells (B) or from unbound bacteria (U). In this particular experiment, only low amounts of double-stranded processed T-DNA intermediates were detected using as a hybridization probe *Hin*dIII fragment 18c. Nevertheless, the double-stranded molecules were more stable than were the single-stranded molecules. Loss of single-stranded molecules from the bound bacteria was rapid, and could be detected within 0.5 hours after the bacteria were incubated with the protoplasts.

and either containing or lacking tobacco protoplasts. Figure 2.2B shows that, in the absence of protoplasts, all forms of processed T-DNA began to disappear at approximately the same rate after 4 hours. The bacteria incubated with protoplasts were divided into two groups: those that bound tightly to plant cells and those that remained unbound. Figure 2.3 shows that the disappearance of all forms of processed T-DNA from *Agrobacterium* that were not tightly associated with plant cells followed closely the same kinetics as the disappearance of the T-DNA from bacteria incubated in the absence of plant cells. The disappearance of T-DNA from the bound bacteria was quite different, however. Single-stranded T-DNA molecules began to disappear rapidly (within the first half hour of incubation with plant cells), whereas the double-stranded T-DNA forms remained relatively stable during the first two hours. These data suggest that, in bacteria in which the T-DNA transfer apparatus has been preinduced, the transfer of T-DNA to plant cells can occur rapidly. In addition, these data suggest that the single-stranded form of the T-DNA is lost most quickly, and may therefore be the true transferred intermediate. Although other explanations for these data are possible (for example, the induction by plant cells of an S1-like nuclease activity in *Agrobacterium*), the relative stability of T-strands in bacteria incubated with but not attached to plant cells suggests that this alternative explanation of the data is not likely.

Although the *vir* genes located on the Ti plasmid play a major role in T-DNA processing and transfer from the bacteria, other genes on the *Agrobacterium* chromosome may play important roles in these processes, both directly and indirectly. A number of *Agrobacterium* chromosomal genes encode functions important in determining the surface characteristics of the bacterium. The gene *chvB* encodes a very large protein responsible for the biosynthesis of a β-1, 2-cyclic glucan that is secreted from the bacterial cell (Zorreguieta and Ugalde, 1986; Zorreguieta *et al.*, 1988). Both the secretion and modification of this glucan are mediated by the product of the *chvA* locus (Cangelosi *et al.*, 1989; Inon de Iannino and Ugalde, 1989; O'Connell and Handelsman, 1989). Mutagenesis of either of these genes resulted in *Agrobacterium* strains that were avirulent or highly attenuated in virulence, depending on the host plant species that was inoculated (Douglas *et al.*, 1982, 1985; Puvanesarajah *et al.*, 1985; Cangelosi *et al.*, 1989, 1990). These bacteria did not bind well to plant cells (Douglas *et al.*, 1982; Hawes and Pueppke, 1989), and the effect of mutations in these genes upon virulence may have resulted from this altered binding behavior. Mutations in the gene *pscA* also resulted in bacterial strains that were severely impaired in virulence (Marks *et al.*, 1987; Thomashow *et al.*, 1987). This gene encodes functions involved in the synthesis and secretion by the bacterial cell of succinoglycan molecules (Uttaro *et al.*, 1990). In addition, genes involved in cellulose biosynthesis by the bacteria (*cel* genes) may play a more subtle role in binding of the bacteria and in virulence. Although *Agrobacterium* strains mutant in these genes showed normal virulence characteristics when the bacteria were inoculated directly into wound sites, these mutant strains demonstrated decreased virulence when the wound was washed subsequent to inoculation (Matthysse, 1983; Deasey and Matthysse, 1984). Matthysse and co-workers suggested that the cellulose fibrils encoded by the products of the *cel* genes may entrap bacteria and plant cells in a loose association. The weak binding of the bacteria to the plant cells resulting from this entrapment may not be important for virulence under laboratory inoculation conditions, but may be important in natural environments.

Several other genes encoded by the *Agrobacterium* chromosome are also involved in virulence. The gene *chvE* encodes a glucose/galactose binding protein (Cornish *et al.*, 1989; Huang *et al.*, 1990a). Mutation of this gene resulted in a severely attenuated induction of the *vir* genes by acetosyringone, paralleled by a decrease in virulence upon certain plant species (Cangelosi *et al.*, 1990; Huang *et al.*, 1990a). This gene will be discussed in greater detail below in the section on the effects of sugars upon *Agrobacterium* gene induction. In addition to *chvE*, this author's laboratory identified two other *Agrobacterium* strains mutant in *virB* induction (Gray *et al.*, 1992). Although the effects of mutations in these strains were not as severe as were mutations in *chvE* (Table 2.1), they nevertheless resulted in *Agrobacterium* strains that showed decreased virulence on tobacco

TABLE 2.1
Induction of *virB::lacZ* in At44, At44#9, and At44#37

Strain[a]	Miller units	Control (%)
At44	445 ± 81	100
At44#9	9 ± 4	2
At44#37	49 ± 9	11

[a] *Agrobacterium tumefaciens* strains were induced for 12 hours in AB medium containing 0.5% sucrose and 100 μ*M* acetosyringone (w/v). These experiments represent replicates of three independent inductions.

leaf disks (Figure 2.4) and *Kalanchoe* stems. These mutations are in the gene *miaA*, which encodes a tRNA isopentenyl transferase (Gray *et al.*, 1992).

In addition to the chromosomal genes described above, a gene termed *ros* (for rough surface) has been identified in *A. tumefaciens* C58. The mutation of *ros* had pleiotropic effects, including altered bacterial growth characteristics, colony morphology, and *vir* gene induction (Close *et al.*, 1985). Unlike the situation in wild-type *Agrobacterium* chromosomal backgrounds, in which the genes *virC* and *virD* were not expressed in the absence of acetosyringone, *Agrobacterium* strains harboring a mutant *ros* gene expressed the *virC* and *virD* genes constitutively. This effect was only upon these two *vir* genes, suggesting alternate mechanisms of regulation of the various *vir* genes (Close *et al.*, 1987). Although the genes *virC* and *virD* were expressed in *ros* strains of *Agrobacterium* in the absence of induction, these genes could be further induced by acetosyringone.

Following the transfer of the T-DNA from *Agrobacterium* to plant cells, the T-DNA covalently integrates into plant nuclear DNA (Chilton *et al.*, 1977, 1980; DeBeuckeleer *et al.*, 1981; Thomashow *et al.*, 1980a,b). Genes encoded by the T-DNA, although expressed weakly in the bacteria (Gelvin *et al.*, 1981; Janssens *et al.*, 1984), are transcribed to a much greater extent in the plant cell. The expression of the T-DNA in the plant is controlled by transcriptional regulatory elements similar to those associated with other eukaryotic genes. The transcription of the T-DNA is mediated by RNA polymerase II (Willmitzer *et al.*, 1981), and typical TATA and CAAT boxes precede most of these genes (Barker *et al.*, 1983). In addition, many T-DNA genes contain upstream transcriptional activating elements similar to UAS sequences preceding many yeast promoters (Bruce and Gurley, 1987; DiRita and Gelvin, 1987; Ebert *et al.*, 1987; Bruce *et al.*,

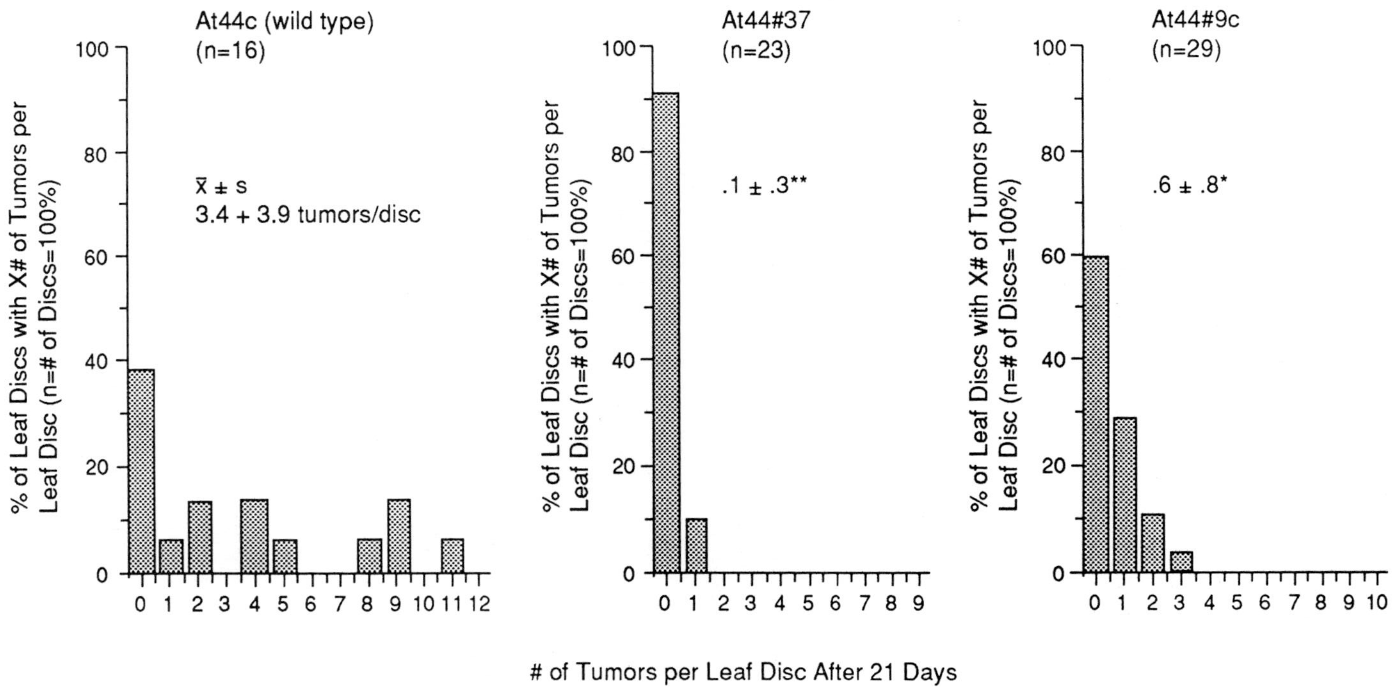

Figure 2.4 Virulence of *A. tumefaciens* At44c (harboring a wild-type C58 chromosome) and *A. tumefaciens* At44#37c and At44#9c (harboring Tn5 insertions in the chromosome) on tobacco leaf disks. The bacteria were grown to a Klett of 100; tumors were scored 21 days after inoculation. *, mean significantly different from wild-type at 95% level of probability; **, mean significantly different from wild-type at 99% level of probability.

1988; Leisner and Gelvin, 1988, 1989; Bouchez *et al.*, 1989; Bandyopadhyay *et al.*, 1989; Mitra and An, 1989; Singh *et al.*, 1989).

2.3 Opines

Crown gall tumor cells proliferate rapidly, both on the plant and in tissue culture. Although these tumors may severely debilitate the plant, the tumorous state per se probably does not benefit *Agrobacterium*. What is of benefit to the inciting strain of *Agrobacterium* most likely is the large quantity of opines synthesized by the tumors. Opines are low molecular weight compounds that are, in general, formed from amino acids, α-keto acids, and sugars. Very high concentrations (up to 5–10 m*M*) can accumulate in crown gall tumors (Firman and Fenwick, 1978; Scott *et al.*, 1979; Saint-Pierre and Dion, 1988). The production of these opines is mediated by opine biosynthetic genes encoded by the T-DNA (Bomhoff *et al.*, 1976; Montoya *et al.*, 1977; Ellis *et al.*, 1984; De Paolis *et al.*, 1985; Komro *et al.*, 1985; Blundy *et al.*, 1986; Hood *et al.*, 1986; Brevet *et al.*, 1988). In addition, the T-DNA from certain Ti plasmids harbors a gene (*ons*) that specifies octopine or nopaline secretion (Messens *et al.*, 1985). Although there have been a few reports of the production of opines by nontransformed plants under certain conditions (e.g., Christou *et al.*, 1986), it is generally considered that opine biosynthesis by plants is a crown gall (or hairy root) tumor-specific process. Plant cells cannot utilize opines. *Agrobacterium* cells, although harboring a Ti plasmid that contains opine biosynthetic genes, do not produce opines. However, the Ti plasmid harbored by the strain of *Agrobacterium* that incites a particular type of crown gall tumor generally carries opine catabolic genes on a non-T-DNA region of the plasmid (Garfinkel and Nester, 1980; Schardl and Kado, 1983). These genes enable particular *Agrobacterium* strains to utilize opines as the sole source of carbon and, in some instances, nitrogen (Bomhoff *et al.*, 1976; Montoya *et al.*, 1977; Klapwijk and Schilperoort, 1979; Dessaux *et al.*, 1987; Sans *et al.*, 1987; Dessaux *et al.*, 1988). Furthermore, only those opines that are directed to be made in the plant by a given Ti plasmid generally are catabolized by *Agrobacterium* strains harboring that Ti plasmid. Although some species of *Pseudomonas* and some gram-positive soil bacteria can also catabolize opines (Saint-Pierre and Dion, 1988), the directed infection of plants for the presumed purpose of the synthesis of opines that will be utilized by the inciting *Agrobacterium* strain has been termed an example of "genetic colonization." The "opine concept" implies that *Agrobacterium* "genetically engineers" plant cells to make compounds, the opines, that (to a first approximation) only it can utilize. *Agrobacterium* thereby is able to establish an ecological niche for itself, with the plant now supplying the nutrients directed to the establishment of this colonization.

Opines can serve at least three functions for the bacterium. First, they can be used as a carbon and, in some cases, nitrogen source for the inciting strain of *Agrobacterium*. Opine catabolism and transport genes are transcriptionally silent in the absence of the appropriate opine inducer, but are activated in the presence of the opine (Gelvin *et al.*, 1981). Genetic data formerly suggested that the genes responsible for octopine or nopaline catabolism and transport were controlled both positively by the opine and negatively by a repressor (Genetello *et al.*, 1977; Kerr *et al.*, 1977; Klapwijk *et al.*, 1978; Petit *et al.*, 1978; Klapwijk and Schilperoort, 1979; Ellis *et al.*, 1982a,b). Some of these authors additionally suggested that the regulation of conjugal Ti plasmid transfer by certain opines (see below) was under the control of this same repressor system (Klapwijk *et al.*, 1978). More recent data suggest, however, that octopine and nopaline catabolism and transport genes are regulated only positively by a protein homologous to the *lysR* family of DNA binding proteins (D. White, data presented at the Fifth International Congress on the Molecular Genetics of Plant–Microbe Interactions, Interlaken, Switzerland, 1990; Habeeb *et al.*, 1991; von Lintig *et al.*, 1991). In an *Agrobacterium* strain that harbors no complete copy of any gene from the octopine catabolism region of the Ti plasmid, the promotor of one of the opine catabolic genes, that directs the expression of a *lacZ* fusion, was silent. This promoter was induced by octopine (von Lintig *et al.*, 1991). Because octopine transport is inducible by octopine, the question arises as to how molecules of octopine initially enter the bacterial cell in order to induce the transport and catabolism genes. Although a possible explanation may be that the opine transport gene promoter is slightly "leaky," resulting in a low level of production of the octopine transport machinery, a likely explanation for this paradox has been proposed by this author's laboratory. Using highly radioactive [^{14}C]octopine or [^{14}C]nopaline, M. Krishnan *et al.*, 1991 found that both of these compounds could be transported into *A. tumefaciens* strains with the C58 chromosomal background even in the absence of a Ti plasmid. This transport required active cellular metabolism in that it was inhibited by incubation of the bacteria with sodium azide (Figure 2.5). The transport of radioactive octopine or nopaline could be competed with by either the heterologous opine or by arginine (Figure 2.6). These data suggest that the Ti plasmid-independent transport of octopine and nopaline into *Agrobacterium* is mediated by a chromosomally encoded arginine transport system. Although this system is much less efficient in transporting octopine or nopaline than are the Ti plasmid-encoded opine transport systems, it is sufficient to permit the initial transport of the opines into the cell to induce the opine catabolic and transport genes.

The Ti plasmid is a conjugal plasmid. Transfer of the entire Ti plasmid between *Agrobacterium* cells is normally repressed. However, in the presence of certain opines, termed conjugal opines, conjugation of the Ti plasmid is induced.

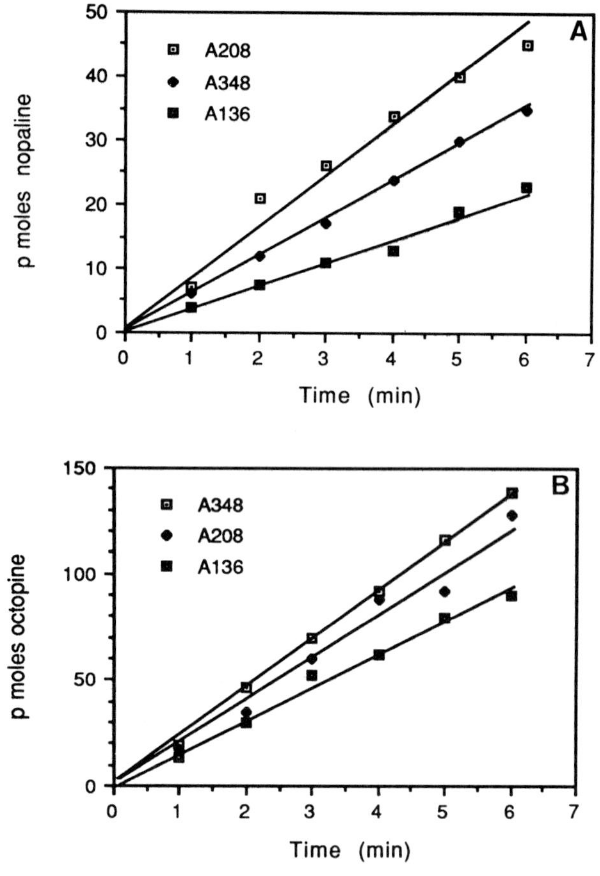

Figure 2.5 Transport of nopaline (A) or octopine (B) by *A. tumefaciens* A136 (lacking a Ti plasmid), A348 (harboring the octopine-type Ti plasmid pTiA6), or A208 (harboring the nopaline-type Ti plasmid pTiT37). The bacteria were grown to a Klett of 100, resuspended at a concentration of 2×10^8 cells/ml in K3 medium containing 5 m*M* MES–NaOH (Veluthambi *et al.*, 1989) and incubated with 50,000 cpm/ml octopine (270 mCi/mmol) plus 1 μ*M* octopine (A), or 50,000 cpm/ml (270 mCi/mmol) nopaline plus 1 μ*M* nopaline (B). Samples were collected by filtration, washed, and counted.

Not all opines are conjugal opines. For example, although octopine-type Ti plasmids encode functions for the catabolism of both octopine and the mannityl opines (mannopine, mannopinic acid, agropine, and agropinic acid), only octopine serves as a conjugal opine (Genetello *et al.*, 1977; Klapwijk *et al.*, 1978; Petit and Tempe, 1978; Petit *et al.*, 1978; Hooykaas *et al.*, 1979). Similarly, nopaline-type Ti plasmids use agrocinopines A and B, but not nopaline, as the

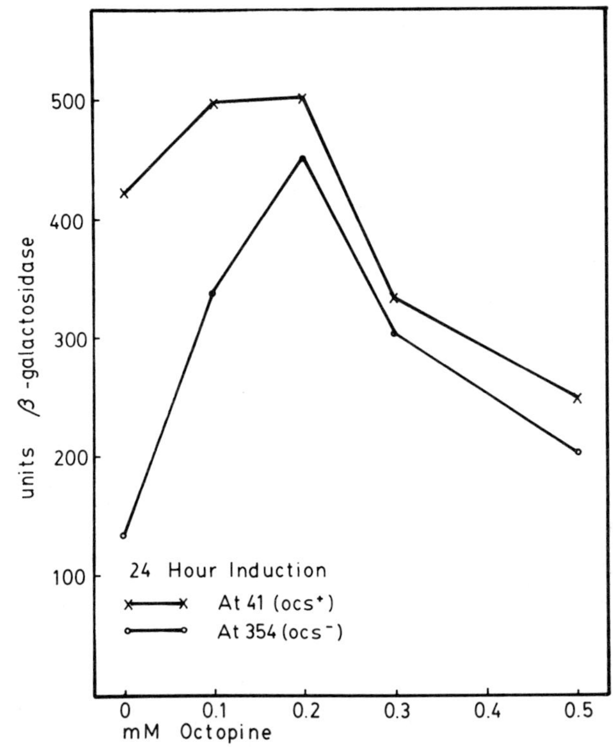

Figure 2.6 Effect of exogenous octopine on the induction of the *pinF* gene in *Agrobacterium* strains that harbor a wild-type (At41) or mutant (At335) octopine synthase gene. Although the maximal hyperinduction by octopine of the *pinF* gene in strain At41 is approximately 15%, octopine at 200 μM hyperinduces the *pinF* gene in strain At335 by 300%. Concentrations of octopine in excess of 200 μM repress the induction of the *pinF* locus. These higher concentrations of octopine match those found in mature crown gall tumors.

conjugal opines (Ellis *et al.*, 1982a). Mannopine/agropine-type Ti plasmids utilize agrocinopines C and D, but not the mannityl opines, as the conjugal opines (Ellis *et al.*, 1982a). The conjugal transfer of a copy of the Ti plasmid from a virulent *Agrobacterium* cell to a cell lacking such a plasmid makes sense in light of the opine hypothesis. Because the opine catabolic genes reside on the Ti plasmid, these *Agrobacterium* cells lacking a Ti plasmid cannot partake in the benefits of "genetic colonization"; that is, they cannot utilize the opines produced by the crown gall tumors. The conjugal receipt of such a plasmid from a virulent donor subsequently allows the recipient to utilize these opines. In addition, the spread of a copy of the Ti plasmid to these bacteria disperses the genetic determinant of tumorigenesis and opine production to other *Agrobacterium* cells, allowing the tumorigenic cycle to continue.

Although Tempe *et al.* (1977) once suggested that the transfer of the T-DNA from *Agrobacterium* to plant cells may utilize the same transfer system as is used in the conjugal transfer of the Ti plasmid between *Agrobacterium* cells, we now know that this is most likely not true. *Agrobacterium* strains in which the conjugal transfer genes were mutated are virulent, and strains lacking T-DNA could still transfer the Ti plasmid conjugally between *Agrobacterium* cells (Ellis *et al.*, 1982b; Klapwijk and Schilperoort, 1979). However, a possible link between T-DNA transfer to plant cells and conjugal Ti plasmid transfer between bacterial cells was recently suggested by Gelvin and Habeck (1990) and Steck and Kado (1990). Using octopine-type and nopaline-type *Agrobacterium* strains, these authors showed that mutations in certain *vir* genes (necessary for T-DNA transfer) could affect the efficiency of conjugal Ti plasmid transfer. Gelvin and Habeck (1990) showed that some mutations in the genes *virA, virB, virC,* and *virG* could depress the frequency of octopine-induced conjugal transfer of the Ti plasmid by up to four orders of magnitude. This effect was not absolute, however, in that it was manifested only under limiting times of octopine induction of the conjugal transfer apparatus (18–24 hours). When the bacteria were induced by octopine for longer times (48–72 hours), the inhibitory effect of mutations in these *vir* genes diminished. These authors suggested that, although *vir* gene functions were not essential for conjugal Ti plasmid transfer, these functions may aid the conjugation process under limiting conditions. These authors therefore established a genetic link between the T-DNA transfer process, mediated by the *vir* genes, and the opine-induced conjugal plasmid transfer process.

A third possible function for certain opines as a signal for mediating tumorigenesis was recently described by Veluthambi *et al.* (1989). We originally posed the question, would *Agrobacterium* cells induce their *vir* genes when in the presence of a crown gall tumor? The bacteria could sense such a tumor by the presence of crown gall tumor-specific compounds, opines, synthesized by the plant cells. Our original hypothesis suggested that, in the presence of opines, the *vir* genes may not be induced by the plant phenolic compounds. To our surprise, the induction of the *vir* genes by acetosyringone, as determined by β-galactosidase activity of *vir*::*lacZ* fusions, was enhanced by certain opines. The concentrations of octopine necessary for this potentiation of *vir* gene induction were high (1–10 m*M*), but within the range of concentration of octopine and octopinic acid found in mature crown gall tumors (Scott *et al.*, 1979; Saint-Pierre and Dion, 1988). Although the strains examined (derivatives of A348, an octopine-type *Agrobacterium* strain) could utilize octopine and the mannityl opines, but not nopaline, nopaline had a greater synergistic effect upon *vir* gene induction by acetosyringone than did octopine. Two other nonmetabolizable opines, leucinopine and succinamopine, also were better potentiators of *vir* gene induction than was octopine. Mannopine, however, did not hyperinduce the *vir* genes in the presence of acetosyringone. Apparently, amino di- and triacid opines (but not arginine, a constituent of octopine and nopaline), were effective compounds in potentiating the induction of the *vir* genes by acetosyringone. The activity of such nonmetabolizable opines as nopaline on *vir* gene induction by an octopine-type *Agrobacterium* strain could in part be explained by the fact that nopaline could be transported into these strains by a non-Ti plasmid-encoded transport system (see above).

The hyperinduction of the *vir* genes by opines and acetosyringone required no Ti plasmid-encoded genes other than *virA* and *virG*. It was possible, however, that chromosomal genes could also mediate this process. The potentiation of *vir* gene induction by opines resulted in the enhanced accumulation of T-strands within the induced *Agrobacterium* cell. The double-stranded cleavage of the T-DNA border repeat sequences was also increased. In addition, the application of both opines and acetosyringone, but not opines or acetosyringone alone, to the cut plant apical meristem resulted in an increased transformation efficiency of cotton, a plant relatively recalcitrant to *Agrobacterium* transformation and regeneration (Veluthambi *et al.*, 1989).

Although the hyperinduction of the *vir* genes by opines increased with opine concentration (up to at least 10–20 m*M*) when the bacteria were induced in a test tube by acetosyringone, the induction of *Agrobacterium* cells by tobacco protoplasts resulted in a different pattern of potentiation by octopine. Using an *Agrobacterium* strain harboring a *lacZ* transcriptional fusion in the gene *pinF* [also known as *virH* (Kanemoto *et al.*, 1989), a gene inducible by acetosyringone but not necessary for the transformation of tobacco (Stachel and Nester, 1986)], Krishnan and Gelvin showed that relatively low concentrations of octopine (100–200 μ*M*) could have a stimulatory effect on induction. Higher concentrations of octopine, however, repressed *vir* gene activity (Figure 2.6). In addition, the *pinF* gene of *Agrobacterium* strains harboring a Tn*5* insertion in the octopine synthase (*ocs*) gene was induced 2- to 3-fold less by protoplasts than was the *pinF* gene in a strain harboring a wild-type *ocs* gene (Figure 2.7). These two strains behaved identically, however, when in the presence of tobacco protoplasts derived from transgenic plants that synthesized octopine (Figure 2.8). Data from Dion's laboratory (Saint-Pierre and Dion, 1988) indicated that in the extracellular fluid of mature crown gall tumors, octopine concentrations reached approximately 200 μ*M*. These results suggested that, in the presence of plant cells, *Agrobacterium* could respond to physiological concentrations of octopine in a complex way. Low concentrations potentiated *vir* gene induction, whereas higher concentrations resulted in a reduced level of induction.

We have hypothesized that the concentration of opines, among other chemicals in the wound environment, may influence the physiological state of the *Agrobacterium* cell. During the initial stages of infection, when the concentra-

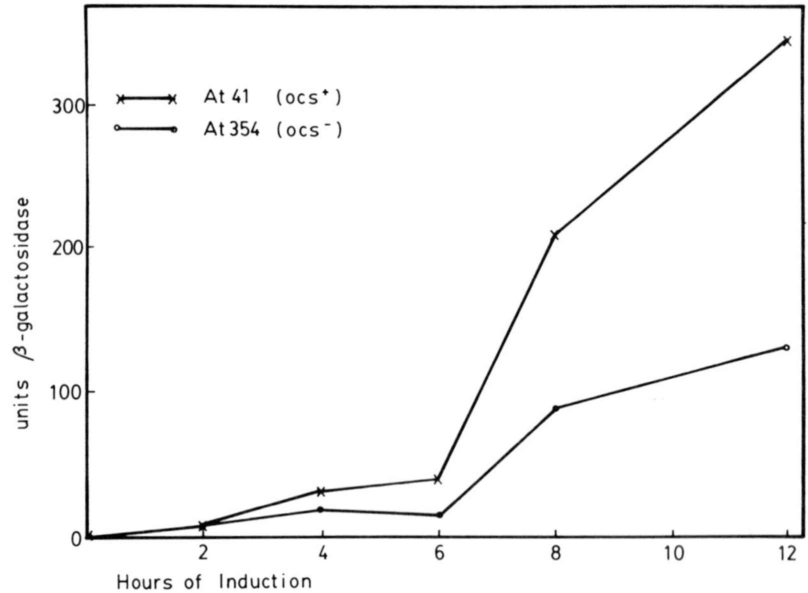

Figure 2.7 Induction by wild-type tobacco protoplasts of the *pinF* gene in *Agrobacterium* strains that harbor a wild-type (At41) or mutant (At335) octopine synthase gene. Although the kinetics of induction of the *pinF* gene are similar in the two bacterial strains, *ocs*$^+$ bacteria induce the *pinF* gene to a 3-fold greater extent than do *ocs*$^-$ bacteria.

tion of phenolic *vir* gene inducing molecules is high, low concentrations of opine synthesized by the nascent tumor may indicate to the bacterium that infection has been successful. The bacteria may be "encouraged" to continue the infection process. As the wound heals and the concentration of phenolic molecules decreases, the limitation of inducing molecules may be partially compensated by increased opine concentrations. We previously showed that, in the presence of limiting acetosyringone concentrations, T-strand production was increased by the addition of certain opines (Veluthambi *et al.*, 1989). Finally, the concentration of phenolic inducing molecules may decrease to such an extent, and the concentration of opines produced by a mature crown gall tumor increase to such a degree, that *vir* gene activity is repressed. The bacteria may sense that infection has been successful enough to produce a tumor and that additional energy need not be expended to process and transfer the T-DNA to plant cells. The "physiology" of the bacterial cell alters from an "inductive" state to a "vegetative" state, whereby *Agrobacterium* now concentrates its efforts upon metabolizing opines. Such a model would predict that *Agrobacterium* cells harboring T-DNA that is unable to direct the production of certain opines may be less virulent than strains that could induce opine-producing tumors. That this may be true is suggested by the data

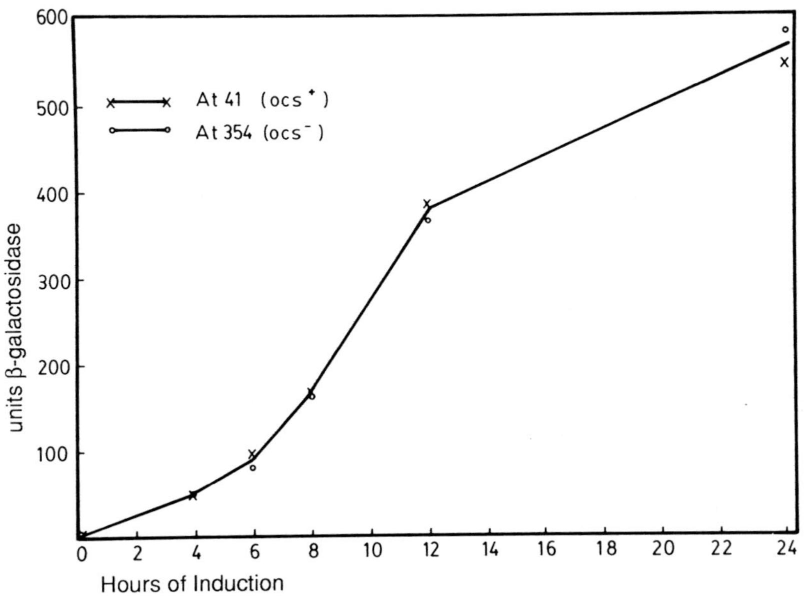

Figure 2.8 Induction by transgenic tobacco protoplasts of the *pinF* gene in *Agrobacterium* strains that harbor a wild-type (At41) or mutant (At335) octopine synthase gene. Note that when the protoplasts synthesize octopine, there is no difference in the induction of the *pinF* gene between the two strains.

presented in Table 2.2. *Agrobacterium* strains harboring a wild-type *ocs* gene are approximately 40% more virulent, as measured by a potato disk assay, than a near-isogenic strain harboring a mutation in the *ocs* gene.

Opines thus may play a number of important roles in crown gall tumorigenesis. As well as metabolites, they may serve as signal molecules impor-

TABLE 2.2
Virulence of *A. tumefaciens* At41 (ocs^+) and At355 (ocs^-) on Potato Tuber Disks[a]

		Average number of tumors/disk	
Experiment	Bacteria/ml	At41	At335
1	10^6	49.8 (*n* = 6)	40.2 (*n* = 6)
2	10^7	14.2 (*n* = 47)	8.4 (*n* = 48)
3	10^6	11.2 (*n* = 49)	7.4 (*n* = 52)

[a]Fifty μl of bacteria were inoculated onto each disk, and the disks scored 18 days after infection. *n*, Number of disks scored.

tant both for inducing conjugal T-DNA transfer and in regulating the T-DNA transfer process in response to environmental signals.

2.4 Genetic Engineering of Plants Using T-DNA

The use of the T-DNA for the genetic engineering of plants depended on four important discoveries and technical developments. The first was the localization in and deletion from the T-DNA of those genes governing tumorigenesis. The second was the development of markers useful for the selection of those plant cells that were genetically transformed by *Agrobacterium*. The third was the development of a number of vectors and genetic procedures that facilitated the integration of foreign genes into the T-DNA. The fourth was the development of various plant tissue culture systems that allowed the efficient infection and regeneration of plant cells into transgenic plants.

There have been few reports of the spontaneous regeneration of plants from crown gall tumors. Crown gall tumors synthesize and secrete large quantities of the phytohormones auxin and cytokinin (Morris, 1986), and the resulting hormone imbalance will not allow the organization of normal plant tissues and organs. The spontaneous regeneration of a small number of plants from crown gall tumors has involved either the deletion of substantial portions of the T-DNA (Yang *et al.*, 1980) or the methylation of T-DNA "oncogenes," resulting in their inactivation (Amasino *et al.*, 1984). Hairy root tumors produced by *Agrobacterium rhizogenes* on a number of different plant species can spontaneously regenerate near-normal looking plants (Tepfer, 1984; Noda *et al.*, 1987). However, tobacco plants displayed some abnormalities, such as altered internode length, abnormal leaf shape and coloration, pleigeotrophy, an altered generation time, and sterility. It became obvious early on, therefore, that the T-DNA genes that regulated the production of abnormal quantities of phytohormones would have to be inactivated in order to produce transgenic plant cells that could regenerate developmentally normal plants.

The identification of the T-DNA regions responsible for phytohormone production (oncogenes) was first accomplished by transposon mutagenesis (Garfinkel and Nester, 1980; Ooms *et al.*, 1981). In a particularly elegant series of experiments, Garfinkel *et al.* (1981) used Tn*5* and Tn*3* mutagenesis to create mutations in the T-DNA that resulted in an altered crown gall tumor morphology on *Kalanchoe daigramontiana* and, in some instances, *Nicotiana tabacum*. These insertions occurred in three regions of the T-DNA. Insertions in the first region, termed *tms* (for tumor morphology shooty) resulted in tumors that resembled shooty teratomas. Tumors that produced large quantities of adventitious roots were incited by *Agrobacterium* strains mutant in the locus *tmr* (for tumor

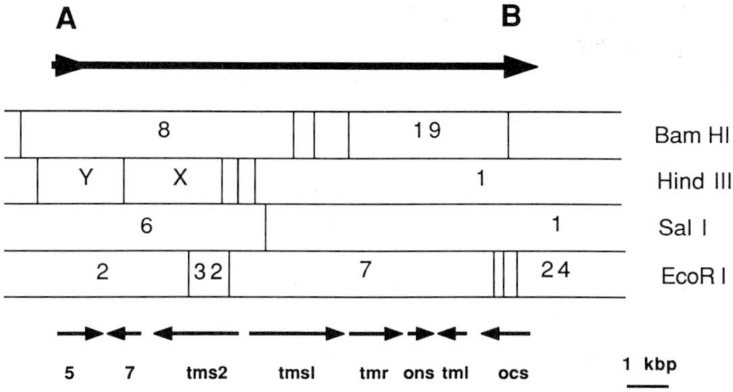

Figure 2.9 Map of the T_L region of a typical wide host range octopine-type Ti plasmid. The arrows beneath the map indicate the positions and orientations of the transcripts. *tmsl* and *tms2,* tumor morphology shooty genes 1 and 2, respectively; *tmr,* tumor morphology rooty; *tml,* tumor morphology large; *ons,* octopine–nopaline secretion; *ocs,* octopine synthase; 5 and 7, transcripts for genes 5 and 7, respectively.

morphology rooty), and very rapidly growing tumors were produced on *Kalanchoe* plants by inoculation of *Agrobacterium* strains mutant in the gene *tml* (for tumor morphology large). Mutations in other regions of the T-DNA, such as the octopine synthase (*ocs*) gene, did not affect the tumor morphology. Such tumors, however, did not produce octopine. A map of the T_L region of the Ti plasmid pTiA6, showing the various T-DNA oncogenes, is shown in Figure 2.9. Based upon the observations of Skoog and Miller (1957) that the morphology of nontransformed tobacco explants could be altered in culture by varying the ratio of auxin to cytokinin in the tissue culture medium, Garfinkel *et al.* (1981) hypothesized that the *tms* locus regulated the production of auxins, and the *tmr* locus regulated the production of cytokinins.

This hypothesis was subsequently proven correct. Akiyoshi *et al.* (1983) studied the ratio of auxins to cytokinins found in tumors incited by *Agrobacterium* strains harboring wild-type oncogenes, or mutations in the genes *tms* or *tmr*. These ratios coincided with the expected phenotypes resulting from the mutation of genes involved in auxin or cytokinin production, respectively. Further molecular, genetic, and biochemical studies identified the *tms* locus as encoding two proteins responsible for indoleacetic acid biosynthesis (Inze *et al.*, 1984; Schroder *et al.*, 1984; Thomashow *et al.*, 1984, 1986; Kemper *et al.*, 1985; Van Onckelen *et al.*, 1985) and the *tmr* locus as encoding an enzyme, isopentenyladenosine transferase, that catalyzed the biosynthesis of cytokinins (Akiyoshi *et al.*, 1984; Barry *et al.*, 1984; Buchmann *et al.*, 1985). The *tml* locus and another locus (gene 5) apparently encode functions that sensitize the plant cell to auxins and cytokinins (Hooykaas *et al.*, 1988; Spanier *et al.*, 1989;

Tinland *et al.*, 1989, 1990). Early genetic engineering experiments generated transgenic plant tumors (e.g., Murai *et al.*, 1983). However, the deletion of the T-DNA oncogenes discussed above led to the development of "disarmed" T-DNA vectors that could deliver novel genes to plant cells yet still permit the regeneration of normal plants from these cells.

The explants of most nontransformed plant tissues require the addition of auxins and cytokinins to the culture medium to proliferate *in vitro*. Because crown gall tumor cells overproduce large quantities of auxins and cytokinins, these cells can be selected on account of their phytohormone autonomy in tissue culture. As mentioned above, however, this phytohormone overproduction precluded the regeneration of the transgenic tumor tissue into normal plants. With the deletion of the T-DNA oncogenes, plants could be regenerated, but another marker was required to select for those cells that had received the T-DNA. Such a need prompted the development of antibiotic resistance genes that could be used to select transgenic plant cells.

Because the proliferation of most plant cells is highly sensitive to aminoglycoside antibiotics, resistance to such antibiotics was used as the first selectable marker for gene transfer to plants. The gene encoding the aminoglycoside-modifying enzyme neomycin phosphotransferase II (NPTII), *nptII* found on the bacterial transposon Tn*5*, had previously been brought into plant cells as part of the transposon Tn*5* during the study of Garfinkel and Nester (1980). Because the *nptII* gene within Tn*5* still contained prokaryotic transcriptional regulatory signals, however, it was not expressed in plants. In the early 1980s several groups developed chimeric genes that could be as selectable markers in plants. The first of these chimeric genes contained the *nptII* coding region under the control of eukaryotic promoters [derived from *Agrobacterium* opine biosynthetic genes or the cauliflower mosaic virus (CaMV) genome] and followed by polyadenylation signal sequences derived from opine biosynthetic genes (Bevan *et al.*, 1983; Fraley *et al.*, 1983; Herrera-Estrella *et al.*, 1983a,b; Yenofsky *et al.*, 1990). The transfer and expression of these chimeric selectable markers rendered transformed plant cells resistant to kanamycin or the related antibiotic G418. Normal (but kanamycin-resistant) calli and plants could subsequently be recovered from a mixed population of nontransformed and transformed cells. The development of numerous other chimeric selectable (and screenable) markers has occurred during the past 10 years. Such markers include resistance to hygromycin (Van den Elzen et al., 1985; Waldron *et al.*, 1985), gentamicin (Hayford *et al.*, 1988), methotrexate (Herrera-Estrella *et al.*, 1983a), and streptomycin (an antibiotic that causes bleaching but not necessarily death of plant cells; Svab *et al.*, 1990). Recently, chimeric genes resulting in resistance to various herbicides have been developed (Comai *et al.*, 1985; Shah *et al.*, 1986; DeBlock *et al.*, 1987; della-Cioppa *et al.*, 1987; Fillatti *et al.*, 1987; Klee *et al.*, 1987a,b; Botterman and Leemans, 1988; Cheung *et al.*, 1988; Haughn *et al.*, 1988; Lee *et al.*, 1988; De

Greef *et al.*, 1989; Gabard *et al.*, 1989; Lyon *et al.*, 1989; Streber and Willmitzer, 1989; Guerineau *et al.*, 1990). In addition, a number of screenable markers useful for the identification (but not necessarily selection) of transformed plant cells have been used in transgenic plants. Among these are opine biosynthetic genes (Dahl and Tempe, 1983), genes directing the production of β-galactosidase (Helmer *et al.*, 1984; Teeri *et al.*, 1989; Matsumoto *et al.*, 1990) or β-glucuronidase (Jefferson *et al.*, 1987; Janssen and Gardner, 1990), genes directing the expression of luciferase (Ow *et al.*, 1986; Koncz *et al.*, 1989; Barnes, 1990), or genes affecting the pigmentation of plant cells (Ludwig *et al.*, 1990). A number of "constitutive" promoters have been used to direct the expression of these selectable markers. These include promoters for the octopine, nopaline, and mannopine synthase (*ocs, nos,* and *mas,* respectively) genes, and the CaMV 35 and 19S gene promoters (Bevan *et al.*, 1983; Herrera-Estrella *et al.*, 1983a,b; Koziel *et al.*, 1984; Velten *et al.*, 1984; Gelvin *et al.*, 1985; Waldron *et al.*, 1985; Kay *et al.*, 1987; Rogers *et al.*, 1987; Rothstein *et al.*, 1987; Comai *et al.*, 1990). In addition, a number of "regulatable" promoters have been used to express chimeric genes in plants. Among these are environmentally and stress-inducible promoters (Simpson *et al.*, 1985; Schoffl and Baumann, 1985; Spena and Schell, 1987; Strittmatter and Chua, 1987; Thornburg *et al.*, 1987; Benfey and Chua, 1989) and developmentally regulated promoters (Koltunow *et al.*, 1990; Mariani *et al.*, 1990).

The development of T-DNA vectors suitable for the introduction of foreign genes into plant cells has advanced considerably since the early 1980s. The Ti plasmids of *Agrobacterium* are very large (200–800 kbp), and this size precludes the direct cloning of genes into the T-DNA region. Early experiments adopted a procedure by which a region of the T-DNA was cloned into a broad host range replicon that could replicate both in *E. coli* and in *Agrobacterium.* Derivatives of the plasmid RK2, such as pRK290 and others (Ditta *et al.*, 1980, 1985; Schmidhauser *et al.*, 1988), were commonly used. Following the cloning of a gene into the T-DNA and transformation of the plasmid into *E. coli,* the plasmids were mated into *A. tumefaciens* by a triparental mating procedure using plasmids such as pRK2013 (Figurski and Helinski, 1979) to supply the *mob* function *in trans.* Double-homologous crossover events could occur, exchanging the "mutagenized" version of the T-DNA with the wild-type T-DNA fragment found on the Ti plasmid. Such "marker exchange" or "homogenotization" events could be selected by the introduction into the *A. tumefaciens* cell of a plasmid of the same incompatibility group as the RK2 derivative [such as pPH1JI (Hirsch and Beringer, 1984)] and plating the cells on the appropriate antibiotics. This protocol, originally devised by Ruvkin and Ausubel (1981) for use in *Rhizobium meliloti,* was adapted by Garfinkel *et al.* (1981) for their transposon mutagenesis study of T-DNA functions, and by others for the insertion of specific genes into the T-DNA (Matzke and Chilton, 1981). Murai *et al.* (1983) used this procedure for the introduction of a bean phaseolin gene into plant tumors, and the technique has been widely used since (e.g., Barton *et al.*, 1983).

Although this homogenotization regime can be efficiently used to replace T-DNA segments with those harboring new genes, the procedure is a bit cumbersome, requiring two separate matings and multiple antibiotic resistance selections. Another method commonly used to introduce foreign DNA into the T-DNA relies upon a single homologous recombination, or cointegration, event. A gene of interest is first cloned into a pBR322-derived vector, and the plasmid transformed into *E. coli.* This *E. coli* strain is mated with the recipient *A. tumefaciens* strain, again using the mobilizing functions provided by pRK2013 during a triparental mating. Plasmids harboring a ColE1 origin of replication, such as pBR322, cannot replicate in *Agrobacterium.* The recipient *Agrobacterium* strain has been engineered to contain a Ti plasmid having sequences homologous to pBR322 within the T-DNA. Such a plasmid may also have the T-DNA oncogenes deleted. Plating the mating mixture on antibiotics that select for the presence of the pBR-derived plasmid will select *Agrobacterium* cells in which the Ti plasmid and the pBR plasmid have cointegrated using the pBR homology in the T-DNA. A number of cointegrate vector systems, including the widely used plasmid pGV3850 and its derivatives (Zambryski *et al.*, 1983), and others (Comai *et al.*, 1983; Deblaere *et al.*, 1985) are based upon this experimental design. Other vectors contained within the pMON split end vector series also rely upon cointegration of a pBR322-derived plasmid with the T-DNA (Fraley *et al.*, 1985). In these cases, however, the T-DNA of the recipient Ti plasmid has had one T-DNA border deleted, and the pBR322-derived plasmid harbors the other T-DNA border. Cointegration occurs between T-DNA sequences on the Ti plasmid and on the pBR plasmid, resulting in a reconstructed T-DNA containing two borders and the foreign gene of interest between them. Although these cointegrate vector systems are relatively convenient and widely used, there is always a risk that the cointegrate can resolve into the two component plasmids in *Agrobacterium.* Such events can be selected against, however, by selection with antibiotics appropriate for the retention of the pBR-derived plasmid in the bacterium.

An observation made by Schilperoort's laboratory (Hoekema *et al.*, 1983) led to the development of an entirely new set of T-DNA vectors. These investigators found that the T-DNA could be separated from the *vir* genes onto two separate replicons in *Agrobacterium.* Such a bacterium was still able to transfer T-DNA to plant cells. Novel sets of T-DNA vectors, called binary vectors, were soon developed by a number of laboratories (e.g., de Frammond *et al.*, 1983; Bevan, 1984; An *et al.*, 1985, 1987; Klee *et al.*, 1985; Simpson *et al.*, 1986; Schardl *et al.*, 1987; Vilaine and Casse-Delbart, 1987a; McBride and Summerfelt, 1990). These vectors contain a set of T-DNA borders between which can be cloned a marker useful for selection in plant cells, and usually a number of unique restriction endonuclease sites. This T-DNA is harbored by a broad host range replicon capable of replication both in *E. coli* and in *A. tumefaciens*. Foreign genes of interest can be inserted into the unique restriction endonuclease sites within the T-DNA, the recombinant plamid transformed into *E. coli,* and subsequently mobilized into an *Agrobacterium* strain harboring a Ti plasmid (either wild-type, or a "disarmed" Ti plasmid in which the T-DNA was deleted). These binary vectors can replicate in *Agrobacterium,* usually with a copy number of 5–10 relative to the Ti plasmid and the bacterial chromosome. The *vir* genes resident within *Agrobacterium* can process and mobilize the T-DNA from the bacterium to a plant cell. Such binary systems are extremely easy to use, requiring merely one mating of the vector from *E. coli* and no subsequent recombination step within *Agrobacterium*. In combination with "disarmed" *Agrobacterium* strains such as LBA4404 (Ooms *et al.*, 1982b), pGV3111 (Zambryski *et al.*, 1983), C58Z707 (Culianez-Marcia and Hepburn, 1988a), or EHA101 (Hood *et al.*, 1986), these binary vectors are commonly used to create and regenerate transgenic plants. A number of specialized binary vectors have been created to select for, for example, promoter and enhancer sequences active in plant cells (An *et al.*, 1988; Ott *et al.*, 1990).

The "disarming" of Ti plasmids by deletion of the T-DNA oncogenes permitted not only the genetic transformation of plant cells, but also the regeneration of these transgenic cells into whole plants. The advent of easily manipulatable T-DNA vector systems gave rise to novel and easy to use plant cell regeneration systems. Initially, tobacco protoplasts were the target cell for transformation and regeneration. Highly refined "cocultivation" regimes were developed (Marton *et al.*, 1979; Fraley *et al.*, 1984; Hain *et al.*, 1985). In these experiments, *Agrobacterium* cells harboring foreign genes in the T-DNA were incubated with protoplasts that had just regenerated cell walls and were starting to divide. Although useful for such relatively easily regenerable plants as tobacco and *Petunia,* these cocultivation systems were not as useful for plant species from which it was difficult to create and regenerate protoplasts. Although the transformation of tobacco cells, rather than protoplasts, has been reported (An, 1985) the protoplast and cell culture technology was somewhat tricky and beyond the capabilities of many plant molecular biology laboratories. Because of the difficulty transforming and regenerating protoplasts of a number of different species, many laboratories are now using a leaf "disk" infection protocol (Horsch *et al.*, 1985; Burow *et al.*, 1990). In this procedure, sterile leaf sections are cut with a scalpel or a cork borer, dipped in *Agrobacterium* cells harboring a genetically engineered T-DNA (using either cointegrate or binary vectors), the leaf sections blotted to remove excess bacteria, and incubated for several days on tissue culture medium lacking antibiotics. The leaf sections are subsequently transferred to shoot-inducing medium containing antibiotics, both to kill the bacterial cells and to select for transgenic plant cells. As a variant of this procedure, stem or cotyledon section have been used for a number of plant species. Tobacco plants regenerated from these leaf disks generally are not chimeric, and, after potting, can set seed within 3–4 months of the initial infection. The leaf disk transformation procedure is currently the most popular method for generating transgenic tobacco or *Petunia* plants.

2.5 Conclusion

Enormous progress has been made during the past 10 years in understanding the molecular basis of crown gall tumorigenesis. The complex interactions involving pathogenesis involve the exchange of numerous chemical signals between the host plant and the bacterium. The resulting efficient and directed transfer of genes from *Agrobacterium* to plant cells is unique in nature. An understanding of the signaling involved in initiating the infection cycle, as well as the molecular events involved in T-DNA processing and transfer, has allowed scientists to harness *Agrobacterium* as a genetic engineering tool that is now routinely used in the laboratory. What is most amazing is the speed with which scientists have been able to develop the *Agrobacterium* system for use in the laboratory. It has been only 13 years since it was shown that the T-DNA region of the Ti plasmid is transferred to plant cells, and only 10 years since the first transgenic tumors (and 8 years since the first transgenic plants) were generated. Although scientists have learned much about the mechanism of crown gall tumorigenesis, a number of aspects of this process, such as the details of the actual T-DNA transfer and integration events, have yet to be elucidated. Several problems in using *Agrobacterium* as a genetic engineering tool must still be resolved. These include the extension of the host range of the bacterium to allow it efficiently to infect important monocot crop species, the ability to utilize *Agrobacterium* for plant gene replacement experiments, and a better control (or elimination) of the position effects involved in T-DNA gene expression. Considering the rapid progress that has been made to date, however, it is likely that these problems also will be solved within the next few years. The resolution of these problems will allow T-DNA-mediated gene transfer to become the universal method for introducing foreign genetic material into plants.

Acknowledgments

The author thanks Susan Karcher, Walt Ream, and Todd Steck for their critical evaluation of earlier versions of this chapter. Research in the author's laboratory has been supported by the National Science Foundation, the United States Department of Agriculture, the Midwest Plant Biotechnology Consortium, and the Dow/Elanco and Dupont companies.

References

Akiyoshi, D. E., Morris, R. O., Hinz, R., Mischke, B. S., Kosuge, T., Garfinkel, D. J., Gordon, M. P., and Nester, E. W. (1983). *Proc. Natl. Acad. Sci. U.S.A.* 80, 407–411.

Akiyoshi, D. E., Klee, H., Amasino, R. M., Nester, E. W., and Gordon, M. P. (1984). *Proc. Natl. Acad. Sci., U.S.A.* 81, 5994–5998.

Albright, L. M., Yanofsky, M. F., Leroux, B., Ma, D., and Nester, E. W. (1987). *J. Bacteriol.* 169, 1046–1055.

Albright, L. M., Huala, E., and Ausubel, F. M. (1989). *Annu. Rev. Genet.* 23, 311–336.

Alt-Moerbe, J., Rak, B., and Schroder, J. (1986). *EMBO J.* 5, 1129–1135.

Amasino, R. M., Powell, A. L. T., and Gordon, M. P. (1984). *Mol. Gen. Genet.* 197, 437–46.

An, G. (1985). *Plant Physiol.* 79, 568–570.

An, G. (1987). *Meth. Enzymol.* 153, 292–305.

An, G., Watson, B. D., Stachel, S., Gordon, M. P., and Nester, E. W. (1985). *EMBO J.* 4, 277–284.

An, G., Ebert, P. R., Mitra, A., and Ha, S. B. (1988). In *Plant Molecular Biology Manual* (Ed. by S. B. Gelvin and R. A. Schilperoort), pp. A3–A19. Dordrecht: Kluwer.

Bandyopadhyay, R. S., Bruce, W. B., and Gurley, W. B. (1989). *J. Biol. Chem.* 264, 19399–19406.

Barker, R. F., Idler, K. B., Thomson, D. V., and Kemp, J. D. (1983). *Plant Mol. Biol.* 2, 335–350.

Barnes, W. M. (1990). *Proc. Natl. Acad. Sci. U.S.A.* 87, 9183–9197.

Barry, G. F., Rogers, S. G., Fraley, R. T., and Brand, L. (1984). *Proc. Natl. Acad. Sci. U.S.A.* 81, 4776–4780.

Barton, K. A., Binns, A. N., Matzke, A. J. M., and Chilton, M.-D. (1983). *Cell* 32, 1033–1043.

Benfey, P. N., and Chua, N.-H. (1989). *Science* 244, 174–181.

Bevan, M. (1984). *Nuc. Acids Res.* 12, 8711–8721.

Bevan, M. W., Flavell, R. B., and Chilton, M.-D. (1983). *Nature (London)* 304, 184–187.

Binns, A. N., and Thomashow, M. F. (1988). *Ann. Rev. Microbiol.* 42, 575–606.

Blundy, K. S., White, J., Firmin, J. L., and Hepburn, A. G. (1986). *Mol. Gen. Genet.* 202, 62–67.

Bolton, G. W., Nester, E. W., and Gordon, M. P. (1986). *Science* 232, 983–985.

Bomhoff, G., Klapwijk, P. M., Kester, H. C. M., Schilperoort, R. A., Hernalsteens, J. P., and Schell, J. (1976). *Mol. Gen. Genet.* 145, 177–181.

Botterman, J., and Leemans, J. (1988). *TIGS* 4, 219–222.

Bouchez, D., Tokuhisa, J. G., Llewellyn, D. J., Dennis, E. S., and Ellis, J. G. (1989). *EMBO J.* 8, 4197–4204.

Bouckaert-Urban, A.-M., and Vendrig, J. C. (1982). *Z. Pflanzenphysiol.* 105, 395–405.

Brevet, J., Borowski, D., and Tempe, J. (1988). *Mol. Plant-Microbe Inter.* 1, 75–79.

Bruce, W. B., and Gurley, W. B. (1987). *Mol. Cell. Biol.* 7, 59–67.

Bruce, W. B., Bandyopadhyay, R., and Gurley, W. B. (1988). *Proc. Natl. Acad. Sci. U.S.A.* 85, 4310–4314.

Buchmann, I., Marner, F.-J., Schroder, G., Waffenschmidt, S., Schroder, J., (1985). *EMBO J.* 4, 853–859.

Burow, M. D., Chlan, C. A., Sen, P., Lisca, A., and Murai, N. (1990). *Plant Mol. Biol. Rep.* 8, 124–139.

Byrne, M. C., Koplow, J., David, C., Tempe, J., and Chilton, M.-D. (1983). *J. Mol. Appl. Genet.* 2, 201–209.

Cangelosi, G. A., Hung, L., Puvanesarajah, V., Stacey, G., Ozga, D. A., Leigh, J. A., and Nester, E. W. (1987). *J. Bacteriol.*, 169, 2086–2091.

Cangelosi, G. A., Martinetti, G., Leigh, J. A., Lee, C. C., Theines, C., and Nester, E. W. (1989). *J. Bacteriol.* 171, 1609–1615.

Cangelosi, G. A., Ankenbauer, R. G., and Nester, E. W. (1990). *Proc. Natl. Acad. Sci. U.S.A.* 87, 6708–6712.

Caplan, A. B., Van Montagu, M., and Schell, J. (1985). *J. Bacteriol.* 161, 655–664.

Cheung, A. Y., Bogorad, L., Van Montagu, M., and Schell, J. (1988). *Proc. Natl. Acad. Sci. U.S.A.* 85, 391–395.

Chilton, M.-D., Drummond, M. H., Merlo, D. J., Sciaky, D., Montoya, A. L., Gordon, M. P., and Nester, E. W. (1977). *Cell* 11, 263–271.

Chilton, M.-D., Saiki, R. K., Yadav, N., Gordon, M. P., Quetier, E. (1980). *Proc. Natl. Acad. Sci. U.S.A.* 77, 4060–4064.

Christie, P. J., Ward, J. E., Winans, S. C., and Nester, E. W. (1988). *J. Bacteriol.* 170, 2659–2667.

Christou, P., Platt, S. G., and Ackerman, M. C. (1986). *Plant Physiol.* 82, 218–221.

Citovsky, V., De Vos, G., and Zambryski, P. (1988). *Science* 240, 501–504.

Citovsky, V., Wong, M. L., and Zambryski, P. (1989). *Proc. Natl. Acad. Sci. U.S.A.* 86, 1193–1197.

Close, T. J., Tait, R. C., and Kado, C. I. (1985). *J. Bacteriol.* 164, 774–781.

Close, T. J., Rogowsky, P. M., Kado, C. I., Winans, S. C., Yanofsky, M. F., and Nester, E. W. (1987). *J. Bacteriol.* 169, 5113–5118.

Comai, L., Schilling-Cordaro, C., Mergia, A., and Houck, C. M. (1983). *Plasmid* 10, 21–30.

Comai, L., Facciotti, D., Hiatt, W. R., Thompson, G., and Rose, R. E. (1985). *Nature (London)* 317, 741–744.

Comai, L., Moran, P., and Maslyar, D. (1990). *Plant Mol. Biol.* 15, 373–381.

Cornish, A., Greenwood, J. A., and Jones, C. W. (1989). *J. Gen. Microbiol.* 135, 3001–3013.

Culianez-Macia, F. A., and Hepburn, A. G. (1988a). *Plant Mol. Biol.* 11, 389–399.

Culianez-Macia, F. A., and Hepburn, A. G. (1988b). *Mol. Plant-Microbe Inter.* 1, 207–214.

Dahl, G. A., and Tempe, J. (1983). *Theor. Appl. Genet.* 66, 233–239.

Das, A. (1988). *Proc. Natl. Acad. Sci. U.S.A.* 85, 2909–2913.

Deasey, M. C., and Matthysse, A. G. (1984). *Phytopath.* 74, 991–994.

DeBeuckeleer, M., Lemmers, M., DeVos, G., Willmitzer, L., Van Montagu, M., and Schell, J. (1981). *Mol. Gen. Genet.* 183, 283–288.

Deblaere, R., Bytebier, B., DeGreve, H., Deboeck, F., Schell, J., and Van Montagu, M. (1985). *Nuc. Acids Res.* 13, 4777–4788.

Deblaere, R., Reynaerts, A., Hofte, H., Hernalsteens, J.-P., Leemans, J., and Van Montagu, M. (1987). *Meth. Enzymol.* 153, 277–292.

DeBlock, M., Botterman, J., Vandewiele, M., Dockx, J., Thoen, C., Gossele, V., Rao Movva, N., Thompson, C., and Van Montagu, M. (1987). *EMBO J.* 6, 2513–2518.

De Frammond, A. J., Barton, K. A., and Chilton, M.-D. (1983). *Bio/Technology* 5, 262–269.

De Greef, W., Delon, R., De Block, M., Leemans, J., and Botterman, J. (1989). *Bio/Technology* 7, 61–64.

della-Cioppa, G., Bauer, S. C., Taylor, M. L., Rochester, D. E., Klein, B. K., Shah, D. M., Fraley, R. T., and Kishore, G. M. (1987). *Bio/Technology* 5, 579–584.

De Paolis, A., Mauro, M. L., Pomponi, M., Carderelli, M., Spano, L., and Costantino, P. (1985). *Plasmid* 13, 1–7.

Dessaux, Y., Tempe, J., and Farrand, S. K. (1987). *Mol. Gen. Genet.* 208, 301–308.

Dessaux, Y., Guyon, P., Petit, A., Tempe, J., Demarez, M., Legrain, C., Tate, M. E., and Farrand, S. K. (1988). *J. Bacteriol.* 170, 2939–2946.

De Vos, G., and Zambryski, P. (1989). *Mol. Plant-Microbe Inter.* 2, 43–52.

DiRita, V. J., and Gelvin, S. B. (1987). *Mol. Gen. Genet.* 207, 233–241.

Ditta, G., Schmidhauser, T., Yakobson, E., Lu, P., Liang, X.-W., Finlay, D. R., Gainey, D., and Helinski, D. R. (1985). *Plasmid* 13, 149–153.

Ditta, G., Stanfield, S., Corbin, D., and Helinski, D. R. (1980). *Proc. Natl. Acad. Sci. U.S.A.* 77, 7347–7351.

Douglas, C. J., Halperin, W., and Nester, E. W. (1982). *J. Bacteriol.* 152, 1265–1275.

Douglas, C. J., Staneloni, R. J., Rubin, R. A., and Nester, E. W. (1985). *J. Bacteriol.* 161, 850–860.

Durrenberger, F., Crameri, A., Hohn, B., and Koukolikova-Nicola, Z. (1989). *Proc. Natl. Acad. Sci. U.S.A.* 86, 9154–9158.

Ebert, P. R., Ha, S. B., and An, G. (1987). *Proc. Natl. Acad. Sci. U.S.A.* 84, 5745–5749.

Ellis, J. G., Kerr, A., Petit, A., and Tempe, J. (1982a). *Mol. Gen. Genet.* 186, 263–269.

Ellis, J. G., Murphy, P. J., and Kerr, A. (1982b). *Mol. Gen. Genet.* 186, 275–281.

Ellis, J. G., Ryder, M. H., and Tate, M. E. (1984). *Mol. Gen. Genet.* 195, 466–473.

Figurski, D. H., and Helinski, D. R. (1979). *Proc. Natl. Acad. Sci. U.S.A.* 76, 1648–1652.

Fillatti, J. J., Kiser, J., Rose, R., and Comai, L., (1987). *Bio/Technology* 5, 726–730.

Firmin, J. L., and Fenwick, G. R. (1978). *Nature (London)* 276, 842–844.

Fraley, R. T., Rogers, S. G., Horsch, R. B., Sanders, P. R., Flick, J. S., Adams, S. P., Bittner, M. L., Brand, L. A., Fink, C. L., Fry, J. S., Galluppi, G. R., Goldberg, S. B., Hoffmann, N. L., and Woo, S. C. (1983). *Proc. Natl. Acad. Sci. U.S.A.* 80, 4803–4807.

Fraley, R. T., Horsch, R. B., Matzke, A., Chilton, M.-D., Chilton, W. S., and Sanders, P. R. (1984). *Plant Mol. Biol.* 3, 371–378.

Fraley, R. T., Rogers, S. G., Horsch, R. B., Eichholtz, D. A., Flick, J. S., Fink, C. L., Hoffmann, N. L., and Sanders, P. R. (1985). *Bio/Technology* 5, 629-635.

Gabard, J. M., Charest, P. J., Iyer, V. N., and Miki, B. L. (1989). *Plant Physiol.* 91, 574–580.

Gallie, D. R., Gay, P., and Kado, C. I. (1988). In *Vectors* (Ed. by R. L. Rodriguez and D. T. Denhardt), pp. 333–342. Boston: Butterworths.

Gardner, R. C., and Knauf, V. C. (1986). *Science* 231, 725–727.

Garfinkel, D. J., and Nester, E. W. (1980). *J. Bacteriol.* 144, 732–743.

Garfinkel, D. J., Simpson, R. B., Ream, L. W., White, F. F., Gordon, M. P., and Nester, E. W. (1981). *Cell* 27, 143–153.

Gasser, C. S., and Fraley, R. T. (1989). *Science* 244, 1293–1299.

Gelvin, S. B. (1990). *Plant Physiol.* 92, 281–285.

Gelvin, S. B., and Habeck, L. L. (1990). *J. Bacteriol.* 172, 1600–1608.

Gelvin, S. B., Gordon, M. P., Nester, E. W., and Aronson, A. I. (1981). *Plasmid* 6, 17–29.

Gelvin, S. B., Karcher, S. J., and Goldsbrough, P. B. (1985). *Mol. Gen. Genet.* 199, 240–248.

Genetello, C., Van Larebeke, N., Holsters, M., Depicker, A., Van Montagu, M., and Schell, J. (1977). *Nature (London)* 265, 561–563.

Ghai, J., and Das, A. (1989). *Proc. Natl. Acad. Sci. U.S.A.* 86, 3109–3113.

Gietl, C., Koukolikova-Nicola, Z., and Hohn, B. (1987). *Proc. Natl. Acad. Sci. U.S.A.* 84, 9006–9010.

Gray, J., Wang, J., and Gelvin, S. B. (1992). *J. Bacteriol.* 174, 1086–1098.

Guerineau, F., Brooks, L., Meadows, J., Lucy, A., Robinson, C., and Mullineaux, P. (1990). *Plant Mol. Biol.* 15, 127–136.

Habeeb, L. F., Wang, L., and Winans, S. C. (1991). *Mol. Plant-Microbe Interact.* 4, 379–385.

Hain, R., Stabel, S., Czernilofsky, A. P., Steinbif, H. H., Herrera-Estrella, L., and Schell, J. (1985). *Mol. Gen. Genet.* 199, 161–168.

Haughn, G. W., Smith, J., Mazur, B., and Somerville, C. (1988). *Mol. Gen. Genet.* 211, 266–271.

Hawes, M. C., and Pueppke, S. G. (1989). *Plant Physiol.* 91, 113–118.

Hayford, M. B., Medford, J. I., Hoffmann, N. L., Rogers, S. B., and Klee, H. J. (1988). *Plant Physiol.* 86, 1216–1222.

Helmer, G., Casadaban, M., Bevan, M., Kayes, L., and Chilton, M.-D. (1984). *Bio/Technology* 2, 520–527.

Hepburn, A. G., and White, J. (1985). *Plant Mol. Biol.* 5, 3–11.

Hepburn, A. G., Clarke, L. E., Blundy, K. S., and White, J. (1983). *J. Mol. Appl. Genet.* 2, 211–224.

Herrera-Estrella, A., Chen, Z.-M., Van Montagu, M., and Wang, K. (1988). *EMBO J.* 7, 4055–4062.

Herrera-Estrella, A., Van Montagu, M., and Wang, K. (1990). *Proc. Natl. Acad. Sci. U.S.A.* 87, 9534–9537.

Herrera-Estrella, L., DeBlock, M., Messens, E., Hernalsteens, J.-P., Van Montagu, M., and Schell, J. (1983a). *EMBO J.* 2, 987–996.
Herrera-Estrella, L., Depicker, A., Van Montagu, M., and Schell, J. (1983b). *Nature (London)* 303, 209–213.
Hirsch, P. R., and Beringer, J. E. (1984). *Plasmid* 12, 139–141.
Hoekema, A., Hirsh, P. R., Hooykaas, P. J. J., and Schilperoort, R. A. (1983). *Nature (London)* 303, 179–180.
Holsters, M., Villarroel, R., Gielen, J., Seurinck, J., DeGreve, H., Van Montagu, M., and Schell, J. (1983). *Mol. Gen. Genet.* 190, 35–41.
Hood, E. E., Chilton, W. S., Chilton, M.-D., and Fraley, R. T. (1986). *J. Bacteriol.* 168, 1283–1290.
Hooykaas, P. J. J. (1988). In *Vectors* (Ed. by R. L. Rodriguez and D. T. Denhardt), pp. 517–538. Boston: Butterworths.
Hooykaas, P. J. J. (1989). *Plant Mol. Biol.* 13, 327–336.
Hooykaas, P. J. J., Roobol, C., and Schilperoort, R. A. (1979). *J. Gen. Microbiol.* 110, 99–109.
Hooykaas, P. J. J., Hofker, M., den Dulk-Ras, H., and Schilperoort, R. A. (1984). *Plasmid* 11, 195–205.
Hooykaas, P. J. J., den Dulk-Ras, H., and Schilperoort, R. A. (1988). *Plant Mol. Biol.* 11, 791–794.
Horsch, R. B., and Klee, H. J. (1986). *Proc. Natl. Acad. Sci. U.S.A.* 83, 4428–4432.
Horsch, R. B., Fry, J. E., Hoffmann, N. L., Eichholtz, D., Rogers, S. G., and Fraley, R. T. (1985). *Science* 227, 1229–1231.
Howard, E., and Citovsky, V. (1990). *BioEssays* 12, 103–108.
Howard, E. A., Winsor, B. A., De Vos, G., and Zambryski, P. (1989). *Proc. Natl. Acad. Sci. U.S.A.* 86, 4017–4021.
Howard, E. A., Zupan, J. R., Citovsky, V., and Zambryski, P. C. (1992). *Cell* 68, 109–118.
Huang, M.-L., Cangelosi, G. A., Halperin, W., and Nester, E. W. (1990a). *J. Bacteriol.* 172, 1814–1822.
Huang, Y., Morel, P., Powell, B., Kado, C. I. (1990b). *J. Bacteriol.* 172, 1142–1144.
Inon de Iannino, N., and Ugalde, R. A. (1989). *J. Bacteriol.* 171, 2842–2849.
Inze, D., Follin, A., Van Lijsebettens, M., Simoens, C., Genetello, C., Van Montagu, M., and Schell, J. (1984). *Mol. Gen. Genet.* 194, 265–274.
Iyer, V. N., Klee, H. J., and Nester, E. W. (1982). *Mol. Gen. Genet.* 188, 418–424.
Janssen, B.-J., and Gardner, R. C. (1990). *Plant Mol. Biol.* 14, 61–72.
Janssens, A., Engler, G., Zambryski, P., and Van Montagu, M. (1984). *Mol. Gen. Genet.* 195, 341–350.
Jayaswal, R. K., Veluthambi, K., Gelvin, S. B., and Slightom, J. L. (1987). *J. Bacteriol.* 169, 5035–5045.
Jefferson, R. A., Kavanagh, T. A., and Bevan, M. W. (1987). *EMBO J.* 6, 3901–3907.
Jen, G. C., and Chilton, M.-D. (1986a). *J. Bacteriol.* 166, 491–499.
Jen, G. C., and Chilton, M.-D. (1986b). *Proc. Natl. Acad. Sci. U.S.A.* 83, 3895–3899.
Jin, S., Prusti, R. K., Roitsch, T., Ankenbauer, R. G., and Nester, E. W. (1990a). *J. Bacteriol.* 172, 4945–4950.
Jin, S., Roitsch, T., Ankenbauer, R. G., Gordon, M. P., and Nester, E. W. (1990b). *J. Bacteriol.* 172, 525–530.
Jin, S., Roitsch, T., Christie, P. J., and Nester, E. W. (1990c). *J. Bacteriol.* 172, 531–537.
John, M. C., and Amasino, R. M. (1988). *J. Bacteriol.* 170, 790–795.
Joos, H., Inze, D., Caplan, A., Sormann, M., Van Montagu, M., and Schell, J. (1983). *Cell* 32, 1057–1067.
Kanemoto, R. H., Powell, A. T., Akiyoshi, D. E., Regier, D. A., Kerstetter, R. A., Nester, E. W., Hawes, M. C., and Gordon, M. P. (1989). *J. Bacteriol.* 171, 2506–2512.
Kay, R., Chan, A., Daly, M., and McPherson, J. (1987). *Science* 236, 1299–1302.

Kemper, E., Waffenschmidt, S., Weiler, E. W., Rausch, T., and Schroder, J. (1985). *Planta* 163, 257–262.
Kerr, A., Manigault, P., and Tempe, J. (1977). *Nature (London)* 265, 560–561.
Klapwijk, P. M., and Schilperoort, R. A. (1979). *J. Bacteriol.* 139, 424–431.
Klapwijk, P. M., Scheuldermon, T., and Schilperoort, R. A. (1978). *J. Bacteriol.* 136, 775–785.
Klee, H. J., White, F. F., Iyer, V. N., Gordon, M. P., and Nester, E. W. (1983). *J. Bacteriol.* 153, 878–883.
Klee, H. J., Yanofsky, M. F., and Nester, E. W. (1985). *Bio/Technology* 3, 637–642.
Klee, H., Horsch, R., and Rogers, S. (1987a). *Ann. Rev. Plant Physiol.* 38, 343–350.
Klee, H. J., Muskopf, Y. M., and Gasser, C. S. (1987b). *Mol. Gen. Genet.* 210, 437-442.
Knauf, V., Yanofsky, M., Montoya, A., and Nester, E. (1984). *J. Bacteriol.* 160, 564–568.
Koltunow, A. M., Truettner, J., Cox, K. H., Wallroth, M., and Goldberg, R. B. (1990). *Plant Cell* 2, 1201–1224.
Komro, C. T., DiRita, V. J., Gelvin, S. B., and Kemp, J. D. (1985). *Plant Mol. Biol.* 4, 253–263.
Koncz, C., Olsson, O., Langridge, W. H. R., Schell, J., and Szalay, A. (1989). *Proc. Natl. Acad. Sci. U.S.A.* 84, 131–135.
Koukolikova-Nicola, Z., Shillito, R. D., Hohn, B., Wang, K., Van Montagu, M., and Zambryski, P. (1985). *Nature (London)* 313, 191–196.
Koziel, M. G., Adams, T. L., Hazlet, M. A., Damm, D., Miller, J., Dahlbeck, D., Jayne, S., and Staskawicz, B. J. (1984). *J. Mol. Appl. Genet.* 2, 549–562.
Krishnan, M., Burgner, J. W., Chilton, W. S., and Gelvin, S. B. (1991). *J. Bacteriol.* 173, 903–905.
Lee, K. Y., Townsend, J., Tepperman, J., Black, M., Chui, C. F., Mazur, B., Dunsmuir, P., and Bedbrook, J. (1988). *EMBO J.* 7, 1241–1248.
Leisner, S. M., and Gelvin, S. B. (1988). *Proc. Natl. Acad. Sci. U.S.A.* 85, 2553–2557.
Leisner, S. M., and Gelvin, S. B. (1989). *Plant Cell* 1, 925–936.
Lemmers, M., deBeuckeleer, M., Holsters, M., Zambryski, P., Depicker, A., Hernalsteens, J. P., Van Montagu, M., and Schell, J. (1980). *J. Mol. Biol.* 144, 353–376.
Leroux, B., Yanofsky, M. F., Winans, S. C., Ward, J. E., Ziegler, S. F., and Nester, E. W. (1987). *EMBO J.* 6, 849–856.
Ludwig, S. R., Bowen, B., Beach, L., and Wessler, S. R. (1990). *Science* 247, 449–450.
Lundquist, R. C., Close, T. J., and Kado, C. I. (1984). *Mol. Gen. Genet* 193, 1–7.
Lyon, B. R., Llewellyn, D. J., Huppatz, J. L., Dennis, E. S., and Peacock, W. J. (1989). *Plant Mol. Biol.* 13, 533–540.
Machida, Y., Usami, S., Yamamoto, A., Niwa, Y., and Takebe, I. (1986). *Mol. Gen. Genet.* 204, 374–382.
Mariani, C., De Beuckeleer, M., Truettner, J., Leemans, J., and Goldberg, R. B. (1990). *Nature (London)* 347, 737–741.
Marks, J. R., Lynch, T. J., Karlinsey, J. E., and Thomashow, M. F. (1987). *J. Bacteriol.* 169, 5835–5837.
Marton, L., Wullems, G. J., Molendijk, L., and Schilperoort, R. A. (1979). *Nature (London)* 277, 129–131.
Matsumoto, S., Takebe, I., and Machida, Y. (1990). *Gene* 66, 19–29.
Matthysse, A. G. (1983). *J. Bacteriol.* 154, 906–915.
Matzke, A. J. M., and Chilton, M.-D. (1981). *J. Mol. Appl. Gen.* 1, 39–49.
McBride, K. E., and Summerfelt, K. R. (1990). *Plant Mol. Biol.* 14, 269–276.
Melchers, L. S., Thompson, D. V., Idler, K. B., Neuteboom, S. T. C., de Maagd, R. A., Schilperoort, R. A., and Hooykaas, P. J. J. (1987). *Plant Mol. Biol.* 9, 635–645.
Melchers, L. S., Thompson, D. V., Idler, K. B., Schilperoort, R. A., and Hooykaas, P. J. J. (1987). *Nucl. Acids Res.* 14, 9933–9942.
Melchers, L. S., Regensburg-Tuink, T. J. G., Bourret, R. B., Sedee, N. J. A., Schilperoort, R. A., and Hooykaas, P. J. J. (1989a). *EMBO J.* 8, 1919–1925.

Melchers, L. S., Regensburg-Tuink, A. J. G., Schilperoort, R. A., and Hooykaas, P. J. J. (1989b). *Mol. Microbiol.* 3, 969–977.

Melchers, L. S., Maroney, M. J., den Dulk-Ras, A., Thompson, D. V., van Vuuren, H. A. J., Schilperoort, R. A., and Hooykaas, P. J. J. (1990). *Plant Mol. Biol.* 14, 249-259.

Merlo, D. J., Nutter, R. C., Montoya, A. L., Garfinkel, D. J., Drummond, M. H., Chilton, M.-D., Gordon, M. P., and Nester, E. W. (1980). *Mol. Gen. Genet.* 177, 637–643.

Messens, E., Lenaerts, A., Van Montagu, M., and Hedges, R. W. (1985). *Mol. Gen. Genet.* 199, 344–348.

Messens, E., Dekeyser, R., and Stachel, S. E. (1990). *Proc. Natl. Acad. Sci. U.S.A.* 87, 4368–4372.

Mitra, A., and An, G. (1989). *Mol. Gen. Genet.* 215, 294–299.

Montoya, A. L., Chilton, M.-D., Gordon, M. P., Sciaky, D., and Nester, E. W. (1977). *J. Bacteriol.* 129, 101–107.

Morel, P., Powell, B. S., Rogowsky, P. M., and Kado, C. I. (1989). *Mol. Microbiol.* 3, 1237–1246.

Morel, P., Powell, B. S., and Kado, C. I. (1990). *C.R. Acad. Sci. Paris* 310, 21–26.

Morris, J. W., and Morris, R. O. (1990). *Proc. Natl. Acad. Sci. U.S.A.* 87, 3614–3618.

Morris, R. O. (1986). *Ann. Rev. Plant Physiol.* 37, 509–538.

Murai, N., Sutton, D. W., Murray, M. G., Slightom, J. L., Merlo, D. J., Reichert, N. A., Sengupta-Gopalan, C., Stock, C. A., Barker, R. F., Kemp, J. D., and Hall, T. C. (1983). *Science* 222, 476–482.

Nixon, B. T., Ronson, C. W., and Ausubel, F. M. (1986). *Proc. Natl. Acad. Sci. U.S.A.* 83, 7850–7854.

Noda, T., Tanaka, N., Mano, Y., Nabeshima, S., Ohkawa, H., and Matsui, C., (1987). *Plant Cell Rep.* 6, 283–286.

O'Connell, K. P., and Handelsman, J. (1989). *Mol. Plant-Microbe Inter.* 2, 11–16.

Ooms, G., Hooykaas, P. J. J., Moolenaar, G., and Schilperoort, R. A. (1981). *Gene* 14, 33–50.

Ooms, G., Bakker, K. A., Molendijk, L., Wullems, G. J., Gordon, M. P., Nester, E. W., and Schilperoort, R. A. (1982a). *Cell* 30, 589–597.

Ooms, G., Hooykaas, P. J. J., Van Veen, R. J. M., Van Beelan, P., Regensburg-Tuink, T. J. G., and Schilperoort, R. A. (1982b). *Plasmid* 7, 15–29.

Ott, R. W., Ren, L., and Chua, N.-H. (1990). *Mol. Gen. Genet.* 221, 121–124.

Ow, D. W., Wood, K. V., DeLuca, M., DeWet, J. R., Helinski, D. R., and Howell, S. H. (1986). *Science* 234, 856–859.

Owens, L. D., and Smigocki, A. C. (1988). *Plant Physiol.* 88, 570–573.

Pazour, G. J., and Das, A. (1990). *J. Bacteriol.* 172, 1241–1249.

Peralta, E. G., and Ream, L. W. (1985). *Proc. Natl. Acad. Sci. U.S.A.* 82, 5112–5116.

Peralta, E. G., Helmiss, R., and Ream, W. (1986). *EMBO J.* 5, 1137–1142.

Petit, A., and Tempe, J. (1978). *Mol. Gen. Genet.* 167, 147–155.

Petit, A., Tempe, J., Kerr, A., Holsters, M., Van Montagu, M., and Schell, J. (1978). *Nature (London)* 271, 570–572.

Powell, B. S., Powell, G. K., Morris, R. O., Rogowsky, P. M., and Kado, C. I. (1987). *Mol. Microbiol.* 1, 309–316.

Powell, B. S., Rogowsky, P. M., and Kado, C. I. (1989). *Mol. Microbiol.* 3, 411–419.

Puvanesarajah, V., Schell, F. M., Stacey, G., Douglas, C. J., and Nester, E. W. (1985). *J. Bacteriol.* 164, 102–106.

Ream, W. (1989). *Ann. Rev. Phytopathol.* 27, 583–618.

Rogers, S. G., Klee, H. J., Horsch, R. B., and Fraley, R. T. (1987). *Meth. Enzymol.* 153, 253–277.

Rogers, S. G., Horsch, R. B., Fraley, R. T., and Klee, H. J. (1988). In *Vectors* (Ed. by R. L. Rodriguez and D. T. Denhardt), pp. 539–557. Boston: Butterworths.

Rogowsky, P. M., Close, T. J., Chimera, J. A., Shaw, J. J., and Kado, C. I. (1987). *J. Bacteriol.* 169, 5101–5112.

Roitsch, T., Wang, H., Jin, S., and Nester, E. W. (1990). *J. Bacteriol.* 172, 6054–6060.
Ronson, C. W., Nixon, B. T., and Ausubel, F. M. (1987). *Cell* 49, 579–581.
Rothstein, S. J., Hahners, K. N., Lotstein, R. J., Carozzi, N. B., Jayne, S. M., and Rice, D. A. (1987). *Gene* 53, 153–161.
Rubin, R. A. (1986). *Mol. Gen. Genet.* 202, 312–320.
Ruvkin, G. B., and Ausubel, F. M. (1981). *Nature (London)* 289, 85–88.
Sahi, S. V., Chilton, M.-D., and Chilton, W. S. (1990). *Proc. Natl. Acad. Sci. U.S.A.* 87, 3879–3883.
Saint-Pierre, B., and Dion, P. (1988). *Can. J. Microbiol.* 34, 793–801.
Sans, N., Schroder, G., and Schroder, J. (1987). *Eur. J. Biochem.* 167, 81–87.
Schafer, W., Gorz, A., and Kahl, G. (1987). *Nature (London)* 327, 529–532.
Schardl, C. L., and Kado, C. I. (1983). *Mol. Gen. Genet.* 191, 10–16.
Schardl, C. L., Byrd, A. D., Benzion, G., Altschuler, M. A., Hildebrand, D. F., and Hunt, A. G. (1987). *Gene* 61, 1–11.
Schell, J. S. (1987). *Science* 237, 1176–1183.
Schmidhauser, T. J., Ditta, G., and Helinski, D. R. (1988). In *Vectors* (Ed. by R. L. Rodriguez and D. T. Denhardt), pp. 287–332. Boston: Butterworths.
Schoffl, F., and Baumann, G. (1985). *EMBO J.* 4, 1119–1124.
Schroder, G., Waffenschmidt, S., Weiler, E. W., and Schroder, J. (1984). *Eur. J. Biochem.* 138, 387–391.
Scott, I. M., Firmin, J. L., Butcher, D. N., Searle, L. M., Sogeke, A. K., Eagles, J., March, J. F., Self, R., and Fenwick, G. R. (1979). *Mol. Gen. Genet.* 176, 57–65.
Sen, P., Pazour, G. J., Anderson, D., and Das, A. (1989). *J. Bacteriol.* 171, 2573–2580.
Shah, D. M., Horsch, R. B., Klee, H. J., Kishore, G. M., Winter, J. A., Tumer, N. E., Hironaka, C. M., Sanders, P. R., Gasser, C. S., Aykent, S., Siegel, N. R., Rogers, S. G., and Fraley, R. T. (1986). *Science* 233, 478–481.
Shaw, C. H., Watson, M. D., Carter, G. H., and Shaw, C. H. (1984). *Nuc. Acids Res.* 12, 6031–6041.
Simpson, J., Timko, M. P., Cashmore, A. R., Schell, J., Van Montagu, M., and Herrera-Estrella, L. (1985). *EMBO J.* 4, 2723–2729.
Simpson, R. B., Spielmann, A., Margossian, L., and McKnight, T. D. (1986). *Plant Mol. Biol.* 6, 403–415.
Singh, K., Tokuhisa, J. G., Dennis, E. S., and Peacock, W. J. (1989). *Proc. Natl. Acad. Sci. U.S.A.* 86, 3733–3737.
Skoog, F., and Miller, C. O. (1957). *Symp. Soc. Exp. Biol.* 11, 118–131.
Slightom, J. L., Durand-Tardif, M., Jouanin, L., and Tepfer, D. (1986). *J. Biol. Chem.* 261, 108–121.
Spanier, K., Schell, J., and Schreier, P. H. (1989). *Mol. Gen. Genet.* 219, 209–216.
Spena, A., and Schell, J. (1987). *Mol. Gen. Genet.* 206, 436–440.
Spencer, P. A., and Towers, G. H. N. (1988). *Phytochem.* 27, 2781–2785.
Stachel, S. E., and Nester, E. W. (1986). *EMBO J.* 5, 1445–1454.
Stachel, S. E., and Zambryski, P. C. (1986a). *Cell* 47, 155–157.
Stachel, S. E., and Zambryski, P. C. (1986b). *Cell* 46, 325–333.
Stachel, S. E., An, G., Flores, C., and Nester, E. W. (1985a). *EMBO J.* 4, 891–898.
Stachel, S. E., Messens, E., Van Montagu, M., and Zambryski, P. (1985b). *Nature (London)* 318, 624–629.
Stachel, S. E., Nester, E. W., and Zambryski, P. C. (1986). *Proc. Natl. Acad. Sci. U.S.A.* 83, 379–383.
Stachel, S. E., Timmerman, B., and Zambryski, P. (1987). *EMBO J.* 6, 857–863.
Steck, T. R., and Kado, C. I. (1990). *J. Bacteriol.* 172, 2191–2193.
Steck, T. R., Morel, P., and Kado, C. I. (1988). *Nuc. Acids Res.* 16, 8736.

Steck, T. R., Close, T. J., and Kado, C. I. (1989). *Proc. Natl. Acad. Sci. U.S.A.* 86, 2133–2137.

Streber, W. R., and Willmitzer, L. (1989). *Bio/Technology* 7, 811–816.

Strittmatter, G., and Chua, N.-H. (1987). *Proc. Natl. Acad. Sci. U.S.A.* 84, 8986–8990.

Svab, Z., Harper, E. C., Jones, J. D. G., and Maliga, P. (1990). *Plant Mol. Biol.* 14, 197–205.

Taylor, B. H., Amasino, R. M., White, F. F., Nester, E. W., and Gordon, M. P. (1985). *Mol. Gen. Genet.* 201, 554–557.

Teeri, T. H., Lehvaslaiho, H., Franck, M., Uotila, J., Heino, P., Palva, E. T., Van Montagu, M., and Herrera-Estrella, L., (1989). *EMBO J.* 8, 343–350.

Tempe, J., Petit, A., Holsters, M., Van Montagu, M., and Schell, J. (1977). *Proc. Natl. Acad. Sci. U.S.A.* 74, 2848–2849.

Tepfer, D. (1984). *Cell* 37, 959–967.

Thomashow, M. F., Nutter, R., Montoya, A. L., Gordon, M. P., and Nester, E. W. (1980a). *Cell* 19, 729–739.

Thomashow, M. F., Nutter, R., Postle, K., Chilton, M.-D., Blattner, F. R., Powell, A., Gordon, M. P., and Nester, E. W. (1980b). *Proc. Natl. Acad. Sci. U.S.A.* 77, 6448–6452.

Thomashow, L. S., Reeves, S., and Thomashow, M. F. (1984). *Proc. Natl. Acad. Sci. U.S.A.* 81, 5071–5075.

Thomashow, M. F., Hugly, S., Buchholz, W. G., and Thomashow, L. S. (1986). *Science* 231, 616–618.

Thomashow, M. F., Karlinsey, J. E., Marks, J. R., and Hurlbert, R. E. (1987). *J. Bacteriol.* 169, 3209–3216.

Thornburg, R. W., An, G., Cleveland, T. E., Johnson, R., and Ryan, C. A. (1987). *Proc. Natl. Acad. Sci. U.S.A.* 84, 744–748.

Timmerman, B., Van Montagu, M., and Zambryski, P. (1988). *J. Mol. Biol.* 203, 373–384.

Tinland, B., Huss, B., Paulus, F., Bonnard, G., and Otten, L. (1989). *Mol. Gen. Genet.* 219, 217–224.

Tinland, B., Rohgritsch, O., Michler, P., and Otten, L. (1990). *Mol. Gen. Genet.* 223, 1–10.

Usami, S., Morikawa, S., Takebe, I., and Machida, Y. (1987). *Mol. Gen. Genet.* 209, 221–226.

Usami, S., Okamoto, S., Takebe, I., and Machida, Y. (1988). *Proc. Natl. Acad. Sci. U.S.A.* 85, 3748–3752.

Uttaro, A. D., Cangelosi, G. A., Geremia, R. A., Nester, E. W., and Ugalde, R. A. (1990). *J. Bacteriol.* 172, 1640–1646.

Van den Elzen, P. J. M., Townsend, J., Lee, K. Y., and Bedbrook, J. R. (1985). *Plant Mol. Biol.* 5, 299–302.

Van Haaren, M. J. J., Pronk, J. T., Schilperoort, R. A., and Hooykaas, P. J. J. (1987). *Plant Mol. Biol.* 8, 95–104.

Van Haaren, M. J. J., Sedee, N. J. A., Krul, M., Schilperoort, R. A., and Hooykaas, P. J. J. (1988). *Plant Mol. Biol.* 11, 773–781.

Van Haaren, M. J. J., Sedee, N. J. A., de Boer, H. A., Schilperoort, R. A., and Hooykaas, P. J. J. (1989). *Plant Mol. Biol.* 13, 523–531.

Van Onckelen, H., Rudelsheim, P., Inze, D., Follin, A., Messens, E., Horemans, S., Schell, J., Van Montagu, M., and De Greef, J. (1985). *FEBS Lett.* 181, 373–376.

Velten, J., Velten, L., Hain, R., and Schell, J. (1984). *EMBO J.* 3, 2723–2730.

Veluthambi, K., Jayaswal, R. K., and Gelvin, S. B. (1987). *Proc. Natl. Acad. Sci. U.S.A.* 84, 1881–1885.

Veluthambi, K., Ream, W., and Gelvin, S. B. (1988). *J. Bacteriol.* 170, 1523–1532.

Veluthambi, K., Krishnan, M., Gould, J. H., Smith, R. H., and Gelvin, S. B. (1989). *J. Bacteriol.* 171, 3696–3703.

Vilaine, F., and Casse-Delbart, F. (1987a). *Gene* 55, 105–114.

Vilaine, F., and Casse-Delbart, F. (1987b). *Mol. Gen. Genet.* 206, 17–23.

von Lintig, J., Zanker, H., and Schroder, J. (1991) *Mol. Plant-Microbe Interact.* 4, 370–378.
Waldron, C., Murphy, E. B., Roberts, J. L., Gustafson, G. D., Armour, S. L., and Malcolm, S. K. (1985). *Plant Mol. Biol.* 5, 103–108.
Wang, K., Herrera-Estrella, L., Van Montagu, M., and Zambryski, P. (1984). *Cell* 38, 455–462.
Wang, K., Stachel, S. E., Timmerman, B., Van Montagu, M., and Zambryski, P. C. (1987). *Science* 235, 587–591.
Wang, K., Herrera-Estrella, A., and Van Montagu, M. (1990). *J. Bacteriol.* 172, 4432–4440.
Ward, E. R., and Barnes, W. M. (1988). *Science* 242, 927–930.
Weiler, E. W., and Schroder, J. (1987). *TIBS* 12, 271–275.
White, F. F. (1989). In *Plant Biotechnology* (Ed. by S.-D. Kung and C. J. Arntzen), pp. 3–34. Boston: Butterworths.
Willmitzer, L., Schmalenbach, W., and Schell, J. (1981). *Nuc. Acids Res.* 9, 4801–4812.
Winans, S. C., Ebert, P. R., Stachel, S. E., Gordon, M. P., and Nester, E. W. (1986). *Proc. Natl. Acad. Sci. U.S.A.* 83, 8278–8282.
Winans, S. C., Jin, S., Komari, T., Johnson, K. M., and Nester, E. W. (1987). In *Plant Molecular Biology* (Ed. by D. von Wettstein and N.-H. Chua). New York: Plenum.
Winans, S. C., Kerstetter, R. A., and Nester, E. W. (1988). *J. Bacteriol.* 170, 4047–4054.
Yadav, N. S., Van der Leyden, J., Bennett, D. R., Barnes, W. M., and Chilton, M.-D. (1982). *Proc. Natl. Acad. Sci. U.S.A.* 79, 6322–6326.
Yamamoto, A., Iwahashi, M., Yanofsky, M. F., Nester, E. W., Takebe, I., and Machida, Y. (1987). *Mol. Gen. Genet.* 206, 174–177.
Yang, F., Montoya, A. L., Merlo, D. J., Drummond, M. H., Chilton, M.-D., Nester, E. W., and Gordon, M. P. (1980). *Mol. Gen. Genet.* 177, 707–714.
Yanofsky, M., Montoya, A., Knauf, V., Lowe, B., Gordon, M., and Nester, E. (1985). *J. Bacteriol.* 163, 341–348.
Yanofsky, M. F., Porter, S. G., Young, C., Albright, L. M., Gordon, M. P., and Nester, E. W. (1986). *Cell* 47, 471–477.
Yenofsky, R. L., Fine, M., and Pellow, J. W. (1990). *Proc. Natl. Acad. Sci. U.S.A.* 87, 3435–3439.
Young, C., and Nester, E. W. (1988). *J. Bacteriol.* 170, 3367–3374.
Zambryski, P. (1988). *Ann. Rev. Genet.* 22, 1–30.
Zambryski, P., Holsters, M., Kruger, K., Depicker, A., Schell, J., Van Montagu, M., and Goodman, H. M. (1980). *Science* 209, 1385–1391.
Zambryski, P., Deplicker, A., Druger, K., and Goodman, H. M. (1982). *J. Mol. Appl. Genet.* 1, 361–370.
Zambryski, P., Joos, P. H., Genetello, C., Leemans, J., Van Montagu, M., and Schell, J. (1983). *EMBO J.* 2, 2143–2150.
Zambryski, P., Tempe, J., and Schell, J. (1989). *Cell* 56, 193–201.
Zerback, R., Dressler, K., and Hess, D. (1989). *Plant Sci.* 62, 83–91.
Zorreguieta, A., and Ugalde, R. A. (1986). *J. Bacteriol.* 167, 947–951.
Zorreguieta, A., Geremia, R. A., Cavaignac, S., Cangelosi, G. A., Nester, E. W., and Ugalde, R. A. (1988). *Mol. Plant-Microbe Inter.* 1, 121–127.

3

Markers for Plant Gene Transfer

Benjamin A. Bowen

Department of Biotechnology Research
Pioneer Hi-Bred International, Inc.
Johnston, Iowa 50131

Genetic analysis would not be possible without marker genes to distinguish between cells, organisms, or gametes. Naturally occurring mutations which can be used as markers range from single base pair differences or restriction fragment length polymorphisms between homologous stretches of DNA to altered genes or chromosomes which affect morphological characters or enzyme-mediated biochemical reactions. In the past, genetic studies in plants have frequently been limited by the availability of suitable mutations which could be used as convenient markers. Now that an increasing number of plant species are amenable to genetic transformation by exogenous DNA, markers can be constructed at will by recombining pieces of DNA from various sources *in vitro*. With these advances in technology, genetic analysis in plants may soon approach the level of sophistication enjoyed by organisms where these tools have been available for some time.

In this review, I have attempted to provide an overview of markers currently used for plant gene transfer and some of the applications where they have proved useful. In order to prolong the utility of the article, I have also included markers that are likely to be used in the near future and highlighted applications for which I anticipate that appropriate markers will soon be developed. Where appropriate, I have discussed problems and limitations that are frequently encountered.

3.1 Reasons for Using Marker Genes

3.1.1 CONFIRMATION OF DNA DELIVERY AND STABLE TRANSFORMATION

Markers play a crucial role in the development of transformation procedures, because they can be used to ascertain how effectively transforming DNA is introduced into plant cells with different treatments (Walden, 1989; Klein *et al.*, 1989a). Selectable markers are generally used for stable transformation. They either allow living cells, tissues, or whole plants to grow under conditions which prevent the growth of untransformed tissues or confer a phenotype which allows transformed and untransformed tissues to be distinguished unequivocally. In contrast, assayable markers direct the expression of a readily quantifiable product (e.g., an enzyme or an antigen), which is not necessarily distinguishable in living cells. Assayable markers are frequently used to optimize DNA delivery in the early stages of developing a transformation procedure. This is because a high level of marker gene expression in the first few days following gene transfer is thought to be essential for stable transformation. However, a high level of transient gene expression may not always be sufficient. Some of the variables which affect transient gene expression (e.g., the number of transformed cells, the number of active templates inside those cells which are transformed, and the viability of the transformed cells following DNA delivery) are undoubtedly important parameters for ensuring stable transformation. Other parameters (e.g., integration into the target cell genome and cell division following integration) are prerequisites for stable transformation, but are not necessarily related to the efficiency of transient marker gene expression. Similarly, the selection of stable transformants can be affected by the efficiency of marker gene expression at the site of integration, and this may be different from the level of expression observed prior to integration (Kay *et al.*, 1987; Odell *et al.*, 1988). In order to maximize stable transformation frequency, therefore, it would be desirable to use a marker which could be detected only when it is integrated into the genome and preferably at the single-cell level. As yet, no such marker is available.

It is important that markers for plant cell transformation can only be expressed inside cells of the target tissue and not inside any contaminating microorganisms which may or may not be deliberately introduced as part of the delivery system (Klein *et al.*, 1989a; Janssen and Gardner, 1989; Vancanneyt *et al.*, 1990). This is not a trivial point, because considerable care must be taken to exclude the possibility of microbial contamination when attempting to confirm transformation for the first time.

3.1.2 QUANTIFICATION OF GENE EXPRESSION

Assayable markers are frequently used to quantify gene expression for reasons other than optimizing DNA delivery. Such markers are also commonly referred to as reporter genes. When fused to plant regulatory sequences *in vitro*, reporter genes are indispensable for the analysis of gene expression in transient assays or in stable transformants. Recombinant gene fusions can be used to determine when, where, and at what level regulatory sequences direct gene expression *in vivo* and which intra- or extracellular signals are involved (Benfey and Chua, 1989, 1990; Tsai and Coruzzi, Chapter 6, this volume). Reporter genes can also be fused to peptide coding sequences to determine whether they are involved in targeting proteins either outside of the cell (Denecke *et al.*, 1990) or into membrane-bound organelles such as chloroplasts (Schreier and Schell, 1986), mitochondria (Boutry *et al.*, 1987), endoplasmic reticulum (Iturriaga *et al.*, 1989), peroxisomes (Gould *et al.*, 1990), or nuclei (Restrepo *et al.*, 1990).

Gene fusions have become *de rigueur* for many applications in plant molecular biology. This is because many genes of interest are more difficult to assay than the markers most commonly used as reporter genes. Without reporter genes, it would be difficult to distinguish the expression of plant genes introduced by transformation from the background level of endogenous gene expression. Similarly, the patterns of gene expression directed by separate members of a gene family encoding the same gene product can be readily ascertained using gene fusions in transgenic plants (for examples see Forde *et al.*, 1989; Edwards *et al.*, 1990; Koes *et al.*, 1990; Tsai and Coruzzi, Chapter 6, this volume).

It is often tacitly assumed that the level of a given marker gene product is directly proportional to the level at which the marker gene is transcribed. However, this assumption should never be taken for granted. Posttranscriptional differences in mRNA and/or protein stability may also play a role (Goldberg *et al.*, 1989; Silverthorne and Tobin, 1990), especially when making comparisons between different plant cell types or treatments which could have pleiotropic effects on cell physiology. Even if these factors are not involved, differences in transcription may not be solely due to the regulatory element under study. For example, in transient expression studies, the effect of adjacent plasmid sequences should be addressed (Alam and Cook, 1990). Similarly, in stable transformants, marker gene expression can be affected by the site of insertion, template copy number, rearrangements, and methylation (Weising *et al.*, 1988). Without analyzing sufficient numbers of transformants, therefore, different patterns of expression should be interpreted cautiously. They may be caused by the pattern of integration in each transformant rather than by any differences between transforming DNA sequences. Many of these problems, however, should be circumvented when efficient procedures for gene targeting are developed in plants (Paszkowski *et al.*, 1988; Lee *et al.*, 1990).

Transient assays are notoriously variable (Hollon and Yoshimura, 1989). With certain techniques used for delivering DNA into plant cells (e.g., polyethylene glycol or electroporation-mediated DNA uptake by protoplasts or microprojectile bombardment of intact tissues), much of this variability results from the degree of physiological stress imposed by the delivery process itself. This can affect the kinetics of transient expression in different cells (Pröls *et al.*, 1988; B. Bowen, unpublished observations). Another major problem is that the amount of DNA taken up by different populations of cells or different tissues can vary significantly. These factors can be controlled to some extent by replication of treatments and/or by including an internal standard in every transfection (M. Fromm *et al.*, 1987; Denecke *et al.*, 1989; Hollon and Yoshimura, 1989; Sheen, 1990). Typically, this consists of a second constitutively expressed marker gene which can be assayed in the same extract used to assay the experimental marker.

Several other potential artifacts should also be considered when evaluating transient assay data. For example, it is possible that trans-acting factors which normally have a low affinity for the template could spuriously activate transcription at high template copy number. It is also conceivable that a reduced level of gene expression may be observed at high template copy number, if one or more trans-acting factors required for transcription are limiting in concentration (Schöler and Gruss, 1984; B. Bowen, unpublished observations). Conversely, when genes encoding trans-acting factors are introduced at high copy number, "squelching" of gene expression can occur (Gill and Ptashne, 1988; Bowen *et al.*, 1991a; see Figure 3.4 and Section 3.3.3). Problems of these sorts are best examined by analyzing transient marker gene expression over a wide range of template copy numbers (Bowen, 1991).

3.1.3 VISUALIZATION OF GENE EXPRESSION

Markers which can be detected in plant tissues directly are extremely powerful tools. In transient expression studies, such markers can indicate the number and location of transfected cells, both of which are important parameters for optimizing stable transformation. In stable transformants, it is possible to determine whether an introduced gene is expressed evenly in all cells of a particular tissue or at higher levels in certain selected cells or cell types. This information often provides greater insight into the functional role of a given gene during plant development, for example, or in response to environmental stimuli than quantification in crude tissue extracts. The β-D-glucuronidase (GUS) reporter gene system has proved remarkably versatile in many investigations of this sort (Jefferson, 1989). One current limitation of this system, however, is that the histochemical procedure for detecting GUS expression usually requires sacrificing the tissue. Markers which can be used to visualize gene expression in living plant tissues are also being developed (Section 3.2.3). Some of these have already proved useful as markers for whole plant transformation (Meyer *et al.*, 1987; Herman and Marks, 1989) or for determining the fate of transformed cells in chimeric tissues (M. Ross, B. Bowen and D. Tomes, unpublished data; see Figure 3.1, color plate). However, the greatest potential for visible markers may lie in selecting transformants which exhibit certain patterns of gene expression. In other organisms, visible markers have been used to tag developmentally regulated genes via an "enhancer-trap" approach (O'Kane and Gehring, 1987; Gossler *et al.*, 1989). Similar approaches could be readily adapted for transformable plants in the near future. Visible markers introduced by gene transfer may also be useful for detecting rare events involving transposition (Jones *et al.*, 1989; Masson and Federoff, 1989; Finnegan *et al.*, 1989; Spena *et al.*, 1989; Frey *et al.*, 1990) or somatic recombination in plants.

3.2 Selectable Markers

3.2.1 CRITERIA FOR CHOOSING SELECTABLE MARKERS

Selectable markers confer a dominant phenotype on transformed cells because they invariably result in the addition of a new trait not normally associated with untransformed cells. Traits which can be used to select transformed cells from those that are untransformed generally fall into two classes: those which confer either cell viability or lethality in the presence of a selective agent; and those which have a negligible effect on cell survival but which confer transformed cells with some distinguishing physical characteristic.

Plant cell transformation frequencies are generally low, so most stable transformation schemes use markers which ensure the survival of transformed cells in the presence of a selective agent. Choosing an appropriate selective agent can depend on several factors. At the very least, the selective agent should be readily available and soluble in plant tissue culture media. Preferably, it should also be inexpensive and comparatively nontoxic to the investigator. If concentrated stock solutions in organic solvents must be added to the culture media, the potential toxicity of these toward the targets of transformation should also be checked. However, the most important parameter to ascertain is whether the frequency at which untransformed cells or plants can survive at a given level of the selective agent is greater than or similar to the expected or observed transformation frequency. If the frequency of "escapes" is high, an alternative selection should be sought. Ideally, the LD_{50} (i.e., the level at which 50% growth inhibition is observed) for untransformed cells should be at least an order of magnitude lower than that for transformed cells (Nutter *et al.*, 1987; Hauptmann *et al.*, 1988; Dekeyser *et al.*, 1989). In practice, this can only be ascertained when transformed cells are available.

One effective way to increase the LD_{50} window between transformed and untransformed cells is to increase the level of selectable marker gene expression (Nutter *et al.*, 1987). Before attempting to obtain stable transformants, therefore, the level of selectable marker gene expression should be optimized by testing different constructs in transient assays (Dekeyser *et al.*, 1989). Promoter strength is of primary importance, but alterations in both coding (Maliga *et al.*, 1988; Yenofsky *et al.*, 1990) and noncoding (Rogers *et al.*, 1985; Nutter *et al.*, 1987; Gallie *et al.*, 1987; Callis *et al.*, 1987; Ingelbrecht *et al.*, 1989; De Almeida *et al.*, 1989) sequences can also have dramatic effects on selectable marker gene expression in plant cells. In some cases, markers only confer a high level of resistance to a given selective agent when correctly targeted to the appropriate cellular compartment, such as the chloroplast (Table 3.1).

The nature of the target tissue and the DNA delivery system should also be considered when developing a selection procedure with a given marker gene. For example, green tissues may be more sensitive to agents which affect plastid-localized processes than callus cells. Furthermore, exposure of callus cells to high levels of these agents during the selection procedure can affect regeneration of green plants or subsequent plastid development. With other selective agents, the survival of transformed cells under selective conditions can be affected by the release of toxic compounds (e.g., phenolics) from surrounding untransformed cells. With bulky explants, transformed cells on the surface may also be starved of nutrients, if secondary thickening is induced by the selective agent in adjacent untransformed cells. In these cases, transformants can often by more easily recovered either by subdividing the target tissue or by using suspension cell cultures or protoplasts, if this is feasible. In addition, large explants or clumps of untransformed cells often survive for longer periods under selective conditions. This can lead to a higher frequency of escapes, especially if transformed cells deplete the level of selective agent in the surrounding medium. In some cases,

Figure 3.1 Anthocyanin regulatory genes as markers for determining the fate of transformed cell lineages in chimeric maize plants. (A) Sector in the cotyledon of a seedling which developed from an immature maize embryo bombarded with 35S::*R* and 35S::C 8–10 days after pollination. (B) Maize plant showing a sector transformed with 35S::*R* in the fourth leaf about 4–5 weeks after bombardment as described above. (C) Same plant as in (B) showing sectored leaves and sectored tassel just prior to anther emergence (i.e., about 15–16 weeks after bombardment). (Figure B from Bowen, 1992.)

Figure 3.2 Transient expression of anthocyanin regulatory genes following microprojectile bombardment of maize tissues. (A) Immature maize leaf (C *r-g B-b pl*) bombarded with 35S::*R* and incubated with illumination. Both epidermal and subepidermal cells accumulating anthocyanin can be seen. (B) Mature aleurone (C *r-g*) bombarded with 35S::*R* and 35S::GUS constructs coprecipitated onto the same batch of microprojectiles. In contrast to the histochemical staining pattern obtained with the chromogenic substrate X-Gluc, anthocyanin accumulation is cell autonomous [see also (A) and (C)]. (C) Embryogenic callus (*c r-r b pl*) bombarded with 35S::*R* and 35S::C and incubated without illumination. (Figures B and C from Bowen, 1992.)

3.1A

3.1B

3.1C

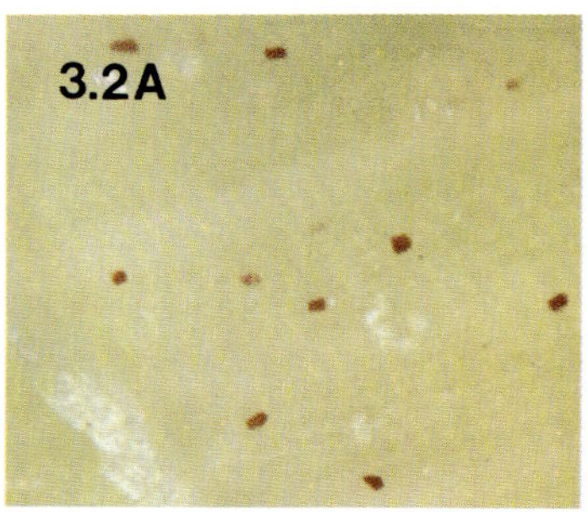
3.2A

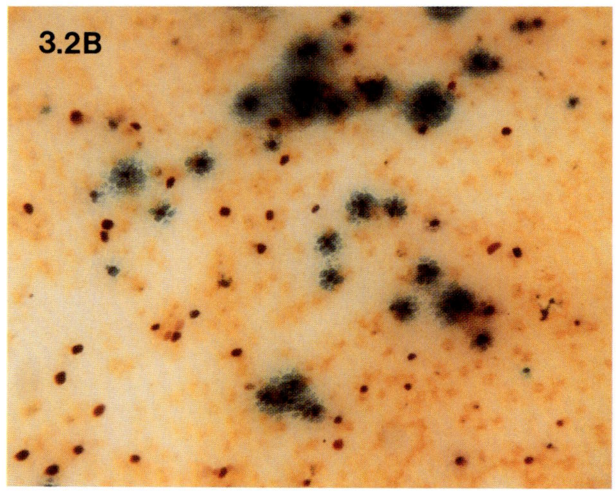
3.2B

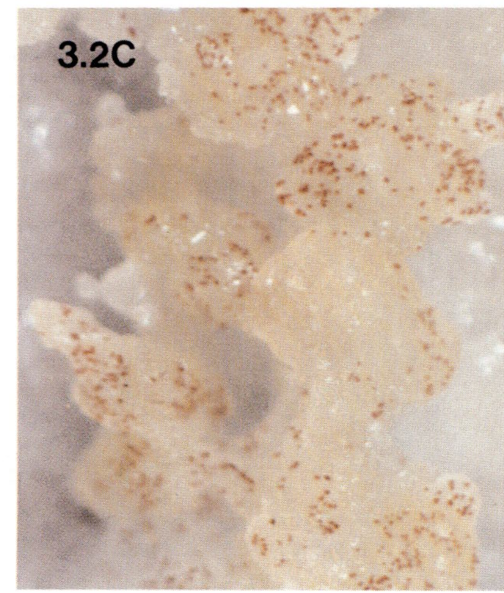
3.2C

such escapes can be easily "weeded" out by screening procedures which involve rooting or germination in the presence of the selective agent.

Selectable markers which require a selective agent may not be very useful with transformation procedures that are likely to develop chimeras at high frequency. This is because transformed sectors of chimeric plants often rely on untransformed cells for the supply of nutrients. Visible markers have been used to select chimeric plants generated by microprojectile bombardment of shoot meristems (McCabe *et al.*, 1988; M. Ross, B. Bowen, and D. Tomes, unpublished data; see Figure 3.1). However, if stable transformation events occur at low frequency, visual selection of transformants is generally impractical, because procedures for culturing and screening large numbers of plants or explants are necessary (Lee *et al.*, 1989). Alternatively, a nondestructive assay which involves spraying intact plants with a selective or chromogenic agent could be employed (Weide *et al.*, 1989). Transgenic plants have also been selected by taste, using a gene encoding the intensely sweet protein thaumatin (Witty, 1989). However, it remains to be seen how this procedure might be legislated in the future.

3.2.2 MARKERS WHICH CONFER VIABILITY OR LETHALITY UNDER SELECTIVE CONDITIONS

3.2.2.1 Herbicide and Antibiotic Resistance Markers

The repertoire of selectable markers which result in resistance to antibiotics, herbicides, or other phytotoxins is now fairly extensive. Table 3.1 lists those that have been successfully used for gene transfer in one or more plant species and a selection of other markers which are likely to be used in the near future. Markers in the latter category which are not listed may also include additional aminoglycoside modifiers (Davies and Smith, 1978), herbicide resistance markers (either mutant target proteins or detoxifying enzymes), and drug resistance markers used in other organisms (see, e.g., Tropschug *et al.*, 1989; Castillo *et al.*, 1990).

Resistance to each selective agent listed in Table 3.1 is conferred by one of three principle mechanisms: detoxification by enzymatic modification or sequestration; expression of a target with reduced affinity for the selective agent; or overexpression of the wild-type target. In general, markers which operate via the first two mechanisms are more effective and consequently in more widespread use. However, for details on which markers have been used most effectively with a given plant species, the reader is referred to other chapters in this treatise and the references contained therein.

TABLE 3.1
Selectable Markers for Plant Gene Transfer

Selective agent(s)	Mode of action[a]	Selectable marker(s)[b]	Mechanism[c]	Source(s)	Refs.[d]
Bialophos, Glufosinate, Phosphinothricin	I/GS	Glutamine synthetase (GS; glutamate–ammonia ligase)[e]	MTP/O	*Medicago sativa*	(*16*, *22*)
		Phosphinothricin acetyltransferase (*bar*; PAT)	D	*Streptomyces hygroscopicus* *Streptomyces viridochromogenes*	(*13*, *23*, *63*, *66*) (*70*)
Blasticidin-S	?	Blasticidin S deaminase (*bsr*)	D	*Aspergillus terreus* *Bacillus cereus*	(*35*) (*35*)
Bleomycin, Phleomycin	Breaks DNA	Bleomycin binding protein (*ble*)	S	*Escherichia coli* *Streptoalloteichus hindustanus*	(*31*) (*48*)
Bromacil	I/PSII	Quinone binding protein (Q_β; *psbA*)[f]	MTP/C	*Anacystis nidulans* *Phalaris paradoxa*	(*32*) (*32*)
Bromoxynil	I/PSII	Bromoxynil nitrilase (*bxn*)	D	*Klebsiella ozaenae*	(*57*)
Cadmium	I/Metalloenzymes	Metallothionein II (MTII)[e]	S	*Cricetulus griseus* *Homo sapiens* *Mus musculus*	(*39*) (*44*) (*41*)
Chloramphenicol	I/Translation	Chloramphenicol acetyltransferase (*cat*; CAT)[e]	D	*Escherichia coli*	(*8*, *12*, *30*, *49*)
Chloropropanoates	I/Vitamin B?	Dehalogenase	D	*Pseudomonas putida*	(*5*)
2,4-Dichlorophenoxyacetate	Auxin analog	2,4-Dichlorophenoxyacetate monooxygenase (*tfdA*; DPAM)	D	*Alcaligenes eutrophus*	(*40*, *58*)
Diuron	I/PSII	Quinone binding protein (Q_β; *psbA*)[f]	MTP/C	*Anacystis nidulans* *Chlamydomonas rheinhardtii*	(*32*) (*32*)
Geneticin/G418	I/Translation	Neomycin 3′-*O*-phosphotransferase [*npt*; *neo*; NPTII; APH(3′)II]	D	*Escherichia coli*	(*50*)
Gentamicin	I/Translation	Aminoglycoside 3-*N*-acetyltransferases [*aacIII*/*IV*; AAC(3)III/IV]	D	*Escherichia coli*	(*27*)

Glyphosate	I/EPSPS	5-Enolpyruvylshikimate-3-phosphate synthase (EPSPS; *aroA*)	MTP/C	*Arabidopsis thaliana*	(*36*)
				Escherichia coli	(*14*)
				Petunia hybrida	(*56*)
				Salmonella typhimurium	(*10*)
Haloxyfop	I/ACCase	Acetyl-CoA carboxylase (ACCase; *Acc*)[f]	MTP/O	*Zea mays*	(*47*)
Hygromycin B	I/Translation	Hygromycin 3′-O-phosphotransferase [*hpt*; APH(3′)IV]	D	*Escherichia coli*	(*64, 65*)
Imidazolinones	I/ALS	Acetolactate synthase (ALS; AHAS)[f]	MTP	*Arabidopsis thaliana*	(*53*)
				Beta vulgaris	(*59*)
				Datura innoxia	(*54*)
				Glycine max	(*60*)
				Gossypium hirsutum	(*60*)
				Nicotiana tabacum	(*60*)
				Zea mays	(*43, 73*)
Isonicotinate hydrazide	I/GD	Glycine decarboxylase (GD)[f]	MTP/M	*Nicotiana tabacum*	(*71*)
Isoxaben	I/Cell wall	? (*IxrA/B*)[f]	?	*Arabidopsis thaliana*	(*28*)
Kanamycin	I/Translation	Neomycin 3′-*O*-phosphotransferase [*npt*; *neo*; NPTII; APH(3′)II]	D	*Escherichia coli*	(*4, 19, 29*)
		Neomycin 3′-*O*-phosphotransferases [NPTI/III; APH(3′)I/III]	D	*Escherichia coli*	(*49*)
Lincomycin	I/Translation	Chloroplast rRNA[f]	MTR/C	*Nicotiana plumbaginifolia*	(*11, 45*)
Lysine/threonine	I/AK	Aspartate kinase (AK; *Ask*)[f]	MTP/O	*Daucus carota*	(*7*)
				Hordeum vulgare	(*2*)
				Zea mays	(*15*)
Methotrexate	I/DHFR	Dihydrofolate reductase (*dhfr*; DHFR)	MTP/O	*Escherichia coli*	(*29*)
				Mus musculus	(*17, 52*)

(*continued*)

TABLE 3.1 *(Continued)*

Selective agent(s)	Mode of action[a]	Selectable marker(s)[b]	Mechanism[c]	Source(s)	Refs.[d]
		Methotrexate binding protein (MTX-BP)[f]	S	*Petunia hybrida*	(*3*)
5-Methyltryptophan	I/AS	Anthranilate synthase (AS)[f]	MTP/O	*Datura innoxia*	(*51*)
				Daucus carota	(*68*)
				Nicotiana tabacum	(*67*)
				Solanum tuberosum	(*6*)
		Tryptophan 2-monooxygenase (*iaaM*; *tms1*)	D	*Agrobacterium tumefaciens*	(*33*)
				Pseudomonas savastanoi	(*33*)
Paromomycin	I/Translation	Neomycin 3′-*O*-phosphotransferase [*npt*; *neo*; NPTII; APH(3′)II]	D	*Escherichia coli*	(*24*)
Puromycin	I/Translation	Puromycin *N*-acetyltransferase (*pac*; PAC)[f]	D	*Streptomyces alboniger*	(*72*)
Pyrimidyloxybenzoates	I/ALS	Acetolactate synthase (ALS; AHAS)[f]	MTP/C	*Gossypium hirsutum*	(*60*)
				Nicotiana tabacum	(*60*)
Sethoxydim	I/ACCase	Acetyl-CoA carboxylase (ACCase; *Acc*)[f]	MTP/O	*Zea mays*	(*47*)
Spectinomycin	I/Translation	Aminoglycoside 3″-*O*-adenylyltransferase [*aadA*; AAD(3″)]	D	*Escherichia coli*	(*61*)
		Chloroplast rRNA	MTR/C	*Nicotiana tabacum*	(*20, 62*)

Streptomycin	I/Translation	Aminoglycoside 3″-*O*-adenylyltransferase [*aadA*; AAD(3″)]	D	*Escherichia coli*	(*61*)
		Chloroplast rRNA	MTR/C	*Nicotiana tabacum*	(*18, 45, 62*)
		Streptomycin 3″-O-phosphotransferase [*spt*; APH(3″)]	D	*Escherichia coli*	(*34, 42*)
Sulfonamides	I/DHPS	Dihydropteroate synthase (DHPS; *sulI*)	MTP/C	*Escherichia coli*	(*25*)
Sulfonylureas	I/ALS	Acetolactate synthase (ALS; AHAS; *SUR*)	MTP/C	*Arabidopsis thaliana*	(*26, 43*)
				Beta vulgaris	(*59*)
				Brassica napus	(*69*)
				Datura innoxia	(*54*)
				Glycine max	(*43, 60*)
				Gossypium hirsutum	(*60*)
				Nicotiana tabacum	(*38, 43, 60*)
				Zea mays	(*21, 73*)
		Sulfonylurea methylhydroxylase (cyt-$P450_{SU1}$; *suaC*)	D	*Streptomyces griseolus*	(*43, 46*)
Tabtoxin, Tabtoxinine-β-lactam	I/GS	Glutamine synthetase (GS)[f]	MTP	*Avena sativa*	(*37*)
		Tabtoxin acetyltransferase (*ttr*)	D	*Pseudomonas syringae* pv *tabaci*	(*1*)
Triazines	I/PSII	Glutathione *S*-transferase (GST)[e]	D	*Zea mays*	(*43, 55*)
		Quinone binding protein (Q_β; *psbA*)[e]	MTP/C	*Amaranthus hybridus*	(*9, 32*)
				Anacystis nidulans	(*32*)
				Brassica campestris	(*32*)
				Chlamydomonas rheinhardtii	(*32*)
				Solanum nigrum	(*32*)

(*continued*)

TABLE 3.1 *(Continued)*

Selective agent(s)	Mode of action[a]	Selectable marker(s)[b]	Mechanism[c]	Source(s)	Refs.[d]
Triazolopyrimidines	I/ALS	Acetolactate synthase (ALS; AHAS)[f]	MTP/C	*Glycine max*	*(60)*
				Gossypium hirsutum	*(60)*
				Nicotiana tabacum	*(60)*

[a]I, Inhibitor; PSII, Photosystem II.

[b]Common abbreviations for genes and/or proteins are indicated in parentheses.

[c]MTP, Mutant target protein; MTR, mutant target RNA; O, overproduction; D, detoxification; S, sequestration; C/M, requires chloroplast (C) or mitochondrial (M) localization.

[d]Key to references: (*1*) Anzai *et al.*, 1989; (*2*) Arruda *et al.*, 1984; (*3*) Barg *et al.*, 1987; (*4*) Bevan *et al.*, 1983; (*5*) Buchanan-Wollaston *et al.*, 1989; (*6*) Carlson and Widholm, 1978; (*7*) Cattoir-Reynaerts *et al.*, 1983; (*8*) Charest *et al.*, 1989; (*9*) Cheung *et al.*, 1988; (*10*) Comai *et al.*, 1985; (*11*) Cseplö *et al.*, 1988; (*12*) De Block *et al.*, 1984; (*13*) De Block *et al.*, 1987; (*14*) Della-Cioppa *et al.*, 1987; (*15*) Dotson *et al.*, 1990; (*16*) Eckes *et al.*, 1989; (*17*) Eichholtz *et al.*, 1987; (*18*) Etzold *et al.*, 1987; (*19*) Fraley *et al.*, 1983; (*20*) H. Fromm *et al.*, 1987; (*21*) Fromm *et al.*, 1990; (*22*) H. Goodman, quoted in (*43*); (*23*) Gordon-Kamm *et al.*, 1990; (*24*) Guerche *et al.*, 1987; (*25*) Guerineau *et al.*, 1990; (*26*) Haughn *et al.*, 1988; (*27*) Hayford *et al.*, 1988; (*28*) Heim *et al.*, 1989, 1990; (*29*) Herrera-Estrella *et al.*, 1983a; (*30*) Herrera-Estrella *et al.*, 1983b; (*31*) Hillé *et al.*, 1986; (*32*) Hirschberg *et al.*, 1987; (*33*) Inzé *et al.*, 1987; (*34*) Jones *et al.*, 1987; (*35*) Kamakura *et al.*, 1990; (*36*) Klee *et al.*, 1987b; (*37*) Knight *et al.*, 1988; (*38*) Lee *et al.*, 1988; (*39*) Lefebvre *et al.*, 1987; (*40*) Lyon *et al.*, 1989; (*41*) Maiti *et al.*, 1989; (*42*) Maliga *et al.*, 1988; (*43*) Mazur and Falco, 1989; (*44*) Misra, 1989; (*45*) Moll *et al.*, 1990; (*46*) Omer *et al.*, 1990, and D. O'Keefe, personal communication; (*47*) Parker *et al.*, 1990; (*48*) Perez *et al.*, 1989; (*49*) Pietrzak *et al.*, 1986; (*50*) Potrykus *et al.*, 1985; (*51*) Ranch *et al.*, 1983; (*52*) Rogers *et al.*, 1987; (*53*) Sathasivan *et al.*, 1990; (*54*) Saxena and King, 1988, 1990; (*55*) Shah *et al.*, 1986a; (*56*) Shah *et al.*, 1986b; (*57*) Stalker *et al.*, 1988; (*58*) Streber and Willmitzer, 1989; (*59*) Stougaard *et al.*, 1990; (*60*) Subramanian *et al.*, 1990; (*61*) Svab *et al.*, 1990a; (*62*) Svab *et al.*, 1990b; (*63*) Thompson *et al.*, 1987; (*64*) van den Elzen *et al.*, 1985; (*65*) Waldron *et al.*, 1985; (*66*) White *et al.*, 1990; (*67*) Widholm, 1972a; (*68*) Widholm, 1972b; (*69*) Wiersma *et al.*, 1989; (*70*) Wohlleben *et al.*, 1988; (*71*) Zelitch and Berlyn, 1982; (*72*) de la Luna *et al.*, 1988; (*73*) Anderson and Georgeson, 1989.

[e]Direct selection for these markers is generally difficult; see references for more details.

[f]Markers characterized but not yet used for plant transformation.

3.2.2.2 Auxotrophic Markers

Auxotrophic mutations which require nutritional supplements are rare in plants, and few can be readily propagated as homozygotes (McCourt and Somerville, 1987). Consequently, documented reports in which auxotrophs have been complemented by transformation with a functional gene are scanty. In one case, the enzyme threonine dehydratase was shown to be deficient in a mutant of *Nicotiana plumbaginifolia* which requires isoleucine. The yeast gene encoding this enzyme, *ILV1*, was fused to plant regulatory sequences and introduced into mutant cells using *Agrobacterium* (Colau *et al.*, 1987). Transformed calli and shoots could grow in medium lacking isoleucine and threonine dehydratase enzyme activity was restored. In another example, the nitrate reductase gene from tobacco (*Nicotiana tabacum*) was used to complement nitrate reductase-deficient mutants of tobacco and *N. plumbaginifolia* using both direct DNA uptake and *Agrobacterium*-mediated transformation (Vaucheret *et al.*, 1990). Transformants could grow in the presence of nitrate, whereas the mutant progenitors could not. Since nitrate reductase also confers sensitivity to chlorate, back mutations to nitrate reductase deficiency should be readily selectable in chlorate-containing media (Caboche and Rouze, 1990). Selection for the gain or loss of nitrate reductase in this fashion could potentially be useful for genetic manipulations analogous to those that are possible in yeast with the *URA3* gene, whose presence or absence can be similarly selected for with different selective agents (Boeke *et al.*, 1987). An analogous genetic system might be feasible using *Arabidopsis* mutants deficient in adenine phosphoribosyltransferase (APRT) which can be selected using 2,6-diaminopurine (Moffatt and Somerville, 1988). Complementation of the deficiency should be associated with reversion to male fertility, which seems to require an active APRT enzyme (Moffatt and Somerville, 1988; Regan and Moffatt, 1990).

Mutants deficient in tryptophan biosynthetic enzymes have been described in *Arabidopsis* (Last and Fink, 1988) and maize (Wright and Neuffer, 1989). Complementation of the *Arabidopsis trp2-1* mutation by the *TSB1* gene encoding the β-subunit of tryptophan synthase (Berlyn *et al.*, 1989) has recently been reported (Last *et al.*, 1990). Attempts are also being made to complement the thiamin-requiring *Arabidopsis th-1* mutant with the *E. coli thiB* gene encoding thiamin-monophosphate pyrophosphorylase (Komeda *et al.*, 1988).

3.2.2.3 Lethal or Conditionally Lethal Markers

It might seem odd to include markers which kill cells in this section, but they could be used for a variety of applications in plant molecular biology, many of which are only in their infancy at present. For example, lethal markers could be used to select for mutations which suppress gene expression or to study cell lineage during development (Evans, 1989). Agents which select for marker loss may also be useful for the enrichment of gene targeting events by "positive–negative" selection (Capecchi, 1989).

Table 3.2 lists a selection of markers which could be used for these purposes in plants. At the time of writing, however, many of these are still in the process of evaluation with respect to their potential use as tools for the kinds of genetic analysis referred to above. Some markers (e.g., diphtheria toxin A-chain or ribonucleases such as RNase T1 or barnase) encode proteins that are extremely toxic to plant cells. These can be made conditionally lethal by fusing them to a promoter which is only expressed in certain tissues (Mariani *et al.*, 1990; Koltunow *et al.*, 1990) or in response to an external stimulus. Other markers (e.g., indoleacetamide hydrolase or sulfonylurea methyl hydroxylase) are lethal only in the presence of an exogenously supplied reagent (C. Dean and J. Jones, personal communication; D. O'Keefe, personal communication). These frequently encode an enzyme that can convert a nontoxic precursor into one that is toxic. Cells that lack the marker do not metabolize the precursor, so they are unaffected. As a variation on this theme, some markers (e.g., T-URF13) may confer sensitivity to a supplied ligand by affecting an essential process such as membrane permeability (Dewey *et al.*, 1988; Huang *et al.*, 1990).

3.2.3 VISIBLE MARKERS

3.2.3.1 Histological Markers

Histological markers confer a distinguishing phenotype in the presence of a substrate or affinity reagent which is supplied exogenously. In many cases, the transformed tissue must be fixed and sectioned prior to histological analysis. Classical histological agents include dyes or chromogenic substrates which react with cellular constituents (Gahan, 1984; Vaughn, 1987). Antibodies can also be used for detecting proteins or low molecular weight antigens at the subfemtomole (10^{-15} mole) level by a histological procedure known as immunocytochemistry (Harlow and Lane, 1988; Harris and Outlaw, 1990). Similarly, gene expression at the mRNA level may be visualized using *in situ* hybridization with a complementary nucleic acid probe (Raikhel *et al.*, 1989). Both approaches can yield valuable information, but they can also be technically demanding and subject to artifacts. As with any histochemical technique, careful controls should be performed to check for background and to ensure that all tissues are made sufficiently and comparably permeable to the supplied reagents. In addition, the signal intensity in different cells should be cautiously interpreted, because it can be affected both by overall cell size and the ratio of vacuolar to cytoplasmic volume.

The GUS reporter gene is probably the most widely used marker in histological studies of transgenic plants. The advantages of this system are manifold and will not be described in detail here, because they have been adequately summarized by Jefferson (1989). The *E. coli lacZ* gene can also be detected histochemically in fixed plant tissues (Teeri *et al.*, 1989), but endogenous β-galactosidase activity has generally discouraged its widespread use (Helmer *et al.*, 1984; Matsumoto *et al.*, 1988). It is likely that other bacterial genes which can be readily detected by using chromogenic substrates (e.g., the *xylE* gene from *Pseudomonas* species: Zukowski *et al.*, 1983) could also be used as markers for gene transfer in plants.

Luciferases from both fireflies and bacteria have been expressed in transgenic plants (Ow *et al.*, 1986; Koncz *et al.*, 1987). Luminescence directed by these enzymes has been detected *in planta* by supplying appropriate luminogenic substrates (Ow *et al.*, 1986; Schneider *et al.*, 1990). The firefly enzyme substrate is poorly taken up by most plant tissues (Howell *et al.*, 1989), whereas the bacterial enzyme substrate appears to traverse membranes more readily (Schauer, 1988). However, detection of either enzyme with high resolution in intact tissues is difficult and usually requires low light camera equipment (Schauer, 1988; Gallie *et al.*, 1989; Schneider *et al.*, 1990; Barnes, 1990). In some cases, x-ray film can also be used (Ow *et al.*, 1986; Fromm *et al.*, 1990). In the future, it may be possible to use other luciferase genes which elicit bioluminescence of different wavelengths (Wood *et al.*, 1989).

Some markers encode enzymes whose products can be detected by histochemical means. For example, amylose produced by genes encoding starch granule-bound starch synthases can be detected in transfected potato tubers and in maize endosperm cells with a solution containing iodine and potassium iodide (Jacobsen *et al.*, 1990; Visser *et al.*, 1990). Similarly, tobacco cells transformed with the *glgC16* gene from *E. coli,* which encodes a feedback insensitive form of ADP-glucose pyrophosphorylase (glucose-1-phosphate adenylyltransferase), contain an increased number of birefringent starch granules as seen by microscopy under polarized light (Timmerman *et al.*, 1991).

3.2.3.2 Morphological Markers

The use of morphological markers for plant gene transfer will probably increase as more genes controlling morphological characters become available. Gene transfer techniques will undoubtedly contribute to a greater understanding of the mechanisms whereby single gene products can affect morphological traits. For example, genes isolated from gall-forming bacteria have dramatic morphogenic effects when expressed in plant tissues (Klee *et al.*, 1987a; Spena *et al.*, 1987; Smigocki and Owens, 1988), and their action is likely to increase our understanding of how plant hormones mediate their effects. Similarly, transgenic plants which overexpress an extracellular peroxidase are susceptible to wilting (Lagrimini *et al.*, 1990), and the overexpression of phytochrome has several pleiotropic effects, including dwarfism (Boylan and Quail, 1989). The morphology of plants which selectively express a toxic gene product in a specific tissue can also be distinctive. For example, tapetum-specific expression of ribonucleases or diphtheria toxin A-chain (Table 3.2) can result in male sterility (Mariani *et al.*, 1990; Koltunow *et al.*, 1990).

In *Arabidopsis,* trichome formation requires the product of the *GL1* gene.

TABLE 3.2
Lethal Markers for Plant Gene Transfer

Lethal marker(s)[a]	Toxigenic agent(s)[b]	Mode of action	Source(s)	Refs.[c]
Adenine phosphoribosyltransferase (APRT; *apt*)[d,e]	2,6-Diaminopurine	Toxic nucleotide production	*Arabidopsis thaliana*	(*20*)
Alcohol dehydrogenase (ADH; *Adh*)[d]	Allyl alcohol	Acrolein production	*Zea mays*	(*6, 24*)
Barnase	Absence of barstar	RNA degradation	*Bacillus amyloliquefaciens*	(*8, 19*)
Diphtheria toxin A-chain (*ctxA*; DT-A)	NA	Ribosome inactivation	*Corynebacterium diphtheriae*	(*2, 4, 16, 23*)
β-Glucuronidase (GUS; *uidA*)	?	Toxic aglycone production	*Escherichia coli*	(*13*)
Indoleacetamide hydrolase (*iaaH*; *tms2*)[f]	Indoleacetamides	Auxin production	*Agrobacterium tumefaciens* *Pseudomonas savastanoi*	(*14, 15*) (*26*)
Nitrate reductase (NR)[d]	Chlorate	Chlorite production	*Nicotiana tabacum*	(*5*)
RAS2[f]	NA	Altered signal transduction?	*Saccharomyces cerevisiae*	(*10*)
Restriction endonucleases (*Eco*RI, etc.)[e,f]	Absence of methylase	DNA degradation	Bacteria	(*1, 21*)
Ricin A chain (R-A)[e]	NA	Ribosome inactivation	*Ricinus communis*	(*17, 25*)
Ribonuclease T1 (RNase T1)	NA	RNA degradation	*Aspergillus oryzae*	(*19*)
S RNases[f]	S RNases?	rRNA degradation	*Nicotiana alata*	(*18*)

Sulfonylurea methylhydroxylase (cyt-$P450_{SU1}$; *suaC*)	Sulfonylurea R7402	ALS inhibitor production	*Streptomyces griseolus*	(*22, 27*)
Thymidine kinase (TK; *tk*; HSV1-TK)[e,f]	ACV, GCV, FIAU, etc.	Toxic nucleotide production	Herpes simplex virus	(*3, 9*)
Tryptophan 2-monooxygenase (*iaaM*; *tms1*)[f]	NA	Indolecetamide overproduction	*Agrobacterium tumefaciens* *Pseudomonas savastanoi*	(*12*) (*12*)
T-URF13 (*urf13-T*; *T-urf13*)	Methomyl; BmT toxin	Mitochondrial permeabilization	*Zea mays* cms-T	(*7, 11*)

[a]Common abbreviations for genes and/or proteins are indicated in parentheses.

[b]NA, Not applicable; other abbreviations as described in the references.

[c]Key to references: (*1*) Barnes and Rine, 1985; (*2*) Behringer *et al.*, 1988; (*3*) Borrelli *et al.*, 1988, 1989; (*4*) Breitman *et al.*, 1987; (*5*) Caboche and Rouze, 1990; (*6*) Callis *et al.*, 1987; (*7*) Dewey *et al.*, 1988; (*8*) Hartley, 1989; (*9*) Heyman *et al.*, 1989; (*10*) Hilson *et al.*, 1990; (*11*) Huang *et al.*, 1990; (*12*) Inzé *et al.*, 1987; (*13*) Jefferson, 1989; (*14*) C. Dean and J. Jones, personal communication; (*15*) Karlin-Neumann and Tobin, 1989; (*16*) Koltunov *et al.*, 1990; (*17*) Landel *et al.*, 1988; (*18*) McClure *et al.*, 1990; (*19*) Mariani *et al.*, 1990; (*20*) Moffatt and Somerville, 1988; (*21*) Kwoh *et al.*, 1988; (*22*) Omer *et al.*, 1990; (*23*) Palmiter *et al.*, 1987; (*24*) Schwartz and Osterman, 1976; (*25*) Stirpe and Barbieri, 1986; (*26*) Yamada *et al.*, 1985; (*27*) D. O'Keefe, personal communication.

[d]Applicable only when recipients of gene transfer have a null phenotype.

[e]Not yet used for plant transformation.

[f]Applicability probably restricted; see references for further information.

This gene has recently been cloned and reintroduced into a *gl1* mutant which completely lacks trichomes (Herman and Marks, 1989). Trichome formation was restored by gene transfer, indicating that the defect was complemented by the introduced gene. Plants with trichomes can be easily distinguished from plants without, so this marker could be very useful for *Arabidopsis* transformation in the future. The *agamous* gene provides another example (Yanofsky *et al.*, 1990).

Plants which are infected with viruses or viroids can frequently be diagnosed by the presence of characteristic mosaics or lesions. Analogous symptoms can be produced by introducing the genetic information normally encoded by a viral genome as part of the T-DNA transferred by *Agrobacterium*. This approach is not only a powerful means of analyzing the various functions encoded by plant viruses, but also provides an extremely sensitive detection mechanism for the transfer of T-DNA to cells at very low frequency (Grimsley and Bisaro, 1987; Hohn *et al.*, 1989). For example, agroinfection of maize by T-DNA constructs containing multiple copies of maize streak virus demonstrated for the first time that *Agrobacterium* could deliver T-DNA to cells in graminaceous monocots (Grimsley *et al.*, 1987). Related gemini viruses have also been used to detect DNA transfer into other monocot tissues (Töpfer *et al.*, 1989; Creissen *et al.*, 1990).

3.2.3.3 Pigmentation Markers

For the plant geneticist, it is fortunate that plants are colorful organisms. Mutations which affect pigment synthesis are rarely detrimental and have been extensively used as markers in plants by geneticists and breeders for hundreds of years (Mol *et al.*, 1989). Many of the most universal (and colorful) pigments unique to plants are anthocyanins or similar flavonoid-related compounds. In recent years, a number of genes associated with anthocyanin production have been cloned (Mol *et al.*, 1988; Ludwig and Wessler, 1990). Some of these are structural genes which encode enzymes involved in anthocyanin biosynthesis, whereas others encode trans-acting factors which regulate expression of the structural genes at the transcriptional level.

Three general approaches have been taken in utilizing cloned anthocyanin genes as dominant markers in transformed plant tissues. From a genetic standpoint, these involve: (1) suppression of endogenous wild-type alleles using antisense RNA (van der Krol *et al.*, 1988), the addition of extra gene copies (Napoli *et al.*, 1990; van der Krol *et al.*, 1990), or dominant negative inhibitor proteins (Bowen *et al.*, 1991a); (2) complementation of mutant alleles (Meyer *et al.*, 1987; Klein *et al.*, 1989b; Nash *et al.*, 1990; Menssen *et al.*, 1990); or (3) transactivation of silent alleles (Ludwig *et al.*, 1990; Goff *et al.*, 1990).

Perhaps the most spectacular example of gene transfer involving a structural gene for anthocyanin biosynthesis was reported by Meyer *et al.* (1987). By transforming a mutant line of petunia with a gene encoding dihydroflavonol reductase from maize (*A1*), the normal anthocyanin biosynthetic pathway was subverted and plants with a novel flower color resulted. The maize *A1* gene has subsequently been used as a reporter gene for petal-specific gene expression in petunia (Benfey *et al.*, 1990).

Trans-activators of anthocyanin biosynthetic genes have recently been exploited as sensitive markers of gene transfer in maize (Ludwig *et al.*, 1990; Goff *et al.*, 1990). Pigmentation can be observed noninvasively in living cells of virtually any maize tissue which has a full complement of the structural genes present, but which is not normally pigmented (Figure 3.2, color plate). Individual cells transfected with genes encoding trans-activators of anthocyanin biosynthesis can be more readily distinguished than cells transfected with genes encoding anthocyanin biosynthetic enzymes or with other visible markers such as GUS or luciferase, because the induced anthocyanin accumulation is cell-autonomous (Ludwig *et al.*, 1990; Goff *et al.*, 1990; Figure 3.2). At the present time, it is unclear why anthocyanin accumulation in cells complemented by wild-type anthocyanin biosynthetic genes is not cell-autonomous (Klein *et al.*, 1989b; Nash *et al.*, 1990; Menssen *et al.*, 1990).

At least two gene products (encoded by the *R* and *C* gene families, respectively) are required for trans-activation of maize anthocyanin biosynthetic genes (Goff *et al.*, 1990; Ludwig and Wessler, 1990). Cells expressing either an *R* gene product or a *C* gene product alone do not synthesize anthocyanins. Similarly, when *R* or *C* genes from maize are introduced individually into cells of other graminaceous monocots, they do not induce the synthesis of flavonoid biosynthetic genes (Bowen *et al.*, 1991b). However, flavonoid pigments can be induced when *R* and *C* genes from maize are introduced together into cells of at least two other cereals (*Sorghum bicolor* and *Saccharum spontaneum*) by microprojectile bombardment (Bowen *et al.*, 1991b). This suggests that the signals recognized by trans-activator proteins from maize are conserved in other graminaceous monocots. It is likely, therefore, that maize *R* and *C* genes will be widely used to develop transformation procedures for other cereals in the near future.

3.3 Assayable Markers

3.3.1 CRITERIA FOR CHOOSING ASSAYABLE MARKERS

Assayable markers most commonly used to quantify gene expression in plants are listed in Table 3.3. Most of the markers listed can only be assayed quantitatively in soluble tissue extracts using one or more of the procedures outlined in Section 3.3.2. Some markers can also be assayed both qualitatively (Section 3.2.3) or semiquantitatively (Section 3.3.3) *in planta*. Choosing the most appropriate assayable marker or reporter gene, therefore, depends on the type of assay which is most suited to the experimental aims of the investigator. For quantitative

TABLE 3.3
Assayable Markers for Plant Gene Transfer

Assayable marker[a]	Available assays[b,c]	Potential limitations[c]
Chloramphenicol acetyltransferase (CAT; *cat*)	C (*43*) F (*28*) I (*13, 33*) R (*11, 34, 41, 43, 50*)	Background activity (*5, 11*) Inhibitors/esterases (*4, 5*)
β-Galactosidase (*lacZ*; β-gal)	C (*27, 34*) F (*28*) G (*16, 47, 50*) H (*26, 27, 28, 47*) I (*34*) L (*23*)	Background activity (*16, 26, 44, 47*) Microbial contamination (*24*) Histochemistry requires fixation (*47*)
β-Glucuronidase (GUS; *gus*; *uidA*)	C (*19*) F (*7, 19, 28, 35*) G (*20*) H (*7, 14, 19, 28, 45*) I (*7*)	Background activity (*1, 18, 32, 40*) Autofluorescence (*3*) Quenching/inhibitors (*35*) Microbial contamination (*4*)
Luciferases (luc; *luxA/B*)	H (*2, 12, 30, 39, 50*) I (*15*) L (*21, 23, 30, 38, 50*)	Quantification requires luminometer (*23*) Substrate uptake *in planta* (*39*)
Neomycin phosphotransferase (*npt*; *neo*; NPTII; APH(3′)II)[d]	G (*37, 50*) I (*10, 33, 50*) R (*17, 25, 31, 46*)	Background activity/kinases (*25, 37, 48, 50*) Mutant constructs (*49*) Inhibitors (*46*) Microbial contamination (*22*) Affinity fractionation improves sensitivity (*17*)
Opine synthases (*ocs*; *nos*; *mas*; *ags*; *acs*)	C (*42, 48*) F (*29, 50*)	Quantification difficult (*42, 48, 50*) Background activity (*6*)

(*continued*)

experiments, the sensitivity and reproducibility of the assay method are probably the most crucial factors (Töpfer *et al.*, 1988; Sheen, 1990; Alam and Cook, 1990). However, the ease with which multiple samples can be assayed in a reasonable time and the cost per assay are also important considerations. Assays which can be performed in a microtiter plate and quantified in the same format are generally the most economical in these respects. Large numbers of assays will be easier to run when robotic devices which can prepare tissue extracts and perform assays in a microtiter plate format are in routine use. Such devices are currently under development (Brumback, 1991).

If tissue extracts are used to quantify reporter gene expression, several biochemical parameters are important in determining the choice of an appropriate

TABLE 3.3
(*Continued*)

Assayable marker[a]	Available assays[b,c]	Potential limitations[c]
Phosphinothricin acetyl-transferase (*bar*; PAT)	C (*8*)	Affinity fractionation improves sensitivity (*36*)
	F (*36*)	
	I (*8*)	
	R (*8, 9*)	

[a]Common abbreviations for genes and/or proteins are indicated in parentheses.

[b]C, Colorimetric; F, fluorometric; G, gel activity; H, histochemical; I, immunochemical; L, luminometric; R, radioactive.

[c]Key to references and commercial sources given in parentheses: (*1*) Alwen *et al.*, 1990; (*2*) Barnes, 1990; (*3*) Bekkaoui *et al.*, 1988; (*4*) B. Bowen, unpublished observations; (*5*) Charest *et al.*, 1989; (*6*) Christou *et al.*, 1986; (*7*) Clontech Laboratories, Inc., Palo Alto, California; (*8*) De Block *et al.*, 1987; (*9*) Denecke *et al.*, 1990; (*10*) Firoozabady *et al.*, 1987; (*11*) M. Fromm *et al.*, 1987; (*12*) Fromm *et al.*, 1990; (*13*) Gendloff *et al.*, 1990; (*14*) Gould and Smith, 1989; (*15*) Gould *et al.*, 1990; (*16*) Helmer *et al.*, 1984; (*17*) Henderson *et al.*, 1991; (*18*) Hu *et al.*, 1990; (*19*) Jefferson, 1987; (*20*) Kavanagh *et al.*, 1988; (*21*) Koncz *et al.*, 1987; (*22*) Konstantinov *et al.*, 1990; (*23*) Kricka, 1988; (*24*) Leifert *et al.*, 1989; (*25*) McDonnell *et al.*, 1987; (*26*) Matsumoto *et al.*, 1988; (*27*) Miller, 1972; (*28*) Molecular Probes, Inc., Eugene, Oregon; (*29*) Otten and Schilperoort, 1978; (*30*) Ow *et al.*, 1986; (*31*) Platt and Yang, 1987; (*32*) Plegt and Bino, 1989; (*33*) 5-Prime 3-Prime, Inc., West Chester, Pennsylvania; (*34*) Promega Corp., Madison, Wisconsin; (*35*) Rao and Flynn, 1990, 1991; (*36*) G. Rao, personal communication; (*37*) Reiss *et al.*, 1984; (*38*) Callis *et al.*, 1987; (*39*) Schneider *et al.*, 1990; (*40*) Schulz and Weissenböck, 1987; (*41*) Seed and Scheen, 1988; (*42*) Shaw *et al.*, 1988; (*43*) Shaw, 1975; (*44*) Simos *et al.*, 1989; (*45*) Sittertz-Bhatkar, 1989; (*46*) Staebell *et al.*, 1990; (*47*) Teeri *et al.*, 1989; (*48*) Yang *et al.*, 1987; (*49*) Yenofsky *et al.*, 1990; (*50*) Gelvin *et al.*, 1990.

[d]Assays may be adapted for other aminoglycoside phosphotransferases (Davies and Smith, 1978).

reporter gene (Rao and Flynn, 1990, 1991; Alam and Cook, 1990). First, the background activity should be as low as possible. In some cases, it may be possible to rectify a high background by crude fractionation (e.g., by using gel filtration, ammonium sulfate precipitation or an affinity matrix), by changing buffer components such as pH, or by heat treatment. Second, if the protein encoded by the reporter gene is available in pure form, it should be determined if an extract contains any factors which inhibit or quench its activity or which unduly affect its stability in short- or long-term storage. Finally, it is important to mention that protein and/or mRNA stability *in vivo* are often neglected as factors in choosing the most appropriate reporter gene for a given application (Alam and Cook, 1990).

Several reporter gene products, such as CAT (Boutry *et al.*, 1987), NPTII (Schreier and Schell, 1986; Denecke *et al.*, 1990), GUS (Kavanagh *et al.*, 1988; Iturriaga *et al.*, 1989; Denecke *et al.*, 1990; Restrepo *et al.*, 1990), and PAT

(Denecke *et al.*, 1990), have been targeted to compartments other than the cytoplasm by using translational fusions. However, the final form of the targeted protein may retain extra amino acids or posttranslational modifications (e.g., glycosyl residues) which can affect protein stability and/or specific activity. It may not always be appropriate to make comparisons between quantitative assays using fusions of this kind. It should also be remembered that our understanding of protein targeting in plant cells is far from complete (Bowles, 1990). At the time of writing, therefore, it is difficult to predict which marker gene may be most effectively targeted to a specific compartment (Denecke *et al.*, 1990).

3.3.2 *IN VITRO* ASSAYS

3.3.2.1 Enzyme Assays

Expression of all the markers listed in Table 3.3 can be quantified by means of one or more types of enzyme assay. These may be colorimetric, fluorometric, or luminometric in nature. Other assays may require radioactive substrates. Assays which can be quantified fluorometrically or luminometrically are generally preferable, but for some investigators, access to the appropriate equipment may be a limiting factor. In most institutions, however, a scintillation counter is usually available for assays which involve radioactive or luminogenic substrates. Most of the enzymes listed in Table 3.3 can be assayed in crude extracts without further fractionation. Affinity gel enrichment has improved the available assays for NPTII and PAT, however (Henderson *et al.*, 1991; G. Rao, personal communication).

Several other assayable markers not listed in Table 3.3 could conceivably be used as reporters for plant gene expression, but they have not been adequately tested. For example, two cassettes encoding alkaline phosphatase from mammalian cells could be used (for references, see Alam and Cook, 1990). One directs alkaline phosphatase activity to the cell surface, while the other results in a secreted form of the enzyme which can accumulate to high levels. Several sophisticated assay techniques for alkaline phosphatase have been developed (Alam and Cook, 1990), and background activity in plant extracts should be minimal, because most plant cell phosphatases have acidic pH optima (Sexton and Hall, 1978). A gene encoding an ice nucleation protein also has potential as a sensitive reporter for quantifying gene expression in plant cells (Lindgren *et al.*, 1989).

3.3.2.2 Gel Activity Assays

A gel activity assay is one in which an enzymatic activity is detected as a band in a native gel following electrophoretic separation from other proteins present in a crude mixture. For example, GUS and *lacZ* can both be detected using fluorometric substrates in polyacrylamide gels (Helmer *et al.*, 1984; Kavanagh *et al.*,

1988; Teeri *et al.*, 1989). Similarly, the NPTII enzyme is often quantified by using a radioactive gel activity assay, in order to distinguish it from other kinases present in many plant cell extracts (Reiss *et al.*, 1984). Colorimetric gel activity assays have been used for detecting patatin acyl hydrolase (Rosahl *et al.*, 1987) and yeast invertase (von Schwaewen *et al.*, 1990) in transgenic plants. In some cases, it has been possible to demonstrate posttranslational modifications of reporter gene products unequivocally by using these kind of assays (Kavanagh *et al.*, 1988). However, Western blots are more commonly used for this purpose (Iturriaga *et al.*, 1989).

Starch gel activity assays are used extensively in the seed industry to screen for isozyme purity in thousands of plant samples at a time (Tanksley and Orton, 1983). Isozyme analysis has so far been used to detect gene transfer in very few instances (Callis *et al.*, 1987; Lagrimini *et al.*, 1990). As transformation procedures improve and gene transfer technology is integrated into the rest of the seed industry, these kinds of assays may be used more extensively than they are at present.

3.3.2.3 Immunochemical Assays

Some of the proteins encoded by genes of potential agronomic interest have no known enzymatic activity. Examples include viral coat proteins, many insecticidal proteins (e.g., from *Bacillus thuringiensis*), or seed storage proteins with a desirable amino acid content. In the initial stages of development, it is often impractical to optimize the expression of these markers by evaluating their agronomic effects. In many instances, sensitive assays can be developed by using antibodies (Harlow and Lane, 1988). Enzyme-linked immunosorbent assay (ELISA) techniques are frequently used, because they can potentially detect antigens in the attomole (10^{-18} mole) range (Stanley *et al.*, 1985; Johannsson *et al.*, 1986; Harris and Outlaw, 1990). Other immunochemical techniques in common use include Western blots and immunolocalization (Harlow and Lane, 1988). These latter techniques are often used for studying posttranslational modifications and protein targeting.

Immunochemical methods can potentially detect both active and inactive (or degraded) proteins which share the same epitope. If possible, therefore, an independent assay for the active protein should be performed as a control. The relative stability of heterologous proteins *in planta* and/or in tissue extracts can also be estimated in this way. In transgenic tobacco, for example, the level of CAT enzyme activity corresponded to the amount of protein detected by ELISA (Gendloff *et al.*, 1990), indicating that this protein was fairly stable in its heterologous host.

Functional antibodies have recently been produced by expressing heavy and light chain immunoglobulin genes in transgenic plants (Hiatt *et al.*, 1989). Just as antigens can be detected in plant extracts by ELISA or other immunochemical

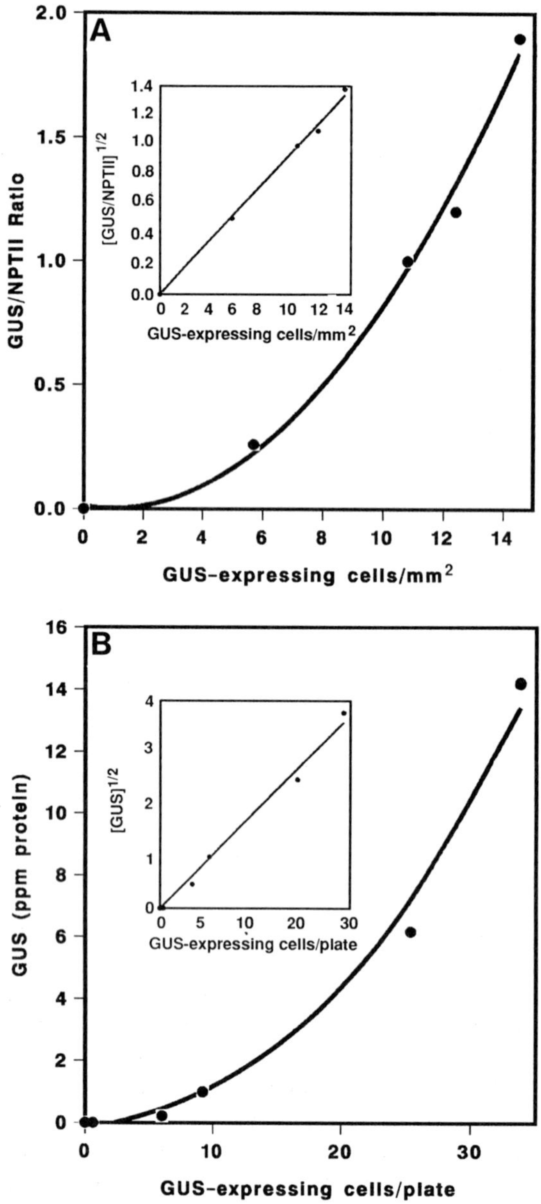
A
GUS/NPTII Ratio
GUS-expressing cells/mm²
[GUS/NPTII]^1/2
GUS-expressing cells/mm²
B
GUS (ppm protein)
GUS-expressing cells/plate
[GUS]^1/2
GUS-expressing cells/plate

methods, antibody expression can be quantified by the binding of an antigen using analogous procedures (Harlow and Lane, 1988).

3.3.3 *IN VIVO* ASSAYS

Most visible markers in current use are qualitative in nature, because it is difficult to quantify their expression in individual cells. For example, differential luciferin uptake is likely to affect the quantification of luciferase in cells of intact plants (Schneider *et al.*, 1990; Barnes, 1990), although it may be possible to quantify enzyme activity in individual protoplasts (Gallie *et al.*, 1989). Similarly, it should be feasible to measure GUS in individual cells (Jefferson, 1989), but this has not been very practical so far (Harkins *et al.*, 1990). Ideally, the quantification of marker gene expression in living cells of intact plants should be noninvasive and require no addition of exogenous reagents. Markers which transactivate anthocyanin biosynthesis hold great promise in this respect (Ludwig *et al.*, 1990; Ludwig and Wessler, 1990). Classical genetic analysis suggests that the rate of pigment accumulation in maize tissues may be directly related to the level of trans-activator expression (Fogel, 1946). Furthermore, the rate of pigment accumulation in individual maize cells can be determined by microspectrophotometry (Commoner and Zucker, 1953).

Figures 3.3 and 3.4 illustrate how visible markers may be used in conjunction with microprojectile bombardment to compare relative levels of marker gene expression without having to perform biochemical assays. Figure 3.3 shows the relationship between the number of cells visibly expressing GUS *in vivo* and the transient level of GUS gene expression assayed *in vitro,* following bombardment of immature soybean cotyledons (Figure 3.3A) or embryogenic maize suspension culture cells (Figure 3.3B). In Figure 3.3A, the data were obtained with equimolar amounts of different constructs expressing GUS at different levels; in Figure 3.3B, a dilution series of a single GUS construct was used. In both experiments, the relationship between the number of GUS-expressing cells and

---

Figure 3.3 Relationship between the level of GUS expression and the number of cells visibly expressing GUS in microprojectile-bombarded tissues. (A) Immature soybean cotyledons were bombarded with a fixed amount of a 35S::NPTII construct and equimolar amounts of different GUS constructs containing various combinations of regulatory sequences. GUS activity (ppm protein) was normalized to NPTII activity (ppm protein) and plotted against the density of cells staining blue with the chromogenic GUS substrate X-Gluc. Each data point (kindly provided by J. Ranch) represents the mean of four replicate treatments. The inset shows the same *x* values plotted against the square root of the *y* values. (B) Embryogenic corn suspension cells were bombarded with various amounts of a 35S::GUS construct serially diluted with carrier DNA (pUC18) such that equal amounts of total DNA were used per bombardment. GUS activity (ppm protein) was plotted against the density of X-Gluc staining cells as in (A). Each data point (kindly provided by B. Drummond) represents the mean of five replicate treatments. The inset shows the *x* values plotted against the square root of the *y* values as in (A).

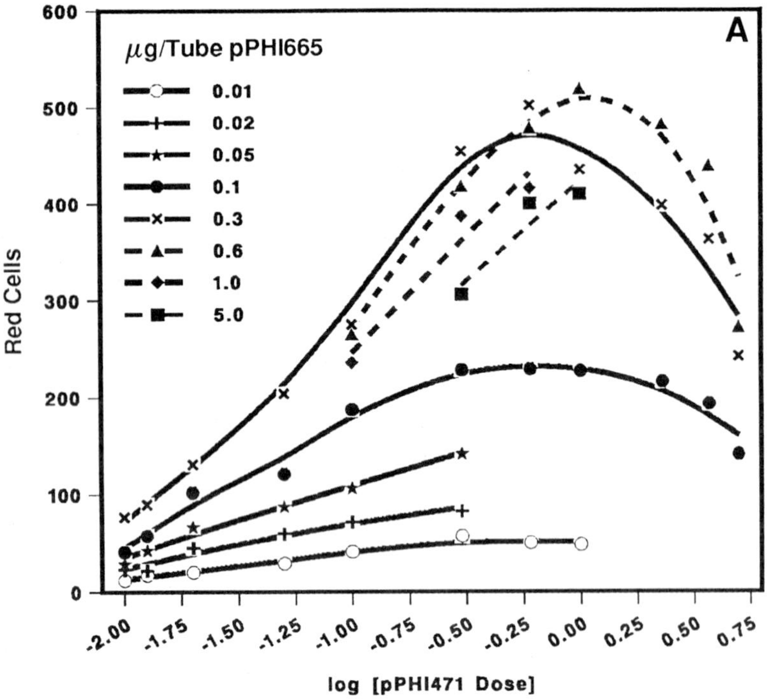

A
μg/Tube pPHI665
0.01
0.02
0.05
0.1
0.3
0.6
1.0
5.0
Red Cells
log [pPHI471 Dose]

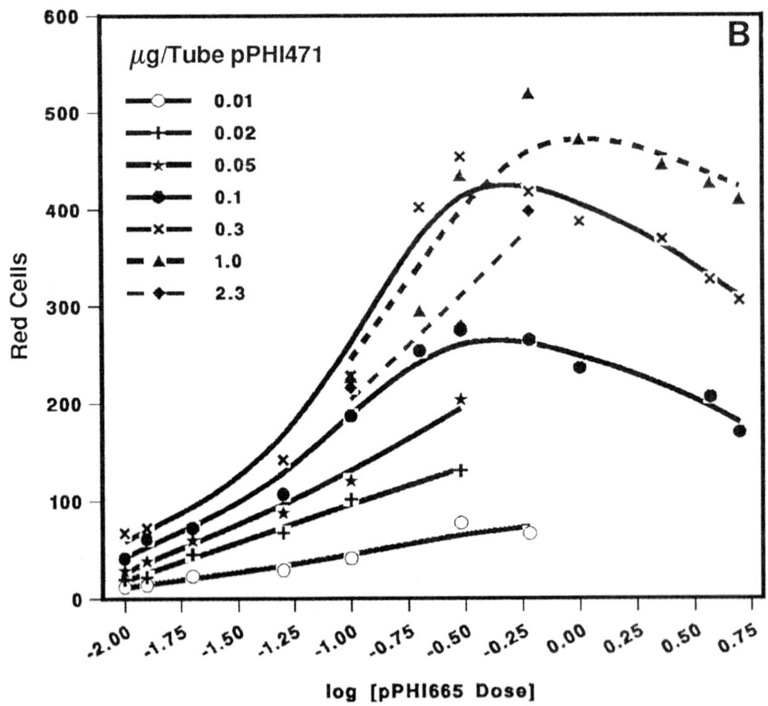

B
μg/Tube pPHI471
0.01
0.02
0.05
0.1
0.3
1.0
2.3
Red Cells
log [pPHI665 Dose]

the level of GUS activity was the same. This demonstrates that gene expression levels can be compared between treatments by counting visibly transfected cells. The insets in Figures 3.3A and 3.3B show that there is a good correlation between the *x* values (i.e., the numbers of GUS-expressing cells) and the square root of the *y* values (i.e., the levels of GUS activity), suggesting that the level of GUS expression in each treatment varies from cell to cell in a Poisson distribution. From the dose–response experiment (Figure 3.3B), it is reasonable to infer that the numbers of active 35S::GUS templates in each cell are distributed similarly.

Figure 3.4 shows two series of dose–response curves depicting the number of red cells seen in embryogenic maize suspension culture cells when 35S::*R* is varied at various fixed doses of 35S::*C*, and when 35S::*C* is varied at various fixed doses of 35S::*R*. Increasing the number of 35S::*R* templates appears to decrease the threshold number of active 35S::*C* templates needed to produce a visible level of anthocyanin inside transfected cells, and vice versa. However, when *R* is in excess over *C*, or vice versa, an apparent decrease in the number of red cells is observed. These observations suggest that *R* and *C* interact cooperatively *in vivo* and that an excess of either inhibits trans-activation, possibly via some kind of "squelching" mechanism (Gill and Ptashne, 1988). For reasons elaborated elsewhere, the threshold level of active templates in each transfected cell is probably quite low (Weintraub, 1988; Armaleo *et al.*, 1990). Interestingly, an independent analysis in mammalian cells has recently shown that quite small changes in the intracellular concentrations of trans-activators can effectively switch gene expression "on" or "off" in a similar fashion to that suggested by the data in Figure 3.4 (Fiering *et al.*, 1990).

3.4 Concluding Remarks

What kinds of markers might be developed in the future? In writing this review, I have been impressed with the wealth of selectable and assayable markers already available for plant gene transfer. In the years to come, I am confident that new selective agents and increasingly sophisticated assays will expand the repertoire of markers listed in Tables 3.1–3.3. However, new versatile markers which

Figure 3.4 Red cell numbers obtained from factorial dose–response experiments with 35S::*R* (A) and 35S::C (B). Embryogenic corn suspension cells (*c r-r b pl*) were bombarded with mixtures of 35S::*R* (pPHI471) and 35S::C (pPHI665) and incubated without illumination. With each treatment, the total amount of DNA precipitated onto the tungsten microprojectiles (4.375 mg/tube) was made up to 10 μg/tube with a 35S::GUS construct and one-tenth of the microprojectiles were used per bombardment. Red cell numbers were summed from three replicate bombardments for each treatment and plotted against the logarithm of either pPHI471 dose (A) or pPHI665 dose (B). All plasmid doses are expressed as μg/tube.

affect either morphological characters or plant pigments may also soon prevail. For example, pigmentation markers may be developed using genes involved in the biosynthesis of melanin (Della-Cioppa *et al.*, 1987), indigo (Mermod *et al.*, 1986), or carotenoids (Buckner *et al.*, 1990), and several genes which affect plant development hold promise as morphological markers (see, e.g., Bleecker *et al.*, 1988; Chory *et al.*, 1989). Totally novel phenotypes may also be possible. In bacteria, for example, new antibiotics have been produced by the subversion of biochemical pathways using genes derived from different streptomycetes (Hopwood *et al.*, 1985). Following the lead of Meyer *et al.* (1987), it may be possible to produce interesting new traits in plants by creative genetic engineering of this sort. However, in many cases, it may also be necessary to turn endogenous genes or pathways off, perhaps via the expression of antisense RNA, dominant inhibitor proteins or antibodies.

In the past, simple genetic experiments with judiciously chosen marker genes have often been instrumental in the discovery of important biological phenomena. For example, the work of Mendel with unlinked genetic markers in peas led to the elucidation of the laws of inheritance. Had he chosen markers which were linked, he might never have made the inferences he did. One can only speculate how Mendel might marvel at the state of plant genetics today. The gene responsible for one of the characters crucial to his interpretation of sexual inheritance has now been cloned (Bhattacharyya *et al.*, 1990), and there are powerful techniques available for manipulating its expression and transferring it to other plant species by nonsexual means. With the aid of such elaborate technology, perhaps plant molecular biologists might one day unravel processes as important and fundamental as those discovered by Mendel with more meager means at his disposal. At the very least, the markers described in this review should be just as crucial for furthering our understanding of plant biology as they were for Mendel in his day.

Acknowledgments

I would like to thank Tom Brumback, Caroline Dean, Bruce Drummond, Lahree Henderson, John Howard, Jonathan Jones, Ganesh Kishore, Dan O'Keefe, Jerry Ranch, Guru Rao, Margit Ross, Kurt Timmerman, and Dwight Tomes for providing unpublished information; Renee Kosslak, Jerry Ranch, Brad Roth, and Erica Unger for comments on the manuscript; and Steve Sargent for graphics expertise.

References

Alam, J., and Cook, J. L. (1990). *Anal. Biochem.* 188, 245–254.

Alwen, W., Vincente, O., and Heberlé-Bors, E. (1990). In *VIth NATO Advanced Study Institute: Plant Molecular Biology,* p. 18.

Anderson, P. C., and Georgeson, M. (1989). *Genome* 31, 994–999.

Anzai, H., Yoneyami, K., and Yamaguchi, I. (1989). *Mol. Gen. Genet.* 219, 492–494.

Armaleo, D., Ye, G. N., Klein, T. M., *et al.* (1990). *Curr. Genet.* 17, 97–103.
Arruda, P., Bright, S. W. J., Kueh, J. S. H., Lea, P. J., and Rognes, S. E. (1984). *Plant Physiol.* 76, 442–446.
Barg, R., Perl, M., and Beckmann, J. S. (1987). *Plant Mol. Biol.* 8, 87–94.
Barnes, G., and Rine, J. (1985). *Proc. Natl. Acad. Sci. U.S.A.* 82, 1354–1358.
Barnes, W. M. (1990). *Proc. Natl. Acad. Sci. U.S.A.* 87, 9183–9187.
Behringer, R. R., Mathews, L. S., Palmiter, R. D., and Brinster, R. L. (1988). *Genes Dev.* 2, 453–461.
Bekkaoui, F., Pilon, M., Laine, E., *et al.* (1988). *Plant Cell Rep.* 7, 481–484.
Benfey, P., and Chua, N. H. (1989). *Science* 244, 174–181.
Benfey, P., and Chua, N. H. (1990). *Science* 250, 959–966.
Benfey, P., Takatsuji, H., Ren, L., Shah, D. M., and Chua, N.-H. (1990). *Plant Cell* 2, 949–856.
Berlyn, M., Last, R. L., and Fink, G. R. (1989). *Proc. Natl. Acad. Sci. U.S.A.* 86, 4604–4608.
Bevan, M., Flavell, R. B., and Chilton, M.-D. (1983). *Nature (London)* 304, 184–187.
Bhattacharyya, M. K., Smith, A. M., Ellis, T. H. N., and Martin, C. (1990). *Cell* 60, 115–122.
Bleecker, A. B., Estelle, M. A., Somerville, C., and Kende, H. (1988). *Science* 241, 1086–1089.
Boeke, J. D., Trueheart, J., Natsoulis, G., and Fink, G. R. (1987). *Meth. Enzymol.* 154, 164–175.
Borrelli, E., Heyman, R., Hsi, M., and Evans, R. M. (1988). *Proc. Natl. Acad. Sci. U.S.A.* 85, 7572–7576.
Borrelli, E., Heyman, R. A., Arias, C., Samchenko, P. E., and Evans, R. M. (1989). *Nature (London)* 339, 538–541.
Boutry, M., Nagy, F., Poulsen, C., Aoyagi, K., and Chua, N. H. (1987). *Nature (London)* 328, 340–342.
Bowen, B. (1992). In *GUS Protocols: Using the GUS Gene as a Reporter of Gene Expression* (Ed. by S. Gallagher), pp. 163–177. New York: Academic Press, Inc.
Bowen, B., Solan, F., Sims, L., *et al.* (1991a). Abstract 400, *3rd International Congress of Plant Molecular Biology*, Tucson, Arizona.
Bowen, B., Bennetzen, J., Nagai, C., Sun, S. and Roth, B. (1991b). Abstract 1093, *3rd International Congress of Plant Molecular Biology*, Tucson, Arizona.
Bowles, D. J. (1990). *Curr. Opin. Cell Biol.* 2, 673–680.
Boylan, M. T., and Quail, P. H. (1989). *Plant Cell* 1, 765–773.
Breitman, M. L., Clapoff, S., Rossant, J., *et al.* (1987). *Science* 238, 1563–1565.
Brumback, T. (1991). In *GUS Protocols: Using the GUS Gene as a Reporter of Gene Expression* (Ed. by S. Gallagher), pp. 77–87. New York: Academic Press, Inc.
Buchanan-Wollaston, V., Gold, B., Naser, A., and Cannon, F. (1989). *J. Cell Biochem.* Suppl. 13D, 330.
Buckner, B., Kelson, T., and Robertson, D. S. (1990). *Plant Cell* 2, 867–876.
Caboche, M., and Rouze, P. (1990). *Trends Genet.* 6, 187–192.
Callis, J., Fromm, M., and Walbot, V. (1987). *Genes Dev.* 1, 1183–1200.
Capecchi, M. R. (1989). *Science* 244, 1288–1292.
Carlson, J. E., and Widholm, J. M. (1978). *Physiol. Plant.* 44, 251–255.
Castillo, G., Vera, J. C., Yang, C.-P. H., Horwitz, S. B., and Rosen, O. M. (1990). *Proc. Natl. Acad. Sci. U.S.A.* 87, 4737–4741.
Cattoir-Reynaerts, A., Degryse, E., Verbruggen, I., and Jacobs, M. (1983). *Biochem. Physiol. Pflanz.* 178, 81–90.
Charest, P. J., Iyer, V. N., and Miki, B. L. (1989). *Plant Cell Rep.* 7, 628–631.
Cheung, A. Y., Bogorad, L., Van Montagu, M., and Schell, J. (1988). *Proc. Natl. Acad. Sci. U.S.A.* 85, 391–395.
Chory, J., Peto, C., Feinbaum, R., Pratt, L., and Ausubel, F. (1989). *Cell* 58, 991–999.
Christou, P., Platt, S. G., and Ackerman, M. C. (1986). *Plant Physiol.* 82, 218–221.
Colau, D., Negrutiu, I., Van Montagu, M., and Hernalsteens, J. P. (1987). *Mol. Cell Biol.* 7, 2552–2557.

Comai, L., Facciotti, D., Hiatt, W. R., *et al.* (1985). *Nature (London)* 317, 741–744.

Commoner, B., and Zucker, M. L. (1953). In *Growth and Differentiation in Plants* (Ed. by W. E. Loomis), pp. 339–392. Ames, Iowa: Iowa State College Press.

Creissen, G., Smith, C., Francis, R., Reynolds, H., and Mullineaux, P. (1990). *Plant Cell Rep.* 8, 680–683.

Cseplö, A., Etzold, T., Schell, J., and Schreier, P. H. (1988). *Mol. Gen. Genet.* 214, 295–299.

Datla, R. S. S., Hammerlindl, J. K., Pelcher, L. E., Selvaraj, G., and Crosby, W. L. (1990). *J. Cell Biochem.* Suppl. 14E, 279.

Davies, J., and Smith, D. I. (1978). *Ann. Rev. Microbiol.* 32, 469–518.

De Almeida, E. R. P., Gossele, V., Muller, C. G., *et al.* (1989). *Mol. Gen. Genet.* 218, 78–86.

De Block, M., Herrera-Estrella, L., Van Montagu, M., Schell, J., and Zambryski, P. (1984). *EMBO J.* 3, 1681–1689.

De Block, M., Botterman, J., Van de Wiele, M., *et al.* (1987). *EMBO J.* 6, 2513–2518.

Dekeyser, R., Claes, B., Marichal, M., Van Montagu, M., and Caplan, A. (1989). *Plant Physiol.* 90, 217–223.

de la Luna, S., Soria, I., Pulido, D., Ortin, J., and Jimenez, A. (1988). *Gene* 62, 121–126.

Della-Cioppa, G., Bauer, S. C., Taylor, M. L., *et al.* (1987). *Bio/Technology* 5, 579–584.

Della-Cioppa, G., Garger, S. J., Sverlow, G. G., Turpen, T. H., and Grill, L. K. (1990). *Bio/Technology* 8, 634–638.

Denecke, J., Gossele, V., Botterman, J., and Cornelissen, M. (1989). *Methods Mol. Cell Biol.* 1, 19–27.

Denecke, J., Botterman, J., and Deblaere, R. (1990). *Plant Cell* 2, 51–59.

Dewey, R. E., Siedow, J. N., Timothy, D. H., and Levings, C. S. (1988). *Science* 239, 293–295.

Dotson, S. B., Frisch, D. A., Somers, D. A., and Gengenbach, B. G. (1990). *Planta* 182, 546–552.

Eckes, P., Schmitt, P., Daub, W., and Wengenmayer, F. (1989). *Mol. Gen. Genet.* 217, 263–268.

Edwards, J. W., Walker, E. L., and Coruzzi, G. M. (1990). *Proc. Natl. Acad. Sci. U.S.A.* 87, 3459–3463.

Eichholtz, D. A., Rogers, S. G., Horsch, R. B., *et al.* (1987). *Somatic Cell Genet.* 13, 67–76.

Etzold, T., Fritz, C. C., Schell, J., and Schreier, P. H. (1987). *FEBS Lett.* 219, 343–346.

Evans, G. A. (1989). *Genes Dev.* 3, 259–263.

Fiering, S., Northrop, J. P., Nolan, G. P., *et al.* (1990). *Genes Dev.* 4, 1823–1834.

Finnegan, E. J., Taylor, B. H., Craig, S., and Dennis, E. S. (1989). *Plant Cell* 1, 757–764.

Firoozabady, E., De Boer, D. L., Merlo, D. J., *et al.* (1987). *Plant Mol. Biol.* 10, 105–116.

Fogel, S. (1946). *The Allelic Variability and Action of the R Gene in Maize.* Ph.D. Dissertation, University of Missouri, Columbia.

Forde, B. G., Day, H. M., Turton, J. F., *et al.* (1989). *Plant Cell* 1, 391–401.

Fraley, R. T., Rogers, S. G., Horsch, R. B., *et al.* (1983). *Proc. Natl. Acad. Sci. U.S.A.* 80, 4803–4807.

Frey, M., Reinecke, J., Grant, S., Saedler, H., and Gierl, A. (1990). *EMBO J.* 9, 4037–4044.

Fromm, H., Edelman, M., Aviv, D., and Galun, E. (1987). *EMBO J.* 6, 3233–3237.

Fromm, M., Callis, J., Taylor, L. P., and Walbot, V. (1987). *Meth. Enzymol.* 153, 351–366.

Fromm, M. E., Morrish, F., Armstrong, C., *et al.* (1990). *Bio/Technology* 8, 833–839.

Gahan, P. B. (1984). *Experimental Botany: an International Series of Monographs Vol. 18a Plant Histochemistry and Cytochemistry: an Introduction* (Ed. by J. F. Sutcliffe and J. Cronshaw). London: Academic Press, Inc.

Gallie, D. R., Sleat, D. E., Watts, J. W., Turner, P. C., and Wilson, T. M. A. (1987). *Nucleic Acids Res.* 15, 8693–8711.

Gallie, D. R., Lucas, W. J., and Walbot, V. (1989). *Plant Cell* 1, 301–311.

Gelvin, S. B., Schilperoort, R. A., and Verma, D. P. S. (1990). *Plant Molecular Biology Manual* (Ed. by S. B. Gelvin, R. A. Schilperoot and D. P. S. Verma), 2nd ed. Boston: Kluwer Academic.

Gendloff, E. H., Bowen, B., and Buchholz, W. G. (1990). *Plant Mol. Biol.* 14, 575–583.
Gill, G., and Ptashne, M. (1988). *Nature (London)* 334, 721–724.
Goff, S. A., Klein, T. M., Roth, B. A., *et al.* (1990). *EMBO J.* 9, 2517–2522.
Goldberg, R. B., Barker, S. J., and Perez-Grau, L. (1989). *Cell* 56, 149–160.
Gordon-Kamm, W. J., Spencer, T. M., Mangano, M. L., *et al.* (1990). *Plant Cell* 2, 603–618.
Gossler, A., Joyner, A. L., Rossant, J., and Skarnes, W. C. (1989). *Science* 244, 463–465.
Gould, J. H., and Smith, R. H. (1989). *Plant Mol. Biol. Rep.* 7, 209–216.
Gould, S. J., Keller, G.-A., Schneider, M., *et al.* (1990). *EMBO J.* 9, 85–90.
Grimsley, N., and Bisaro, D. (1987). In *Plant DNA Infectious Agents* (Ed. by T. Hohn and J. Schell), pp. 87–107. Vienna: Springer Verlag.
Grimsley, N., Hohn, T., Davies, J. W., and Hohn, B. (1987). *Nature (London)* 325, 177–179.
Guerche, P., Bellin, C., Le Moullec, J. M., and Caboche, M. (1987). *Biochimie* 69, 621–628.
Guerineau, F., Brooks, L., Meadows, J., *et al.* (1990). *Plant Mol. Biol.* 15, 127–136.
Harkins, K. R., Jefferson, R. A., Kavanagh, T. A., Bevan, M. W., and Galbraith, D. W. (1990). *Proc. Natl. Acad. Sci. U.S.A.* 87, 816–820.
Harlow, E., and Lane, D. (1988). *Antibodies.* New York: Cold Spring Harbor Laboratory.
Harris, M. J., and Outlaw, W. H. (1990). *Physiol. Plant* 78, 495–500.
Hartley, R. W. (1989). *Trends Biochem. Sci.* 14, 450–454.
Haughn, G. W., Smith, J., Mazur, B., and Somerville, C. (1988). *Mol. Gen. Genet.* 211, 266–271.
Hauptmann, R. M., Vasil, V., Ozias-Akins, P., *et al.* (1988). *Plant Physiol.* 86, 602–606.
Hayford, M. B., Medford, J. I., Hoffman, N. L., Rogers, S. G., and Klee, H. J. (1988). *Plant Physiol.* 86, 1216–1222.
Heim, D. R., Roberts, J. L., Pike, P. D., and Larrinua, I. M. (1989). *Plant Physiol.* 90, 146–150.
Heim, D. R., Roberts, J. L., Pike, P. D., and Larrinua, I. M. (1990). *Plant Physiol.* 92, 858–861.
Helmer, G., Casadaban, M., Bevan, M., Kayes, L., and Chilton, M. D. (1984). *Bio/Technology* 2, 520–527.
Henderson, L., Rao, A. G., and Howard, J. (1991). *Anal. Biochem.* 194, 64–68.
Herman, P. L., and Marks, M. D. (1989). *Plant Cell* 1, 1051–1055.
Herrera-Estrella, L., De Block, M., Messens, E., *et al.* (1983a). *EMBO J.* 2, 987–995.
Herrera-Estrella, L., Depicker, A., Van Montagu, M., and Schell, J. (1983b). *Nature (London)* 303, 209–213.
Heyman, R. A., Borrelli, E., Lesley, J., *et al.* (1989). *Proc. Natl. Acad. Sci. U.S.A.* 86, 2698–2702.
Hiatt, A. H., Cafferkey, R., and Bowdish, K. (1989). *Nature (London)* 342, 76–78.
Hillé, J., Verheggen, F., Roelvink, P., *et al.* (1986). *Plant Mol. Biol.* 7, 171–176.
Hilson, P., Dewulf, J., Delporte, F., *et al.* (1990). *Plant Mol. Biol.* 14, 669–686.
Hirschberg, J., Yehuda, A. B., Pecker, I., and Ohad, N. (1987). In *NATO Advanced Science Institutes Series: Series A; Life Sciences,* Vol. 140, pp. 357–366. New York: Plenum.
Hohn, B., Koukolikova-Nicola, Z., Bakkeren, G., and Grimsley, N. (1989). *Genome* 31, 987–993.
Hollon, T., and Yoshimura, F. K. (1989). *Anal. Biochem.* 182, 411–418.
Hopwood, D. A., Malpartida, F., Kieser, H. M., *et al.* (1985). *Nature (London)* 314, 642–644.
Howell, S. H., Ow, D. W., and Schneider, M. (1989). In *Plant Molecular Biology Manual* (Ed. by S. B. Gelvin and R. A. Schilperoort), pp. 1–11. Boston: Kluwer Academic.
Hu, C.-Y., Chee, P. P., Chesney, R. H., *et al.* (1990). *J. Cell Biochem.* Suppl. 14E, 282.
Huang, J., Lee, S.-H., Lin, C., *et al.* (1990). *EMBO J.* 9, 339–347.
Ingelbrecht, I. L. W., Herman, L. M. F., Deykeser, R. A., Van Montagu, M. C., and Depicker, A. G. (1989). *Plant Cell* 1, 671–680.
Inzé, D., Follin, A., Velten, J., *et al.* (1987). *Planta* 172, 555–562.
Iturriaga, G., Jefferson, R. A., and Bevan, M. W. (1989). *Plant Cell* 1, 381–390.
Jacobsen, E., Krijgsheld, H. T., Hovenkamp-Hermelink, J. H. M., *et al.* (1990). *Plant Sci.* 67, 177–182.

Janssen, B. J., and Gardner, R. C. (1989). *Plant Mol. Biol.* 14, 61–72.
Jefferson, R. A. (1987). *Plant Mol. Biol. Rep.* 5, 387–405.
Jefferson, R. A. (1989). *Nature (London)* 342, 837–838.
Johannsson, A., Ellis, D. H., Bates, D. L., Plumb, A. M., and Stanley, C. J. (1986). *J. Immunol. Meth.* 87, 7–11.
Jones, J. D. G., Svab, Z., Harper, E. C., Hurwitz, C. D., and Maliga, P. (1987). *Mol. Gen. Genet.* 210, 86–89.
Jones, J. D. G., Carland, F. M., Maliga, P., and Dooner, H. K. (1989). *Science* 244, 204–207.
Kamakura, T., Yoneyama, K., and Yamaguchi, I. (1990). *J. Cell Biochem.* Suppl. 14E, 335.
Karlin-Neumann, G., and Tobin, E. (1989). *J. Cell Biochem.* Suppl. 13D, 320.
Kavanagh, T. A., Jefferson, R. A., and Bevan, M. W. (1988). *Mol. Gen. Genet.* 215, 38–45.
Kay, R., Chan, A., Daly, M., and McPherson, J. (1987). *Science* 236, 1299–1302.
Klee, H. J., Horsch, R. B., Hinchee, M., Hein, M. B., and Hoffman, N. L. (1987a). *Genes Dev.* 1, 86–96.
Klee, H. J., Muskopf, Y. M., and Gasser, C. S. (1987b). *Mol. Gen. Genet.* 210, 437–442.
Klein, T. M., Roth, B. A., and Fromm, M. E. (1989a). In *Genetic Engineering, Vol. 11* (Ed. by J. K. Setlow), pp. 13–31. New York: Plenum.
Klein, T. M., Roth, B. A., and Fromm, M. E. (1989b). *Proc. Natl. Acad. Sci. U.S.A.* 86, 6681–6685.
Knight, T. J., Bush, D. R., and Langston-Unkefer, P. J. (1988). *Plant Physiol.* 88, 333–339.
Koes, R. E., van Blokland, R., Quatrocchio, F., van Tunen, A. J., and Mol, J. N. M. (1990). *Plant Cell* 2, 379–392.
Koltunow, A. M., Truettner, J., Cox, K. H., Wallroth, M., and Goldberg, R. B. (1990). *Plant Cell* 2, 1201–1224.
Komeda, Y., Tanako, M., and Nishimune, T. (1988). *Plant Physiol.* 88, 248–250.
Koncz, C., Olsson, O., Langridge, W. H. R., Schell, J., and Szalay, A. A. (1987). *Proc. Natl. Acad. Sci. U.S.A.* 84, 131–135.
Konstantinov, K. L., Mladenovic, S. D., Denic, M. P., and Plecas, M. (1990). In *VIth NATO Advanced Study Institute: Plant Molecular Biology,* p. 99.
Kricka, L. J. (1988). *Anal. Biochem.* 175, 14–21.
Kwoh, T. J., Obermiller, P. S., McCue, A. W., *et al.* (1988). *Nucleic Acids Res.* 16, 11489–11506.
Lagrimini, L. M., Bradford, S., and Rothstein, S. (1990). *Plant Cell* 2, 7–18.
Landel, C. P., Zhao, J., Bok, D., and Evans, G. A. (1988). *Genes Dev.* 2, 1168–1178.
Last, R. L., and Fink, G. R. (1988). *Science* 240, 305–310.
Last, R. L., Bissinger, P. M., and Fink, G. R. (1990). In *VIth NATO Advanced Study Institute: Plant Molecular Biology,* p. 109.
Lee, B., Murdoch, K., Kreis, M., and Jones, M. G. K. (1989). *Plant Mol. Biol. Rep.* 7, 129–134.
Lee, K. Y., Townsend, J., Tepperman, J., *et al.* (1988). *EMBO J.* 7, 1241–1248.
Lee, K. Y., Lund, P., Lowe, K., and Dunsmuir, P. (1990). *Plant Cell* 2, 415–425.
Lefebvre, D. D., Miki, B. L., and Laliberte, J. F. (1987). *Bio/Technology* 5, 1053–1056.
Leifert, C., Waites, W. M., and Nicholas, J. R. (1989). *J. Applied Bacteriol.* 67, 353–361.
Lindgren, P. B., Frederick, R., Govindarajan, A. G., *et al.* (1989). *EMBO J.* 8, 1291–1301.
Ludwig, S. R., and Wessler, S. R. (1990). *Cell* 62, 849–851.
Ludwig, S. R., Bowen, B., Beach, L., and Wessler, S. R. (1990). *Science* 247, 449–450.
Lyon, B. R., Llewellyn, D. J., Huppatz, J. L., Dennis, E. S., and Peacock, W. J. (1989). *Plant Mol. Biol.* 13, 533–540.
Maiti, I. B., Wagner, G. J., Yeargan, R., and Hunt, A. G. (1989). *Plant Physiol.* 91, 1020–1024.
Maliga, P., Svab, Z., Harper, E. C., and Jones, J. D. G. (1988). *Mol. Gen. Genet.* 214, 456–459.
Mariani, C., De Beuckeleer, M., Truettner, J., Leemans, J., and Goldberg, R. (1990). *Nature (London)* 347, 737–741.
Masson, P., and Federoff, N. V. (1989). *Proc. Natl. Acad. Sci. U.S.A.* 86, 2219–2223.

Matsumoto, S., Takebe, I., and Machida, Y. (1988). *Gene* 66, 19–29.
Mazur, B. J., and Falco, S. C. (1989). *Ann Rev. Plant Physiol. Plant Mol. Biol.* 40, 441–470.
McCabe, D. E., Swain, W. F., Martinell, B. J., and Christou, P. (1988). *Bio/Technology* 6, 923–926.
McClure, B. A., Gray, J. E., Anderson, M. A., and Clarke, A. E. (1990). *Nature (London)* 347, 757–760.
McCourt, P., and Somerville, C. R. (1987). In *The Biochemistry of Plants* Vol. 15 (Ed. by D. Davies), pp. 32–64. New York: Academic Press, Inc.
McDonnell, R. E., Clark, R. D., Smith, W. A., and Hinchee, M. A. (1987). *Plant Mol. Biol. Rep.* 5, 380–386.
Menssen, A., Hohmann, S., Martin, W., *et al.* (1990). *EMBO J.* 9, 3051–3057.
Mermod, N., Harayama, S., and Timmis, K. N. (1986). *Bio/Technology* 4, 321–324.
Meyer, P., Heidman, I., Forkman, G., and Saedler, H. (1987). *Nature (London)* 330, 677–678.
Miller, J. H. (1972). *Experiments in Molecular Genetics* (Ed. by J. H. Miller). New York: Cold Spring Harbor Laboratory.
Misra, S. (1989). *J. Cell Biochem.* Suppl. 13D, 339.
Moffatt, B., and Somerville, C. (1988). *Plant Physiol.* 86, 1150–1154.
Mol, J. N. M., Stuitje, A. R., Gerats, A. G. M., and Koes, R. E. (1988). *Plant Mol. Biol. Rep.* 6, 274–279.
Mol, J., Stuitje, A., Gerats, A., van der Krol, A., and Jorgensen, R. (1989). *Trends Biotechnol.* 7, 148–153.
Moll, B., Polsby, L., and Maliga, P. (1990). *Mol. Gen. Genet.* 221, 245–250.
Napoli, C., Lemieux, D., and Jorgensen, R. (1990). *Plant Cell* 2, 279–289.
Nash, J., Luehrsen, K. R., and Walbot, V. (1990). *Plant Cell* 2, 1039–1049.
Nutter, R., Everett, N., Pierce, D., *et al.* (1987). *Plant Physiol.* 84, 1185–1192.
Odell, J. T., Knowlton, S., Lin, W., and Mauvais, C. J. (1988). *Plant Mol. Biol.* 10, 263–272.
O'Kane, C. J., and Gehring, W. J. (1987). *Proc. Natl. Acad. Sci. U.S.A.* 78, 7634–7638.
Omer, C. A., Lenstra, R., Litle, P. J., *et al.* (1990). *J. Bacteriol.* 172, 3335–3345.
Otten, L. A. B. M., and Schilperoort, R. A. (1978). *Biochim. Biophys. Acta* 527, 497–500.
Ow, D. W., Wood, K. W., De Luca, M., *et al.* (1986). *Science* 234, 856–859.
Palmiter, R. D., Behringer, R. R., Quaife, C. J., *et al.* (1987). *Cell* 50, 435–443.
Parker, W. B., Marshall, L. D., Burton, J. D., *et al.* (1990). *Proc. Natl. Acad. Sci. U.S.A.* 87, 7175–7179.
Paszkowski, J., Baur, M., Bogucki, A., and Potrykus, I. (1988). *EMBO J.* 7, 4021–4026.
Perez, P., Tiraby, G., Kallerhoff, J., and Perret, J. (1989). *Plant Mol. Biol.* 13, 365–373.
Pietrzak, M., Shillito, R. D., Hohn, T., and Potrykus, I. (1986). *Nucleic Acids Res.* 14, 5857–5868.
Platt, S. G., and Yang, N.-S. (1987). *Anal. Biochem.* 162, 529–535.
Plegt, L., and Bino, R. J. (1989). *Mol. Gen. Genet.* 216, 321–327.
Potrykus, I., Saul, M. W., Petruska, J., Paszkowski, J., and Shillito, R. D. (1985). *Mol. Gen. Genet.* 199, 178–182.
Pröls, M., Töpfer, R., Schell, J., and Steinbiss, H. (1988). *Plant Cell Rep.* 7, 221–224.
Raikhel, N. V., Bednarek, S. Y., and Lerner, D. R. (1989). In *Plant Molecular Biology Manual* (Ed. by S. B. Gelvin and R. A. Schilperoort), pp. 1–32. Boston: Kluwer Academic.
Ranch, J. P., Rick, S., Brotherton, J. E., and Widholm, J. M. (1983). *Plant Physiol.* 71, 136–140.
Rao, A. G., and Flynn, P. (1990). *Biotechniques* 8, 38–40.
Rao, A. G., and Flynn, P. (1991). In *GUS Protocols: Using the GUS Gene as a Reporter of Gene Expression* (Ed. by S. Gallagher), pp. 89–99. New York: Academic Press, Inc.
Regan, S. M., and Moffatt, B. (1990). *Plant Cell* 2, 877–889.
Reiss, B., Sprengel, R., Will, H., and Schaller, M. (1984). *Gene* 30, 211–218.
Restrepo, M. A., Freed, D. D., and Carrington, J. C. (1990). *Plant Cell* 2, 987–998.
Rogers, S. G., Fraley, R. T., Horsch, R. B., *et al.* (1985). *Plant Mol. Biol. Rep.* 3, 111–116.

Rogers, S. G., Klee, H. J., Horsch, R. B., and Fraley, R. (1987). *Meth. Enzymol.* 153, 253–277.
Rosahl, S., Schell, J., and Willmitzer, L. (1987). *EMBO J.* 6, 1155–1159.
Sathasivan, K., Haughn, G. W., and Murai, N. (1990). *Nucleic Acids Res.* 19, 2188.
Saxena, P. K., and King, J. (1988). *Plant Physiol.* 86, 863–867.
Saxena, P. K., and King, J. (1990). *Plant Physiol.* 94, 1111–1115.
Schauer, A. T. (1988). *Trends Biotechnol.* 6, 23–27.
Schneider, M., Ow, D., and Howell, S. H. (1990). *Plant Mol. Biol.* 14, 935–937.
Schöler, H. R., and Gruss, P. (1984). *Cell* 36, 403–411.
Schreier, P. H., and Schell, J. (1986). *Philos. Trans. R. Soc. London,* Ser. B 313, 429–432.
Schulz, M., and Weissenböck, G. (1987). *Phytochem.* 26, 933–937.
Schwartz, D., and Osterman, J. (1976). *Genetics* 83, 63–65.
Seed, B., and Sheen, J.-Y. (1988). *Gene* 67, 271–277.
Sexton, R., and Hall, J. L. (1978). In *Electron Microscopy and Cytochemistry of Plant Cells* (Ed. by J. L. Hall), pp. 63–147. Amsterdam: Elsevier.
Shah, D. M., Hironaka, C. M., Wiegand, R. C., *et al.* (1986a). *Plant Mol. Biol.* 6, 203–211.
Shah, D. M., Horsch, R. B., Klee, H., *et al.* (1986b). *Science* 233, 478–481.
Shaw, M. L., Conner, A. J., Lancaster, J. E., and Williams, M. K. (1988). *Plant Mol. Biol. Rep.* 6, 155–164.
Shaw, W. V. (1975). *Meth. Enzymol.* 53, 737–755.
Sheen, J. (1990). *Plant Cell* 2, 1027–1038.
Silverthorne, J., and Tobin, E. M. (1990). *Plant Cell* 2, 1181–1190.
Simos, G., Giannakouros, T., and Georgatsos, J. G. (1989). *Phytochem.* 28, 2587–2592.
Sittertz-Bhatkar, H. (1989). *Biotechniques* 7, 922–924.
Smigocki, A. C., and Owens, L. D. (1988). *Proc. Natl. Acad. Sci. U.S.A.* 85, 5131–5135.
Spena, A., Schmülling, T., Koncz, C., and Schell, J. S. (1987). *EMBO J.* 6, 3891–3899.
Spena, A., Aalen, R. B., and Schulze, S. C. (1989). *Plant Cell* 1, 1157–1164.
Staebell, M., Tomes, D., Weissinger, A., *et al.* (1990). *Anal. Biochem.* 185, 319–323.
Stalker, D. M., McBride, K. E., and Malyj, L. D. (1988). *Science* 242, 419–423.
Stanley, C. J., Johannsson, A., and Self, C. H. (1985). *J. Immunol. Meth.* 83, 89–95.
Stirpe, F., and Barbieri, L. (1986). *FEBS Lett.* 195, 1–8.
Stougaard, P., Bojsen, K., and Christensen, T. (1990). *J. Cell Biochem.* Suppl. 14E, 310.
Streber, W. R., and Willmitzer, L. (1989). *Bio/Technology* 7, 811–816.
Subramanian, M. V., Hung, H. Y., Dias, J. M., *et al.* (1990). *Plant Physiol.* 94, 239–244.
Svab, Z., Harper, E. C., Jones, J. D. G., and Maliga, P. (1990a). *Plant Mol. Biol.* 14, 197–205.
Svab, Z., Hajdukiewvicz, P., and Maliga, P. (1990b). *Proc. Natl. Acad. Sci. U.S.A.* 87, 8526–8530.
Tanksley, S. D., and Orton, T. J. (1983). *Isozymes in Plant Genetics and Breeding.* Amsterdam: Elsevier.
Teeri, T. H., Lehväslaiho, H., Franck, M., *et al.* (1989). *EMBO J.* 8, 343–350.
Thompson, C. J., Rao Movva, N., Tizard, R., *et al.* (1987). *EMBO J.* 6, 2519–2523.
Timmerman, K. P., Stark, D. M., Barry, G. F., Preiss, J., and Kishore, G. M. (1991). Abstract 1846, *3rd International Congress of Plant Molecular Biology*, Tucson, Arizona.
Töpfer, R., Pröls, M., Schell, J., and Steinbiss, H.-H. (1988). *Plant Cell Rep.* 7, 225–228.
Töpfer, R., Gronenborn, B., Schell, J., and Steinbiss, H. H. (1989). *Plant Cell* 1, 133–139.
Tropschug, M., Barthelmess, I. B., and Neupert, W. (1989). *Nature (London)* 342, 953–955.
Vancanneyt, G., Schmidt, R., O'Connor-Sanchez, A., Willmitzer, L., and Rocha-Sosa, M. (1990). *Mol. Gen. Genet.* 220, 245–250.
van den Elzen, P. J. M., Townsend, J., Lee, K. Y., and Bedbrook, J. R. (1985). *Plant Mol. Biol.* 5, 299–302.
van der Krol, A. R., Lenting, P. J., Veenstra, J. G., *et al.* (1988). *Nature (London)* 333, 866–869.

van der Krol, A. R., Mur, L. A., Beld, M., and Stuitje, A. R. (1990). *Plant Cell* 2, 291–299.
Vaucheret, M., Chabaud, M., Kronenberger, J., and Caboche, M. (1990). *Mol. Gen. Genet.* 220, 468–474.
Vaughn, K. C. (1987). *CRC Handbook of Plant Cytochemistry* (Ed. by K. C. Vaughn). Boca Raton, Florida: CRC Press.
Visser, G. F., Feenstra, W. J., and Jacobsen, E. (1990). *J. Cell Biochem.* Suppl. 14E, 271.
von Schwaewen, A., Stitt, M., Schmidt, R., Sonnerold, V., and Willmitzer, L. (1990). *EMBO J.* 9, 3033–3044.
Walden, R. (1989). *Genetic Transformation in Plants.* Englewood Cliffs, New Jersey: Prentice Hall.
Waldron, C., Murphy, E. B., Roberts, J. L., *et al.* (1985). *Plant Mol. Biol.* 5, 103–108.
Weide, R., Koornneef, M., and Zabel, P. (1989). *Theor. Appl. Genet.* 78, 169–172.
Weintraub, H. (1988). *Proc. Natl. Acad. Sci. U.S.A.* 85, 5819–5823.
Weising, K., Schell, J., and Kahl, G. (1988). *Ann. Rev. Genet.* 22, 421–477.
White, J., Chang, S. Y. P., Bibb, M. J., and Bibb, M. J. (1990). *Nucleic Acids Res.* 18, 1062.
Widholm, J. M. (1972a). *Biochim. Biophys. Acta* 261, 52–58.
Widholm, J. M. (1972b). *Biochim. Biophys. Acta* 279, 48–57.
Wiersma, P. A., Schmiemann, M. G., Condie, J. A., Crosby, W. L., and Moloney, M. M. (1989). *Mol. Gen. Genet.* 219, 413–420.
Witty, M. (1989). *Nucleic Acids Res.* 17, 3312.
Wohlleben, W., Arnold, W., Broer, I., *et al.* (1988). *Gene* 70, 25–37.
Wood, K. V., Lam, Y. A., Seliger, H. H., and McElroy, W. D. (1989). *Science* 244, 700–702.
Wright, A. D., and Neuffer, M. G. (1989). *J. Hered.* 80, 229–233.
Yamada, T., Palm, C. J., Brooks, B., and Kosuge, T. (1985). *Proc. Natl. Acad. Sci. U.S.A.* 82, 6522–6526.
Yang, N. S., Platt, S. G., and Christou, P. (1987). *Anal. Biochem.* 160, 342–345.
Yanofsky, M., Ma, H., Bowman, J. L., *et al.* (1990). *Nature (London)* 346, 35–39.
Yenofsky, R. L., Fine, M., and Pellow, J. W. (1990). *Proc. Natl. Acad. Sci. U.S.A.* 87, 3435–3439.
Zelitch, I., and Berlyn, M. B. (1982). *Plant Physiol.* 69, 198–204.
Zukowski, M. M., Gaffney, D. F., Speck, D., *et al.* (1983). *Proc. Natl. Acad. Sci. U.S.A.* 80, 1101–1105.

4

Techniques for Gene Transfer

Barnabás Jenes[1], Helen Moore,
Jun Cao[2], Wanggen Zhang,
and Ray Wu

Section of Biochemistry
Molecular and Cell Biology
Biotechnology Building
Cornell University
Ithaca NY, 14853

4.1 Gene Transfer and Genetic Transformation

The revolutionary progress in recombinant DNA technology makes it necessary to continue to improve the techniques needed for the genetic engineering of plants. Improvement of these techniques requires better tissue culture methodologies and gene transfer techniques. Before describing gene transfer and genetic transformation, we will adopt the following definitions to eliminate the ambiguity often found in the literature.

4.1.1 GENE TRANSFER

Gene transfer (or DNA uptake) refers to a process which moves a specific piece of DNA (usually a foreign gene ligated to a bacterial plasmid) into protoplasts or

[1]Present address: Institute for Plant Sciences, Agricultural Biotechnology Center, Gödöllö, Szent-Györgyi A. út 4. P. O. Box 170, Hungary 2101
[2]Plant Science Center, Biotechnology Building, Cornell University, Ithaca, NY, 14853, USA

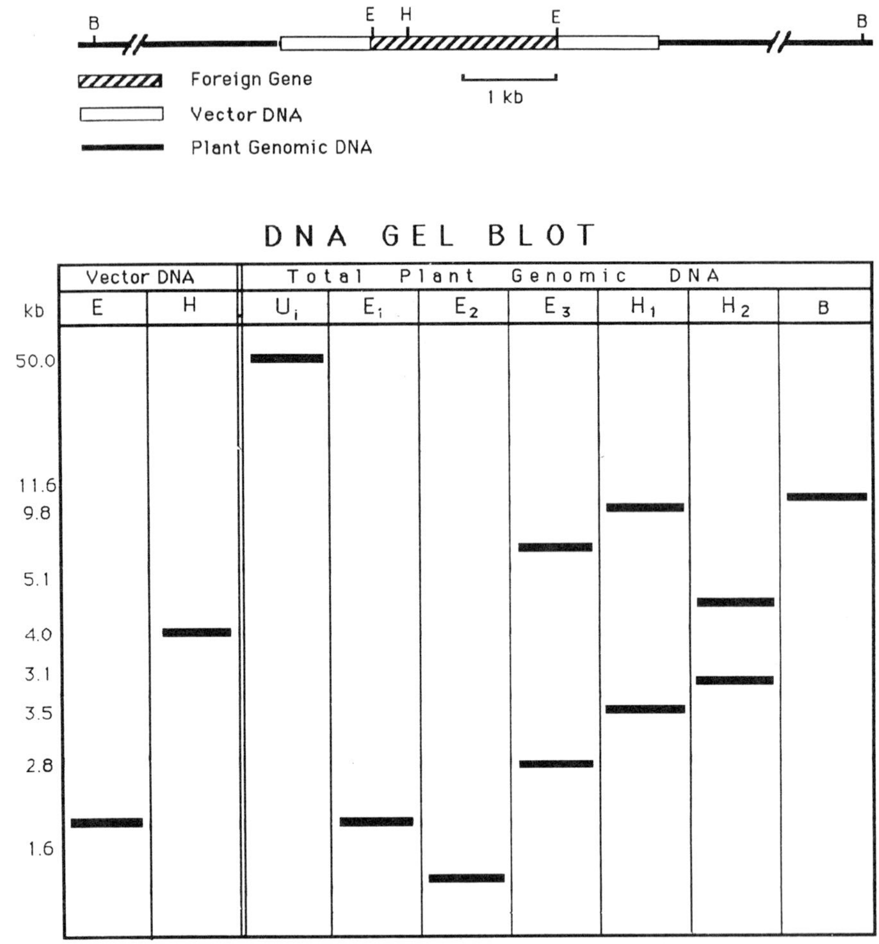

Figure 4.1 Proof of integration of a foreign gene in the plant genome. **B, E,** and **H** are restriction enzymes; the subscripts **1, 2,** and **3** represent total genomic DNA samples from different transgenic plants that are uncut or digested with a restriction enzyme. The Southern blot is probed with the 2-kb foreign DNA fragment. **U_1,** Uncut total genomic DNA. The hybridizing band is usually a broad band (around 50 kb) which corresponds to the average size of the genomic DNA prepared by common methods. **E_1,** Genomic DNA from plant 1 cut with restriction enzyme E. If the hybridizing band is the same size as the probe, as in lanes E and E_1, it means that there is no rearrangement of the foreign gene in the plant genome. It also means that integration occurred outside of this region. If integration occured within this region, there would be two hybridizing bands of different sizes. **E_2,** Genomic DNA from plant 2 cut with E. A hybridizing band smaller in size than that in lane E suggests that there is a partial deletion of the foreign DNA in the plant genome. **E_3,** Genomic DNA from plant 3 cut with E. Two hybridizing bands with sizes different from that of the foreign gene indicate that integration occurred within the foreign gene. **H_1,** Genomic DNA from plant 1 restricted with H. Two hybridizing bands indicate

cells. If the gene transfer is efficient and the foreign gene is introduced sufficient number of cells, transient gene expression can be quantitatively measured (12–48 hours after gene transfer). Usually the level of the protein encoded by the foreign gene is measured because it is simpler than measuring the level of mRNA to show the success of gene transfer. For study of transient gene expression, stable integration of the introduced gene is not necessary, but a reproducible and efficient gene transfer method is needed. Examples are PEG-mediated gene transfer, electroporation, and the biolistic method (see details in Section 4.2 and 4.3).

4.1.2 GENETIC TRANSFORMATION

Genetic transformation (transformation) refers to stable integration of a foreign gene into the genome of a plant regenerated from DNA-treated protoplasts or intact cells. Integration can be demonstrated by DNA gel blot (Southern blot) analysis. The analysis should include the use of several restriction enzymes to provide proof for integration of the foreign gene. As shown in Figure 4.1, for example, it is useful to include (1) an enzyme that cuts twice within the plasmid to show the presence of the foreign gene and to show that there is no rearrangement of the foreign gene on the plasmid in the transgenic plants; (2) an enzyme that cuts once within the plasmid to yield restriction fragments which vary in size depending on the nature of the flanking DNA at the site of insertion of the plasmid in the transformed plants; and (3) an enzyme that does not cut within the plasmid to distinguish between the free plasmid (in monomeric or oligomeric form) and the integrated plasmid.

4.1.3 TRANSGENIC PLANTS

Transgenic plants refers to plants carrying the stably integrated foreign genes. R_0 plants are those regenerated from DNA-treated protoplasts or cells in gene transfer experiments. R_1 plants are derived from seeds produced by R_0 plants. Although the R_0 plants are useful for analysis of tissue-specific or developmental stage-specific gene expression, the production of R_1 plants is essential for crop improvement. Proof of R_1 transgenic plants also requires physical and phe-

---

that one copy of the plasmid DNA has integrated into the plant genome at one insertion site. **H_2,** Genomic DNA from plant 2 restricted with H. Two hybridizing bands of different size from those in H_1 reflect different sites of integration of one copy of the foreign gene into the genome in plants 1 and 2. **B,** Genomic DNA restricted with B. The presence of one hybridizing band (10 kb) that differs in size from that of multimers of the 4-kb plasmid DNA provides evidence for integration of the plasmid DNA into the plant genome.

notypic evidence for the inheritance of the foreign gene to the offspring, including data on genetic segregation of the foreign gene.

4.1.4 GOALS OF GENE TRANSFER

The ability to transfer DNA sequences into plant cells suggests a wide variety of potential practical uses. The most obvious potential application is in crop improvement. Often mentioned goals are changes in oil and protein contents of crops, herbicide resistance, insect resistance, and virus tolerance (Walden, 1989).

For basic research, genetic transformation is essential for functional analysis of gene regulatory elements (Kuhlemeier *et al.* 1987; McElroy *et al.*, 1990, 1991) and for studies of protein function (Cardarelli *et al.*, 1987; Boutry *et al.*, 1987), molecular biology of viruses (Deom *et al.*, 1987), and hormonally regulated genes (Ou-Lee *et al.*, 1988; Nolan and Ho, 1988; Huttly and Baulcombe, 1989), to cite but a few examples.

For detecting or quantitating the efficiency of the gene transfer it is convenient to use reporter genes such as the chloramphenicol acetyltransferase (CAT) gene or β-glucuronidase (GUS) gene. The presence of the enzymes encoded by these genes in the cells demonstrates that the foreign gene is taken up by the cells and is expressed.

For stable integration and the production of transgenic plants it is helpful to make use of dominant selectable marker genes such as the neomycin phosphotransferase gene (NPTII, or *neo*), dihydrofolate reductase gene (DHFR), phosphinothricin acetyltransferase gene (*bar*), and hygromycin phosphotransferase gene (*hph*). The use of selectable markers makes it possible to select directly for transgenic cells or colonies which can survive and grow under selective conditions and finally regenerate into fertile transgenic plants.

4.1.5 TECHNIQUES FOR GENE TRANSFER

In Section 4.2 of this chapter a general review of the methods for transformation of plant protoplasts, cells, and tissues is presented. In Section 4.3, the more useful of these methods are discussed in detail for dicots and for monocots.

4.2 Methods for Transformation of Plant Protoplasts, Cells, and Tissues

4.2.1 TRANSFORMATION OF PROTOPLASTS

Each single cell of a plant is potentially capable of developing into a whole plant. This phenomenon is known as cell totipotence. Significant advances have been made by scientists during the past decade in plant regeneration from protoplasts (isolated single plant cells with cell walls removed) in both dicots and monocots. This not only has confirmed the cell totipotency hypothesis, but also has made it possible to transfer foreign genes into plants using protoplasts as the starting material. There are two major reasons to use protoplasts instead of cells or tissues for genetic manipulation. One is that protoplasts are separated single cells which have been stripped of their cell wall, facilitating the transfer of foreign genes through the plasma membrane. The other is that all cells of a transgenic plant regenerated from a protoplast will contain the foreign genes of interest; thus the transgenic plant will have a uniform (nonchimeric) genetic makeup.

Protoplasts capable of gene integration and plant regeneration may be referred to as competent. Protoplasts may be induced to become competent by alteration of specific experimental conditions. Several methods have been used for stimulating protoplast competence, such as heat shock (Abdullah *et al.*, 1986), low-dose irradiation (Kohler *et al.*, 1989), and media ultrafiltration (Davies *et al.*, 1989). These treatments, in addition to enzymatic and mechanical treatments used in the protoplast isolation procedure, can increase gene transformation frequency and protoplast division. When competent protoplasts with high viability are available, several methods such as polyethylene glycol (PEG) treatment, electroporation, and *Agrobacterium* infection can be used for gene transfer.

4.2.1.1 Gene Transfer into Protoplasts Using Chemical Agents

Various chemical treatments have been used to stimulate DNA uptake by protoplasts. At present, PEG is the most common chemical treatment used to stimulate DNA uptake into protoplasts (Negrutiu *et al.*, 1987) of both dicots and monocots.

PEG acts to increase the permeability of cell membranes, and has been used as an efficient protoplast fusion agent in somatic cell hybridization of many plant and animal species. Various parameters important for gene transfer with PEG have been identified. First, PEG concentration can affect protoplast viability and gene transfer efficiency. The optimal concentration of PEG is 15–25%. If PEG concentration is too high, cell viability will decrease; if it is too low, the gene transfer efficiency will decrease (Hayashimoto *et al.*, 1990). Second, the components of the buffer will influence the gene transfer efficiency greatly. Our group, working with rice protoplasts, found that 15 m*M* $MgCl_2$ and 400 m*M* mannitol at pH 5.6 increased gene transfer and transformation efficiency compared with other buffers (Zhang and Wu, 1988). Finally, culture media for protoplasts will directly affect protoplast division and plant regeneration (Jenes and Pauk, 1989). Currently two major methods, solid culture and nurse culture, are commonly used for efficient protoplast culture after gene transformation (Cao *et al.*, 1991).

4.2.1.2 Gene Transfer into Protoplasts by Electroporation

The electroporation method is based on the use of short electrical pulses of high field strength to facilitate DNA uptake by increasing the permeability of protoplast membranes. It is known that the gene transfer efficiency and protoplast viability are influenced by the amplitude and duration of the electric pulse and by the composition of the electroporation medium. Fromm *et al.* (1985) first reported gene transfer into maize protoplasts and Langridge *et al.* (1985) first reported the stable transformation of carrot protoplasts with DNA by electroporation. Application of high voltage either directly (both electrodes immersed in the sample solution) or indirectly (without anode contact) to a solution containing plasmid DNA and protoplasts of rice, wheat, or sorghum resulted in different degrees of protoplast viability and gene transfer efficiency (Ou-Lee *et al.*, 1986). Protoplast size and competent state differ from one species to another, and therefore the optimal electroporation conditions may vary among species. For example, Shimamoto *et al.* (1989) used a long electrical pulse (10 msec) generated by a large capacitor and a moderate voltage (300 V/cm) for gene transformation of embryogenic protoplasts of rice (10–20 μm in diameter). Rhodes *et al.* (1988) applied a shorter pulse and higher voltage to maize protoplasts (25–40 μm in diameter). As in the case of the PEG method, the buffer also greatly influences the gene transfer efficiency as well as the protoplast survival rate. A recent report demonstrates that the addition of an organic acid, aspartic acid monopotassium salt, replacing the chloride ions in the buffer increases the transformation frequency at least 10-fold and increases plating efficiency (Tada *et al.*, 1990). After the electroporation procedure, protoplasts are cultured in much the same way as for the PEG method.

4.2.1.3 Gene Transfer into Protoplasts Using *Agrobacterium*

Agrobacterium tumefaciens is the bacterial agent of crown gall disease, manifested in nature as tumors of root and stem tissue of more than one hundred susceptible plant species (Agrios, 1978). The bacterium enters the plant at wound tissue sites, and infection results in a local proliferation of undifferentiated plant tissue. *Agrobacterium tumefaciens* has the unique ability to transfer some of its DNA into the nuclear DNA of an infected host plant by virtue of functions encoded on its "Ti" (tumor-inducing) plasmid. The transferred DNA, or T-DNA, of this plasmid encodes substances which act as plant hormones, causing the proliferation of tissue seen in the infected plant.

The Ti plasmid of *A. tumefaciens* has been altered through genetic engineering to create "disarmed" plasmids or vectors which can carry any DNA sequence into an *A. tumefaciens*-infected plant, without tumorous growth of the host plant. Ti plasmid-derived cloning vectors have been designed for replication in both *A.*

tumefaciens and *Escherichia coli* bacterial strains. This facilitates the cloning of foreign DNA sequences into the vectors using an *E. coli* host, and subsequent transfer of the completed vector to an *A. tumefaciens* host. *Agrobacterium tumefaciens* genetics is discussed more thoroughly in Chapter 10 of this volume, and in many other reviews (Schell, 1987; Gelvin and Schilperoort, 1988).

Protoplasts of many dicot and some monocot (nongraminaceous) species can be transformed using *A. tumefaciens* strains harboring foreign gene(s) of interest. In general, *A. tumefaciens*-mediated transformation of plant tissues (see Section 4.3.1.1 of this chapter) requires less effort and expertise in tissue culture than transformation of protoplasts. However, transformation of protoplasts using *A. tumefaciens* is an appropriate method in species where *A. tumefaciens*-mediated transformation of tissue explants is difficult or has not been achieved, or in experiments where expression in regenerated transgenic plants is not the primary goal (Thomzik and Hain, 1990).

The foreign DNA to be expressed in the transformed protoplasts must be cloned into an appropriate vector (some version of Ti plasmid) and the vector transferred to an *A. tumefaciens* strain capable of infecting protoplasts of the host plant of interest. Protoplasts must be prepared using optimal conditions for competency for regeneration, cocultivated with the bacteria, and plated on appropriate media for protoplast culture and regeneration, if desired. Growth of *A. tumefaciens* in the protoplast culture after cocultivation is arrested or retarded by the use of antibiotics in the culture media.

4.2.1.4 Gene Transfer into Protoplasts by Sonication

An alternative transformation method has been reported recently by Joersbo and Brunstedt (1990) and by Jia's group (Zhang *et al.*, 1991). Joersbo and Brunstedt introduced DNA into sugar beet (*Beta vulgaris* L.) and tobacco (*Nicotiana tabacum* L.) protoplasts by applying a brief exposure of 20 kHz ultrasound in the presence of a plasmid containing the CAT gene fused to the 35S promoter. A high level of CAT activity was detected in the transformed cells. The protoplasts surviving the exposure to ultrasound were found to have long-term viability and ability to regenerate to microcalli similar to the untreated protoplasts.

Zhang *et al.* (1991) succeeded in regeneration of transgenic tobacco plants from protoplasts transformed by sonication. They used dimethyl sulfoxide (DMSO) to raise the transient transformation frequency to 86% and the shoot differentiation frequency to 60%. They also found that salmon sperm DNA as carrier DNA was essential for obtaining stable transformants.

One possible advantage of the method is that the system may be simpler than electroporation. The efficiency of transformation is comparable to other, often used methods. It remains to be seen whether this method can be extended to a number of plant species.

4.2.1.5 Microinjection into Plant Protoplasts

Microinjection is one of the most precise techniques for delivering macromolecules into specific intracellular compartments of living cells (Reich *et al.*, 1986). Studies with animal cells have shown the utility of intracellular microinjection for efficient transformation (Capecchi, 1980; Brinster *et al.*, 1985). This has been confirmed in experiments using plant protoplasts (Crossway *et al.*, 1986; Reich *et al.*, 1986).

Basically, microcapillaries and microscopic devices (micromanipulators) are used to deliver DNA into the protoplasts; if competent, the injected cells can survive and divide. No other transformation method could compete with microinjection if competent cells could be visually identified in plant tissue (Potrykus, 1990). However, it is not yet possible to identify competent cells visually. The disadvantage of this method is that it is very time consuming and is not amenable when large numbers of transformants are desired. Moreover, it requires expensive instruments and considerable skill to carry out (Wu, 1989).

4.2.2 TRANSFORMATION OF INTACT CELLS OR TISSUES

4.2.2.1 Gene Transfer to Plant Tissue Using *Agrobacterium*

The mechanism of *Agrobacterium tumefaciens* transformation is briefly described in Section 4.2.1 above. A more complete description of the system may be found in Chapter 10 of this volume.

The availability of a wide range of Ti plasmid-derived cloning vectors and *A. tumefaciens* strains for transformation has made it possible, and increasingly popular, to transform susceptible plant species using *A. tumefaciens* which harbors a specific DNA sequence of interest. The presence of a selectable marker gene in the T-DNA region of the vector makes it possible to directly select transformed plant cells after cocultivation with *A. tumefaciens*. Resistant cells and tissues can then be regenerated to plantlets and mature plants. Transformation using disarmed (nontumorigenic) *A. tumefaciens* plasmid vectors can result in transgenic plants of normal phenotype which express the introduced gene(s).

The methods for transformation of intact cells or tissues with *A. tumefaciens* have been developed most extensively using excised tissue of *Nicotiana* and *Petunia* species (Horsch *et al.*, 1985; Rogers *et al.*, 1986). Transformation experiments with these species established rapid and reproducible procedures which have been applied to transformation of other species. *A. tumefaciens*-mediated transformation has not proven useful for the agriculturally important graminaceous monocots. This is apparently due to the lack of evocation by *A. tumefaciens* of a wound response in these species (Potrykus, 1990). Although *A.*

tumefaciens-mediated transformation of graminaceous monocot species has been reported (Raineri *et al.*, 1990), the method has not been widely reproducible. More information regarding species which can be transformed using *A. tumefaciens* strains may be found in this volume and elsewhere (Uchimiya *et al.*, 1989).

4.2.2.2 Biolistic Gene Transfer

Biolistic gene transfer is a relatively new approach to plant transformation. The term "biolistic" (biological ballistics) was coined to describe the nature of the delivery of foreign DNA into living cells or tissues through "bombardment" with a biolistic device (a particle gun). The biolistic process has been described in detail (Klein *et al.*, 1987; Wang *et al.*, 1988; Klein *et al.*, 1989; Sanford, 1990). The process involves high-velocity acceleration of microprojectiles carrying foreign DNA, penetration of the cell wall and membrane by microprojectiles, and delivery of the DNA inside plant cells. Figure 4.2 shows a diagram of a biolistic device.

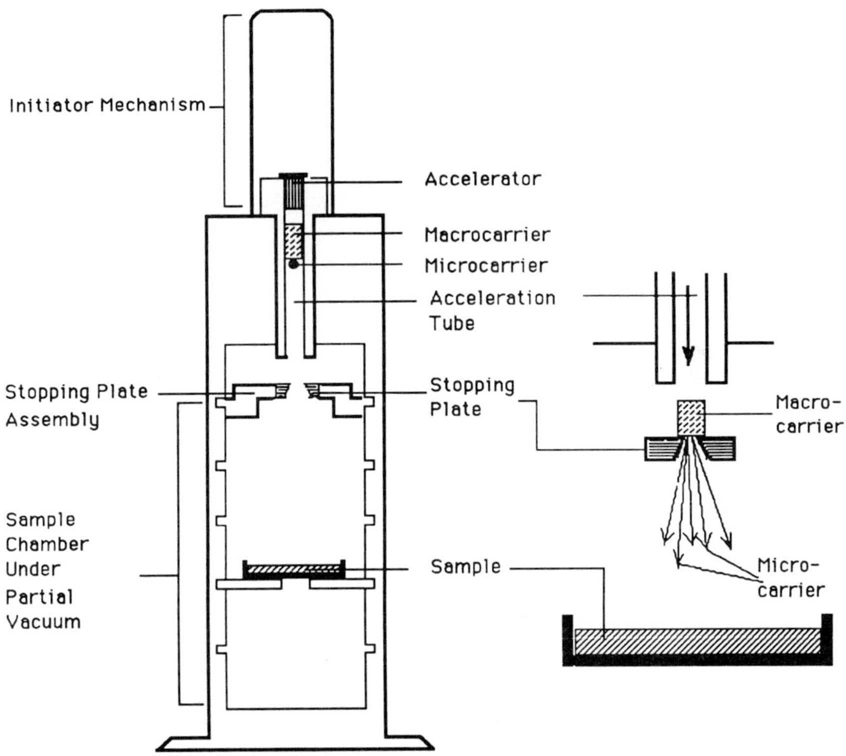

Figure 4.2 Principle of operation for the biolistic device (adapted from the **BIOLISTIC** Particle Delivery System of Du Pont Company).

The biolistic process has demonstrated broad utility and appears to be effective among all plant species tested thus far. In plants, the process has created possibilities to transform many important crop species which have been difficult to transform by other methods. In the past few years it has been reported that foreign genes have been delivered and expressed in both dicots and monocots, including economically important crop species such as soybean, wheat, corn, and rice (Wang *et al.*, 1988; Wu, 1989; Cao *et al.*, 1990). Virtually any type of cells or tissues can be used as targets for gene transfer. Some target tissues such as embryogenic suspension culture (Fromm *et al.*, 1990) and meristematic tissue (McCabe *et al.*, 1988) have proven to be transformable and able to give rise to transgenic plants.

DNA has been reproducibly delivered into mitochondria in yeast (Johnston *et al.*, 1988) and chloroplasts in *Chlamydomonas* (Boynton *et al.*, 1988) through the biolistic process. In recent years the biolistic process has been the only effective means for transformation of plant organelles (Svab *et al.*, 1990; Daniell *et al.*, 1991).

Use of the biolistic process for transient assay of gene constructs in specialized plant cells or tissues is a valuable application. Any developmentally regulated or tissue specific promoter can be fused to a reporter gene and delivered into specialized cells or tissues. On visualization or quantitation of expression of the introduced genes, one can study developmentally regulated gene functions in specific plant cells or tissues (Cao *et al.*, 1991; Lee *et al.*, 1991; Goldfarb *et al.*, 1991; Seki *et al.*, 1991).

Efficiency of the biolistic gene transfer depends on both biological and physical factors. Using the GUS gene as a reporter, 400 to 2000 blue spots (products of the GUS enzyme assay) per filter paper (5 cm diameter) carrying plant cell suspension were reported (Wang *et al.*, 1988; Sanford, 1990). The frequency of gene transfer for transient assay was estimated to be one out of 10^3 or 10^4. About 2–5% of transient transformants became stably transformed. The frequency of gene transfer has been improved 5- to 10-fold with recent improvement of the particle gun and method of sample preparation (Ye *et al.*, 1990). So far, the delivery efficiency into intact cells is still low in comparison to that obtained through other well established transformation methods using protoplasts. However, it can be expected that with improvement of the biolistic process, transformation frequency will continue to increase.

When the biolistic process is applied to plant tissues, plants regenerated from such tissues usually are chimeric in terms of introduced foreign genes due to random bombardment of a small number of cells in a multiple cell system. However, transformants can be sorted out and stabilized in their progenies, especially when selectable marker genes are used (Finer and McMullen, 1990; Fromm *et al.*, 1990; Gordon-Kamm *et al.*, 1990; McCabe *et al.*, 1988; Tomes *et al.*, 1990; Christou *et al.*, 1991; McCown *et al.*, 1991).

Invention of the biolistic process provides additional and alternative means for transforming higher plants. The biolistic process is still at its early stage of development. Biological and physical factors remain to be studied, understood, and improved. However, its versatility and practical simplicity make it a valuable transformation technique for higher plants.

4.2.2.3 Other Methods

Microinjection into Plant Tissue. Beginning with the assumption that few celled zygotic proembryos contain competent cells, Potrykus's group established plant regeneration from isolated zygotic proembryos of soybean (*Glycine soya* L.), cotton (*Gossipium* hybrid), sunflower (*Helianthus annuus* L.), tobacco (*Nicotiana tabacum* L.) and *Arabidopsis* (Potrykus, 1990). Following multiple microinjections of cells of zygotic embryos with marker genes they analyzed the putative primary transgenic chimeras and sexual offspring, finding indicative evidence for putative transgenic chimeras. The method is more suited to the goal of achieving stable transformants than for application in transient expression experiments.

Macroinjection. Zhou *et al.* (1983) injected total DNA from one cotton variety (donor) into the embryos of another cotton variety (recipient) and obtained a large number of plants. Several of the recipient plants (R_1 through R_4 generations) showed morphological characteristics that resemble those of the donor plant. The same research group injected DNA isolated from a wilt disease (*Fusarium*)-resistant cotton variety into the embryos of a wilt-sensitive variety; wilt-resistant offspring were obtained (Zhou *et al.*, 1988). This method of gene transfer has been used successfully by several groups in China.

De la Peña *et al.* (1987) injected DNA carrying a dominant selectable marker gene (APH[3′]II) into rye tillers at a very early stage of development. The seeds which developed on the injected tillers were screened for kanamycin resistance (10 μg/ml kanamycin sulfate). Seven seedlings survived the kanamycin selection and two of these showed neomycin phosphotransferase activity. The results were confirmed by genomic DNA gel blot analysis. However, these experiments have not been successfully reproduced thus far.

Pollen Tube Pathway. A method for transferring foreign DNA into recently pollinated rice florets was first reported by Duan and Chen (1985). They transferred total DNA from a rice variety with purple coloration to florets of a common variety, by cutting the stigma off the recipient floret and applying a drop of DNA solution to the cut end of the style. Seeds which developed from the treated florets were planted and some of the resulting plants showed purple coloration in progeny at the second through the fourth generations. The method was repeated by Luo and Wu (1988) using a plasmid-encoded gene, neomycin phosphotransferase, as a reporter. They presented preliminary molecular evidence of gene transfer by both genomic blot hybridization and NPTII enzyme assay in transgenic plants. The transformation efficiency was around 2% based on the number of rice florets treated with DNA. More recent experiments showed that the transformation efficiency was much lower (Duan and Wu, unpublished results). The method has proven to be difficult to reproduce and needs further improvement.

Electroporation of Tissues. A procedure was reported by Dekeyser *et al.* (1990) to electroporate DNA into intact leaf tissue of rice. Optimization of the system mainly involved eliminating explant-released nucleases, prolonging the DNA-explant incubation time and expanding the electrical pulse time. Using a GUS gene fused to constitutive promoters they demonstrated that all cell types within a leaf base were susceptible to electroporation-mediated DNA uptake. The technique was reported to be applicable to other monocots, including wheat, maize, and barley, and other rice tissue explants, such as etiolated and green sheath and lamine tissues. This procedure was published very recently and has not yet been reproduced in other laboratories.

Dry Embryo Incubation in DNA. The uptake of DNA by dry plant tissue through membranes, whose physiochemical characteristics change while natural desiccation occurs, was thought to be an alternative and very simple way of gene transfer (Simon, 1974). Researchers used this approach decades ago (Ledoux and Huart, 1969; Ledoux *et al.*, 1971, 1974; Soyfer, 1980). However, their results have been subject to much controversy (Kleinhofs *et al.*, 1975; Soyfer and Titov, 1981). Töpfer's group isolated embryos of seven cereals and three legumes by blowing off the bran of the cereals and subsequent separation of endosperm and embryos using cyclohexane and carbon tetrachloride (Töpfer *et al.*, 1989). The intact embryos were selected manually and imbibed in a solution of plasmid DNA carrying a chimeric NPTII gene. Molecular evidence was presented for the transient expression of the foreign gene. The authors suggest that the method is suitable for transient gene expression studies.

Additional Methods. Other methods for transfer of DNA into plant cells are discussed in the book by Walden (1989).

4.3 Preferred Methods for Dicots and Monocots

4.3.1 PREFERRED METHODS FOR DICOTS

4.3.1.1 *Agrobacterium*-Mediated Gene Transfer

Introduction of foreign genes into plant tissues by *Agrobacterium tumefaciens* cocultivation is at present the major method for study of plant gene expression in dicotyledonous plants. The foreign genes studied in *A. tumefaciens*-transformed plants have included light regulated genes (Lamppa *et al.*, 1985; Kuhlemeier *et *al.*, 1987a; Stockhaus *et al.*, 1987), flower-specific genes (Mariani *et al.*, 1990; Moore and Nasrallah, 1990), seed storage protein genes (Chen *et al.*, 1986; Colot *et al.*, 1987), insect inhibitor genes (Perlak *et al.*, 1990), fruit ripening genes (Giovannoni *et al.*, 1989), and maize transposon sequences (Baker *et al.*, 1986; Yoder, 1990), to name but a few. *A. tumefaciens*-mediated gene transformation of both heterologous and homologous host plants will continue to be one of the most effective and widely used methods for study of plant gene expression.

Methods for cocultivation of *A. tumefaciens* with leaf disc tissue have been described in detail (Horsch *et al.*, 1985; Rogers *et al.*, 1986). Cocultivation with other excised plant tissues follows the same basic procedure. A brief description of the general protocol is as follows:

1. *Preparation of* A. tumefaciens *strain harboring the foreign DNA sequence of interest.* An appropriate TI plasmid-derived cloning vector, preferably one containing a chimeric selectable marker gene for expression in plants, must be chosen to carry the DNA of interest. Many such vectors have been created. The DNA of interest is cloned into the TI vector and replicated in *E. coli*. The vector is then transferred to an *A. tumefaciens* host strain ("helper" strain) carrying a Ti plasmid which is "disarmed" (nontumorigenic) but still carries the virulence functions necessary for infection of plant tissue. The transfer to *A. tumefaciens* may be accomplished by a triparental mating procedure (involving the *E. coli* strain harboring the vector, the *A. tumefaciens* strain, and an *E. coli* strain carrying mobilization functions) (Ditta *et al.*, 1980; Rogers *et al.*, 1986) or by direct transfer (Holsters *et al.*, 1978; An *et al.*, 1988). The final *A. tumefaciens* strain carrying the DNA of interest is grown in an overnight culture for the next day's cocultivation experiment.
2. *Preparation of plant tissue for cocultivation.* Plant tissues for cocultivation may be excised from fresh greenhouse material and sterilized before cocultivation, or may be excised from axenic seedlings and used directly. Leaf tissue may be cut as disks or simply as small strips, and other tissues treated similarly, to prepare wounded surfaces for the bacterial infection.
3. *Cocultivation of plant tissue with* A. tumefaciens. The prepared tissues are generally dipped briefly in the *A. tumefaciens* overnight culture, blotted on sterile filter paper, and plated on nonselective media for a cocultivation period of a few days.
4. *Tissue culture and regeneration of plants.* After the cocultivation period, the plant tissue is plated on media containing general antibiotics (carbenicillin or cefotaxime) to kill or retard the growth of the residual *A. tumefaciens,* and containing the appropriate antibiotic, herbicide, or other chemical to which resistance is encoded by the selectable marker gene on the vector. Tissue culture is continued until plantlets are regenerated, the plantlets are transferred to soil, and the plants are maintained in a growth chamber or greenhouse.
5. *Study of transgenic plants.* The plants are analyzed by genomic DNA

gel blot analysis to confirm the presence of the foreign DNA, and may be further analyzed for expression of the foreign DNA in plant tissues. To confirm that stable integration of the foreign DNA has occurred, extensive DNA gel blot analysis needs to be carried out (see Section 4.1.2 above). To show seed transmission of the foreign DNA the transgenic plants should be self-crossed, the progeny seed planted, and the resulting plants analyzed for presence of the foreign DNA.

4.3.1.2 Biolistic Gene Transfer in Dicots

Biolistic gene transfer has been applied to many species in dicots. Tissues to which foreign genes have been successfully transferred using the biolistic process include meristems, young leaves, and embryogenic suspension culture cells.

For dicot plants which are easily micropropagated, leaf sections serve as good targets for the biolistic gene transfer, as was demonstrated in transformation of tobacco leaf (Tomes *et al.*, 1990). Leaf pieces in thin strips (1 cm long and 1 mm wide) placed on filter paper dampened with liquid medium were exposed to bombardment to receive GUS and NPTII genes. Following culture of bombarded leaf pieces on medium containing 100 or 200 μg/ml kanamycin for 6–8 weeks, plants were regenerated. About 160 plants were found to be putatively transgenic from 16 independent transformation events, with 76% of the plants showing NPTII activity and 50% of the plants showing GUS activity. Enzymatic and DNA hybridization analyses of two transformants and their progenies indicated Mendelian segregation of the introduced genes in one transformant.

If plants can be regenerated from embryogenic cell suspension culture, it can also be used as a target for the biolistic gene transfer. In cotton transformation experiments (Finer and McMullen, 1990), approximately 0.5 ml packed cell volume of embryogenic suspension culture was dispersed in a 3.5-cm diameter petri dish, which was placed in the center of a 9-cm petri dish and covered with a nylon screen. A chimeric hygromycin resistance gene was delivered into cotton cells through bombardment. The bombarded cells were cultured on a proliferation medium supplemented with hygromycin (100 μg/ml). About 30 stable transgenic colonies were recovered from each bombardment. The transgenic colonies were transferred to regeneration medium and plants were regenerated. Genomic DNA blot analysis of transgenic colonies and resulting plants confirmed the presence of the introduced *hph* gene.

DNA-coated gold particles were successfully introduced into soybean meristematic tissues using an electric discharge apparatus (McCabe *et al.*, 1988; Christou *et al.*, 1989). Immature seed-derived embryogenic axes, after removal of primary leaves to expose the meristem, were placed on target plates in the presence of 1% agar and subjected to bombardment with GUS and NPTII chimeric genes. The bombarded explants were maintained on a medium enriched with a high level of cytokinin for organogenesis. Multiple shoots per explant were regenerated one month following bombardment and grown to mature plants. Histochemical GUS assays were carried out for leaf disks collected from R_0 plants. Most of the R_0 plants were chimeric (composed of a mixture of transformed and nontransformed tissues). Two plants which were GUS positive in all leaves examined were further studied in detail. Genomic DNA blot analysis of these two plants indicated that more than one copy of foreign DNA was integrated into the soybean genome. Inheritance of the GUS and NPTII genes in one plant followed a Mendelian segregation pattern with a ratio of 3 : 1 in R_1 and R_2.

Stable transformation of *Populus* species was successfully carried out by McCown *et al.* (1991). They used protoplast-derived cells, nodules, and stems as target tissues. Gene combinations including NOS-NPTII, CaMV 35S-GUS, and CaMV 35S-BT (modified endotoxin gene from *Bacillus thuringiensis*) were delivered into the target tissues by electric discharge particle acceleration. Four transformed plants of one hybrid genotype containing all three genes were recovered and analyzed.

Development of the biolistic process has made it possible for foreign DNA to be transferred to organelles. Daniell *et al.* (1990) first reported the introduction and expression of foreign genes in chloroplasts of tobacco suspension cells. Two plasmids were constructed expressing the chloroplast CAT gene with chloroplast replicon (pHD 407) and without it (pHD 312). These vectors plus a nuclear vector (a 35S-CAT gene fusion construct) and pUC 118 as a negative control were biolistically delivered into tobacco suspension cells. Transient assay of CAT activity was carried out. Cells bombarded with pHD 407 reached a high level of CAT activity after 120 hr, whereas at that time the cells bombarded with the nuclear vector (35S-CAT) or the control vector (pUC 118) showed no activity. Cells bombarded with pHD 312 gave lower activity than those obtained with pHD 407 at all time intervals. The high-level expression shown in the pHD 407-bombarded cells suggests that this plasmid was delivered to and expressed in the chloroplasts. No expression of the chloroplast/CAT vectors was observed when the vectors were used in gene transfer experiments using electroporation.

4.3.2 PREFERRED METHODS FOR MONOCOTS

4.3.2.1 PEG-Mediated Direct Transformation

PEG-mediated direct transformation is one of the major gene transfer systems in monocots. The method has been widely used for transformation of various monocot species including *Pennisetum purpureum* L., *Panicum maximum* L., *Triticum monococcum* L., and *Saccharum officinarum,* as well as important crops including *Oryza sativa* L., *Zea mays* L. and *Sorghum bicolor* (Hauptmann *et al.,* 1988).

PEG-mediated protoplast transformation has been employed frequently in our laboratory for japonica rice (*Oryza sativa* L.), for transient assay and stable transformation. To stimulate DNA uptake, we incubated protoplasts with linearized or circular plasmid DNA (pAI$_1$*gus*N with ADH1 gene promoter) together with 18% PEG. The results showed that a foreign gene could be stably integrated into rice genomic DNA and correctly expressed under anaerobic conditions (Zhang and Wu, 1988). In addition, we used this system for transient assay and found that a rice *Act*-1 (actin 1) promoter, fused to the GUS reporter gene, strongly and constitutively expressed GUS (McElroy *et al.*, 1990, 1991). We have recently investigated GUS gene expression directed by the rice *Act*-1 5′ region in different cell types of transgenic rice plants and found that the gene is strongly expressed in phloem, meristem, and mesophyll cells but is not or is weakly expressed in cortex and xylem cells (Zhang *et al.*, 1991). Extensive genomic DNA analysis of these plants was carried out, according to the scheme outlined in Figure 4.1 of this text, and proved that integration of *Act*-1/*gus* occurred.

Hayashimoto *et al.* (1990) recently reported that they transformed rice protoplasts with a hygromycin resistance gene using 25% PEG and recovered two fertile transgenic plants. These plants both transmitted the *hph* gene into the second generation according to a genetic analysis.

Recently, Lee *et al.* (1991) successfully transformed small rice cell groups (50–100 cells) using 20% PEG and *Adh*/*gus*. They regenerated about 70 transgenic plants.

Advances in transformation of Indica rice have also been made using PEG-mediated direct transformation. Peng *et al.* (1990) cotransformed plasmids pKAN (containing the CaMV 35S promoter, the NPTII gene, and a T-DNA polyadenylation sequence) and pPUR (containing the CaMV 35S promoter, the *gus* coding sequence and a *nos* polyadenylation sequence) into Indica rice protoplasts using the PEG system. They produced 104 kanamycin-resistant lines of rice of which 44 lines were cotransformed with *gus*. Datta *et al.* (1990) also have reported PEG-mediated transformation and regeneration of Indica rice protoplasts, using a hygromycin resistance gene.

For other monocot species the PEG method is also widely used, both for transient gene expression and for stable transformation. Potrykus *et al.* (1985) transformed protoplasts of *Lolium multiflorum* (Italian ryegrass) with an aminoglycoside (neomycin) phosphotransferase gene and obtained stably transformed colonies. Lörz *et al.* (1985) used the PEG method to transform *Triticum monococcum* L. cells with the same construct, and obtained stable transformed colonies. These results show that the PEG method is a reliable technique and can be developed into a major system for direct gene transfer in monocot species. However, it must be noted that considerable skill in tissue culture technique is needed to produce fertile transgenic plants.

4.3.2.2 Electroporation of Protoplasts

The electroporation method was first applied to maize protoplast gene transfer by Walbot's group (Fromm *et al.*, 1985) and was used in gene transfer experiments with rice, wheat, and sorghum protoplasts by Ou-Lee *et al.* (1986). Now this method is widely used to produce transgenic plants (Toriyama *et al.*, 1988). Horn *et al.* (1988) reported the success of gene transformation and regeneration by electroporation and PEG methods in a monocot species, orchard grass (*Dactilis glomerata* L.). Shimamoto *et al.* (1989) used this method to transfer a bacterial *hph* gene, encoding hygromycin B resistance, into rice protoplasts and developed 14 fertile transgenic plants. They also transferred a maize transposable element *Ac* into rice plants by electroporation. They found that the *Ac* element appears to be active during development of transgenic rice plants from calli. Some *Ac* elements in transgenic rice plants are stable whereas others are able to transpose further during development of leaves (Izawa *et al.*, 1991). Their results open an opportunity to develop a transposon tagging system in rice.

Tada *et al.* (1991) introduced the promoter of rice light-harvesting chlorophyll a/b binding protein of photosystem II (LHCP), linked to *gus*, to rice by electroporation. They found that in green organs of transgenic plants, the LHCP/GUS activity was 10-fold higher than that with CaMV 35S/GUS, but much lower in transformed protoplasts and cells.

Rhodes *et al.* (1988) obtained transgenic maize plants (although sterile) from protoplasts electroporated with chimeric NPTII genes. These results confirmed that the electroporation method is another useful system which can be applied to direct gene transfer in monocot species. Again, more investigation is needed to define all the experimental parameters of the method, from the protoplast stage to the regeneration of fertile transgenic plants.

4.3.2.3 Biolistic Gene Transfer in Monocots

Biolistic gene transfer has proven effective for delivery of foreign DNA into monocots, including the crop species rice, wheat, maize, and sorghum (Cao *et al.*, 1990; Vasil *et al.*, 1991; Gordon-Kamm *et al.*, 1990; Hagio *et al.*, 1991). Embryogenic suspension cells and calli as well as embryo sections have been shown to be biolistically transformable and able to produce transgenic plants.

Gordon-Kamm *et al.* (1990) first reported success in biolistic introduction of foreign genes into maize embryogenic suspension cells followed by regeneration of transgenic maize plants. They made use of the *bar* selectable marker gene, which encodes the enzyme phosphinothricin acetyltransferase (PAT), providing resistance to the herbicide bialaphos. Embryogenic maize cells were bombarded with plasmid pDPG 165 containing 35S-*bar* alone, or in combination with pDPG 208 carrying *gus*. Bombarded cells, after being maintained in nonselective liquid medium for one or two weeks, were dispersed evenly over a filter overlaying medium containing 1 or 3 μg/ml bialaphos. Resistant colonies from the bombarded suspension cells appeared on the selective medium 6–7 weeks after bombardment. DNA and enzymatic analyses demonstrated introduction and expression of the *bar* gene in transformants. About 52% of transformants grown on regeneration medium became regenerable transformed callus lines, giving rise to 295 plants. Enzymatic and DNA analyses of R_1 plants confirmed the presence of the introduced *bar* genes. Segregation of PAT-positive and -negative plants followed a Mendelian pattern. When the herbicide was sprayed onto leaves of transgenic R_0, R_1, and control plants, transgenic plants showed no necrosis, while nontransgenic plants did.

Fromm *et al.* (1990) also made successful use of the biolistic process for delivery of foreign DNA into maize embryogenic cells and regeneration of transgenic maize plants. In their experiments they found that the use of selective growth media during the regeneration process was essential to the recovery of transgenic plants carrying the selectable marker gene. DNA and enzymatic assay of R_1 plants showed that the foreign gene was stably maintained in the progeny.

Foreign DNA carrying the GUS gene has been biolistically introduced into intact cells of embryo sections dissected from mature seeds of rice and maize (Cao *et al.*, 1990). When bombarded embryo sections were cultured, about 80% of the embryo sections developed into plantlets 2–3 weeks later. Histochemical assay of GUS activity in the plantlets showed that the transgenic plants (around 20 R_0 plants) were chimeric and DNA gel blot analysis demonstrated the presence of the introduced gene in leaves. However, the GUS gene was not found in R_1 plants. Presumably, if we could generate a much larger number of R_0 transgenic plants, several of these may transmit the foreign gene to the next generation.

Christou reported success in recovery of transgenic rice plants from both japonica and indica varieties using electric discharge particle acceleration and gold particles (Christou *et al.*, 1991). Immature embryos were isolated 12–15 days after flowering using a dissecting microscope, and placed on a water–agar plate with the adaxial side in contact with the medium. Scutellar regions of the embryos were subjected to electric discharge gold particle bombardment using plasmids carrying *gus* and *bar,* or *gus* and *hph.* Transformed calli and plants were recovered either under nonselection or selection with hygromycin at 50 mg/liter or bialaphos at 10 mg/liter. Genomic DNA gel blot analysis of R_0 plants and their progenies demonstrated stable integration of the introduced genes into the rice genome. Genetic study of R_1 progenies of transgenic rice plants showed segregation in a Mendelian ratio (3 : 1).

Vasil *et al.* (1991) reported stably transformed callus lines after microprojectile bombardment of wheat cell suspension with NPTII, GUS, and EPSP (5-enolpyruvylshikamate phosphate) synthase reporter gene combinations on different plasmids. Southern blot analysis proved that the genes had been stably integrated in all of the selected lines.

Recently, Daniell was successful in biolistic delivery of the GUS gene into chloroplasts of intact wheat cells (Daniell *et al.*, 1991). When nuclear and chloroplast GUS vectors were separately introduced into wheat cells, transient histochemical assay of cells bombarded with the nuclear vector showed the insoluble dye, a product of GUS cleavage, evenly distributed throughout the cytosol, while the dye product in cells bombarded with the chloroplast vector appeared within the chloroplasts. The chloroplast GUS vector was delivered and expressed in both albino and green leaves of anther culture-derived transgenic wheat plants.

It has been difficult to study developmentally regulated gene expression in specialized cells or tissues in monocots due to the lack of an effective transformation technology or difficulties in plant regeneration systems. The biolistic method for DNA delivery proved to be a useful tool for such gene expression studies. For example, to investigate the seed-specific expression of rice α-amylase/GUS gene constructs, aleurone tissue was isolated from mature oat seeds and bombarded with deletion constructs. Three days later the transformed material was examined by a fluorometric assay using methylumbelliferin glucoronide (MUG) as the substrate (Jenes, Goldman and Wu, unpublished data). This method is simpler than the oat aleurone protoplast system (Huttly and Baulcombe, 1989) because it involves less tissue culture manipulation. It has been also reported that the biolistic process was applied for transient expression study of regulation of anthocyanin biosynthetic genes in maize aleurone and embryo tissues (Klein *et al.*, 1989).

We have also used the biolistic process for delivery of other developmentally regulated gene constructs into various rice tissues, and performed transient assay of their expression activities (Cao *et al.*, 1991). The system allows for accurate assessment of promoter strength in the specific tissue *in vivo*. In principle this method can be used in any tissue to study specific regulation on a quantitative level.

In summary, three preferred methods for transforming monocots are described. However, additional work is needed to improve protoplast regeneration techniques for the PEG and electroporation methods, and intact cell transformation efficiency for the biolistic method.

Acknowledgments

This work was supported in part by a research Grant RF 86058, #60, from the Rockefeller Foundation and by Grant No. 593 0009-03 from the Midwest Plant Biotechnology Consortium. B. Jenes was supported by a fellowship from the Ministry of Agriculture of Hungary.

References

Abdullah, R., Cocking, E. C., and Thompson, J. A. (1986). *Bio/Technology* 4, 1087–1090.

Agrios, G. (1978). *Plant Pathology,* 2nd ed. New York: Academic Press.

An, G., Ebert, P. R., Mitra, A., and Ha, S. B. (1988). In *Plant Molecular Biology Manual* (Ed. by S. B. Gelvin and R. Schilperoort), pp. 1–19. Dordrecht, The Netherlands: Martinus Nijhoff Publishers.

Baker, B., Schell, J., Lörz, H., and Federoff, N. (1986). *Proc. Natl. Acad. Sci. U.S.A.* 83, 4844–4848.

Boutry, M., Nagy, F., Poulsen, C., Aoyagi, K., and Chua, N.-H. (1987). *Nature (London)* 328, 340–342.

Boynton, J. E., Gillham, N. W., Harris, E. H., Hossler, J. P., Johnson, A. M., Jones, A. R., Randolphe-Anderson, B. L., Robertson, D., Klein, T. M., Shark, K. B., and Sanford, J. C. (1988). *Science* 240, 1534–1538.

Brinster, R. L., Chen, H. Y., Trumbauer, M., Yagle, M. K., and Palmiter, R. D. (1985). *Proc. Natl. Acad. Sci. U.S.A.* 82, 4438–4442.

Cao, J., Wang, Y. C., Klein, T. M., Sanford, J. C., and Wu, R. (1990). In *Plant Gene Transfer* (Ed. by C. J. Lamb and R. N. Beachy), pp. 21–33. New York: Wiley-Liss.

Cao, J., Zhang, W. G., McElroy, D., and Wu, R. (1991). In *Rice Biotechnology* (Ed. by G. S. Khush and G. Toenniessen). Manila: IRRI.

Capecchi, M. R. (1980). *Cell* 22, 479–488.

Cardarelli, M., Mariotti, D., Pomoni, M., Spanò, L., Capone, I., and Constantino, P. (1987). *Mol. Gen. Genet.* 209, 475–480.

Chen, Z. L., Schuler, M. A., and Beachy, R. N. (1986). *Proc. Natl. Acad. Sci. U.S.A.* 83, 8560–8564.

Christou, P., Swain, W. F., Yang, N. S., and McCabe, D. E. (1989). *Proc. Natl. Acad. Sci. U.S.A.* 86, 7500–7504.

Christou, P., Ford, T. M., and Kofron, M. (1991). *Bio/Technology* 9, 957–962.

Cocking, E. C., and Davey, M. R. (1987). *Science* 236, 1259–1262.

Colot, V., Robert, L. S., Kavanagh, T. A., Bevan, M. W., and Thompson, R. D. (1987). *EMBO J.* 6, 3559–3564.

Crossway, A., Oake, J. V., Irvine, J. M., Ward, B., Knauf, V. C., and Shewmaker, C. K. (1986). *Mol. Gen. Genet.* 202, 179–185.

Daniell, H., Vivekananda, J., and Nielsen, B. L. (1990). *Proc. Natl. Acad. Sci. U.S.A.* 87, 88–92.

Daniell, H., Krishnan, M., and McFadden, B. A. (1991). *Plant Cell Rep.* 9, 615–621.

Datta, S. K., Peterhans, A., Datta, K., and Potrykus, I. (1990). *Bio/Technology* 8, 736–740.

Davies, P. A., Larkin, P. J., and Tanner, G. J. (1989). *Plant Sci.* 60, 237–244.

Dekeyser, R. A., Claes, B., De Rycke, R. M. U., Habets, M. E., Van Montagu, M. C., and Caplan, A. B. (1990). *The Plant Cell* 2, 591–602.

De la Peña, A., Lörz, H., and Schell, J. (1987). *Nature (London)* 325, 274–276.

Deom, C. M., Oliver, M. J., and Beachy, R. N. (1987). *Science* 237, 389–394.

Ditta, G., Stanfield, S., Corbin, D., and Helinski, D. (1980). *Proc. Natl. Acad. Sci. U.S.A.* 77, 7347–7351.

Duan, X., and Chen, S. (1985). *China. Agricult. Sci.* 3, 6–9.

Finer, J. J., and McMullen, M. D. (1990). *Plant Cell Rep.* 8, 586–589.

Fromm, M., Taylor, L. P., and Walbot, V. (1985). *Proc. Natl. Acad. Sci. U.S.A.* 82, 5824–5828.

Fromm, M. E., Morrish, F., Armstrong, C., Williams, R., Thomas, J. and Klein, T. M. (1990). *Bio/Technology* 8, 833–839.

Gelvin, S. B., and Schiperoort, R., eds. (1988). Plant Molecular Biology Manual. Dordrecht, The Netherlands: Martinus Nijhoff Publishers.

Goldfarb, B., Strauss, S. H., Howe, G. T., and Zaerr, J. B. (1991). *Plant Cell Rep.* 10, 517–521.
Gordon-Kamm, W. J., Spencer, T. M., Mangano, M. L., Adams, T. R., Daines, R. J., Start, W. G., O'Brien, J. V., Chambers, S. A., Adams, W. R., Willetts, N. G., Rice, T. B., Mackey, C. J., Kreuger, R. W., Kausch, A. P., and Lemaux, P. G. (1990). *The Plant Cell* 2, 603–618.
Giovannoni, J. J., DellaPenna, D., Bennett, A. B., and Fischer, R. L. (1989). *The Plant Cell* 1, 53–63.
Hagio, T., Blowers, A. D., and Earle, E. D. (1991). *Plant Cell Rep.* 10, 260–264.
Hauptmann, R. M., Vasil, V., Ozias-Akins, P., Tabaeizadeh, Z., Rogers, S. G., Fraley, R. T., Horsh, R. B., and Vasil, I. K. (1988). *Plant Physiol.* 86, 602–606.
Hayashimoto, A., Li, Z., and Murai, N. (1990). *Plant Physiol.* 93, 857–863.
Holsters, M., de Waele, D., Depicker, A., Messens, M., Van Montagu, M., and Schell, J. (1978). *Mol. Gen. Genet.* 163, 181–187.
Horn, M. E., Shillito, R. D., Conger, B. V., and Harms, C. T. (1988). *Plant Cell Rep.* 7, 469–472.
Horsch, R. B., Fry, J. E., Hoffmann, N. L., Eichholtz, D., Rogers, S. G., and Fraley, R. T. (1985). *Science* 227, 1229–1231.
Huttly, A. K., and Baulcombe, D. C. (1989). *EMBO J.* 8, 1907–1913.
Izawa, T., Miyazaki, C., Yamamoto, M., Terada, R., Iida, S., and Shimamoto, K. (1991). *Mol. Gen. Genet.* 227, 391–396.
Jenes, B., and Pauk, J. (1989). *Plant Sci.* 63, 187–198.
Joersbo, M., and Brunstedt, J. (1990). *Plant Cell Rep.* 9, 207–210.
Johnston, S. A., Anziano, P. Q., Shark, K., Sanford, J. C., and Butow, R. A. (1988). *Science* 240, 1538–1541.
Klein, T. M., Wolf, E. D., Wu, R., and Sanford, J. C. (1987). *Nature (London)* 327, 70–73.
Klein, T. M., Roth, B., and Fromm, M. E. (1989). *Proc. Natl. Acad. Sci. U.S.A.* 86, 6681–6685.
Kleinhofs, A., Eden, F. C., Chilton, M.-D., and Bendich, A. J. (1975). *Proc. Natl. Acad. Sci. U.S.A.* 72, 2748–2752.
Kohler, F., Cardon, G., Pohlman, M., Gill, R., and Schieder, O. (1989). *Plant Molec. Biol.* 12, 189–199.
Kuhlemeier, C., Fluhr, R., Green, P. J., and Chua, N.-H. (1987a). *Genes and Dev.* 1, 247–255.
Kuhlemeier, C., Green, P. J., and Chua, N.-H. (1987b). *Annu. Rev. Plant Physiol.* 38, 221.
Lamppa, G., Nagy, F., and Chua, N.-H. (1985). *Nature (London)* 316, 750–752.
Langridge, W. H., Li, B. J., and Szalay, A. A. (1985). *Plant Cell Rep.* 4, 355–359.
Ledoux, L., and Huart, R. (1969). *J. Mol. Biol.* 43, 243–262.
Ledoux, L., Huart, R., and Jacobs, M. (1971). *Eur. J. Biochem.* 23, 96–108.
Ledoux, L., Huart, R., and Jacobs, M. (1974). *Nature (London)* 249, 17–21.
Lee, B. T., Murdoch, K., Topping, J., Jones, M. G. K., and Kreist, M. (1991). *Plant Sci.* 78, 237–246.
Lee, N., Wang, Y., Yang, J., Ge, K., Huang, S., and Tan, J. (1991). *Proc. Natl. Acad. Sci. U.S.A.* 88, 6389–6393.
Lörz, H., Baker, B., and Schell, J. (1985). *Mol. Gen. Genet.* 199, 178–182.
Luo, Z., and Wu, R. (1988). *Plant Molec. Biol. Rep.* 6(3), 165–174.
Mariani, C., De Beuckeleer, M., Truettner, J., Leemans, J., and Goldberg, R. B. (1990). *Nature (London)* 347, 737–741.
McCabe, D. E., Swain, W. F., Martinell, B. J., and Christou, P. (1988). *Bio/Technology* 6, 923–926.
McCown, B. H., McCabe, D. E., Russell, D. R., Robinson, D. J., Barton, K. A., and Raffa, K. F. (1991). *Plant Cell Rep.* 9, 590–594.
McElroy, D., Zhang, W., Cao, J., and Wu, R. (1990). *The Plant Cell* 2, 163–174.
McElroy, D., Blowers, A. D., Jenes, B., and Wu, R. (1991). *Mol. Gen. Genet.* 231, 150–160.
Moore, H. M., and Nasrallah, J. B. (1990). *The Plant Cell* 2, 29–38.

Negrutiu, I., Shillito, R., Potrykus, I., Biasini, G., and Sala, F. (1987). *Plant Molec. Biol.* 81, 363–373.
Nolan, R. C., and Ho, T.-H. D. (1988). *Planta* 174, 551–560.
Ou-Lee, T. M., Turgeon, R., and Wu, R. (1986). *Proc. Natl. Acad. Sci. U.S.A.* 83, 6815–6819.
Ou-Lee, T.-M., Turgeon, R., and Wu, R. (1988). *Proc. Natl. Acad. Sci. U.S.A.* 85, 6366–6369.
Peng, J., Lyznik, L. A., Lee, L., and Hodges, T. K. (1990). *Plant Cell Rep.* 9, 168–172.
Perlak, F. J., Deaton, R. W., Armstrong, T. A., Fuchs, R. L., Sims, S. R., Greenplate, J. T., and Fischhoff, D. A. (1990). *Bio/Technology* 8, 939–943.
Potrykus, I. (1990). *Bio/Technology* 6, 535–543.
Potrykus, I., Saul, M. W., Peruska, J., Paszkowski, J., and Shillito, R. (1985). *Mol. Gen. Genet.* 199, 183–188.
Raineri, D. M., Bottino, P., Gordon, M. P., and Nester, E. W. (1990). *Bio/Technology* 8, 33–38.
Reich, T. J., Iyer, V. N., and Miki, B. L. (1986). *Bio/Technology* 4, 1001–1004.
Rhodes, C. A., Pierce, D. A., Mettler, I. J., Mascarenhas, D., and Detmer, J. J. (1988). *Science* 240, 204–207.
Rogers, S. G., Horsch, R. B., and Fraley, R. T. (1986). *Meth. Enzymol.* 118, 627–640.
Sanford, J. C. (1990). *Physiol. Plant.* 79, 206–209.
Schell, J. (1987). *Science* 237, 1176–1183.
Seki, M., Komeda, Y., Iida, A., Yamada, Y., and Morikawa, H. (1991). *Plant Mol. Biol.* 17, 259–263.
Shimamoto, K., Terada, R., Izawa, T., and Fujimoto, H. (1989). *Nature (London)* 338, 274–276.
Şimon, E. W. (1974). *New Phytol.* 73, 377–420.
Soyfer, V. N. (1980). *Theor. Appl. Genet.* 58, 225–235.
Soyfer, V. N., and Titov, Y. B. (1981). *Mol. Gen. Genet.* 182, 361–363.
Stockhaus, J., Eckes, P., Rocha-Sosa, M., Schell, J., and Willmitzer, L. (1987). *Proc. Natl. Acad. Sci. U.S.A.* 84, 7943–7947.
Svab, Z., Hajdukiewiez, P., and Maliga, P. (1990). *Proc. Natl. Acad. Sci. U.S.A.* 87, 8526–8530.
Tada, Y., Sakamoto, M., and Fujimura, T. (1990). *Theor. Appl. Genet.* 80, 475–480.
Tada, Y., Sakamoto, M., Matsuoka, M., and Fujimura, T. (1991). *EMBO J.* 10, 1803–1808.
Thomzik, J. E., and Hain, R. (1990). *Plant Cell. Rep.* 9, 233–236.
Tomes, S. T., Weissinger, A. K., Ross, M., Higgins, R., Drummond, B. J., Schaaf, S., Malone-Schoneberg, J., Staebell, M., Flynn, P., Anderson, J., and Howard, J. (1990). *Plant. Molec. Biol. Rep.* 14, 261–268.
Töpfer, R., Gronenborn, B., Schell, J., and Steinbiss, H.-H. (1989). *The Plant Cell* 1, 133–139.
Toriyama, K., Arimoto, Y., Uchimiya, H., and Hinata, K. (1988). *Bio/Technology* 6, 1072–1074.
Uchimiya, H., Hauda, T., and Brar, D. S. (1989). *J. Biotechnology* 12, 1–20.
Vasil, V., Brown, S. M., Re, D., Fromm, M. E., and Vasil, I. K. (1991). *Bio/Technology* 9, 743–747.
Walden, R. (1989). *Genetic Transformation of Plants.* Englewood Cliffs, New Jersey: Prentice Hall.
Wang, Y. C., Klein, T. M., Fromm, M., Cao, J., Sanford, J. C., and Wu, R. (1988). *Plant Mol. Biol.* 11, 433–439.
Wu, R. (1989). Methods for transforming plant cells. In *Plant Biotechnology* (Ed. by S. D. Kung and C. J. Arntzen), pp. 35–51. Boston: Butterworths.
Ye, G. N., Daniell, H., and Sanford, J. C. (1990). *Plant Mol. Biol.* 15, 809–819.
Yoder, J. I. (1990). *The Plant Cell* 2, 723–730.
Zhang, L.-J., Cheng, L.-M., Xu, N., Zhao, N.-M., Li, C.-G., Yuan, J., and Jia, S.-R. (1992). *Bio/Technology* 9, 996–997.
Zhang, W., and Wu, R. (1988). *Theor. Appl. Genet.* 76, 835–840.
Zhang, W., McElroy, D., and Wu, R. (1991). *The Plant Cell* 3, 1155–1165.
Zhou, G. Y., Weng, J., Zeng, Y., Huang, J., Qian, S., and Liu, G. (1983). *Meth. Enzymol.* 101, 433–481.
Zhou, G. Y., Huang, J., and Chen, S. (1988). *Scientia Agric. Sinica* 21(3), 1–6.

5

Cell Culture and Regeneration of Transgenic Plants

Steven W. Ritchie and Thomas K. Hodges

Department of Botany and Plant Pathology
Purdue University
West Lafayette, Indiana 47907

To review the subject of regeneration of transgenic plants first requires a comment on the importance that a plant regeneration system holds within a successful transformation system. With the exception of the *Agrobacterium*-mediated seed transformation of *Arabidopsis* (Feldmann and Marks, 1987), all successful and reproducible transformation systems require an *in vitro* plant regeneration system. The regeneration of transgenic plants encompasses an integration of (1) a transformation technique, (2) a functional plant regeneration system for a given species and cultivar, and (3) a selectable or screenable marker gene system. Due to the diversity of transformation and regeneration techniques, not all transformation techniques are compatible with all regeneration systems. Furthermore, regeneration systems are generally species and often cultivar specific, and not all selectable markers work well with all species. Thus, simply having a regeneration system established and a transformation technique and selectable marker genes available, do not guarantee success at regeneration of transgenic plants.

Potrykus (1990) has recently outlined qualifications for a truly successful transformation system. Probably the most convincing evidence for plant transformation would be to show physiological, molecular, and genetic analysis of the progeny in which the foreign gene is (1) integrated into the plant genome, (2) correctly expressed, and (3) inherited in a near-Mendelian ratio. Unfortunately, progeny analysis may not be easily obtained depending on the number of R_0 plants or the time required for species of interest to breed (e.g., woody

TRANSGENIC PLANTS, VOLUME 1
Engineering and Utilization

Considering the potential plant chimerism obtained with use of the particle bombardment technique, however, molecular progeny analysis is now essential. In Table 5.1 we have compiled a list of species that have been transformed with engineered foreign genes. We have included information in Table 5.1 for the type of explant and the selectable agent used for the plant regeneration protocol. Table 5.1 does not cover all papers that have produced transgenic plants, but includes select papers that were unique with regard to either the genotype, explant, transformation procedure, or the selection system employed.

Many excellent recent reviews have been written concerning *in vitro* regeneration of plants. These reviews have covered both the specific experimental manipulations required for plant regeneration for many species (Evans *et al.*, 1981; Flick *et al.*, 1983; Ammirato, 1983), and the developmental biology of *in vitro* plant regeneration or morphogenesis (Ammirato, 1986; Raghavan, 1986; Williams and Maheswaran, 1986; Halperin, 1986; Brown and Thorpe, 1986; Vasil and Vasil, 1986; Ammirato, 1987; Christianson, 1987; Morrish *et al.*, 1987; Christianson and Warnick, 1988; Ranch and Pace, 1988; Vasil, 1988). This review will focus on *in vitro* plant regeneration and morphogenesis at both the practical and conceptual level. Where possible, we will attempt to blend the conceptual with the practical, especially with regard to successful plant transformation systems. This review will not detail the species and cultivar-specific manipulations for plant regeneration because the number of plant species that can be regenerated by somatic embryogenesis, organogenesis, and micropropagation is quite extensive, and because specific protocols have been outlined in previous reviews, in the following chapters within this book, and in the papers cited within this chapter. For general information on media preparation, sterile technique, and general protocols, consult Biondi and Thorpe (1981), Gamborg and Shyluk (1981), Reinert and Yeoman (1982), Gamborg (1982), Brown and Thorpe (1984), Constabel (1984), Potrykus and Shillito (1986), or Smith (1986).

5.1 General Characteristics of Plant Cell Cultures

5.1.1 CELLULAR TOTIPOTENCY

The foundation for *in vitro* plant regeneration is that certain individual somatic plant cells, or protoplasts from these cells, are capable of proliferation and organization into tissues, and eventually development into a complete plant or organism. This capability is termed totipotency, and has been experimentally shown at the single-cell level in the model species tobacco (Vasil and Hildebrandt, 1965) and carrot (Nomura and Komamine, 1985). It was initially thought all plant cells were capable of expressing totipotency. However, some

TABLE 5.1
Techniques Used for Transformation and Regeneration of Plants[a]

Plant	Transformation technique[b]	Tissue source	Selectable agent[c]	Reference
I. Dicots				
Apium graveolens (celery)	At	Petiole pieces	km	Catlin *et al.* (1988)
Arabidopsis thaliana (arabidopsis)	At	Leaf disks	hm	Lloyd *et al.* (1986)
	At	Root pieces	km	Valvekens *et al.* (1988)
	At	Mature seeds	km, G418	Feldmann and Marks (1987)
Beta vulgaris (sugar beet)	At	Shoot-base tissues	km	Lindsey and Gallois (1990)
Brassica napus (oilseed rape)	At	Cotyledonary petioles	km	Moloney *et al.* (1989)
	Ar(At)	Cotyledonary nodes	Hrp, km	Boulter *et al.* (1990)
	At	Inflorescence stalks	km	Boulter *et al.* (1990)
	At	Hypocotyl protoplasts	km	Thomzik and Hain (1990)
	At	Thin cell layers	km, cm	Charest *et al.* (1988)
	At	Hypocotyl segments	km, pt	De Block *et al.* (1989)
	At	Hypocotyl pieces	km	Radke *et al.* (1988)
	At	Microspores, proembryos	km, hm	Pechan (1989)
	MiI	Microspore embryoids	km	Neuhaus *et al.* (1987)
	At	Stem sections	km	Fry *et al.* (1987)
	At	Stem sections	mtx Screen	Pua *et al.* (1987)
	Ep	Leaf protoplasts	km	Guerche *et al.* (1987)
Brassica oleracea	At	Hypocotyl segments	km, pt	De Block *et al.* (1989)
Brassica oleracea (cauliflower)	At	Leaf disks	km	Srivastava *et al.* (1988)
Citrus jambhiri (rough lemon)	PEG	Nucellar callus protoplasts	pm	Vardi *et al.* (1990)

(*continued*)

TABLE 5.1 *(Continued)*

Plant	Transformation technique[b]	Tissue source	Selectable agent[c]	Reference
Cucumis melo (muskmelon)	At	Cotyledonary pieces	km	Fang and Grumet (1990)
Cucumis sativus (cucumber)	Ar	Hypocotyl segments	km	Trulson *et al.* (1986)
	At	Cotyledonary pieces	km	Chee (1990)
Daucus carota (carrot)	At	Suspension cells	km	Scott and Draper (1987)
	At	Callus cells	km	Wurtele and Bulka (1989)
	At	Hypocotyl sections	km	Thomas *et al.* (1989)
Fragaria × *ananassa* (strawberry)	At	Leaf disks	km	Nehra *et al.* (1990)
	At	Leaf pieces, petiole segments	km	James *et al.* (1990)
Glycine canescens	Ar	Hypocotyls	km	Rech *et al.* (1989)
Glycine max (soybean)	Pg	Zygotic embryos	km	Christou *et al.* (1988)
	At	Immature embryos	None, km	Parrott *et al.* (1989)
	At	Germinating seeds	None	Chee *et al.* (1989)
	Pg	Seed apical meristems	None	McCabe *et al.* (1988)
	At	Cotyledons	km	Hinchee *et al.* (1988)
Gossypium hirsutum (cotton)	Pg	Suspension cells	hm	Finer and McMullen (1990)
	At	Apical meristems	km	Ulian *et al.* (1988)
	At	Hypocotyl pieces	km	Umbeck *et al.* (1987)
	At	Cotyledon pieces	km	Firoozabady *et al.* (1987)
Helianthus annuus (sunflower)	At	Hypocotyls	km	Everett *et al.* (1987)
	At	Shoot apical meristems	km	Schrammeijer *et al.* (1990)
Kalanchoe laciniata (kalanchoe)	At	Leaf explants	km	Jia *et al.* (1989)

Juglans regia (walnut)	At	Somatic embryos	km	McGranahan *et al.* (1988, 1990)
Lactuca sativa (lettuce)	At	Cotyledon pieces	km	Michelmore *et al.* (1987)
	Ep	Leaf protoplasts	pm, km	Chupeau *et al.* (1989)
Linum usitatissimum (flax)	At	Hypocotyl sections	km	Basiran *et al.* (1987)
Lotus corniculatus (bird's-foot trefoil)	Ar	Seedlings	Hrp	Jensen *et al.* (1986)
Lycopersicon esculentum (tomato)	At	Leaf disks	km	Horsch *et al.* (1985)
	Ar, At	Cotyledons	km	Shahin *et al.* (1986b)
	At	Cotyledon pieces	km, gly	Fillatti *et al.* (1987a)
	At	Leaf pieces	km	McCormick *et al.* (1986)
	PEG	Callus protoplasts	km	Koornneef *et al.* (1986)
Lycopersicon peruvianum	Ep	Leaf protoplasts	cs, pm	Bellini *et al.* (1989)
Malus pumila (apple)	At	Leaf disk or pieces	km	James *et al.* (1989)
Medicago sativa (alfalfa)	At	Stem segments	km	Shahin *et al.* (1986a)
	At	Stem and petiole segments	pt, km, hm	D'Halluin *et al.* (1990)
	Ar (At)	Stem segments	Hrp, nos Screen	Sukhapinda *et al.* (1987)
Medicago varia (*M. sativa* × *M. falcata*)	At	Stem segments	km	Deak *et al.* (1986)
	At	Leaflets	km	Chabaud *et al.* (1988)
Nicotiana alata	At	Hypocotyl pieces	km	Ebert and Clarke (1990)
Nicotiana edwardsonii	At	Leaf disks	km, G418	Kiernan *et al.* (1989)
Nicotiana plumbaginifolia	At	Leaf protoplasts	km	Horsch *et al.* (1984)
	Mg/PEG	Leaf protoplasts	km	Negrutiu *et al.* (1987)
	At	Leaf protoplasts	km, stm	Jones *et al.* (1987)
	At	Leaf pieces	km	Ellis *et al.* (1988)
	PEG	Leaf protoplasts	km	Horth *et al.* (1987)
	At	Epidermal peels	km	Trinh *et al.* (1987)
	At	Leaf protoplasts	cm, km, mtx	De Block *et al.* (1984)

(*continued*)

TABLE 5.1 *(Continued)*

Plant	Transformation technique[b]	Tissue source	Selectable agent[c]	Reference
Nicotiana tabacum (tobacco)	At	Leaf protoplasts	cm, km, mtx	De Block *et al.* (1984)
	At	Leaf disks	km	Horsch *et al.* (1985)
	At	Leaf disks	km, brm	Stalker *et al.* (1988)
	At	Leaf disks	km, stm	Jones *et al.* (1987)
	At	Leaf disks	km, ppt	De Block *et al.* (1987)
	Ar	Leaf disk	km, gly	Comai *et al.* (1985)
	Ar(At)	Leaf disks	Hrp, km	Hatamoto *et al.* (1990)
	At	Leaf disks	cs, km	Haughn *et al.* (1988)
	At	Leaf disk	km, hm	Landsmann *et al.* (1988)
	Calcium phosphate	Leaf protoplasts	km	Hain *et al.* (1985)
	Mg/PEG	Leaf protoplasts	km	Negrutiu *et al.* (1987)
	PEG	Leaf protoplasts	km	Paszkowski *et al.* (1984)
	Ep	Leaf protoplasts	km	Riggs and Bates (1986)
	PEG/Ep	Leaf protoplasts	km	Shillito *et al.* (1985)
	PEG/Ep	Leaf protoplasts	km, cm	Pietrzak *et al.* (1986)
	Liposomes	Leaf protoplasts	km	Deshayes *et al.* (1985)
	Liposomes	Leaf protoplasts	km	Zhu *et al.* (1990)
	Pg	Leaf pieces	km	Tomes *et al.* (1990)
	Pg	Leaves	km	Klein *et al.* (1988)
	At	Suspension cells	km	An (1985)
	At	Plantlet protoplasts	km, cs	Lee *et al.* (1988)
Petunia hybrida (petunia)	At	Leaf disks	km	Horsch *et al.* (1985)
	At	Leaf protoplasts	km	Nagy *et al.* (1985)
	At	Leaf disks	mtx	Eichholtz *et al.* (1987)
	PEG/Ep	Plantlet protoplasts	km	Tagu *et al.* (1988)
Populus alba × *P. grandidentata* (poplar)	At	Leaf segments	km	Fillatti *et al.* (1987b)

Solanum melongena (eggplant)	At, co-integrate	Leaf strips	km	Guri and Sink (1988)
	At, binary	Leaf strips	km	Rotino and Gleddie (1990)
Solanum tuberosum (potato)	At	Intact shoots	km	Ooms *et al.* (1987)
	At	Tuber disks	km	Stockhaus *et al.* (1987)
	At	Tuber disks	km	Sheerman and Bevan (1988)
	At	Leaf pieces	km, hm	Knapp *et al.* (1988)
	At	Microtubers	km	Ishida *et al.* (1989)
	At	Leaf strips	km	Wenzler *et al.* (1989)
	At	Excised leaves	km	De Block (1988)
Stylosanthes humilis	At	Leaf pieces	km	Manners (1988)
	Ar	Leaf and stem pieces	km	Manners and Way (1989)
Vigna aconitifolia (moth bean)	At	Leaf protoplasts	km	Eapen *et al.* (1987)
II. Monocots				
Asparagus officinalis (asparagus)	At	Stem segments	km	Bytebier *et al.* (1987)
Dactylis glomerata (orchard grass)	PEG, Ep	Suspension protoplasts	hm	Horn *et al.* (1988)
Oryza sativa (indica rice)	PEG	Suspension protoplasts	km	Peng *et al.* (1990)
Oryza sativa (type II rice)	PEG	Suspension protoplasts	hm	Datta *et al.* (1990)
Oryza sativa (japonica rice)	Ep	Suspension protoplasts	km	Zhang *et al.* (1988)
	Ep	Suspension protoplasts	G418	Toriyama *et al.* (1988)
	PEG	Suspension protoplasts	hm	Hayashimoto *et al.* (1990)
	Ptp	Flowering spikes	None or km	Lou and Wu (1988)
Secale cereale (rye)	Mal	Floral tillers	km	de la Peña *et al.* (1987)
Triticum aestivum (wheat)	At, Ptp	Preflowering spikelets	km	Hess *et al.* (1990)
Zea mays (ssp. *mays*) (corn)	Ptp	Silks	None	Ohta (1986)

(*continued*)

TABLE 5.1 *(Continued)*

Plant	Transformation technique[b]	Tissue source	Selectable agent[c]	Reference
	Ep	Suspension protoplasts	km	Rhodes *et al.* (1988)
	Pg	Suspension cells	pt	Gordon-Kamm *et al.* (1990)
	Pg	Suspension cells	pt, cs	Fromm *et al.* (1990)

[a]Intended as a reference of different techniques that have been used to regenerate transgenic plants. It does not include all papers to date in which transgenic plants were obtained. Multiple listings for a specific species are given only when sufficiently different techniques for regenerating transgenic plants were used. Transformation studies in which only transgenic callus or tissue was obtained are generally not included.

[b]At, *Agrobacterium tumefaciens*; Ar, nononcogenic *Agrobacterium rhizogenes*; Ar(At), oncogenic *A. rhizogenes* plus *A. tumefaciens* T-DNA; Ep, electroporation; MaI, macroinjection; MiI, microinjection; Pg, particle gun; Ptp, pollen-tube pathway; PEG, polyethylene glycol.

[c]bial, Bialaphos; brm, bromoxynil; cm, chloramphenicol; cs, chlorsulfuron; G-418 sulfate; gly, glyphosate; Hrp, hairy root phenotype; hm, hygromycin B; km, kanamycin sulfate; mtx, methotrexate; nos, nopaline synthase (nopaline dehydrogenase); pm, paromomycin sulfate; pt, phosphinotricin; stm, streptomycin.

cells in an explant or within a culture will irreversibly differentiate, rendering them incapable of cellular proliferation and totipotency (Vasil and Vasil, 1986; Ammirato, 1987). In addition, chromosomal abberations (e.g., polyploidization) in long-term cultures may also render cultured cells incapable of expressing totipotency although they may be capable of proliferation (Halperin, 1969; Halperin, 1986; Lindsey and Yeoman, 1985; Lee and Phillips, 1988). Thus, it is generally recognized that not all individual cells in an *in vitro* culture are capable of expressing totipotency, even if the culture as a whole is regenerable (Raghavan, 1986; Ammirato, 1987; Ranch and Pace, 1988; Vasil, 1988; Potrykus, 1990).

Research directed toward understanding the nature of totipotency, especially in nonmodel species, is extremely important from both a practical and scientific perspective. The first step toward understanding totipotency is identifying and characterizing which cells (or cell clusters) in a culture are totipotent. This knowledge would be helpful in the establishment of highly totipotent suspension cultures. Although the distinction between totipotent and nontotipotent cells has been achieved to some degree morphologically and cytologically, the biochemical and molecular basis for totipotency is poorly understood. Until we understand the nature and characteristics of totipotency, rather than just identifying its presence or absence morphologically, the science of tissue culture will continue to carry an artistic component.

5.1.2 COMPETENCE

Because tissue explants are a mixture of cells varying physiologically, biochemically, and developmentally (Lindsey and Yeoman, 1985), it is not surprising that exposure of this tissue to a specific *in vitro* environment will elicit various reactions from the different cell types. The result of this cellular heterogeneity is that only certain cells in an explant will be capable of or "competent" to respond to the *in vitro* culture conditions (Halperin, 1969; Ammirato, 1986; Vasil and Vasil, 1986; Potrykus, 1990). The term competence refers to the capability of a cell or group of cells to respond to an inductive stimulus for a developmental process (Meins and Binns, 1979). It is possible for a cell to be totipotent, but not competent to express totipotency under a particular set of experimental inductive conditions. Competence (or lack thereof) has been used to explain cellular response for most *in vitro* developmental processes including (1) why some cells of explants are capable of dedifferentiating and resuming mitotic activity; (2) why some explant or cultured cells are capable of responding to morphogenic stimuli [e.g., shoot- or root-inducing media (Christianson and Warnick, 1988)]; and (3) why some explant or cultured cells can regenerate following protoplast isolation and culture (Potrykus and Shillito, 1986). Presumably, the competent state represents, for each individual developmental process or series of processes, some unique genetic, epigenetic, and physiological status of the responding cell. Because many developmental processes in plants are poorly understood, the term competence has unfortunately become a catch-all term to describe the status of those cells that respond to *in vitro* inductive environments. Interestingly, competence has also been used to describe cells capable of being transformed (Potrykus, 1990).

5.1.3 PLANT CELL CULTURE MORPHOLOGIES

All plant cell cultures are a heterogeneous mixture of cell types (Murashige, 1974; Lindsey and Yeoman, 1985; Smith, 1986). Culture heterogeneity would be expected in organized cultures (see below) as the presence of any morphogenesis in the culture requires heterogeneity in cell type. However, even in relatively homogeneous cultures, the microenvironment within different sized cell clusters could conceivably contribute to variation in cellular character (Lindsey and Yeoman, 1985). Additionally, there are at least two types of cells in any proliferating culture: meristematic and differentiated. Differentiated cells are characterized by being large, variably shaped, and highly vacuolated cells with a thin cytoplasm and inconspicuous nuclei. In contrast, meristematic cells are characterized by having a small isodiametric shape, dense cytoplasm, little to no vacuolization, many starch grains, and large nuclei with prominent nucleoli (Murashige, 1974; Brown and Thorpe, 1986; Smith, 1986). In addition to culture heterogeneity resulting from cellular maturation, other sources of heterogeneity may be traced back to the original tissue explant, or genetic aberrations and epigenetic variations that arise during culture (Lindsey and Yeoman, 1985; Halperin, 1986; Lee and Phillips, 1988).

There are basically two types of *in vitro* cultures with respect to the morphologies in an established culture. These morphologies can simply be catagorized as unorganized and organized (Potrykus and Shillito, 1986; Ranch and Pace, 1988). Beyond this simple and somewhat subjective classification, numerous specific organized morphologies may arise depending on the species, cultivar, explant source, and culture conditions (physical and chemical). The recognition of and ability to distinguish and select between different morphologies for given situations (e.g., species and culture conditions) is a skill that must be acquired through experience. Unorganized cultures are nonmorphogenic and relatively homogeneous in appearance, although single differentiated cells and variable sized cell clusters may exist in the same culture (Halperin, 1966; Torrey, 1966). In addition, these culture types can be subdivided into regenerable and nonregenerable unorganized cultures. Regenerable but unorganized cultures are capable of being induced to organize and regenerate plants through organogenesis or embryogenesis (see Sections 5.3.1 and 5.3.2 below). Most cultures that regenerate by indirect organogenesis (e.g., tobacco cultures) fit into this category as do carrot suspension cultures maintained at a high 2,4-D concentration and high cell density. The second type of unorganized culture is the nonregenerable unorganized culture which does not have the capability for morphogenic induction and will always remain in a mode of cellular proliferation. An example of a nonregenerable unorganized culture is the Black Mexican Sweet maize line that has been an important model system in maize protoplast and transformation work (Chourey and Zurawski, 1981; Ludwig *et al.*, 1985; Fromm *et al.*, 1986; Spencer *et al.*, 1990). In contrast to the unorganized cultures, some type and level of morphogenesis can be recognized in organized cultures. The morphogenic organization is generally for embryogenesis but may include both embryogenesis and organogenesis. The level of organization may be slight, or may be advanced as for soybean (Finer, 1988a) or walnut (McGranahan *et al.*, 1988) cultures in which secondary somatic embryos proliferate from existing somatic embryos. Organized cultures may contain a mixture of single cells, unorganized cell groups, and organized structures.

Finally, unorganized or slightly organized cultures have been described as friable. Friability refers to the structural property of a culture in which the cells or cell clusters tend to separate easily. The basis for friability is believed to be a cell wall property that limits cell to cell contact. Auxin concentrations and other components (e.g., nitrogenous compounds) in medium have been shown to have an affect on friability (Armstrong and Green, 1985). Callus culture friability is important for suspension culture initiation as friable cultures will disperse readily in liquid (Evans *et al.*, 1981; Kamo and Hodges, 1986; Smith, 1986). In general, suspension cultures are less morphogenic and less organized than callus cultures maintained on semisolid media.

5.2 Factors Affecting *in Vitro* Culture

The three factors affecting *in vitro* plant regeneration are genotype, explant source, and culture conditions (including culture medium and environment). Successful *in vitro* culture initiation and regeneration is dependent upon the correct decisions for all of these factors. In fact, because we do not yet understand plant hormonal mechanisms or plant developmental processes, the best approach for developing an *in vitro* regeneration system is still based on empirically testing variations of these three factors until some combination proves successful (Halperin, 1986; Ranch and Pace, 1988). However, as seen in Section 5.2.1, one can take advantage of known genotype responsiveness by crossing recalcitrant genotypes to regenerable genotypes (Hodges *et al.*, 1986).

5.2.1 SPECIES AND CULTIVAR

The choice of species and cultivar will obviously depend on one's particular experimental objectives. However, if one simply wishes to regenerate a plant by *in vitro* culture, the choice of genotype is probably the most important factor to consider. Using a given protocol, some species or cultivars within a species simply have a greater capacity to regenerate *in vitro* than other species or cultivars within a species. This statement must be tempered with the realization that poor culturability and regenerability of some species or genotypes can be overcome by alterations in the culture protocol. Nevertheless, tobacco (*Nicotiana tabacum*) has been the model species for organogenic studies and carrot (*Daucus carota*) has been the model species for the study of embryogenesis (Flick *et al.*, 1983; Ammirato, 1983). These species have probably arisen as model systems because of the relative ease and repeatability of culture initiation and plant regeneration, and possibly due to the relatively wider range of culture conditions that can be successfully used with these species (Flick and Evans, 1984). General protocols for culture initiation, maintenance, and regeneration of these species are numerous (Evans *et al.*, 1981; Reinert and Yeoman, 1982; Flick *et al.*, 1983; Ammirato, 1983; Smith, 1986). Other species within the same family as tobacco (Solanaceae) and within the same family as carrot (Umbelliferae) have also been easily amenable to *in vitro* culture and regenerability, whereas other families (e.g., Gramineae) have not been as amenable to *in vitro* culture (Flick *et al.*, 1983; Ammirato, 1983). The finding that particular cultivars within a species respond differently in *in vitro* culture (see examples in Brown and Thorpe, 1986), and studies that show *in vitro* plant regeneration can be enhanced by breeding for this trait (Bingham *et al.*, 1975; Koornneef *et al.*, 1986; Hodges *et al.*, 1986), illustrate that the genetic component is highly influential on success for *in vitro* culture and plant regeneration. However, this conclusion is based on evidence derived using one or a few specific experimental protocols, and different protocols can have pronounced effects on the response of a given cultivar. Vasil and Vasil (1986) argue that the genotypic component can be accounted for if the correct explant and culture medium are used. The corollary to this statement is that all species and cultivars are amenable to *in vitro* culture if the correct combination of explant and culture conditions are used. In a general sense this corollary is reasonable, but it is difficult at this time to imagine a species such as maize having the range and flexibility of successful *in vitro* regeneration protocols (i.e., use of different explant sources and culture conditions) as that seen for tobacco. We would agree with Potrykus (1990) that the type of wound response (both in terms of callus proliferation and irreversible loss of competence for resuming mitotic activity) produced by a species and explant is a primary reason for the differences seen in culturability and regenerability between species such as maize and tobacco.

5.2.2 EXPLANT SOURCE AND ONTOGENETIC STAGE

For the carrot and tobacco model systems, plants have been regenerated from explants originating from almost every part of the plant (Ammirato, 1983; Flick *et al.*, 1983; Flick and Evans, 1984). Thus, the choice of explant source for these species is not as critical as for nonmodel species. For all species, however, generally younger or vigorously growing tissues from healthy plants should be used as the explant source, although differentiated tissue (e.g., tobacco pith) can be used in many cases for dicots (Flick *et al.*, 1983). Successful *in vitro* regeneration for dicot species has been achieved using leaf pieces (disks or strips), leaf petiole segments, cotyledonary petioles, cotyledonary pieces, hypocotyl segments, root segments, stem segments, various floral and inflorescence structures, storage root and tuber pieces, embryos, and immature embryos (Flick *et al.*, 1983; Table 5.1). The use of young tissue is especially important for cereal monocots such as maize where parenchyma cells *in vivo* quickly lose the ability to redifferentiate as they mature (Aloni and Plotkin, 1985). Although plants have been regenerated *in vitro* from many different tissues of monocots (Flick *et al.*, 1983; Ammirato, 1983), the immature organs or meristematic and undifferentiated tissues are generally the most responsive and reliable explant sources (Green and Phillips, 1975; Kamo *et al.*, 1985; Vasil and Vasil, 1986). The reason for this stage-specific response may be due to genetic, epigenetic, or physiological changes that occur in mature cells (Halperin, 1986; Vasil, 1988; Lee and Phillips, 1988). Other factors of the explant to consider include size, orientation in culture, pretreatment, and inoculation density (Brown and Thorpe, 1986).

5.2.3 DEDIFFERENTIATION AND CALLUS INDUCTION

One of the first steps required for *in vitro* culture establishment by the indirect organogenic and embryogenic pathways (Figures 5.1 and 5.2) must be a dedifferentiation of tissue explant cells and a resumption of cell division (Evans *et al.*, 1981; Ammirato, 1986; Christianson and Warnick, 1988). As differentiation and dedifferentiation are terms describing visual developmental or morphogenic processes (Christianson, 1987), the process of dedifferentiation is not apparent until we see callus proliferation. It can be assumed that explant cells are probably determined to some extent for a developmental pathway. Thus, another possible initial developmental process that occurs upon initial culturing would be a loss to some degree of the previously determined state (i.e., dedetermination) of the explant cells. Presumably, factors that initiate dedifferentiation and callus proliferation (e.g., growth regulator application or tissue wounding) also initiate dedetermination of the cells. There is no real evidence that complete dedetermination occurs on initial culture of explants although dedifferentiation may seem complete due to nonmorphogenic callus proliferation. However, if a root or leaf explant proliferates callus followed by redifferentiation to a new developmental pathway of shoot organogenesis, there is no question that the previous determined pathway has been changed. Thus, the time that is required to attain competence to respond to organogenic induction is perhaps the time required to

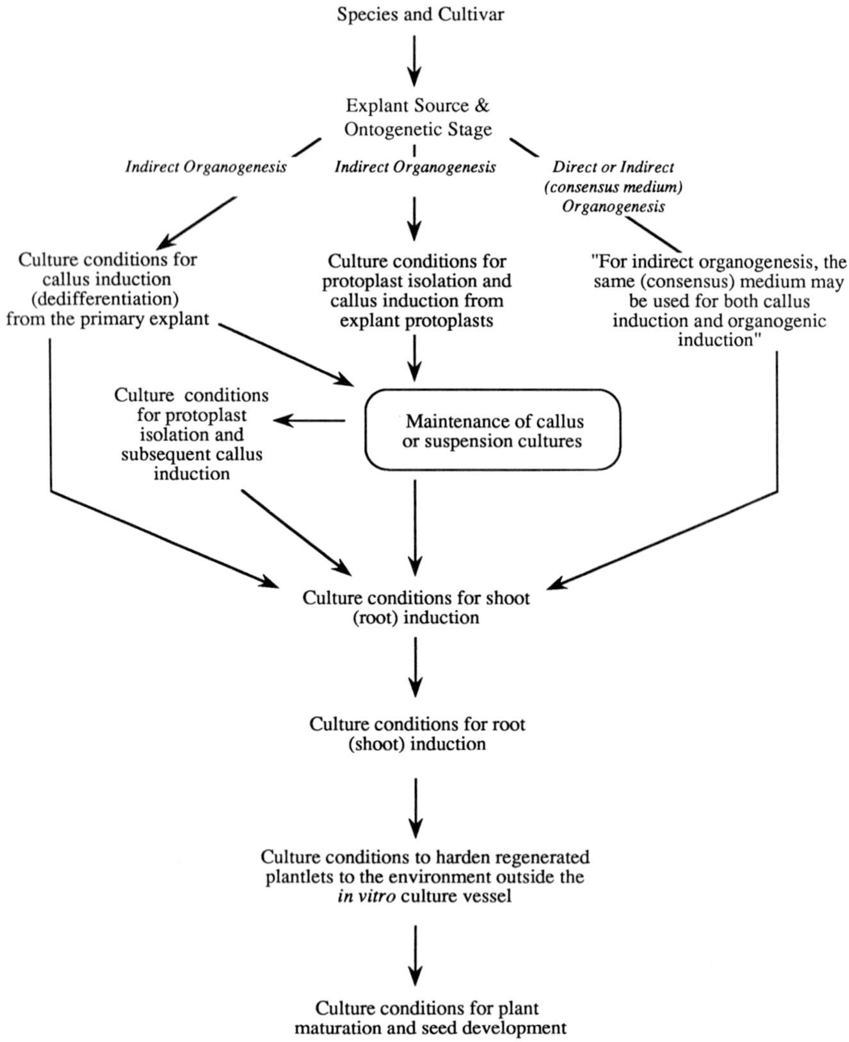

Figure 5.1 *In vitro* organogenesis.

attain "complete" dedifferentiation (Coleman and Ernst, 1990) or dedetermination. There are several accounts where tissues grown *in vitro* retain a portion of the "character" or "history" of the original explant for an extended time (Meins and Binns, 1979). Indeed, epigenetic cellular differences are believed to be one of the sources of cellular heterogeneity originating from the explant upon initiation of cultures (Lindsey and Yeoman, 1985; Halperin, 1986). It has also been

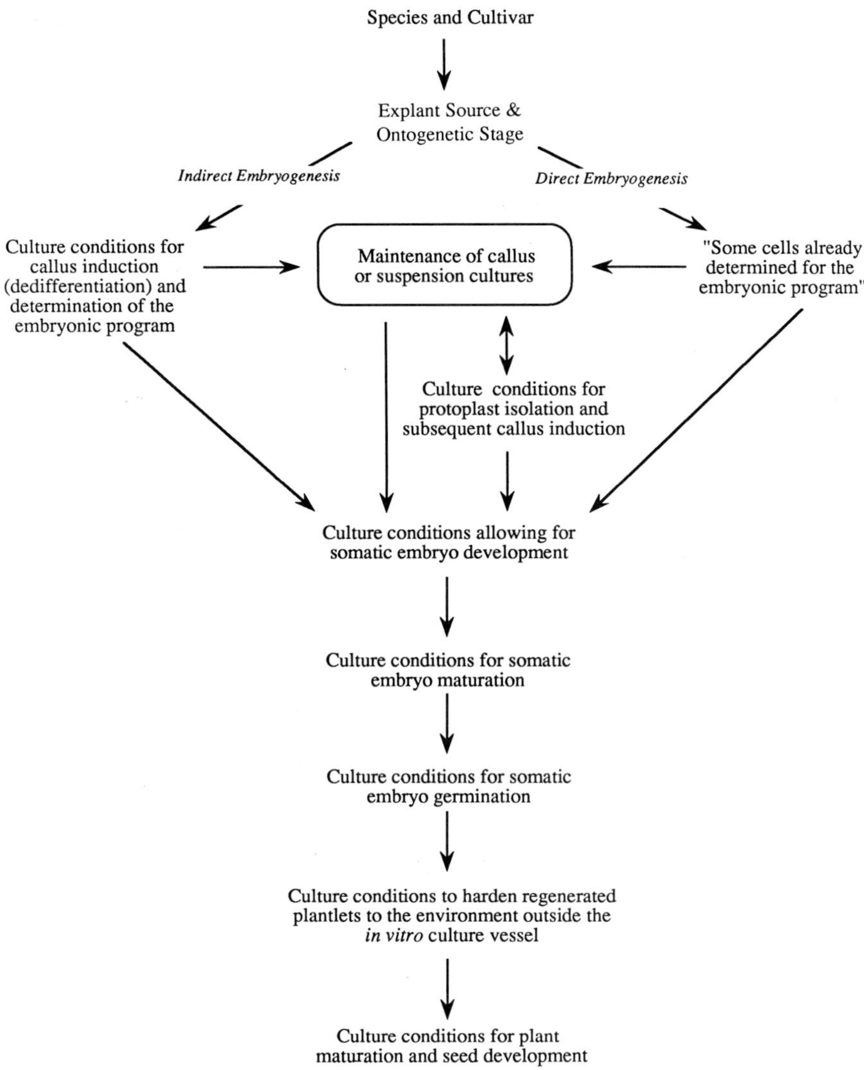

Figure 5.2 *In vitro* somatic embryogenesis.

hypothesized for cereals that establishment of embryogenic callus from immature embryo explants is successful because the explant consists of many predetermined embryogenic cells, or cells that have retained the embryogenic program despite their becoming determined as scutellar cells (Williams and Maheswaran, 1986).

Both practical and conceptual questions about cell culture still remain regarding determination, dedifferentiation, and dedetermination. For example, when the explant is initially cultured, what degree of determination is carried over; how much dedetermination occurs? Can the degree of dedetermination or retained epigenetic character be manipulated or controlled? How much control do the initial endogenous hormonal concentrations have on dedetermination? Is dedetermination merely a reversal of determination, but at an accelerated rate? How closely is dedetermination coupled to dedifferentiation and callus proliferation? When callus proliferation *in vitro* occurs, does this show that a certain level of dedetermination has been attained? How often is the problem of apparent recalcitrance caused by incomplete dedetermination? What are the molecular mechanisms that control determination and dedetermination? Are determination and dedetermination possibly controlled by gene methylation and/or chromatin structure? Is gene methylation the major control mechanism for developmental processes *in vivo* (Poethig, 1990) and *in vitro* (LoSchiavo *et al.*, 1989)? All of these questions will become answered as scientists gain an understanding of plant developmental processes at the molecular level.

5.2.4 WOUND CALLUS AND WOUND RESPONSE

Because the initiation of *in vitro* cultures almost invariably requires some injury to the explanted tissue, wound response is an unavoidable consequence during the initial culture. The type of wound response exhibited by the explant will depend on the species, cultivar, tissue type, and tissue age (Kahl, 1982). When tissues from many dicot species (e.g., tobacco stems) are injured, they produce a callus in response to the injury at the site of the wound. This wound callus initially consists of nonmorphogenic cells that will fill in the wounded area and may eventually differentiate into cells and tissues representative of those that were destroyed by the injury (Kahl, 1982). In addition, the cell walls of this wound callus will become impregnated with phenolics and lignin, and many injury-specific and defense-related gene activations will occur (Lamb *et al.*, 1989). Thus, wounding in these cases initiates a series of complex developmental processes involving dedifferentiation from the existing determined and quiescent state, induction of cell division, and finally redifferentiation and a return to quiescence (Bloch, 1952). The first steps in this simplified sequence of wound healing (i.e., dedifferentiation and callus proliferation) are exactly those required when initiating a callus culture. In addition, the morphological similarities between wound callus and callus cultures were noticed long ago (Bloch, 1952). It has been suspected or speculated, therefore, that callus cultures from species and explants that exhibit this type of wound response are cultures that maintain and amplify the wound response (Smith, 1986; Potrykus and Shillito, 1986; Potrykus, 1990). Recently, Grosset *et al.* (1990) have shown that the newly synthsized mRNA following tobacco mesophyll protoplast isolation corresponds, to a large degree, to newly synthesized mRNA found following wounding of the whole leaf. These mRNAs were shown to be synthesized in response to the initial wounding before protoplast isolation, and not in response to the protoplast isolation procedure per se. Furthermore, these wound-induced mRNA were also present in an established tobacco suspension culture. From their results, and from the observation that *in vivo* wound callus and *in vitro* callus both have the potential for shoot or root organogenesis, they conclude that *in vitro* cultivated cells are actually wound callus. Whether this general conclusion by Grosset *et al.* (1990) is true for all explants and all species is questionable, although it probably applies to species that exhibit a wound response similar to tobacco. A second type of wound response is seen in potato tuber slices where cellular proliferation and wound healing are less random due to the establishment of a wound periderm (Lipetz, 1970; Esau, 1977; Kahl, 1982). Because callus proliferation normally occurs with this type of wound response, *in vitro* cultures of these tissues also may represent an amplified and maintained wound response making it similar to tobacco. In contrast, monocots typically show a poor wound response by increasing cell wall lignification of cells adjacent to the wound and by showing little or no proliferating wound callus in the wound site (Swamy and Sivaramakrishna, 1972; Esau, 1977; Kahl, 1982). Additionally, monocot cells may lose their ability to dedifferentiate and resume meristematic activity upon maturity (Wernicke and Brettell, 1980; Wernicke *et al.*, 1981; Aloni and Plotkin, 1985). To what extent (if any) wounding contributes to callus induction and *in vitro* culture of monocot species or dicots that also show a limited wound response is unknown.

The type of wound response produced by a particular species and explant is of practical significance, and is perhaps the single most important factor contributing to successful plant transformations at present. Most successful transformation systems to date utilize the *Agrobacterium*-mediated transformation technique (Table 5.1). Success with this technique is reported to be dependent on cell divisions taking place in cells adjacent to the wound (Binns, 1990). If a species such as a cereal monocot shows as poor wound response (i.e., a small amount of cell division near the wound site by only a few cells), this widely used transformation technique would not be expected to work (Potrykus, 1990). However, it is possible that we have not yet learned the most efficient means for maintaining and amplifying the limited wound response of cereal monocots. Efforts along these lines could be important for genetic engineering of monocots using *Agrobacterium*.

5.2.5 CULTURE CONDITIONS

Vasil (1988) pointed out that success with regeneration of grasses and cereals was not achieved initially because the applied culture conditions (in particular the

2,4-D concentration) were the same as those used for successful regeneration of dicots. Thus, the culture conditions (both chemical and physical) are highly influential in developing a regeneration system, and culture conditions optimized for one species or genotype may not be optimal or even usable for another. When developing a regeneration system, optimization of the different culture conditions must be determined individually and, unfortunately, empirically for each particular genotype and explant of choice.

Of all the various culture conditions, the composition of the medium is probably the most important. However, the amount of time that an explant or callus is exposed to a particular medium has also proved to be critical for several species (Christianson and Warnick, 1988; Valvekens *et al.*, 1988; Flinn *et al.*, 1988; Coleman and Ernst, 1990). Typically, the components of a medium will include (1) inorganic macro- and micronutrients, (2) reduced nitrogen, (3) an energy or carbon source, (4) vitamins, and (5) growth regulators (Gamborg and Shyluk, 1981). Of these components the growth regulator concentrations and ratios have proven to be the most critical for culture initiation and morphogenesis (Skoog and Miller, 1957; Vasil, 1988). In addition to their nutritional value, other components such as the reduced nitrogen (Halperin, 1966) and sugar content (Ammirato, 1986; Brown and Thorpe, 1986) may significantly affect morphogenesis.

Cultures can be initiated and/or maintained on semisolid medium or as suspensions in liquid with shaking. The decision to use one or the other mode of culture may depend on the eventual use of a particular transformation technique. Other culture conditions that need individual attention include light (including intensity, quality, and photoperiod), temperature, and culture container (including container size and permeability to gas exchange). Aspects of various culture conditions are discussed in greater detail by Evans *et al.* (1981), Ammirato (1986), Raghavan (1986), and Morrish *et al.* (1987).

5.3 Avenues for *in Vitro* Plant Regeneration

Transgenic plants have been regenerated by organogenesis, somatic embryogenesis, or by *in vitro* culture (micropropagation) of existing isolated apical meristems (Table 5.1). A general and simplified flow diagram depicting the steps in *in vitro* plant regeneration by organogenesis and somatic embryogenesis are provided in Figures 5.1 and 5.2, respectively.

5.3.1 ORGANOGENESIS

Organogenesis is a developmental pathway in which shoots or roots (i.e., organs) have been induced to differentiate from a cell or group of cells. *In vitro* plant regeneration by organogenesis usually involves induction and development of a shoot from the explant tissue (with or without an intervening callus stage), followed by transfer to a different medium to induce root formation and development (see Figure 5.1). If the shoot or root is induced and develops directly from a pre-existing cell(s) in the explant without undergoing an initial callus phase, this is termed direct or adventitious organogenesis (Ammirato, 1986; Brown and Thorpe, 1986; Christianson and Warnick, 1988). Typical examples of this type of organogenesis is when shoot cuttings for horticultural plants are rooted for further macropropagation, or when shoot segments are maintained aseptically for *Agrobacterium*-mediated transformations. An example of direct organogenesis *in vitro* is found with tobacco epidermal peels (Tran Thanh Van, 1981) or leaf disks of strawberry (Nehra *et al.*, 1989). Indirect organogenesis involves an initial phase of callus proliferation and growth, followed by shoot or root induction and development from this proliferated callus tissue (see Figure 5.1).

The first major advancement in our control of organogenesis can be attributed to Skoog and Miller (1957) who reported that alterations of the auxin and cytokinin ratios were sufficient to control morphogenesis in tobacco. High cytokinin-to-auxin ratios were found to produce shoots, low cytokinin-to-auxin ratios produced roots, and more equal concentrations of these phytohormones were found to cause callus proliferation. Since this initial report, media formulations for callus, shoot, and root induction have been devised for many plant species (Evans *et al.*, 1981; Flick *et al.*, 1983). In general, these media formulations have been derived empirically due to a lack of our understanding of plant development (*in vivo* and *in vitro*) and mechanisms of hormone action, and to the inability to successfully apply plant regeneration protocols from one species or cultivar within a species to other species or cultivars.

By using experimental procedures of sequential media transfers, *in vitro* studies using *Convolvulus arvensis* (Christianson and Warnick, 1983, 1984, 1985), *Medicago sativa* (Walker *et al.*, 1979), *Arabidopsis thaliana* (Feldmann and Marks, 1986; Valvekens *et al.*, 1988), *Pinus strobus* (Flinn *et al.*, 1988), and *Populus deltoides* (Coleman and Ernst, 1990), it has been demonstrated that a certain (but defined) amount of time is required before a tissue is competent for morphogenic induction. Following the attainment of competence, a certain additional amount of exposure time to induction medium is required for the tissue to then become determined for the developmental pathway (Christianson and Warnick, 1983). Several important practical lessons can be gleaned from all of these organogenesis studies. First, not only is the medium composition important in morphogenesis (Skoog and Miller, 1957), but also the amount of time in which the tissue is exposed to a particular medium is critical (Christianson and Warnick, 1988). Second, the amount of time required for dedifferentiation and for attainment of competence for induction is surprisingly short in many instances but varies considerably with different genotypes. Finally, several studies show that competence for induction is transient. There is a window of competence where induction for morphogenesis is possible; on either side of this window, only callus proliferation occurs (Valvekens *et al.*, 1988; Coleman and Ernst, 1990).

For organogenesis to occur from a single medium, this medium must cause dedifferentiation (callus induction), attainment of competence, induction for the organogenic pathway, determination for the pathway, and finally, not interfere with the morphogenic expression of the developmental pathway. This represents a considerable amount of developmental biology to expect of a single medium; nevertheless, such consensus media do exist (Christianson and Warnick, 1985). The use of such a consensus medium was the basis for the regeneration system used in the *Agrobacterium*-mediated leaf disk transformation system (Horsh *et al.*, 1985; see Figure 5.1).

Organogenesis is the most widely used avenue of *in vitro* plant regeneration in transformation systems at the present time (Table 5.1). All *Agrobacterium*-mediated transformation systems utilizing the leaf disk procedure, or variations thereof, base their regeneration system on direct or indirect organogenesis. Furthermore, transformation systems utilizing protoplasts may ultimately lead to plants regenerated by indirect organogenesis (see Figure 5.1) as well as embryogenesis (see Section 5.3.2). Due to potential problems with somaclonal variation in callus cultures, it is more desirable to regenerate transgenic plants by direct rather than indirect organogenesis, or to at least minimize callus proliferation before regeneration.

5.3.2 EMBRYOGENESIS

Somatic embryogenesis is a developmental pathway in which embryos have been induced to form from a somatic cell or group of somatic cells. As with organogenesis, somatic embryogenesis (see Figure 5.2) can occur directly from cells of the explant tissue without an intervening callus phase (Konar and Nataraja, 1965; Conger *et al.*, 1983). However, the indirect embryogenesis pathway, where somatic embryos are induced and develop from a proliferated callus, is generally more common (Williams and Maheswaran, 1986).

Embryogenic cells appear very similar to meristematic cells in that they are small, densely cytoplasmic, have a large nuclei and prominent nucleoli, and contain many small vacuoles, lipid droplets, and starch grains (Williams and Maheswaran, 1986; Vasil, 1988). Embryogenic cells in culture (suspension or callus) commonly form small compact clumps that Williams and Maheswaran (1986) have termed proembryonal complexes. These complexes have also been referred to as proembryogenic masses (PEM's) in carrot suspension cultures (Halperin, 1966) and represent the initially organized structures from which one to several somatic embryos form (Nomura and Komamine, 1985; deVries *et al.*, 1988). Nonembryogenic cells resembling the previously described differentiated cells (i.e., highly vacuolated, large variable shape) will also form in embryogenic cultures. Macroscopic morphologies of established embryogenic cultures (suspension or callus) may vary from seemingly unorganized cultures (e.g., carrot cultures containing PEM's) to cultures containing somatic embryos with nodular shapes and/or structures strikingly similar to zygotic embryos (Halperin, 1969; Raghavan, 1986; Williams and Maheswaran, 1986; Ammirato, 1987; Morrish *et al.*, 1987). In addition, many monocot cultures may be a mix of organogenic and embryogenic structures (McCain and Hodges, 1986; Morrish *et al.*, 1987).

During the initiation of embryogenic cultures, the supplied auxin (usually 2,4-D) will induce both cellular proliferation (callus induction) and the embryogenic pathway of development, if not already determined (see Figure 5.2). It is generally thought that the embryogenic pathway is induced and becomes determined very early in embryogenic cultures (Evans *et al.*, 1981; Nomura and Komamine, 1985), and this clearly seems to be the case in the model species carrot (Wilde *et al.*, 1988; de Vries *et al.*, 1988). Similar to organogenesis, a certain period of time is apparently required for newly initiated carrot suspension cultures to dedifferentiate and attain competence for the embryogenic pathway (de Vries *et al.*, 1988). Considering that the induction and determination for embryogenesis occurs quickly, and that the high maintenance levels of auxin repress embryogeny, established carrot cultures are thus embryogenically determined cultures maintained at some very early stage of embryo development (see Figure 5.2; Williams and Maheswaran, 1986; Ammirato, 1987; Wilde *et al.*, 1988; de Vries *et al.*, 1988). Once the embryogenic pathway is determined, there is no difference in maintenance of cultures and developmental potential between induced embrogenically determined cells (indirect embryogenesis pathway) and cells that were embryogenically predetermined (direct embryogenesis pathway; Williams and Maheswaran, 1986). For species other than carrot, embryogenic cultures are probably also comprised of determined cells in which some level of embryo development is maintained in culture. The level of morphogenic expression of embryonic development can vary considerably as described above. The assumption that embryogenic cultures of other species are similar to those of carrot is reasonable, but the issue of when cellular determination occurs in other species is still speculative.

When embryogenic suspension cultures are initiated (see Section 5.3.4 also), they usually contain a large percentage of elongated and highly vocuolated cells. Over the next several months, however, the culture will gradually become "established" as the number of proembryonal complexes (small cell clusters) increases and the number of nonembryogenic cells declines. This process may be accelerated by selection techniques such as seiving (Prioli and Sondahl, 1989), selecting the best of several flasks (Shillito *et al.*, 1989), or by addition of filter-sterilized conditioned media from already established embryogenic cultures (de

Vries *et al.*, 1988). Most established embryogenic suspensions will appear unorganized; however, very organized suspension cultures of some species have been reported [e.g., cotton (Finer, 1988b)]. Once established, the embryogenic suspension culture can be used for protoplast isolation and culture (see Figure 5.2) or for studies of embryo development. Several species have been transformed using embryogenic suspension cultures as starting material including carrot, corn, rice, and cotton (Table 5.1).

In established embryogenic cultures, the exogenously supplied auxin maintains cellular proliferation and, therefore, to some degree represses morphogenic expression of embryogenicity (Ammirato, 1983, 1987). The degree of morphogenic repression will depend primarily on the auxin concentration or cultural practices that affect the auxin concentration (e.g., extended time between subculturing), but may also depend on the cell density for suspension cultures. It is hypothesized that the auxin represses morphogenesis by disrupting important cell-to-cell interaction. This allows some embryogenic cells in the clump to develop autonomously and break away from the clump, rather than remain as part of the developmentally integrated primordial complex. This fragmentation of single cells or small groups of cells from the clump is believed to be the mechanism for proliferation of embryogenic suspension cultures (Williams and Maheswaran, 1986; Ammirato, 1987).

Following removal or reduction of the auxin supply, the embryonic development in the culture can proceed to the maturation and germination steps (see Figure 5.2). The induction and maturation steps may occur on the same medium or on different media; however, the germination step generally requires removal of the exogenously supplied phytohormone (Ammirato, 1983, 1987). Following removal of the auxin, the amount of embryonic development that can occur in liquid suspension will vary from complete somatic embryo development in carrot suspensions (Ammirato, 1987) to blockage at the late globular or early scutellar stage for most cereals and grasses (Vasil and Vasil, 1986).

5.3.3 MICROPROPAGATION AND MACROPROPAGATION

Because micropropagation of active or arrested meristems does not involve *de novo* initiation of the meristematic tissue, Ranch and Pace (1988) pointed out that this practice, and the practice of embryo rescue, should not be considered as true *in vitro* plant regeneration. Nevertheless, micropropagation of apical meristems has been used as the regeneration system for transformation of cotton (Ulian *et al.*, 1988) and sunflower (Schrammeijer *et al.*, 1990; Table 5.1). Regeneration by micropropagation has the advantage that plants are regenerated directly from an organized tissue without an intervening callus stage. This not only saves time, but also eliminates undesirable somaclonal variation associated with long callus culture periods. Plant regeneration by organogenesis, somatic embryogenesis, and micropropagation eventually require a macropropagation phase once plantlets have been produced (see Figures 5.1 and 5.2). This generally involves exposing the plantlet to an outside environment that has a slightly lower (yet still high) relative humidity and a higher light level for acclimation or hardening. After sustained growth in this halfway environment, the hardened plantlet can be placed under more normal greenhouse or growth chamber conditions.

5.3.4 SUSPENSION CULTURES AND PROTOPLASTS

Although a suspension cell culture is not a separate pathway toward *in vitro* plant regeneration, it is an important component of indirect organogenesis and embryogenesis. Suspension cell cultures are usually initiated from friable callus cultures that have been established from plant explants. However, haploid cell suspension cultures of some species can be initiated by culturing isolated anthers or microspores directly into liquid medium as with carrot hypocotyl segments (de Vries *et al.*, 1988). Either way, callus induction and cellular proliferation will occur and must be maintained by defined methods. Suspension cultures have been, and continue to be, of major importance for *in vitro* plant regeneration and plant biotechnology for two reasons. First, carrot cell suspension cultures have been an important model system for the study of somatic embryogenesis in plants. Much of our knowledge on somatic embryogenesis has been compiled by studying this system at the morphological, cytological, biochemical, and molecular levels. Second, cell suspension cultures are an important, and sometimes exclusive, source of totipotent protoplasts. For monocots, suspension cells and suspension protoplasts have been the primary target tissues for plant transformation (Table 5.1) even though most regeneration protocols are for primary explant tissues (Vasil and Vasil, 1986). Suspension cells have recently been used to obtain transgenic maize via particle bombardment (Table 5.1). In addition, suspension protoplasts combined with electroporation or PEG-mediated transformation techniques have been utilized to obtain transgenic rice, corn, and orchard grass (Table 5.1). The number of species that have regenerated completely to plants following protoplast culture (starting from suspensions or directly from explants, see Figures 5.1 and 5.2) is quite large and continues to grow (Roest and Gilissen, 1989). The versatility of protoplasts is illustrated by the fact that they have been utilized in transformation systems using *Agrobacterium*, PEG, and electroporation (Table 5.1), and also for protoplast fusion (Potrykus and Shillito, 1986) and microinjection studies (Neuhaus *et al.*, 1987).

5.4 Selectable Markers and Transformation Techniques

5.4.1 SELECTABLE MARKERS

Selectable markers are also discussed in Chapters 1 and 9 in this volume and will only be briefly discussed here. There are two types of selectable marker systems that have been used to obtain transgenic plants (Tables 5.1 and 5.2). The most common type is one in which a gene for antibiotic or herbicide resistance is inserted into the plant genome (Table 5.2). In this situation, resistance to the selectable agent may be conferred by (1) overproduction of the protein that is targeted by the selectable agent; (2) production of a target protein that is resistant to the selectable agent; or (3) production of a protein that detoxifies the selectable agent (Gasser and Fraley, 1989). The objective of using the chemical type of selectable marker is to distinguish, by survivability or by growth rate, transgenic cells from nontransgenic cells during the plant regeneration process. The selectable agent is usually applied early in the plant regeneration system to allow better elimination of the nontransgenic cells or allow better selection and continued culture of transgenic cells. In some cases, exposure to the selectable agent may need to continue throughout the regeneration process to increase the regeneration

TABLE 5.2
Selectable Markers Used Successfully for Plant Transformation

Selectable marker gene and (coded protein)	Selectable agent[a]
Chemical	
Antibiotics	
nptII (neomycin phosphotransferase II)	km, pm, G-418
nptI (neomycin phosphotransferase I)	km, pm, G-418
cat (chloramphenicol acetyltransferase)	cm
hpt (hygromycin phosphotransferase)	hm
dhfr (dihydrofolate reductase)	mtx
stp (streptomycin phosphotransferase)	stm
Herbicides	
bar (phosphinothricin acetyltransferase)	pt, bial
aroA [5-enol-pyruvylshikimic acid 3-phosphate (EPSP) synthase]	gly
als (acetolactase synthase)	cs
bxn (bromoxynil-specific nitrilase)	brm
Nonchemical	
Hairy root phenotype	(−) Hormone media

[a]Same abbreviations as in Table 5.1.

frequency of transgenic plants and reduce the number of escapes (Fromm *et al.*, 1990).

The chemical selectable marker system, as with most parameters in a transformation system, must be determined empirically by first exposing nontransformed regenerating tissue to different concentrations of the chemical selectable agent. The sensitivity of some species or tissues to different selectable agents will vary (Hauptmann *et al.*, 1988; Dekeyser *et al.*, 1989; Catlin, 1990). Using the appropriate concentrations of the selectable agent that kills or inhibits growth of nontransgenic tissues, transgenic tissues must then be tested for resistance to the selectable agent. The resistance of the transgenic tissue will ultimately depend on the strength and tissue specificity of the promoter sequences driving the selectable marker gene (Dekeyser *et al.*, 1989; Thomzik and Hain, 1990).

In the nonchemical type of selectable marker system (Table 5.2) the hairy root phenotype (Hrp) has been used to screen (rather than select) for plants that have been cotransformed using both the *Agrobacterium rhizogenes* oncogenic Ri plasmid plus a separate *A. tumefaciens* binary vector carrying the gene of interest. The abnormal hairy root phenotype found in the R_0 plants is then removed by outcrossing with a normal phenotype. The Hrp and cotransformed T-DNA gene of interest will independently segregate in the progeny and screening for the gene of interest can then be made on the resulting normal phenotypic R_1 plants (Shahin *et al.*, 1986b; Sukhapinda *et al.*, 1987). Obviously, the attainment of phenotypically normal transgenic plants with this selectable marker system takes much longer to complete than with a chemical system. This Hrp system is similar to reporter gene systems such as luciferase (Ow *et al.*, 1986; Koncz *et al.*, 1987) or β-glucuronidase (Jefferson, 1987) in which transgenic plants are obtained by screening rather than selection.

5.4.2 TRANSFORMATION TECHNIQUES

The transformation techniques used in Table 5.1 include (1) *Agrobacterium*-mediated techniques with *A. tumefaciens* and/or *A. rhizogenes;* (2) direct DNA uptake methods utilizing electroporation, polyethylene glycol, liposomes, microinjection, and particle bombardment; and (3) nontissue culture approaches such as pollen-tube pathway, macroinjection, and seed transformation. As discussed in the above wound response and organogenesis sections, the *Agrobacterium*-mediated methods are the dominant transformation techniques to date. *Agrobacterium*-mediated transformation has not worked well in the past for transient expression studies because the genes were expressed in the bacteria as well as in the plant cells. However, *gusA* expression vectors are now available that have little or no background expression from the bacteria and will be useful for transient analysis of *Agrobacterium*-targeted or susceptible tissues (Janssen and Gardner, 1989; Vancanneyt *et al.*, 1990; Castle and Morris, 1990). This should be useful in screening for *Agrobacterium*-targeted tissues in species that have previously not responded to *Agrobacterium* transformation (e.g., cereal monocots). Given the knowledge that T-DNA transfer into monocot species is efficient (Grimsley, 1990), and the reported success using *Agrobacterium* to infect rice (Raineri *et al.*, 1990), the development of regeneration techniques for the *Agrobacterium*-targeted tissues will be an important area of research in the near future.

Although the direct DNA uptake methods are also applicable to dicots, they are especially important for monocots due to the poor success with monocots using *Agrobacterium*-mediated methods (Table 5.1). Of particular interest for cereal monocots is the particle bombardment technique that has recently been successful with maize suspension cells (Gordon-Kamm *et al.*, 1990; Fromm *et al.*, 1990). Given that regenerable wheat (Redway *et al.*, 1990) and rice (Table 5.1) suspension cultures are available, it is reasonable to assume that this method may also be used for transformation of other cereal monocots. Furthermore, the particle gun has been shown to deliver DNA into a wide variety of plant cells which are capable of expressing these genes transiently (Sanford, 1990). The biggest hurdle for success with this transformation technique for monocots or dicots may be the regeneration and selection of the target tissues. However, this technique is still developing and new instruments for delivery of the particles may be more compatible with existing *in vitro* regeneration or micropropagation protocols.

Finally, because the potential for somaclonal variation is an undesirable component of transformation systems that utilize *in vitro* regeneration systems, the development of nontissue culture approaches is a welcome alternative (Feldmann and Marks, 1987). Although there have been single accounts of transgenic plants obtained by the nontissue culture methods of seed transformation for soybean, pollen-tube pathway for rice, wheat, and corn, and macroinjection for rye (see Table 5.1), duplication of these procedures has not been reported by others. A detailed discussion of various transformation techniques is given in Chapter 4 in this volume.

5.5 Summary

In this review, we have attempted to combine the practical and conceptual aspects of *in vitro* regeneration systems. In the rush to obtain transgenic plants over the past ten years, we have perhaps forgotten that *in vitro* plant regeneration systems involve many basic plant developmental processes and that perhaps more attention should be given to developmental biology when devising an *in vitro* regeneration system. However, a good understanding of the concepts of dedifferentiation, competence, and determination at the molecular or mechanistic level is still lacking. The use of transgenic plants actually holds great promise for eventually clarifying these concepts. In addition, many practical problems still exist in the science of *in vitro* plant regeneration. We are still strongly limited by species-, cultivar-, and tissue-specific responses by explanted tissues under specific culture conditions. Undoubtedly, a major challenge for biotechnologists in the future will be the development of genotype-independent regeneration and transformation systems for the cereal monocots and other important food crops. As our understanding of hormone action mechanisms and plant developmental processes improves, these practical problems will be overcome. Advances in understanding of developmental biology will not only help *in vitro* regeneration systems, but will also lead to a better understanding of which genes control important agronomic traits. This knowledge will aid in the design of new and useful transgenic crops in the future.

Acknowledgment

This is journal paper number 13326 of the Purdue University Agricultural Experimental Station.

References

Aloni, R., and Plotkin, T. (1985). *Planta* 163, 126–132.
Ammirato, P. V. (1983). In *Handbook of Plant Cell Culture, Vol. 1* (Ed. by D. A. Evens, W. R. Sharp, P. V. Ammirato, and Y. Yamada), pp. 82–123. New York: Macmillian.
Ammirato, P. V. (1986). In *Plant Tissue Culture and its Agricultural Applications* (Ed. by L. A. Withers and P. G. Alderson), pp. 23–45. Boston: Butterworths.
Ammirato, P. V. (1987). In *Plant Tissue and Cell Culture* (Ed by C. E. Green, D. A. Somers, W. P. Hackett, and D. D. Biesboer), pp. 57–81. New York: Alan R. Liss.
An, G. (1985). *Plant Physiol.* 79, 568–570.
Armstrong, C. L., and Green, C. E. (1985). *Planta* 164, 207–214.
Basiran, N., Armitage, P., Scott, R. J., and Draper, J. (1987). *Plant Cell Rep.* 6, 396–399.
Bellini, C., Chupeau, M.-C., Guerche, P., Vastra, G., and Chupeau, Y. (1989). *Plant Sci.* 65, 63–75.
Bingham, E. T., Hurley, L. V., Kaatz, D. M., and Saunders, J. M. (1975). *Crop Sci.* 15, 719–721.
Binns, A. N. (1990). *Physiol. Plant.* 79, 135–139.
Biondi, S., and Thorpe, T. A. (1981). In *Plant Tissue Culture, Methods and Applications in Agriculture* (Ed. by T. A. Thorpe), pp. 1–20. New York: Academic Press, Inc.
Bloch, R. (1952). *Bot. Rev.* 18, 655–677.
Brown, D. C. W., and Thorpe, T. A. (1984). In *Cell Culture and Somatic Cell Genetics of Plants, Vol. 1* (Ed. by I. K. Vasil), pp. 1–12. New York: Academic Press, Inc.
Brown, D. C. W., and Thorpe, T. A. (1986). In *Cell Culture and Somatic Cell Genetics of Plants, Vol. 3* (Ed. by I. K. Vasil), pp. 49–65. New York: Academic Press, Inc.
Boulter, M. E., Croy, E., Simpson, P., *et al.* (1990). *Plant Sci.* 70, 91–99.
Bytebier, B., Deboeck, F., De Greve, H., Van Montagu, M., Hernalsteens, J-P. (1987). *Proc. Natl. Acad. Sci. U.S.A.* 84, 5345–5349.
Castle, L. A., and Morris, R. O. (1990). *Plant Mol. Biol. Rep.* 8(1), 28–39.

Catlin, D. (1990). *Plant Cell Reports* 9, 285–288.
Catlin, D., Ochoa, O., McCormick, S., and Quiros, C. F. (1988). *Plant Cell Rep.* 7, 100–103.
Chabaud, M., Passiatore, J. E., Cannon, F., and Buchanan-Wollaston, V. (1988). *Plant Cell Rep.* 7, 512–516.
Charest, P. J., Holbrook, L. A., Gabard, J., Iyer, V. N., and Miki, B. L. (1988). *Theor. Appl. Genet.* 75, 438–445.
Chee, P. P. (1990). *Plant Cell Rep.* 9, 245–248.
Chee, P. P., Fober, K. A., and Slightom, J. L. (1989). *Plant Physiol.* 91, 1212–1218.
Chourey, P. S., and Zurawski, D. B. (1981). *Theor. Appl. Genet.* 59, 341–344.
Christianson, M. L. (1987). In *Plant Tissue and Cell Culture* (Ed. by C. E. Green, D. A. Somers, W. P. Hackett, and D. D. Biesboer), pp. 45–55. New York: Alan R. Liss.
Christianson, M. L., and Warnick, D. A. (1983). *Develop. Biol.* 95, 288–293.
Christianson, M. L., and Warnick, D. A. (1984). *Develop. Biol.* 101, 382–390.
Christianson, M. L., and Warnick, D. A. (1985). *Develop. Biol.* 112, 494–497.
Christianson, M. L., and Warnick, D. A. (1988). *Hort. Sci.* 23(3), 515–519.
Christou, P., McCabe, D. E., and Swain, W. F. (1988). *Plant Physiol.* 87, 671–674.
Chupeau, M.-C., Bellini, C., Guerche, P., Maisonneuve, B., Vastra, G., and Chupeau, Y. (1989). *Bio/Technology* 7, 503–508.
Coleman, G. D., and Ernst, S. G. (1990). *Plant Sci.* 71, 83–92.
Comai, L., Facciotti, D., Hiatt, W. R., *et al.* (1985). *Nature (London)* 317, 741–744.
Conger, B. V., Hanning, G. E., Gray, D. J., and McDaniel, J. K. (1983). *Science* 221, 850–851.
Constabel, F. (1984). In *Cell Culture and Somatic Cell Genetics of Plants, Vol. 1* (Ed. by I. K. Vasil), pp. 27–35. New York: Academic Press, Inc.
Datta, S. K., Peterhans, A., Datta, K., and Potrykus, I. (1990). *Bio/Technology* 8, 736–740.
Deak, M., Kiss, G. B., Konez, C., and Dudits, D. (1986). *Plant Cell Rep.* 5, 97–100.
De Block, M. (1988). *Theor. Appl. Genet.* 76, 767–774.
De Block, M., Botterman, J., Vandewiele, M., *et al.* (1987). *EMBO J.* 6, 2513–2518.
De Block, M., Herrera-Estrella, L., Van Montagu, M., Schell, J., and Zambryski, P. (1984). *EMBO J.* 3, 1681–1689.
De Block, M., De Brouwer, D., and Tenning, P. (1989). *Plant Physiol.* 91, 694–701.
Dekeyser, R., Claes, B., Marichal, M., Montagu, M. V., and Caplan, A. (1989). *Plant Physiol.* 90, 217–223.
de la Peña, A., Lörz, H., and Schell, J. (1987). *Nature (London)* 325, 274–276.
Deshayes, A., Herrera-Estrella, L., and Caboche, M. (1985). *EMBO J.* 4, 2731–2737.
de Vries, S. C., Booij, H., Meyerink, P., *et al.* (1988). *Planta* 176, 196–204.
D'Halluin, K., Botterman, J., and Greef, W. D. (1990). *Crop Sci.* 30, 866–871.
Eapen, S., Köhler, F., Gerdemann, M., and Schieder, O. (1987). *Theor. Appl. Genet.* 75, 207–210.
Ebert, P. R., and Clarke, A. E. (1990). *Plant Mol. Biol.* 14, 814–824.
Eichholtz, D. A., Rogers, S. G., Horsch, R. B., *et al.* (1987). *Somatic Cell Mol. Genet.* 13, 67–76.
Ellis, J. R., Shirsat, A. H., Hepher, A., *et al.* (1988). *Plant Mol. Biol.* 10, 203–214.
Esau, K. (1977). In *Anatomy of Seed Plants, Second Edition.* New York: John Wiley.
Evans, D. A., Sharp, W. R., and Flick, C. E. (1981). In *Plant Tissue Culture, Methods and Applications in Agriculture* (Ed. by T. A. Thorpe), pp. 45–113. New York: Academic Press, Inc.
Everett, N. P., Robinson, K. E. P., and Mascarenhas, D. (1987). *Bio/Technology* 5, 1201–1204.
Fang, G., and Grumet, R. (1990). *Plant Cell Rep.* 9, 160–164.
Feldmann, K. A., and Marks, M. D. (1986). *Plant Sci.* 47, 63–69.
Feldmann, K. A., and Marks, M. D. (1987). *Mol. Gen. Genet.* 208, 1–9.
Fillatti, J. J., Kiser, J., Rose, R., and Comai, L. (1987a). *Bio/Technology* 5, 726–730.
Fillatti, J. J., Sellmer, J., McCown, B., Haissig, B., and Comai, L. (1987b). *Mol. Gen. Genet.* 206, 192–199.

Finer, J. J. (1988a). *Plant Cell Rep.* 7, 238–241.
Finer, J. J. (1988b). *Plant Cell Rep.* 7, 399–402.
Finer, J. J., and McMullen, M. D. (1990). *Plant Cell Rep.* 8, 586–589.
Firoozabady, E., DeBoer, D. L., Merlo, D. J., *et al.* (1987). *Plant Mol. Biol.* 10, 105–116.
Flick, C. E., and Evans, D. A. (1984). In *Handbook of Plant Cell Culture, Vol. 2* (Ed. by D. A. Evans, W. R. Sharp, P. V. Ammirato, and Y. Yamada), pp. 606–630. New York: Macmillian.
Flick, C. E., Evans, D. A., and Sharp, W. R. (1983). In *Handbook of Plant Cell Culture, Vol. 1* (Ed. by D. A. Evans, W. R. Sharp, P. V. Ammirato, and Y. Yamada), pp. 13–81. New York: Macmillian.
Flinn, B. S., Webb, D. T., and Newcomb, W. (1988). *Can J. Bot.* 66, 1556–1565.
Fromm, M. E., Taylor, L. P., and Walbot, V. (1986). *Nature (London)* 319, 791–793.
Fromm, M. E., Morrish, F., Armstrong, C., *et al.* (1990). *Bio/Technology* 8, 833–839.
Fry, J., Barnason, A., and Horsch, R. B. (1987). *Plant Cell Rep.* 6, 321–325.
Gamborg, O. L. (1982). In *Plant Tissue Culture Methods, Second Edition* (Ed. by L. R. Wetter and F. Constabel), pp. 1–9. National Research Council of Canada, Saskatoon.
Gamborg, O. L., and Shyluk, J. P. (1981). In *Plant Tissue Culture, Methods and Applications in Agriculture* (Ed. by T. A. Thorpe), pp. 21–44. New York: Academic Press, Inc.
Gasser, C. S., and Fraley, R. T. (1989). *Science* 244, 1293–1299.
Gordon-Kamm, W. J., Spencer, T. M., Mangano, M. L., *et al.* (1990). *The Plant Cell* 2, 603–618.
Green, C. E., and Phillips, R. L. (1975). *Crop. Sci.* 15, 417–421.
Grimsley, N. H. (1990). *Physiol. Plant.* 79, 147–153.
Grosset, J., Marty, I., Chartier, Y., and Meyer, Y. (1990). *Plant Mol Biol.* 15, 485–496.
Guerche, P., Charbonnier, M., Jouanin, L., *et al.* (1987). *Plant Sci.* 52, 111–116.
Guri, A., and Sink, K. C. (1988). *J. Plant Physiol.* 133, 52–55.
Hain, R., Stabel, P., Czernilofsky, A. P., *et al.* (1985). *Mol. Gen. Genet.* 199, 161–168.
Halperin, W. (1966). *Am. J. Bot.* 53(3), 443–418.
Halperin, W. (1969). *Annu. Rev. Plant Physiol.* 20, 395–418.
Halperin, W. (1986). In *Cell Culture and Somatic Cell Genetics of Plants, Vol. 3* (Ed. by I. K. Vasil), pp. 3–47. New York: Academic Press, Inc.
Hatamoto, H., Boulter, M. E., Shirsat, A. H., Croy, E. J., and Ellis, J. R. (1990). *Plant Cell Rep.* 9, 88–92.
Haughn, G. W., Smith, J., Mazur, B., Somerville, C. (1988). *Mol. Gen. Genet.* 211, 266–271.
Hauptmann, R. M., Vasil, V., Ozias-Akins, P., *et al.* (1988). *Plant Physiol.* 86, 602–606.
Hayashimoto, A., Li, Z., and Murai, N. (1990). *Plant Physiol.* 93, 857–863.
Hess, D., Dressler, K., and Nimmrichter, R. (1990). *Plant Sci.* 72, 233–244.
Hinchee, M. A. W., Connor-Ward, D. V., Newell, C. A., *et al.* (1988). *Bio/Technology* 6, 915–922.
Hodges, T. K., Kamo, K. K., Imbrie, C. W., and Becwar, M. R. (1986). *Bio/Technology* 4, 219–223.
Horn, M. E., Shillito, R. D., Conger, B. V., and Harms, C. T. (1988). *Plant Cell Rep.* 7, 369–472.
Horsch, R. B., Fraley, R. T., Rogers, S. G., *et al.* (1984). *Science* 223, 496–498.
Horsch, R. B., Fry, J. E., Hoffmann, N. L., *et al.* (1985). *Science* 227, 1229–1231.
Horth, M., Negrutiu, I., Burny, A., Van Montagu, M., and Herrera-Estrella, L. (1987). *EMBO J.* 6, 2525–2530.
Ishida, B. K., Snyder, G. W., Jr., and Belknap, W. R. (1989). *Plant Cell Rep.* 8, 325–328.
James, D. J., Passey, A. J., Barbara, D. J., and Bevan, M. (1989). *Plant Cell Rep.* 7, 658–661.
James, D. J., Passey, A. J., and Barbara, D. J. (1990). *Plant Sci.* 69, 79–94.
Janssen, B.-J., and Gardner, R. C. (1989). *Plant Mol. Biol.* 14, 61–72.
Jefferson, R. J. (1987). *Plant Mol. Biol. Rep.* 5, 387–405.
Jensen, J. S., Marcker, K. A., Otten, L., and Schell, J. (1986). *Nature (London)* 2321, 669–674.
Jia, S.-R., Yang, M.-Z., Ott, R., and Chua, N.-H. (1989). *Plant Cell Rep.* 8, 336–340.

Jones, J. D. G., Svab, Z., Harper, E. C., Hurwitz, C. D., and Maliga, P. (1987). *Mol. Gen. Genet.* 210, 86–91.
Kahl, G. (1982). In *Molecular Biology of Plant Tumors* (Ed. by B. Horecker, N. O. Kaplan, J. Marmur, and H. A. Scheraga), pp. 211–267. New York: Academic Press, Inc.
Kamo, K. K., and Hodges, T. K. (1986). *Plant Sci.* 45, 111–117.
Kamo, K. K., Becwar, M. R., and Hodges, T. K. (1985). *Bot. Gaz.* 146, 327–334.
Kiernan, J. M., Goldberg, K.-B., Young, M. J., Schoelz, J. E., and Shepherd, R. J. (1989). *Plant Sci.* 64, 67–78.
Klein, T. M., Harper, E. C., Svab, Z., *et al.* (1988). *Proc. Natl. Acad. Sci. U.S.A.* 85, 8502–8505.
Knapp, S., Coupland, G., Uhig, H., Starlinger, P., and Salamini, F. (1988). *Mol. Gen. Genet.* 213, 285–290.
Konar, R. N., and Nataraja, K. (1965). *Phytomorph.* 15, 132–137.
Koncz, C., Olsson, O., Langridge, W. H. R., Schell, J., and Szalay, A. A. (1987). *Proc. Natl. Acad. Sci. U.S.A.* 84, 131–135.
Koornneef, M., Hanhart, C., Jongsma, M., *et al.* (1986). *Plant Sci.* 45, 201–208.
Lamb, C. J., Lawton, M. A., Dron, M., and Dixon, R. A. (1989). *Cell* 56, 215–224.
Landsmann, J., Llewellyn, D., Dennis, E. S., and Peacock, W. J. (1988). *Mol. Gen. Genet.* 214, 68–73.
Lee K. Y., Townsend, J., Tepperman, J., *et al.* (1988). *EMBO J.* 7, 1241–1248.
Lee, M., and Phillips, R. L. (1988). *Annu. Rev. Plant Physiol. Plant Mol. Biol.* 39, 413–437.
Lindsey, K., and Gallois, P. (1990). *J. Expt. Bot.* 41, 529–536.
Lindsey, K., and Yeoman, M. M. (1985). In *Cell Culture and Somatic Cell Genetics of Plants, Vol. 2* (Ed. by I. K. Vasil), pp. 61–101. New York: Academic Press, Inc.
Lipetz, J. (1970). *Int. Rev. Cytol.* 27, 1–28.
Lloyd, A. M., Barnason, A. R., Rogers, S. G., *et al.* (1986). *Science* 234, 464–466.
LoSchiavo, F., Pitto, L., Giuliano, G., *et al.* (1989). *Theor. Appl. Genet.* 77, 325–331.
Lou, Z-X., and Wu, R. (1988). *Plant Mol. Biol. Rep.* 6, 165–174.
Ludwig, S. R., Somers, D. A., Peterson, W. L., *et al.* (1985). *Theor. Appl. Genet.* 71, 344–350.
Manners, J. M. (1988). *Plant Sci.* 55, 61–68.
Manners, J. M., and Way, H. (1989). *Plant Cell Rep.* 8, 341–345.
McCabe, D. E., Swain, W. F., Martinell, B. J., and Christou, P. (1988). *Bio/Technology* 6, 923–926.
McCain, J. W., and Hodges, T. K. (1986). *Bot. Gaz.* 147, 453–460.
McCormick, S., Niedermeyer, J., Fry, J., *et al.* (1986). *Plant Cell Rep.* 5, 81–84.
McGranahan, G. H., Leslie, C. A., Uratsu, S. L., Martin, L. A., and Dandekar, A. M. (1988). *Bio/Technology* 6, 800–804.
McGranahan, G. H., Leslie, C. A., Uratsu, S. L., Martin, L. A., and Dandekar, A. M. (1990). *Plant Cell Rep.* 8, 512–516.
Meins, F., Jr., and Binns, A. N. (1979). *BioScience* 29, 221–225.
Michelmore, R., Marsh, E., Seely, S., and Landry, B. (1987). *Plant Cell Rep.* 6, 439–442.
Moloney, M. M., Walker, J. M., and Sharma, K. K. (1989). *Plant Cell Rep.* 8, 238–242.
Morrish, F., Vasil, V., and Vasil, I. K. (1987). *Adv. Genet.* 24, 431–499.
Murashige, T. (1974). *Annu. Rev. Plant Physiol.* 25, 135–166.
Nagy, F., Morelli, G., Fraley, R. T., Rogers, S. G., and Chua, N.-H. (1985). *EMBO J.* 4, 3063–3068.
Negrutiu, I., Shillito, R., Potrykus, I., Biasini, G., and Sala, F. (1987). *Plant Mol. Biol.* 8, 363–373.
Nehra, N. S., Stushnoff, C., and Kartha, K. K. (1989). *J. Am. Soc. Hort. Sci.* 114, 1014–1018.
Nehra, N. S., Chibbar, R. N., Kartha, K. K., Datla, R. S. S., Crosby, W. L., and Stushnoff, C. (1990). *Plant Cell Rep.* 9, 293–298.

Neuhaus, G., Spangenberg, G., Mittelsten Scheid, O., and Schweiger, H.-G. (1987). *Theor. Appl. Genet.* 75, 30–36.

Normura, K., and Komamine, A. (1985). *Plant Physiol.* 79, 988–991.

Ohta, Y. (1986). *Proc. Natl. Acad. Sci. U.S.A.* 83, 715–719.

Ooms, G., Burrell, M. M., Krap, A., Bevan, M., and Hille, J. (1987). *Theor. Appl. Genet.* 73, 744–750.

Ow, D. W., Wood, K. V., DeLuca, M., *et al.* (1986). *Science* 234, 856–859.

Parrott, W. A., Hoffman, L. M., Hildebrand, D. F., Williams, E. G., and Collins, G. B. (1989). *Plant Cell Rep.* 7, 615–617.

Paszkowski, J., Shillito, R. D., Saul, M., *et al.* (1984). *EMBO J.* 3, 2717–2722.

Pechan, P. M. (1989). *Plant Cell Rep.* 8, 387–390.

Peng, J., Lyznik, L. A., Lee, L., and Hodges, T. K. (1990). *Plant Cell Rep.* 9, 168–172.

Pietrzak, M., Shillito, R. D., Hohn, T., and Potrykus, I. (1986). *Nuc. Acids Res.* 14, 5857–5869.

Poethig, R. S. (1990). *Science* 250, 923–930.

Potrykus, I. (1990). *Bio/Technology* 8, 535–542.

Potrykus, I., and Shillito, R. D. (1986). *Meth. Enzymol.* 118, 549–579.

Prioli, L. M., and Sondahl, M. R. (1989). *Bio/Technology* 7, 589–594.

Pua, E.-C., Mehra-Palta, A., Nagy, F., and Chua, N.-H. (1987). *Bio/Technology* 5, 815–817.

Radke, S. E., Andrews, B. M., Moloney, M. M., *et al.* (1988). *Theor. Appl. Genet.* 75, 685–694.

Raghavan, V. (1986). *Embryogenesis in Angiosperms, A Developmental and Experimental Study,* pp. 115–151. Cambridge: Cambridge University Press.

Raineri, D. M., Bottino, P., Gordon, M. P., and Nester, E. W. (1990). *Bio/Technology* 8, 33–38.

Ranch, J., and Pace, G. M. (1988). *Iowa State Journal of Research* 62, 537–569.

Rech, E. L., Golds, T. J., Husnain, T., *et al.* (1989). *Plant Cell Rep.* 8, 33–36.

Redway, F. A., Vasil, V., and Vasil, I. K. (1990). *Plant Cell Rep.* 8, 714–717.

Reinert, J., and Yeoman, M. M. (1982). *Plant Cell and Tissue culture, A Laboratory Manual.* New York: Springer-Verlag.

Riggs, C. D., and Bates, G. W. (1986). *Proc. Natl. Acad. Sci. U.S.A.* 83, 5602–5606.

Rhodes, C. A., Pierce, D. A., Mettler, I. J., Mascarenhas, D., and Detmer, J. J. (1988). *Science* 240, 204–207.

Roest, S., and Gilissen, L. J. W. (1989). *Acta Bot. Neerl.* 38(1), 1–23.

Rotino, G. L., and Gleddie, S. (1990). *Plant Cell Rep.* 9, 26–29.

Sanford, J. C. (1990). *Physiol. Plant.* 79, 206–209.

Schrammeijer, B., Sijmons, P. C., van den Elzen, P. J. M., and Hoekema, A. (1990). *Plant Cell Rep.* 9, 55–60.

Scott, R. J., and Draper, J. (1987). *Plant Mol. Biol.* 8, 265–274.

Shahin, E. A., Spielmann, A., Sukhapinda, K., Simpson, R. B., and Yashar, M. (1986a). *Crop Sci.* 26, 1235–1239.

Shahin, E. A., Sukhapinda, K., Simpson, R. B., and Spivey, R. (1986b). *Theor. Appl. Genet.* 72, 770–777.

Sheerman, S., and Bevan, M. W. (1988). *Plant Cell Rep.* 7, 13–16.

Shillito, R. D., Saul, M. W., Paszkowski, J., Müller, M., and Potrykus, I. (1985). *Bio/Technology* 3, 1099–1103.

Shillito, R. D., Carswell, G. K., Johnson, C. M., DiMaio, J. J., and Harms, C. T. (1989). *Bio/Technology* 7, 581–587.

Skoog, F., and Miller, C. O. (1957). *Symp. Soc. Exp. Biol.* 11, 118–140.

Smith, R. H. (1986). *Meth. Enzymol.* 118, 539–549.

Spencer, T. M., Gordon-Kamm, W. J., Daines, R. J., Start, W. G., and Lemaux, P. G. (1990). *Theor. Appl. Genet.* 79, 625–631.

Srivastava, V., Reddy, A. S., and Guha-Mukherjee, S. (1988). *Plant Cell Rep.* 7, 504–507.

Stalker, D. M., McBride, K. E., and Malyj, L. D. (1988). *Science* 242, 419–423.
Stockhaus, J., Eckes, P., Blau, A., Schell, J., and Willmitzer, L. (1987). *Nuc. Acids Res.* 15, 3479–3491.
Sukhapinda, K., Spivey, R., and Shahin, E. A. (1987). *Plant Mol. Biol.* 8, 209–216.
Swamy, B. G. L., and Sivarmakrishna, D. (1972). *Phytomorph.* 22, 305–314.
Tagu, D., Bergounioux, C., Cretin, C., Perennes, C., and Gadal, P. (1988). *Protoplasma* 146, 101–105.
Thomas, J. C., Guiltinan, M. J., Bustos, S., Thomas, T., and Nessler, C. (1989). *Plant Cell Rep.* 8, 354–357.
Thomzik, J. E., and Hain, R. (1990). *Plant Cell Rep.* 9, 233–236.
Tomes, D. T., Weissinger, A. K., Ross, M., *et al.* (1990). *Plant Mol. Biol.* 14, 261–268.
Toriyama, K., Arimoto, Y., Uchimiya, H., and Hinata, K. (1988). *Bio/Technology* 6, 1072–1074.
Torrey, J. G. (1966). *Adv. Morphogen.* 5, 39–91.
Tran Thanh Van, K. M. (1981). *Annu. Rev. Plant Physiol.* 32, 291–311.
Trinh, T. H., Mante, S., Pua, E.-C., and Chua, N.-H. (1987). *Bio/Technology* 5, 1081–1084.
Trulson, A. J., Simpson, R. B., and Shahin, E. A. (1986). *Theor. Appl. Genet.* 73, 11–15.
Ulian, E. C., Smith, R. H., Gould, J. H., McKnight, T. D. (1988). *In Vitro Cell Devel. Biol.* 24, 951–954.
Umbeck, P., Johnson, G., Barton, K., and Swain, W. (1987). *Bio/Technology* 5, 263–266.
Valvekens, D., Van Montagu, M., and Van Lijsebettens, M. (1988). *Proc. Natl. Acad. Sci. U.S.A.* 85, 5536–5540.
Vancanneyt, G., Schmidt, R., O'Connor-Sanchez, A., Willmitzer, L., and Rocha-Sosa, M. (1990). *Mol. Gen. Genet.* 220, 245–250.
Vardi, A., Bleichman, S., and Aviv, D. (1990). *Plant Sci.* 69, 199–206.
Vasil, I. K. (1988). *Bio/Technology* 6, 397–402.
Vasil, I. K., and Hildebrandt, A. C. (1965). *Science* 150, 889–890.
Vasil, I. K., and Vasil, V. (1986). In *Cell Culture and Somatic Cell Genetics of Plants, Vol. 3* (Ed. by I. K. Vasil), pp. 121–150. New York: Academic Press, Inc.
Walker, K. A., Wendeln, M. L., and Jaworski, E. G. (1979). *Plant Sci. Letters* 16, 23–30.
Wenzler, H., Mignery, G., May, G., and Park, W. (1989). *Plant Sci.* 63, 79–85.
Wernicke, W., and Brettell, R. (1980). *Nature (London)* 287, 138–139.
Wernicke, W., Brettell, R., Wakizuka, T., and Potrykus, I. (1981). *Z. Pflanzenphysiol.* 103, 361–365.
Wilde, H. D., Nelson, W. S., Booij, H., de Vries, S. C., and Thomas, T. L. (1988). *Planta* 176, 205–211.
Williams, E. G., and Maheswaran, G. (1986). *Ann. Bot.* 57, 443–462.
Wurtele, E. S., and Bulka, K. (1989). *Plant Sci.* 61, 253–262.
Zhang, H. M., Yang, H., Rech, E. L., *et al.* (1988). *Plant Cell Rep.* 7, 379–384.
Zhu, Z., Hughes, K. W., and Huang, L. (1990). *Plant Cell. Tissue Organ Cul.* 22, 135–145.

Part II

Transgenic Plants and Basic Research

6

Transgenic Plants for Studying Genes Encoding Amino Acid Biosynthetic Enzymes

Fong-Ying Tsai and Gloria M. Coruzzi

Department of Biology
New York University
New York, New York 10003

In bacteria and yeast, amino acid biosynthetic pathways have been analyzed using combined genetic and biochemical approaches. The ability to isolate amino acid auxotrophs in these microorganisms and to complement the deficiency with gene rescue has led to the cloning of genes encompassing entire amino acid biosynthetic pathways. In plants, the study of amino acid biosynthetic enzymes has been hampered by problems with both biochemical and genetic approaches. Although amino acid biosynthesis occurs largely in plastids, many reactions also occur in the cytosol and mitochondria. The existence of multiple isoenzymes in multiple subcellular compartments has complicated biochemical studies due to immunological cross-reactivity, inadequate subcellular fractionation, and enzymatic instabilities *in vitro*. Genetic approaches to the selection of plant amino acid auxotrophs have also been hampered by the problem of multiplicity, as molecular data now reveal that many amino acid biosynthetic enzymes are encoded by multiple genes. Despite this caveat, tryptophan auxotrophs have recently been identified in *Arabidopsis thaliana* (Last and Fink, 1988). However, until techniques have been developed to rescue plant mutants using genome complementation, the identification of the affected genes will require the tedious process of gene mapping and chromosome walking.

To date, only a handful of plant genes encoding amino acid biosynthetic enzymes have been cloned using a variety of approaches which include heter-

TRANSGENIC PLANTS, VOLUME 1
Engineering and Utilization

ologous hybridization, bacterial complementation, and screening with synthetic oligonucleotides designed using amino acid sequences derived from purified peptide fragments. The genes cloned include: glutamine synthetase (GS; glutamate–ammonia ligase) (Cullimore *et al.*, 1984; Tingey and Coruzzi, 1987; Tingey *et al.*, 1987; Donn *et al.*, 1984; Hirel *et al.*, 1987; Snustad *et al.*, 1988); asparagine synthetase (AS) (Tsai and Coruzzi, 1990); enolpyruvylshikimic acid-3-phosphate synthase (EPSPs) (Gasser *et al.*, 1988; Klee *et al.*, 1987; Padgette *et al.*, 1987); acetolactate synthase (ALS) (Mazur *et al.*, 1987; Wiersma *et al.*, 1989); tryptophan synthase-β (TRPB) (Berlyn *et al.*, 1989); and 3-deoxy-D-*arabino*-heptulosonate-7-phosphate synthase (DAHPs) (Dyer *et al.*, 1990). Since some of these genes (e.g., ALS, EPSPs, GS) are the targets for commercial herbicides, some studies involving transgenic plant technology have been directed at creating herbicide-resistant plants (see Chapter 10, this volume). Other studies on amino acid biosynthesis genes in plants have focused on correlating gene expression patterns during development with the physiology of nitrogen metabolism and will be the subject of this chapter.

The ability to isolate an individual gene encoding a particular amino acid biosynthetic isoenzyme and to study its specific pattern of expression in wild-type and transgenic plants has enabled molecular biologists to determine the role of a particular isoenzyme in plant nitrogen metabolism. In particular, the molecular analysis of GS and AS genes as reviewed below has answered questions concerning the roles of individual enzymes, which could not be obtained using a biochemical approach alone.

6.1 Roles of Glutamine and Asparagine in Nitrogen Assimilation and Transport

Plants obtain nitrogen from their environment mainly in the form of inorganic compounds, namely nitrate and ammonia taken up from roots, and atmospheric N_2 reduced to ammonia in nitrogen-fixing root nodules. Although some of the nitrate and a small amount of ammonia can be detected in the transporting vessels (xylem and phloem), the majority of nitrogen is transported in organic form. Among all the nitrogen-containing organic compounds made in plants cells, only a few with specific characteristics (including high N:C ratio, high stability, and/or high solubility) are able to serve as nitrogen transport compounds in plants. Some nitrogen transport compounds, such as the amides glutamine and asparagine, are general nitrogen transport compounds made in all types of plants; others, such as ureides and canavanine are important nitrogen transport compounds in only a few species (Lea and Miflin, 1980).

In addition to their roles as building blocks of proteins, the amino acids glutamine and asparagine are the two major nitrogen transport compounds in many higher plants. Ammonia assimilated into organic form by glutamine synthetase (GS; EC 6.3.1.2, glutamate–ammonia ligase) is subsequently converted to asparagine by asparagine synthetase (AS; EC 6.3.5.4) Glutamine is very metabolically active and is the nitrogen donor in the biosynthesis of virtually all nitrogen-containing compounds in plants. In contrast, asparagine is a relatively inert nitrogen transport compound which can be stored in vacuoles. Glutamine and asparagine are transported at high levels in situations where large amounts of nitrogen must be mobilized, such as in nitrogen-fixing root nodules, and also serve to transport stored nitrogen out of cotyledons during germination. The ratio of glutamine to asparagine transported is also affected by the plant growth conditions. For example, glutamine is the predominant form of transported nitrogen in light-grown plants, while asparagine (which has a higher N:C ratio) is the preferred nitrogen transport compound in dark-grown plants (Urquhart and Joy, 1981). Physiological studies reviewed below demonstrated that the GS and AS enzyme activities modulate during plant development in accordance with the synthesis of glutamine and asparagine.

6.2 Physiological Studies on Glutamine Synthetase: Insights and Limitations

Early biochemical studies demonstrated that there are multiple isoforms of GS whose activities vary in numerous organs examined (McNally and Hirel, 1983). In leaves, it was demonstrated that one isoform (GS2) was present in preparations of isolated chloroplasts (Hirel and Gadal, 1981), and furthermore that there was a light-induced increase in chloroplast GS2 activity (Hirel *et al.*, 1982). The cytosolic forms of GS were examined thoroughly in legumes where certain GS isoforms were shown to accumulate during the development of nitrogen-fixing root nodules (Lara *et al.*, 1983; Robertson *et al.*, 1975). In some species (e.g., *Phaseolus*) a novel GS isoform appeared in nitrogen-fixing root nodules (Lara *et al.*, 1983).

While these physiological studies demonstrated the existence of multiple GS isoforms, the reason for the multiplicity and relative function of the various cytosolic GS isoenzymes and the chloroplast GS2 isoform was unknown. For example, while analysis of photorespiratory mutants in barley demonstrated that chloroplast GS2 was essential for the reassimilation of photorespiratory ammonia, it was not clear why the normal levels of cytosolic GS which existed in the mutants could not compensate for the loss of chloroplast GS to assimilate lost ammonia (Wallsgrove *et al.*, 1987). Some of these questions have now been answered using molecular approaches.

6.3 Molecular Studies on Glutamine Synthetase Gene Expression *in Vivo* and in Transgenic Plants

GS cDNAs were originally cloned from *Phaseolus vulgaris* using antibodies to immunoprecipitate GS mRNA from polysomes for cDNA synthesis (Cullimore *et al.*, 1984) and from *Nicotiana plumbaginifolia* by heterologous hybridization to a Chinese Hamster GS clone (Tingey and Coruzzi, 1987). GS genes were subsequently analyzed in a number of legume species including *Pisum sativum* (Tingey *et al.*, 1987; Tingey *et al.*, 1988), *Glycine max* (Hirel *et al.*, 1987), and *Medicago sativa* (Donn *et al.*, 1984). More recently, GS cDNAs were cloned from *Zea mays* by virtue of their ability to complement an *Escherichia coli* *glnA*$^-$ mutant (Snustad *et al.*, 1988).

Pisum sativum and *P. vulgaris* are the two species in which all of the GS genes have been cloned and their pattern of expression examined in the homologous plants and in heterologous transgenic plants. Both legume species form nitrogen-fixing root nodules, however, the fixed nitrogen is transported predominantly as the amides glutamine and asparagine in *P. sativum,* and as ureides in *P. vulgaris*. The similarities and differences in GS expression between these two legume species will be outlined below.

6.3.1 GS GENE FAMILY IN *PISUM SATIVUM*

In pea, there are three types of GS genes which encode three classes of GS polypeptides. One nuclear gene encodes chloroplast GS2, and the expression of this gene has been shown to occur preferentially in leaves in a light-dependent fashion (Tingey *et al.*, 1987; Tingey *et al.*, 1988). The light-regulated expression of chloroplast GS2 is due in part to a phytochrome-mediated response and in part to changes in chloroplast metabolism (Edwards and Coruzzi, 1989). In particular, it has been shown that the mRNA for chloroplast GS2 accumulates preferentially in plants grown under photorespiratory conditions (Edwards and Coruzzi, 1989). For cytosolic GS, there are two classes of genes which encode homologous but distinct cytosolic GS polypeptides. The GS1 gene (also called GS299) encodes a GS polypeptide of 38 kDa which is expressed at low levels in leaves and roots, and is induced (5-fold) in nitrogen-fixing root nodules (Tingey *et al.*, 1988; Walker and Coruzzi, 1989). The other two genes for cytosolic GS (GS3A and GS3B; also known as GS341 and GS132, respectively) are nearly identical in sequence and encode 37-kDa polypeptides which are expressed at highest levels in nitrogen-fixing root nodules and in cotyledons of germinating seedlings (Tingey *et al.*, 1987; Walker and Coruzzi, 1989).

The genes encoding chloroplast GS2 and cytosolic GS3A were introduced into transgenic tobacco plants to examine their specific patterns of expression (Edwards *et al.*, 1990). For these experiments, the promoter for chloroplast GS2 or cytosolic GS3A was fused to a β-glucuronidase (GUS) reporter gene in a Ti binary vector and introduced into tobacco using *Agrobacterium*-mediated DNA transfer. Organs from the resulting transgenic tobacco were isolated, sectioned, and stained for GUS activity. These studies revealed that the promoter for chloroplast GS2 functions predominantly in photosynthetic cell types (e.g., palisade parenchymal cells of the leaf blade, chlorenchymal cells of the midrib and stem, photosynthetic cells of the tobacco cotyledons) (Edwards *et al.*, 1990). These patterns of expression reflect the role of chloroplast GS2 in the assimilation of ammonia generated by nitrate reduction and photorespiratory ammonia.

In contrast, the promoter for cytosolic GS3A directs expression of GUS specifically in the phloem elements of all organs of mature plants and in germinating seedlings (Edwards *et al.*, 1990). This phloem-specific expression pattern conferred by the GS3A promoter indicates that the GS3A cytosolic isoform functions to generate glutamine for intercellular nitrogen transport. The intense expression of the GS3A-GUS transgene in the vasculature of cotyledons of germinating seedlings underscores a role for the GS3A isoform in the synthesis of glutamine for mobilization of seed storage nitrogen reserves (Edwards *et al.*, 1990). The distinct cell-specific patterns of expression conferred by the promoters for chloroplast GS2 and cytosolic GS3A indicate that the corresponding GS isoforms perform separate functions in nitrogen assimilation and transport (Edwards *et al.*, 1990).

These transgenic studies on GS have therefore helped to assign roles to each GS isoform and can explain certain questions which have remained unanswered. For example, the inability of cytosolic GS to compensate for the loss of chloroplast GS2 activity in photorespiratory mutants of barley (Wallsgrove *et al.*, 1987) may be explained by the fact that chloroplast GS and cytosolic GS are expressed in separate cell types, as demonstrated for pea GS genes in the transgenic tobacco (Edwards *et al.*, 1990). Furthermore, the phloem-specific expression pattern of the pea GS3A gene indicates that at least one isoform of cytosolic GS from pea functions exclusively to generate glutamine for intercellular transport (Edwards *et al.*, 1990). This fits well with the high-level expression of the GS3A gene in nodules and in cotyledons of germinating seedlings, two contexts where large amounts of nitrogen must be mobilized to the rest of the plant (Walker and Coruzzi, 1989).

6.3.2 GS GENE FAMILY OF *PHASEOLUS VULGARIS*

In *P. vulgaris* there are four types of GS polypeptides: α, β, γ, and δ (Cullimore and Bennett, 1988). One GS gene (gln-δ) encodes the chloroplast form of GS

(Lightfoot *et al.*, 1988), while the other three (gln-α, gln-β, gln-γ) encode cytosolic GS polypeptides (Bennett *et al.*, 1989; Gebhardt *et al.*, 1986). The GSnl isoenzyme which was found to be "nodule specific" is made up of a mixture of the gln-β and gln-γ polypeptides (Cullimore and Bennett, 1988; Robert and Wong, 1986). Gene expression studies have shown that the gln-β gene (also known as GS-R1) is expressed preferentially in roots while the gln-γ gene (also known as GS-N1) is expressed strongly in nodules and termed "nodule-specific GS" (Cullimore *et al.*, 1984; Gebhardt *et al.*, 1986). While the gln-γ gene was previously thought to be expressed exclusively in nodules (Cullimore *et al.*, 1984; Gebhardt *et al.*, 1986), recent experiments demonstrate that the gln-γ mRNA accumulates to a lesser extent in other organs such as stems, petioles, and green cotyledons (Bennett *et al.*, 1989).

In transgenic studies, the gln-γ and gln-β gene promoters were fused to the GUS reporter gene and were introduced into the legume *Lotus corniculatus* (Forde *et al.*, 1989). These studies showed that the gln-γ promoter conferred high-level GUS expression in nodules, predominantly in the *Rhizobium*-infected cells. The gln-β gene which was expressed at highest levels in roots, was also expressed in both the infected region and cortical cells of young nodules and became restricted to the vasculature tissue as the nodule matured (Forde *et al.*, 1989).

The study of the cell-specific expression patterns of the gln-γ and gln-β genes has provided insight into the assembly of the GS holoenzyme in nodules. The fact that the GS holoenzyme in nodules is sometimes composed of a mixture of γ and β subunits and sometimes is composed of all β subunits (Robert and Wong, 1986) can be explained by the fact that some nodule cells express both gln-γ and gln-β genes while others express gln-β exclusively (Forde *et al.*, 1989).

6.4 Physiological Studies on Asparagine Levels and Asparagine Synthetase Activity

6.4.1 PHYSIOLOGICAL ROLE OF ASPARAGINE IN PLANTS

Asparagine, which was the first amino acid discovered, was initially characterized from asparagus (Vauquelin and Robiquet, 1806). Asparagine is a more efficient nitrogen transport compound compared to glutamine because it has a higher N:C ratio (2:4) and is relatively inert in nitrogen metabolism (Sieciechowicz *et al.*, 1988). In peas, the phloem exudates isolated from leaves of light-grown plants contain high amounts of the amides glutamine and asparagine. However, when pea plants are transferred to the dark, the asparagine levels increase dramatically and become the predominant nitrogen compound in the phloem exudate (Urquhart and Joy, 1981). The preferred synthesis of asparagine in the dark is consistent with the hypothesis that asparagine is a more economical nitrogen transport compound and is used preferentially when supplies of carbohydrate are limited. Asparagine has also been found to be the major nitrogen transport compound used in contexts where large amounts of nitrogen need to be mobilized. For example, when nitrogen fixation takes place in root nodules of certain legume species (e.g., peanut), up to 75% of ^{15}N fed into plants is recovered in nodule extracts and xylem exudates in the form of asparagine (Peaples *et al.*, 1986). During germination, high amounts of asparagine detected in cotyledons indicates that asparagine is the major nitrogen transport amino acid which mobilizes seed storage protein nitrogen to the growing seedling (Dilworth and Dure, 1978; Capdevila and Dure, 1977). Another developmental stage in which large amounts of nitrogen can be actively transported from nitrogen sources to sinks is during seed formation. It has been revealed in white lupin that asparagine represents 50–70% of the nitrogen carriers that mobilize nitrogen from senescing leaves to developing seeds (Atkins *et al.*, 1975). All these findings support the conclusion that asparagine plays a critical role in nitrogen transport in plants.

6.4.2 BIOCHEMICAL AND MOLECULAR BIOLOGICAL STUDIES ON PROCARYOTIC AND ANIMAL ASPARAGINE SYNTHETASE

Asparagine synthetase (AS) is the enzyme responsible for the biosynthesis of asparagine in most living cells. In *E. coli* (Felton *et al.*, 1980; Humbert and Simoni, 1980) and *Klebsiella aerogenes* (Reitzer and Magasanik, 1982), two genes (asnA and asnB) have been identified to encode the AS enzymes. The asnA gene codes for an AS enzyme which uses ammonia as the only nitrogen donor, while the AS enzyme encoded by asnB can use both glutamine and ammonia as substrates and has a preference for glutamine. In eukaryotic cells, only the glutamine-dependent AS activity has been identified. The genetic and biochemical studies in yeast reveal that yeast has two distinct genes encoding cytosolic glutamine-dependent AS with similar molecular weights (Jones, 1978; Ramos and Wiame, 1980). In animal systems, both AS-deficient cell lines (Andrulis *et al.*, 1987) and AS-overproducing cell lines (Gantt *et al.*, 1980; Andrulis *et al.*, 1983) have been identified. These AS-overproducing cell lines have been shown to result from gene amplification and have allowed the isolation of AS genes (Ray *et al.*, 1984; Andrulis, 1985). The cDNA clones for AS have been isolated from Chinese Hamster (Ray *et al.*, 1984) and human (Andrulis *et al.*, 1987) and the AS polypeptides have been purified from rat liver (Hongo and Sato, 1983) and bovine pancreas (Gantt and Arfin, 1981; Luehr and Schuster,

1985). Although animal AS can use both glutamine and ammonia as substrates (Mehlhaff *et al.*, 1985; Luehr and Schuster, 1985; Pfeiffer *et al.*, 1987), glutamine is the preferred nitrogen donor for animal AS (Hongo and Sato, 1983). In addition, the polyclonal and monoclonal antibodies against animal AS have been generated (Pfeiffer *et al.*, 1986) and used to localize the critical sites for the enzyme activity (Pfeiffer *et al.*, 1987).

6.4.3 BIOCHEMICAL STUDIES ON PLANT ASPARAGINE SYNTHETASE

Despite the important role that AS plays in synthesizing a major nitrogen transport compound in plants, a biochemical understanding of the plant AS enzyme is relatively limited. An AS activity which catalyzes the transfer of the amide nitrogen of glutamine to aspartate in the presence of ATP and Mg^{2+} has been characterized and partially purified in lupin (Rognes, 1975; Lea and Fowden, 1975), soybean (Huber and Streeter, 1984; Huber and Streeter, 1985), alfalfa (Egli *et al.*, 1989; Snapp and Vance, 1986), pea (Ta *et al.*, 1984), mung bean (Kern and Chrispeels, 1978), cotton (Dilworth and Dure, 1978), corn (Oaks and Ross, 1984; Oaks *et al.*, 1980), and wheat (Ramnefjell and Rognes, 1981). Glutamine has proven to be the preferred nitrogen donor for all plant AS enzyme studied (Sieciechowicz *et al.*, 1988; Rognes, 1975). However, the results of *in vitro* AS assays in extracts from various plants have revealed that ammonia can replace glutamine as nitrogen donor, albeit at a much higher K_m value (Sieciechowicz *et al.*, 1988; Oaks and Ross, 1984; Huber and Streeter, 1985; Huber and Streeter, 1984; Lea and Fowden, 1975). Based on these findings, it has been assumed that AS might be involved in ammonia detoxification by using ammonia as direct substrate when plant cells accumulate large amounts of ammonia (Givan, 1979; Sieciechowicz *et al.*, 1988). However, it is still unclear whether the ammonia-dependent AS reaction actually occurs in plants *in vivo*.

In general, high AS activities have been detected in nitrogen-rich organs such as cotyledons of germinating seedlings (Dilworth and Dure, 1978; Boland *et al.*, 1982; Lea and Fowden, 1975) and nitrogen-fixing nodules (Scott *et al.*, 1976; Kern and Chrispeels, 1978; Rognes, 1975; Huber and Streeter, 1984; Huber and Streeter, 1985), where large amounts of nitrogen need to be transported to the rest of the growing plant. In peas, AS activity in leaves can be enhanced by dark treatment (Joy *et al.*, 1983). These results for AS activities are consistent with the physiological role of asparagine. However, further studies on plant AS enzymes (i.e., purification, antibody production) have been hampered due to the difficulties encountered in biochemical approaches. The main difficulty has been the instability of the plant AS enzyme *in vitro* (Sieciechowicz *et al.*, 1988). In addition, the existence of a contaminating asparaginase activity (Huber and Streeter, 1985), as well as a nonprotein AS inhibitor (Kern and Chrispeels, 1978; Joy *et al.*, 1983) have made it almost impossible to purify AS to homogeneity from plants. These difficulties also prevent the detection of low AS activities in some organs. Although plant AS enzymes have been studied for more than a decade, some of the basic questions such as the molecular weight, the isozyme number, the cellular localization, and the regulation of AS biosynthesis are still unanswered. The molecular analysis of AS genes has begun to address some of these questions as outlined below.

6.5 Molecular Studies on Plant Asparagine Synthetase Genes *in Vivo* and in Transgenic Plants

6.5.1 ISOLATION AND CHARACTERIZATION OF ASPARAGINE SYNTHETASE cDNAs FROM PEAS

To overcome the difficulties encountered in biochemical studies of plant AS, a molecular approach has been utilized to directly isolate AS genes. The plant AS genes can subsequently be used to address questions about the encoded proteins, number of AS genes, their expression patterns, and the regulatory mechanisms involved in AS gene expression. In addition, the plant AS cDNA can also be used in a "reverse-biochemical" approach to characterize the encoded AS enzymes.

Two classes of AS cDNAs (AS1 and AS2) were isolated from pea nodule and root cDNA libraries when a human AS cDNA was used as a heterologous probe (Tsai and Coruzzi, 1990). The full-length cDNAs for AS1 and AS2 encode proteins with predicted molecular weights of 66,300 and 65,600 respectively, and are very close to the predicted size of the human AS polypeptide (Andrulis *et al.*, 1987). The AS1 and AS2 cDNAs share high homology (81% at the nucleotide level and 86% at the amino acid level) within their coding regions and are completely divergent in 5′ and 3′ noncoding regions. Both full-length cDNAs of AS1 and AS2 contain at their NH_2 termini a conserved peptide sequence which has been shown to be the glutamine binding site in animal AS (Heeke and Schuster, 1989). This finding suggests that both AS1 and AS2 encode glutamine-dependent AS enzymes. The absence of a transit peptide at the N termini of AS1 and AS2 polypeptides strongly suggests that the AS1 and AS2 polypeptides are each cytosolic in nature (Tsai and Coruzzi, 1990).

6.5.2 EXPRESSION PATTERNS OF THE AS GENES IN PEAS

In order to determine the expression patterns of the AS1 and AS2 genes, DNA fragments containing the 3′ noncoding sequence of the AS1 and AS2 cDNAs were used as gene-specific probes to detect specific AS mRNAs. The results of gene-specific Northern blot analysis reveal that both AS1 and AS2 mRNAs accumulate to high levels in cotyledons of germinating pea seedlings and in nitrogen-fixing pea root nodules (Tsai and Coruzzi, 1990) where high AS activities have also been detected in previous biochemical studies (Dilworth and Dure, 1978; Boland *et al.*, 1982; Lea and Fowden, 1975; Scott *et al.*, 1976; Kern and Chrispeels, 1978; Rognes, 1975; Huber and Streeter, 1984; Huber and Streeter, 1985). Since AS activity in mature pea leaves has been shown to be enhanced by dark treatment (Joy *et al.*, 1983), AS mRNA levels were also examined in leaves from peas grown in different light regimes. The steady-state levels of mRNA for both AS1 and AS2 are almost undetectable in light-grown leaves, and dark treatment can induce the accumulation of AS mRNAs greater than 30-fold (Tsai and Coruzzi, 1990). Although AS1 and AS2 share similar expression patterns in cotyledons, nodules, and leaves, the mRNA levels for AS2 were 2- to 5-fold lower than that for AS1 in all those organs (Tsai and Coruzzi, 1990).

6.5.3 REGULATORY MECHANISMS INVOLVED IN AS GENE EXPRESSION

The dark-induced or light-repressed expression of AS1 has been demonstrated to be regulated via the chromophore phytochrome (Tsai and Coruzzi, 1990). The expression of some phytochrome-mediated light-regulated genes such as genes encoding the chlorophyll *a*/*b* binding proteins (Cab) have been shown to be regulated by circadian rhythm, while others such as genes encoding ribulose-1, 5-bisphosphate carboxylase small subunit (rbcS) are not. It was therefore of interest to examine whether circadian rhythm is involved in the apparent light-regulated expression of plant AS genes. When pea plants were grown in a normal day/night cycle, the AS1 mRNA levels fluctuated dramatically in response to the light/dark cycle. These results demonstrated that the light/dark regulation of AS1 is physiologically significant within a normal day/night cycle. Further experiments on pea plants grown in a variety of light versus dark regimes reveal that AS1 expression is simply regulated by light/dark conditions, and not by a biological clock or circadian rhythm (Tsai and Coruzzi, 1991).

The results of nuclear run-on experiments reveal that the light regulation of AS1 and AS2 expression is regulated at the transcriptional level. This response is rapid and a decrease in the rate of transcription for the AS1 gene can be detected 20 min after pea plants were transferred from the dark to the light. Consistent with the differences in the levels of steady-state mRNA, the transcriptional rate of the AS1 gene is 4- to 6-fold higher than that of the AS2 gene in pea leaves (Tsai and Coruzzi, 1991). These results suggest that the promoter of the AS1 gene may be stronger than that of the AS2 gene. Interestingly, the light-repressed expression of AS genes in pea plants has been shown to be independent of organ type. For the AS1 gene, the light-repressed expression occurs at equal levels in leaves, stems, and roots. Further studies showed that the limited amount of light which penetrates through the soil is sufficient to repress the expression of AS1 gene. These results indicate that repression of AS1 gene expression by light is a sensitive response (Tsai and Coruzzi, 1991).

6.5.4 USING TRANSGENIC PLANTS TO STUDY CELL-SPECIFIC EXPRESSION PATTERNS OF AS1 AND AS2 GENES

Transgenic plants have been widely used to study the regulatory elements involved in plant gene expression. The use of the bacterial β-glucuronidase (GUS) gene as a reporter makes it possible to detect the cell type in which a promoter is active by using a GUS histochemical assay. Studies in which the AS1 and AS2 promoters are fused to GUS gene and introduced into transgenic plants will demonstrate whether the cell- and tissue-specific expression patterns of these genes are identical or distinct (Tsai and Coruzzi, work in progress). In addition, these constructs should enable the dissection of *cis*-acting DNA elements involved in light-repressed expression.

To determine whether such transgenic studies could be conducted in tobacco, the regulation of AS by light was examined in tobacco plants. Heterologous Northern blot analysis using DNA fragments containing pea AS coding regions as probes was performed on RNA samples isolated from *Nicotiana plumbaginifolia* and *Nicotiana tabacum*. These studies showed that the accumulation of AS mRNA in both species can be induced by dark treatment, and the results ensure the feasibility of using transgenic tobacco plants to study the light regulation of AS gene expression (Tsai and Coruzzi, 1991).

Both AS1 and AS2 are encoded by single genes in pea and the genomic clones for AS1 and AS2 genes have been isolated and characterized. Constructs containing different lengths of the AS promoters fused to a GUS reporter gene have been constructed and introduced into tobacco plants (Tsai and Coruzzi, work in progress). The analysis of the regulated expression of the AS-GUS fusion in these transgenic tobacco plants should identify the cell types in which each AS promoter is functional. In addition, these studies will allow the dissection of the *cis*-elements responsible for light/dark regulation of AS genes.

6.6 Conclusions

The studies on the GS and AS genes reviewed here have served to illustrate how a molecular analysis of gene expression and the determination of the cell-specific patterns of expression in transgenic plants have elucidated the physiological role of individual GS and AS isoenzymes. Future analysis of other amino acid biosynthesis genes along common nitrogen metabolic pathways should help to elucidate the molecular mechanisms which coordinate the expression of entire amino acid biosynthetic pathways.

Acknowledgments

This work was supported by grants to G. Coruzzi from NIH (GM32877) and DOE (DEFG0289ER14034). F.-Y. Tsai is supported by the Lucille P. Markey Charitable Trust, Miami, Florida.

References

Andrulis, I. L. (1985). In *Molecular Cell Genetics* (Ed. by M. M. Gottesman), pp. 489–518.
Andrulis, I. L., Duff, C., Evans-Blackler, S., Worton, R., and Siminovitch, L. (1983). *Mol. Cell. Biol.* 3, 391–398.
Andrulis, I. L., Chen, J., and Ray, P. N. (1987). *Mol. Cell. Biol.* 7, 2435–2443.
Atkins, C. A., Pate, J. S., and Sharkey, P. J. (1975). *Plant Physiol.* 56, 807–812.
Bennett, M. J., Lightfoot, D. A., and Cullimore, J. V. (1989). *Plant Mol. Biol.* 12, 553–565.
Berlyn, M. B., Last, R. L., and Fink, G. R. (1989). *Proc. Natl. Acad. Sci. U.S.A.* 86, 4604–4608.
Boland, M. J., Hanks, J. F., Reynolds, P. H. S., Blevins, D. G., Tolbert, N. E., and Schubert, K. R. (1982). *Planta* 155, 45–51.
Capdevila, A. M., and Dure, L., III (1977). *Plant Physiol.* 59, 268–273.
Cullimore, J. V., and Bennett, M. J. (1988). *J. Plant Physiol.* 132, 387–393.
Cullimore, J. V., Gebhardt, C., Saarelainen, R., Miflin, B. J., Idler, K. B., and Barker, R. F. (1984). *J. Mol. Appl. Genet.* 2, 589–599.
Dilworth, M. F., and Dure, L., III (1978). *Plant Physiol.* 61, 698–702.
Donn, G., Tischer, E., Smith, J., and Goodman, H. (1984). *J. Mol. Appl Genet.* 2, 621–635.
Dyer, W. E., Weaver, L. M., Zhao, J., Kuhn, D. N., Weller, S. C., and Herrmann, K. M. (1990). *J. Biol. Chem.* 265, 1608–1614.
Edwards, J. W., and Coruzzi, G. M. (1989). *Plant Cell* 1, 241–248.
Edwards, J. W., Walker, E. L., and Coruzzi, G. M. (1990). *Proc. Natl. Acad. Sci. U.S.A.* 87, 3459–3463.
Egli, M., Griffith, S. M., Miller, S. S., Anderson, M. P., and Vance, C. P. (1989). *Plant Physiol.* 91, 898–904.
Felton, J., Michaelis, S., and Wright, A. (1980). *J. Bacteriol.* 142, 221–228.
Forde, B. G., Day, H. M., Turton, J. F., Shen, W.-J., Cullimore, J. V., and Oliver, J. E. (1989). *Plant Cell* 1, 391–401.
Gantt, J. S., and Arfin, S. M. (1981). *J. Biol. Chem.* 256, 7311–7315.
Gantt, J. S., Chiang, C.-S., Hatfield, G. W., and Arfin, S. M. (1980). *J. Biol. Chem.* 255, 4808–4813.
Gasser, C. S., Winter, J. A., Hironake, C. M., and Shah, D. M. (1988). *J. Biol. Chem.* 263, 4280–4287.

Gebhardt, C., Oliver, J. E., Forde, B. G., Saarelainen, R., and Miflin, B. J. (1986). *EMBO J.* 5, 1429–1435.
Givan, C. V. (1979). *Phytochem.* 18, 375–382.
Heeke, G. V., and Schuster, M. (1989). *J. Biol. Chem.* 264, 5503–5509.
Hirel, B., and Gadal, P. (1981). *Z. Pflanzenphysiol.* 102, 315–319.
Hirel, B., Vidal, J., and Gadal, P. (1982). *Planta* 155, 17–23.
Hirel, B., Bouet, C., King, B., Layzell, B., Jacobs, F., and Verma, D. P. S. (1987). *EMBO J.* 6, 1167–1171.
Hongo, S., and Sato, T. (1983). *Biochim. Biophys. Acta* 742, 484–489.
Huber, T. A., and Streeter, J. G. (1984). *Plant Physiol.* 74, 605–610.
Huber, T. A., and Streeter, J. G. (1985). *Plant Science* 42, 9–17.
Humbert, R., and Simoni, R. D. (1980). *J. Bacteriol.* 142, 212–220.
Jones, G. E. (1978). *J. Bacteriol.* 134, 200–207.
Joy, K. W., Ireland, R. J., and Lea, P. J. (1983). *Plant Physiol.* 73, 165–168.
Kern, R., and Chrispeels, M. J. (1978). *Plant Physiol.* 62, 815–819.
Klee, H. J., Muskopf, Y. M., and Gasser, C. S. (1987). *Mol. Gen. Genet.* 210, 437–442.
Lara, M., Cullimore, J. V., Lea, P. J., Miflin, B. J., Johnston, A. W. B., and Lamb, J. W. (1983). *Planta* 157, 254–258.
Last, R. L., and Fink, G. R. (1988). *Science* 240, 305–310.
Lea, P. J., and Fowden, L. (1975). *Proc. Roy. Soc. Lond.* 192, 13–26.
Lea, P. J., and Miflin, B. J. (1980). In *The Biochemistry of Plants, Vol. 5* (Ed. by P. F. Stumpf and E. E. Conn), pp. 569–607. New York: Academic Press, Inc.
Lightfoot, D. A., Green, N. K., and Cullimore, J. V. (1988). *Plant Mol. Biol.* 11, 191–202.
Luehr, C. A., and Schuster, S. M. (1985). *Arch. Biochem. Biophys.* 237, 335–346.
Mazur, B. J., Chui, G. F., and Smith, J. K. (1987). *Plant Physiol.* 85, 1110–1117.
McNally, S., and Hirel, B. (1983). *Physiol. Veg.* 21, 761–774.
Mehlhaff, P. M., Luehr, C. A., and Schuster, S. M. (1985). *Biochem.* 24, 1104–1110.
Oaks, A., and Ross, D. W. (1984). *Can. J. Bot.* 62, 68–73.
Oaks, A., Stulen, I., Jones, K., Winspear, M. J., Misra, S., and Boesel, I. L. (1980). *Planta* 148, 447–484.
Padgette, S. R., Huynh, Q. K., and Borgmeyer, J. (1987). *Arch. Biochem. Biophys.* 258, 564–573.
Peaples, M. B., Pate, J. S., Atkins, C. A., and Bergersen, F. J. (1986). *Plant Physiol.* 82, 946–951.
Pfeiffer, N. E., Mehlhaff, P. M., Wylie, D. E., and Schuster, S. M. (1986). *J. Biol. Chem.* 261, 1914–1919.
Pfeiffer, N. E., Mehlhaff, P. M., Wylie, D. E., and Schuster, S. M. (1987). *J. Biol. Chem.* 262, 11565–11570.
Ramnefjell, O., and Rognes, S. E. (1981). *Plant Physiol.* 67 (Suppl.), 45.
Ramos, F., and Wiame, J.-M. (1980). *Eur. J. Biochem.* 108, 373–377.
Ray, P. N., Siminovitch, L., and Andrulis, I. L. (1984). *Gene* 30, 1–9.
Reitzer, L. J., and Magasanik, B. (1982). *J. Bacteriol.* 151, 1299–1313.
Robert, F. M., and Wong, P. P. (1986). *Plant Physiol.* 81, 142–148.
Robertson, J. G., Farnden, K. J. F., Warburton, M. P., and Banks, J. M. (1975). *Aust. J. Plant Physiol.* 2, 265–272.
Rognes, S. E. (1975). *Phytochemistry* 14, 1975–1982.
Scott, D. B., Farnden, K. J. F., and Robertson, J. G. (1976). *Nature (London)* 263, 703–705.
Sieciechowicz, K. A., Joy, K. W., and Ireland, R. J. (1988). *Phytochem.* 27, 663–671.
Snapp, S. S., and Vance, C. P. (1986). *Plant Physiol.* 82, 390–395.
Snustad, D. P., Hunsperger, J. P., Chereskin, B. M., and Messing, J. (1988). *Genetics* 120, 1111–1124.

Ta, T. C., Joy, K. W., and Ireland, R. J. (1984). *Plant Physiol.* 74, 822–826.
Tingey, S. V., and Coruzzi, G. M. (1987). *Plant Physiol.* 84, 366–373.
Tingey, S. V., Walker, E. L., and Coruzzi, G. M. (1987). *EMBO J.* 6, 1–9.
Tingey, S. V., Tsai, F.-Y., Edwards, J. W., Walker, E. L., and Coruzzi, G. M. (1988). *J. Biol. Chem.* 263, 9651–9657.
Tsai, F.-Y., and Coruzzi, G. M. (1990). *EMBO J.* 9, 323–332.
Urquhart, A. A., and Joy, K. W. (1981). *Plant Physiol.* 68, 750–754.
Vauquelin, L. N., and Robiquet, P. J. (1806). *Ann. Chim. (Paris)* 1, 57.
Walker, E. L., and Coruzzi, G. M. (1989). *Plant Physiol.* 91, 702–708.
Wallsgrove, R. M., Turner, J. C., Hall, N. P., Kendally, A. C., and Bright, S. W. J. (1987). *Plant Physiol.* 83, 155–158.
Wiersma, P. A., Schmiemann, M. G., Condie, J. A., Crosby, W. L., and Moloney, M. M. (1989). *Mol. Gen. Genet.* 219, 413–420.

7

Transgenic Plants for Studying Responses to the Hormones Auxin and Cytokinin

Stephen H. Howell

Boyce Thompson Institute for Plant Research
Ithaca, New York 14853

7.1 Introduction

Plant hormones are key regulators of plant development. Hormones direct normal plant growth and abnormal development in certain pathogenic states. Genes that govern activities involved in the biosynthesis, metabolism, perception, or response to plant hormones have been isolated from plants and pathogenic bacteria. The study of these genes may not only help us to understand plant hormone action, but also may aid in the production of plants with beneficial characteristics derived from altered hormone biosynthesis or responses.

In this chapter, we will focus on two major hormones involved in plant development: auxins and cytokinins. We will discuss what has been learned about plant hormones by introducing genes into transgenic plants which can alter the supply, location, or the response to plant hormones.

7.1.1 AUXIN METABOLISM

General descriptions of auxins, their metabolism,, sites of synthesis, and effects are found elsewhere (see, e.g., Davies, 1987). A brief introduction is given here for the purposes of discussion. The principal auxin in plants is indole-3-acetic acid (IAA) and its derivatives. It has been shown that tryptophan is a precursor for IAA synthesis in bacterial systems (Figure 7.1), but the precursor in plants has not been conclusively demonstrated (for review see Sembdner *et al.*, 1980). One of several problems in studying the pathway is that the tryptophan pool is usually as much as three orders of magnitude larger than the IAA pool, (Epstein *et al.*, 1980). The relative size of the tryptophan precursor pool and the instability of intermediates make it difficult to identify bona fide intermediates in the biosynthesis pathway using classical biochemical techniques. Nonetheless, evidence from enzymatic studies and *in vitro* systems supports the contention that tryptophan is converted to IAA via a general aromatic transaminase reaction with indole-3-pyruvic acid and indole-3-acetaldehyde accumulating as intermediates (Truelsen, 1973; Wightman, 1973). A problem with the indole-3-pyruvic acid pathway, as conceived, is the lack of control that a general aromatic transaminase might provide to IAA biosynthesis. Other pathways, such as ones found in crown gall tumors involving tryptamine as an intermediate, will be described in a later section.

Another view of the IAA biosynthetic pathway has emerged from recent studies of *Arabidopsis* mutants with defects in tryptophan biosynthesis (summarized in Leader, 1991). These studies point to indole or a related compound as a possible precursor of IAA (Figure 7.1). Mutants that are defective in the conversion of tryptophan form indole, such as the *trp2*-1 mutant (Last *et al.*, 1991), have nonetheless, elevated levels of IAA. Another mutant, the *trp1*-1 mutant, is defective in a much earlier step in the tryptophan pathway and has a short, bushy phenotype expected for a plant deprived of auxin. These findings suggest that tryptophan may not be a direct precursor of IAA, but that a compound further upstream on the tryptophan biosynthetic pathway can be the precursor of IAA.

IAA levels in plants could be controlled by means other than the regulation of *de novo* IAA biosynthesis, such as by hydrolysis of conjugates, by transport from one site to another, or by catabolism. Most IAA in seeds and plant tissues is tied up as conjugates and the controlled hydrolysis of conjugates could influence IAA levels. It will be described in a subsequent section that *Agrobacterium rhizogenes* genes appear to alter plant hormone levels by hydrolysis of auxin-glucoside conjugates (Estruch *et al.*, 1991a). However, hydrolysis of auxin conjugates has never been reported as a normal process in uninfected plants.

7.1.2 CYTOKININ METABOLISM

Cytokinins affect a wide range of growth and developmental processes in higher plants including cell division, chloroplast differentiation, bud development, leaf enlargement, and delay in leaf senescence (Davies, 1987). Cytokinins are classically defined as compounds that stimulate the division of cultured plant cells in the presence of auxin. The definition is broad and includes a large number of

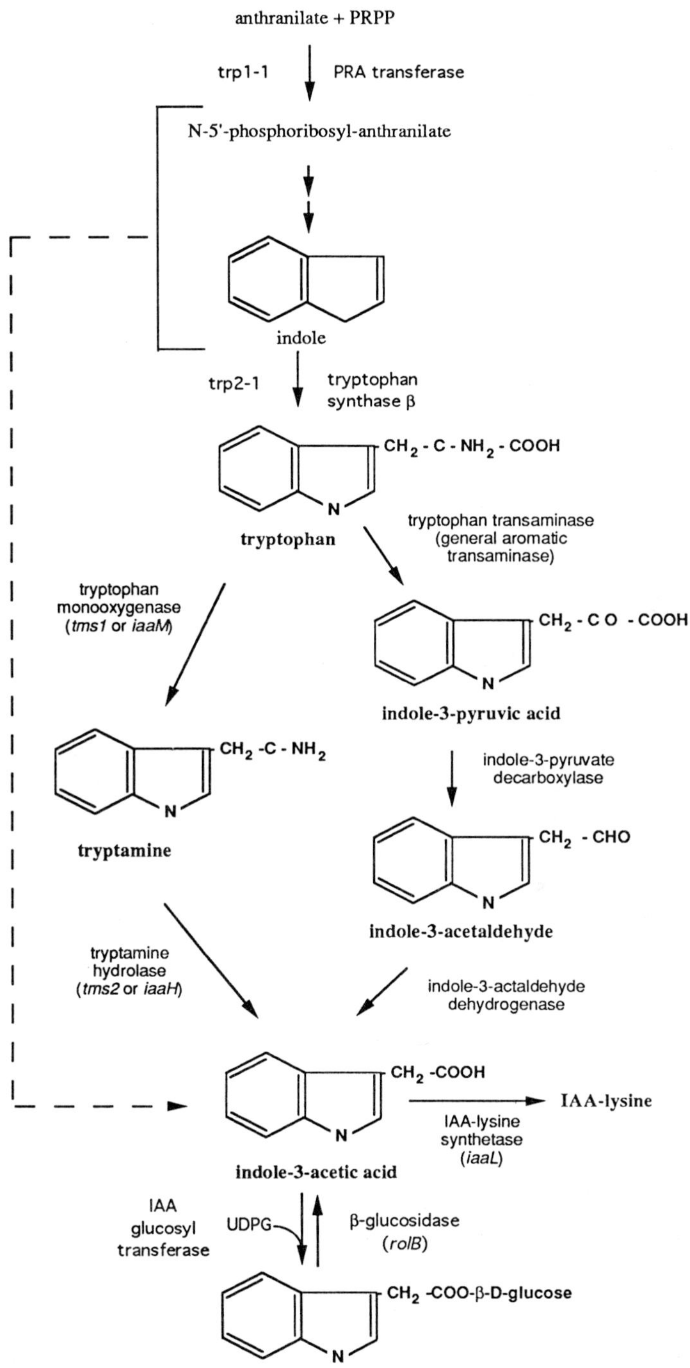

Figure 7.1 Key steps in the biosynthesis of indole-3-acetic acid (IAA). Shown is the procaryotic pathway for the synthesis of IAA via tryptamine and a proposed pathway in higher plants via indole-3-pyruvic acid. Adapted from Sembdner *et al.*, 1980. Dashed line represents a possible pathway for conversion of indole or its precursors to IAA inferred from genetic data (see text).

compounds that have this biological activity (see, e.g., Matsubara, 1990). Typical cytokinins are N^6-substituted adenine derivatives. The first naturally occurring plant cytokinin was isolated from *Zea mays* kernels and termed zeatin by Letham (Letham, 1963). The N^6-substituted adenine derivatives fall into three main groups, depending on the composition of the side chains: the zeatin, isopentenyl adenine, and dihydrozeatin types (Figure 7.2).

Some basic features about cytokinin biosynthesis are still not well understood in plants. Cytokinins are thought to be synthesized in roots of growing plants, then transported and accumulated in the shoot meristem (Kende, 1965; Short and Torrey, 1972). The biosynthesis and metabolism of cytokinins in plants (Letham and Palni, 1983; McGaw, 1987; McGaw *et al.*, 1984) and in phytopathogens (Morris, 1986) have been reviewed. Putative enzymes involved in the pathway have only been partially purified and tracer studies have been plagued by large precursor pools and small pools of intermediates. Controversy still persists as to whether all cytokinins are synthesized *de novo* or whether some are derived from the breakdown of tRNA. Studies on the biosynthesis of cytokinins in microorganisms or in plants infected with *Agrobacterium tumefaciens* have aided in understanding the *de novo* synthesis pathway in plants. As will be discussed in a later section, *A. tumefaciens* has two genes, *tmr*(*ipt*) and *tzs*, both of which encode enzymes that synthesize cytokinins. The *ipt* gene encodes isopentenyltransferase that catalyzes the synthesis of N^6-(Δ^2-isopentenyl) adenosine 5′-monophosphate, (9R-5′P)iP, from AMP and Δ^2-isopentenyl pyrophosphate (Figure 7.2). A similar enzyme has been characterized in the slime mold, *Dictyostelium discoideum* (Taya *et al.*, 1978) and tobacco calli (Chen, 1982).

Although the presence of isopentenyltransferase activity in plant tissues gives substance to the notion of a *de novo* pathway, the immediate product, (9R-5′P)iP, is rarely found in plant tissue (Palni *et al.*, 1983; Stuchbury *et al.*, 1979). From this it is assumed that (9R-5′P)iP is rapidly and stereospecifically hydroxylated to the more abundant cytokinins, *trans*-zeatin riboside 5′-monophosphate (9R-5′P)Z which is a precursor of *trans*-zeatin riboside (9R)Z and *trans*-zeatin (Z). (9R-5′P)Z and (9R)Z can be synthesized from zeatin by a salvage pathway because labeled Z is readily converted to riboside and ribotide forms (McGaw *et al.*, 1984).

The riboside and ribotide conjugates of zeatin are relatively unstable, and in certain tissues other conjugates such as, *N*- and *O*-glucosides are the more abundant and presumably more stable forms. The *N*-glucosides are thought to be important in the regulation of cytokinin levels because the purine ring derivatives, the 7- and 9-glucosides, are biologically inactive (Letham *et al.*, 1983) and resistant to hydrolysis in normal plant growth (Parker and Letham, 1973). As will be described in a subsequent section, the *rolC* gene in *Agrobacterium rhizogenes* encodes an activity with the ability to hydrolyze cytokinin

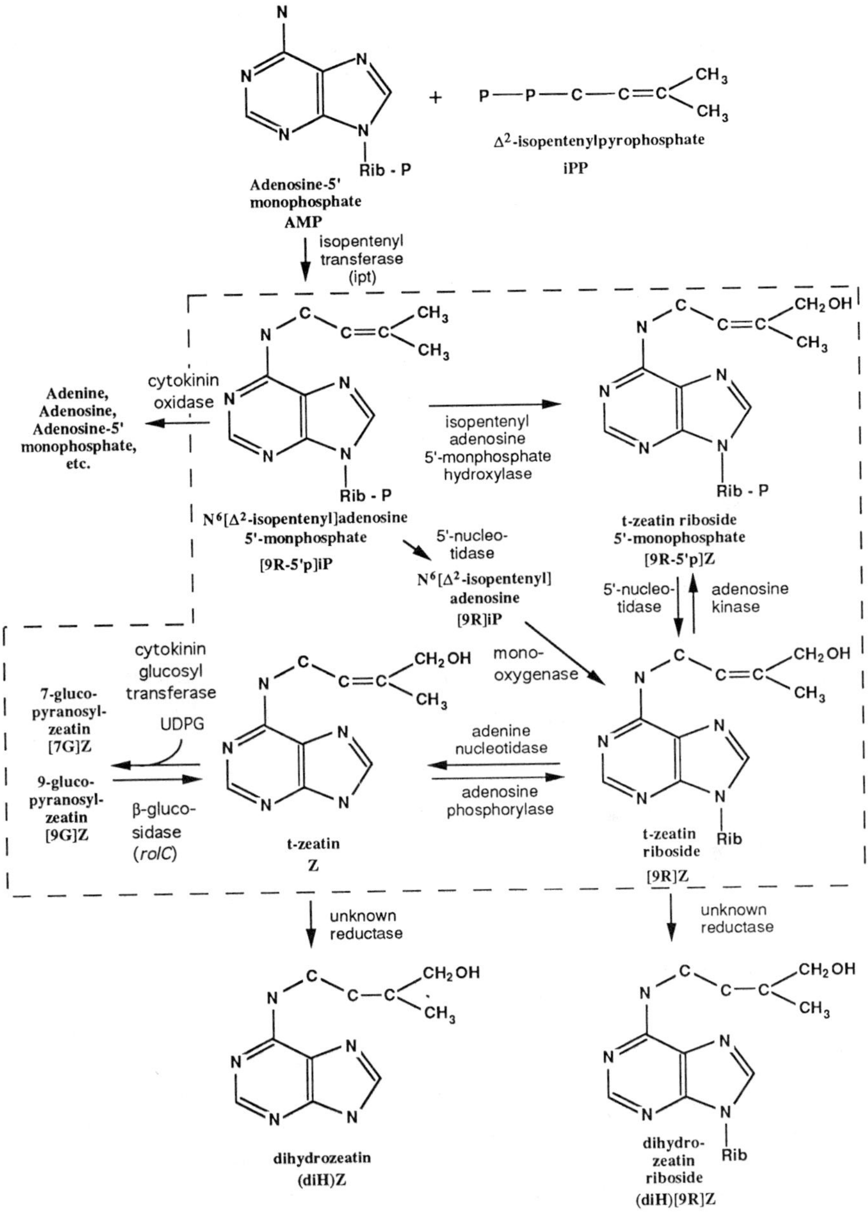

Figure 7.2 Key steps in the biosynthesis and breakdown of cytokinins from adenosine 5′-monophosphate (AMP) and Δ^2-isopentenyl pyrophosphate. Compounds enclosed by dashed line are substrates most readily subject to degradation by cytokinin oxidase. Adapted from McGaw, 1987.

N-glucoside conjugates and release biologically active cytokinin (Estruch *et al.*, 1991a). *O*-glucosides are biologically active either because the compounds themselves are active or because they are rapidly hydrolyzed to their aglycone forms. Nonetheless, *O*-glucosides can be the most abundant forms found in leaves such as in *Phaseolus vulgaris* (Palmer *et al.*, 1981). The amino acid conjugates, like the *N*-glucoside forms, are biologically inactive and very stable (Letham *et al.*, 1983; Parker *et al.*, 1978). The *rolC* gene product in *Agrobacterium rhizogenes* which can break down cytokinin-glucoside conjugates is apparently unable to hydrolyze amino acid conjugates of cytokinins (Estruch *et al.*, 1991a).

Cytokinins and their riboside and ribotide derivatives are subject to hydrolysis by 5′-ribonucleotidases and 5′-ribonucleosidases, side-chain reduction [e.g., to form dihydrozeatin (diH)Z,] and oxidative side-chain cleavage. The latter is mediated by cytokinin oxidase which has been partially purified from a variety of species (McGaw and Horgan, 1983a; McGaw and Horgan, 1983b). *O*-glucoside substitution appears to render cytokinins relatively resistant to oxidation.

7.2 Plant Genes

Relatively few plant genes have been cloned that are involved in auxin and cytokinin biosynthesis, perception, and signal transduction. However, there is great interest in this area, and, so, the following section covers the plant genes which have been cloned and/or genetically characterized so far.

7.2.1 AUXIN

7.2.1.1 Auxin-Binding Proteins

The first step in the auxin response pathway is thought to involve the binding of an auxin to a receptor(s). Much progress has been made in the past few years in identifying auxin-binding proteins which might be potential auxin receptors. A number of excellent reviews have summarized the progress in this field (see Libbenga *et al.*, 1986; Napier and Venis, 1990; Venis, 1985). This section will deal with the analysis of auxin-binding proteins which ultimately led to the cloning of a putative auxin receptor. The cloning of a putative auxin receptor was made possible from protein sequence information obtained by purification of auxin-binding proteins. High affinity auxin-binding proteins were first described in *Zea mays* by Hertel *et al.* (1972). It was recognized that auxin-binding activities were associated with several subcellular fractions in maize, but the bulk of the activity was located in endoplasmic reticulum membranes, called site I (Dohrmann *et al.*, 1978). High-level purification of auxin-binding activities from maize (primarily site I activity) was achieved in several different laboratories using a variety of techniques including affinity chromatography and the use of antibodies (Hesse *et al.*, 1989; Löbler and Klämbt, 1985; Shimomura *et al.*, 1986). The auxin-binding protein isolated from *Zea mays* coleoptiles by Hesse *et al.* (1989) was called auxin-binding protein 1 or abp1. Abp1 bound one mole of [^{3}H]naphthaleneacetic acid (NAA) per mole of 22-kDa protein with a $k_D \sim 2 \times 10^{-7}$ *M*. Other fractions contained two additional binding proteins called abp2 and abp3. Both Hesse *et al.* (1989) and Inohara *et al.* (1989) obtained *N*-terminal sequence information from their respective purified auxin-binding protein preparations. From the sequence, oligonucleotide probes were constructed and used to select cDNA clones. The clones from both groups encoded a protein of 201 amino acids and corresponded to the partial sequence of abp1.

The sequence encoding abp1 had a leader sequence of 38 hydrophobic amino acid residues that is thought to serve as a signal peptide from secretion. The protein had one potential *N*-glycosylation site (Asp-Thr-Thr) to which was linked a high-mannose type glycan. The deduced amino acid sequence terminated in a KDEL sequence which, along with the high-mannose side-chain, is characteristic of proteins found in the lumen of the endoplasmic reticulum (ER). The protein was, indeed, enriched in ER fractions and released using Na_2CO_3 or nonionic detergents. With the exception of the signal peptide, abp1 did not have a transmembrane spanning domain, a property which was inconsistent with prior observations (Dohrmann *et al.*, 1978). However, the membrane association of the major auxin-binding proteins could be explained if the protein was not actually bound to membranes but was sequestered in the lumen compartment of the ER and purified with the membranes. Most vexing was the observation that abp1 was associated with ER membranes and not with plasma membranes, because other experiments indicate that at least some of the biologically active receptors are found on the cell surface.

Using an antibody raised against a purified maize auxin-binding protein fraction, Barbier-Brygoo *et al.* (1989) demonstrated that the antibody blocked the NAA-induced hyperpolarization of tobacco plasma membranes, that is, the antibody increased the auxin dose required to elicit the same membrane potential response. In these experiments, protoplasts were impaled with a microelectrode, and antibody was added to the medium containing tobacco protoplasts. The antiserum appeared to block the auxin receptor and not the proton channel because the antibody did not affect membrane potential changes induced by fusicoccin. (Fusicoccin activates the proton pump by a route independent of auxin.) However, both NAA and fusicoccin-induced hyperpolarization were blocked by an antiserum directed at the membrane H^+-pumping ATPase. These studies indicate that components of an auxin-stimulated response system with epitopes similar to the maize auxin-binding protein are present and biologically active on the surface of tobacco protoplasts. Most recently it has been demonstrated by Venis *et al.* (1990) that impermeable auxin analogs will elicit an auxin response in pea epicotyls and tobacco protoplasts. The analogs were made impermeable by conjugation to carrier proteins. The hormone–protein complexes were made accessible to pea epicotyls by abrading the plant surface during application. These results reinforce the argument that auxin receptors are located on the outer surface of the cell.

The diageotropica (*dgt*) mutant in tomato further complicated the auxin receptor story. Homozygous mutants have diagravitropic shoot growth, abnormal vascular tissue, altered leaf morphology, and unbranched lateral roots (Zobel, 1972; Zobel, 1973). *Dgt* mutants are insensitive to auxin in that the auxin failed to stimulate typical responses such as stem elongation (Kelley and Bradford, 1986) and ethylene evolution (Bradford and Yang, 1980; Jackson, 1979; Zobel, 1973). Using a photoaffinity auxin analog [^{3}H]5N_3-IAA, Hicks *et al.* (1989) demonstrated the auxin-binding proteins found in membrane fractions of the normal parental variety VFN8 tomatoes were absent in similar fractions of the stems of the *dgt* mutant. However, two facts complicate matters with respect to findings in other systems: (1) binding proteins in the normal parental tomato did not correspond in size to maize abp1, and (2) the photoaffinity label was thought to label proteins at the membrane surface, not in the lumen of the ER. The photoaffinity analog labeled two tomato proteins of 40 and 42-kDa in the normal parental variety. Curiously, membranes from the roots of the *dgt* mutant contained the 40- and 42-kDa proteins, although membranes from the stems did not. If the 40- and 42 kDa proteins are responsible for the *dgt* phenotype, then the distribution of the proteins was consistent with the observation that shoots from normal tomatoes rescued the *dgt* phenotype when grafted onto *dgt* roots (Zobel, 1974).

Despite the complications, the *dgt* mutant provided some of the first genetic evidence implicating auxin-binding protein(s) in the auxin response. However, the 40- and 42-kDa proteins do not correspond in size to abp1 and differ from abp1 in other properties. For example, the 40- and 42-kDa proteins presumably are surface exposed, plasma membrane proteins, while abp1 is a soluble protein found in the lumen of the ER. One could account for these diverse observations in a number of ways. First, the auxin binding proteins might constitute a family of related proteins, some of which have ER and plasma membrane protein equivalents. Second, the same proteins have different sizes and different membrane associations in different plants.

Recently, another auxin-binding protein has been identified using an elegant approach employing anti-idiotypic antibodies (Prasad and Jones, 1991). These are antibodies which are produced against other antibodies that have been raised in response to auxins. The idea is that the binding pocket in the original antibody may have the same structure as the plant receptor, and therefore, the anti-idiotypic antibody would recognize the plant receptor. Using this technique

Prasad and Jones have identified a 65-kDa protein that is found in nuclear fractions.

Why are there so many candidates for the auxin receptor? It may be because there are many different receptors and that these different receptors may be tied into different auxin signal transduction pathways. Also, it may be that some of these auxin-binding proteins are not bona fide receptors, although some may still have important functions in auxin preception. It has been suggested by Barbier-Brygoo *et al.* (1990, 1991) that the maize abp1-type protein could be a plasma membrane surface protein that binds to a transmembrane protein which is responsible for transducing the auxin signal to the interior of the cell or to other plasma membrane proteins. They have been able to demonstrate that purified abp1 can be added to suspensions of tobacco protoplasts and under these conditions abp1 sensitizes the protoplasts to auxins, that is, abp1 reduces the auxin dose required to obtain the same membrane potential change response. It is curious that abp1 lowers the dose required for auxin response rather than increasing the magnitude of the response. It might be expected that addition of receptors would increase the magnitude of the response at a given hormonal dose, but that is not what is found. The authors speculate that abp1 might bind to plasma membrane transmembrane proteins creating "active signal transducing units" from inactive ones (Barbier-Brygoo *et al.*, 1991). The recruitment of these active units would produce greater hormone responses at lower hormone doses.

7.2.1.2 Auxin Signal Transduction Pathways

It is not known what pathways are used in auxin signal transduction. It is possible that auxin responses utilize several different signaling pathways and that the pathway that leads to proton efflux may be different than the pathway that leads to auxin-induced gene expression (to be discussed below). In animal systems the transduction of intracellular signals elicited by membrane impermeable hormones often involves many components in addition to plasma membrane-located receptors. These components are required to transmit the hormone signal across the cell membrane, to convert a signal into another form, and to amplify a small signal to a powerful response. One major signal response pathway in animal cells is the G protein-mediated phosphoinositide signaling system which involves the production of inositol triphosphate and diacylglycerol as second messengers (Berridge, 1987). The components of the phosphoinositide signaling system have been found in plants (Boss, 1989; Heim and Wagner, 1986).

It is not known whether auxin utilizes a G protein-mediated pathway, however, preliminary studies suggest that signaling pathways involving phospholipases may be involved in certain auxin responses. For example, it has been reported that auxin accelerates the turnover of both phosphatidylinositol phosphate (PIP) and phosphatidylinositol bisphosphate (PIP_2) in response to auxin in carrot plasma membrane vesicles (Zbell and Walter-Beck, 1988). Furthermore, it has been reported that auxin causes a transient rise in inositol 1,4,5-triphosphate (IP_3) in *Catharanthus roseus* cell suspension cultures blocked in the G_1 phase of the cell cycle (Ettlinger and Lehle, 1988). As in animal cells, IP_3 has been shown to release Ca^{2+} from isolated vacuolar vesicles (Schumaker and Sze, 1987). It is not known what reported enzymatic steps in phosphoinositide metabolism are sensitive to the auxin signal—whether it is the synthesis of PIP and PIP_2 by plant phosphoinositide kinases (Sandelius and Sommarin, 1986; Sommarin and Sandelius, 1988) or the breakdown of PIP_2 by plant phospholipase C (Helsper *et al.*, 1987; Irvine *et al.*, 1980; Melin *et al.*, 1987; Pfaffmann *et al.*, 1987).

Another phospholipase, phospholipase A_2, which hydrolyzes membrane phospholipids to lysophospholipids and free fatty acids, may be a better candidate for a component of the pathway in which auxin stimulates proton pumping. It has been reported that auxins enhance phospholipase A_2 activity *in vivo* and *in vitro* and that induction of the phospholipase activity in isolated microsomal membranes can be blocked by an antibody to abp1 (André and Scherer, 1991; Scherer and André, 1989). Others have demonstrated that lysophospholipids stimulate proton pumping in isolated oat root plasma membranes (Palmgren and Sommarin, 1989). In coupling various findings together, it could be argued that lysophospholipids activate a protein kinase that may, in turn, phosphorylate and regulate the activity of the plasma membrane proton pump (Martiny-Baron and Scherer, 1989; Schaller and Sussman, 1988).

7.2.1.3 Auxin-Induced Gene Expression

Auxin stimulates a number of different responses in plant cells, including the activation of expression of auxin-induced genes [reviewed by Guilfoyle (1986) and Theologis (1986)]. Auxin-induced changes in gene expression have been most thoroughly studied in organs such as soybean hypocotyls in which other physiological responses to auxin, such as cell elongation, have been extensively documented. Auxin-induced changes in gene expression were first demonstrated in soybean hypocotyl (Zurfluh and Guilfoyle, 1982b), maize coleoptile (Zurfluh and Guilfoyle, 1982a), and pea epicotyls (Theologis and Ray, 1982) by isolating mRNAs at various times after auxin addition, translating the RNAs, and examining the translation products on two-dimensional gels. These studies demonstrated that a small number of mRNAs, in one case, 10 out of 500 mRNAs identifiable as translation products on two-dimensional gels (Zurfluh and Guilfoyle, 1982b), were induced by auxin. Through cDNA cloning techniques, cDNAs representing some of the auxin-induced mRNA were cloned from soybean (Ainley *et al.*, 1988; Hagen *et al.*, 1984; McClure *et al.*, 1989; Walker and Key, 1982), pea (Theologis *et al.*, 1985), *Arabidopsis* (Alliotte *et al.*, 1989), and tobacco mesophyll protoplasts (Takahashi *et al.*, 1989).

Activation of some auxin-induced genes was very rapid and occurred before other physiological responses such as cell elongation. For several auxin-induced genes, activation was transcriptional and did not require additional protein synthesis (Hagen and Guilfoyle, 1985; Hagen *et al.*, 1988). At issue in these studies was the cause and effect relationship between auxin-induced gene expression and physiological responses; that is, was auxin-induced gene activation required for certain physiological responses, such as cell elongation, or vice versa? Clearly the onset of expression of some auxin-induced genes (described by some investigators as "early" genes) preceded that of the first stage of cell elongation which lagged about 10–12 min in soybean hypocotyls (Vanderhoef, 1980). A possibility that has been explored was whether proton efflux, a step in cell elongation which locally acidifies the cell wall and alkalinizes the cytoplasm, triggered the expression of the auxin-induced genes. Theologis *et al.* (1985) demonstrated that cerulenin, an inhibitor of auxin-induced proton translocation, did not block auxin-induced gene expression. Furthermore, fusiococcin which promoted proton translocation did not, in itself, stimulate auxin-induced gene expression in pea epicotyls. These experiments demonstrated that auxin-induced proton pumping did not trigger early gene expression events. However, the data did not address the possibility that proton translocation and auxin-induced gene expression respond to a common second messenger signal.

A special class of auxin-induced genes are those encoding the SAUR RNAs (small auxin up RNAs) (McClure *et al.*, 1989). These genes represent a kinetic class of mRNAs that are most rapidly activated by auxins. Time course measurements have demonstrated that these genes are transcriptionally activated within 2–3 min after auxin addition. The group of genes is unusual in that they are clustered and divergently transcribed at the so-called SAUR locus in the soybean genome. Three of five reported SAUR genes were highly induced by auxin in soybean hypocotyls. The inducible expression was tissue specific with highest levels of induction in the elongating segments of the hypocotyls, tissues which undergo rapid cell elongation in response to auxin. Lower, but yet significant levels of induction were observed in other organs such as leaves. The different SAUR genes encoded similar proteins of 9–10.5 kDa of unknown function.

A 7 kilobase (kb) DNA fragment containing the soybean SAUR locus has been transferred to petunia (Wright *et al.*, 1989). In transgenic petunias, the auxin-induced genes in the SAUR locus were transcribed in explanted young leaves in response to auxins. Hence, the *cis*-acting elements necessary for activation of these auxin-induced genes were present in the 7-kb soybean fragment. The gene transfer experiment also demonstrated that the soybean genes were able to respond to auxin signals in petunia. In more recent studies, the promoter from SAUR gene 10A was linked to the β-glucuronidase gene and introduced into tobacco plants where it was expressed in a number of tissues including roots (Li *et al.*, 1991). In roots, the chimeric SAUR gene appeared to respond to endogenous gradients of auxin established by gravitropism. GUS activity was expressed asymmetrically on the downward side of roots in transgenic plants oriented horizontally. The asymmetric expression of the chimeric GUS was abolished by inhibitors of auxin transport.

The functions of very few auxin-regulated genes are known. In fact, no functions have been ascribed to genes which have been selected by differential screening (+/− hormone). However, genes identified by other criteria have been shown to be auxin regulated. For example, 1-aminocyclopropane 1-carboxylate (ACC) synthase was cloned by immunoscreening a cDNA library with an antibody raised against ACC synthase (Sato and Theologis, 1989). Using the cDNA as a probe, Sato and Theologis (1989) found that the accumulation of ACC synthase RNA in young zucchini fruit was stimulated by auxin.

A cDNA for an *Arabidopsis* gene, *dbp*, that is auxin regulated was selected on the basis of its rather uniform organ-specific distribution, except that levels were higher in apical shoot and root segments (Alliotte *et al.*, 1989). Induction of the mRNA was comparatively slow (between 4 and 8 hr) and so the gene was not thought to be directly induced by auxins. Interestingly, the gene product appeared to be a non-sequence-specific DNA-binding protein. The protein was quite basic and had certain features in common with histone H1 protein, however, it was not thought to be a histone.

7.2.1.4 Auxin Response Mutants

Mutants with altered responses to auxin have been identified in a number of plants including the diageotropica (*dgt*) mutant in tomato discussed in a previous section. Auxin-resistant mutants have been obtained in the moss *Physcomitrella* (Ashton *et al.*, 1979), *Arabidopsis* (Estelle and Sommerville, 1987; Maher and Martindale, 1980; Mirza *et al.*, 1984), *Glycine max* (Davidonis *et al.*, 1982), *Nicotiana tabacum* (Chaleff and Parsons, 1978), and *N. plumbaginifolia* (Bitoun *et al.*, 1990; Muller *et al.*, 1985). The *N. plumbaginifolia* mutants isolated by Muller *et al.* (1985) were obtained by selecting for NAA-resistant clones from mutagenized haploid mesophyll protoplasts. Plants regenerated from the clones failed to produce roots although the plants could be rescued by grafting or dipping the base of the plant in auxin-containing solutions. The inheritance of the trait was consistent with the mutation being a single dominant nuclear mutation. In later studies, Bitoun *et al.* (1990) isolated *N. plumbaginifolia* mutants by selection for auxin resistance at the seedling stage. The authors obtained recessive mutants which defined three different complementation groups *iba1*, *iba2*, and *iba3*. The *iba1*/*iba1* plants also displayed cross resistance to ABA and to an inhibitor of gibberellic acid biosynthesis, paclobutrazol. The authors suggest that the cross resistance is a product of the interaction of the hormones in promoting normal growth.

Maher and Martindale (1980) isolated 30 auxin-resistant lines of *Arabidopsis* and examined one in detail. This line, which was due to a single recessive mutation, produced roots which were not positively geotropic on agar medium. In a further analysis of these lines, Mirza *et al.* (1984) focused on two recessive mutants *aux-1* and *aux-2* and a dominant mutant *Dwf*. All were resistant to exogenous auxin but *aux-1* and *Dwf* were ageotropic, while *aux-2* was not.

Estelle and Sommerville (1987) isolated a number of auxin-resistant mutants in *Arabidopsis* in which root growth was resistant to high levels of auxin. Mutants in one group were recessive at a locus called *axr-1* (Lincoln *et al.*, 1990). Although one *axr-1* mutant line was less severe than the others, all affected leaf and flower morphology. Mutant plants were shorter than nonmutants and displayed reduced apical dominance. Another mutant line carried a dominant mutation at a locus called *axr-2*. These mutants were dwarfed, had abnormal gravitropic responses, and were resistant to growth-inhibiting concentrations of ethylene and abscisic acid.

Although no mutant genes have been isolated from auxin-resistant plants, the opportunity to undertake map-based cloning or cloning by transposon mutagenesis is at hand.

7.2.2 CYTOKININ

7.2.2.1 Cytokinin-Binding Proteins

As with auxin, the first step in the cytokinin response is thought to be the binding of cytokinin to a receptor. In the search for cytokinin receptors, several groups have studied cytokinin-binding proteins in extracts from wheat germ (Brinegar *et al.*, 1988; Erion and Fox, 1981; Keim *et al.*, 1981; Moore, 1979; Polya and Davies, 1983; Polya and Davis, 1978; Polya and Bowman, 1979), barley embryos and leaves (Reddy *et al.*, 1983; Romanov *et al.*, 1988), tobacco leaves or cultured cells (Chen *et al.*, 1980; Sussman and Kende, 1978; Takegami and Yoshida, 1975), and cucumber cotyledons (Jayabaskaran, 1990). Recently a binding protein from barley leaves has been described that has a K_d for zeatin $\sim 1.5 \times 10^{-8}$ *M* (Romanov *et al.*, 1988). The protein is 40–45 kDa and binds various cytokinins in accordance with their biological activity. An unexpected feature about the protein is that binding requires high salt. Although this protein has many features expected of a cytokinin receptor, in the absence of a functional test, it has been difficult to attach physiological roles to any of the cytokinin-binding proteins. Hence, there has been less progress to date in the search for a cytokinin receptor than for an auxin receptor.

7.2.2.2 Cytokinin-Induced Gene Expression

Although cytokinin responses have been studied in many different plant systems, the mode of action of cytokinin is unknown (Davies, 1987). The addition of cytokinins to cultured plant material results in the appearance of a number of new mRNAs and proteins (Bauw *et al.*, 1987; Chatfield and Armstrong, 1986; Chen *et al.*, 1987; Chen and Leisner, 1985; Fosket and Tepfer, 1978). Certain tobacco cell suspension cultures respond to cytokinin by greening in the light and accumulating the light-harvesting chlorophyll *a*/*b*-binding (cab) proteins and their mRNAs (Axelos *et al.*, 1984; Flores and Tobin, 1986; Flores and Tobin, 1988;

Teyssendier de la Serve *et al.*, 1985). *Cab* gene expression is regulated by light and cytokinin in other greening systems such as in duckweed, *Lemna*. Flores and Tobin (Flores and Tobin, 1986; Flores and Tobin, 1988) have argued that in *Lemna,* cytokinin stimulation of *cab* gene expression is regulated primarily at the posttranscriptional level. Cytokinins also induce cytokinin oxidase in *Phaseolus* cell cultures (Chatfield and Armstrong, 1986). Cytokinin oxidase is presumably involved in the breakdown of cytokinin and cleaves the N^6 side chains from cytokinins (Figure 7.2). The enhancement of cytokinin oxidase was blocked by RNA synthesis inhibitors, suggesting that cytokinin regulation of cytokinin oxidase expression is transcriptional.

Cytokinins down-regulate the expression of other proteins, most notably β-glucanase. Cytokinin or auxin reduced levels of this protein and its mRNA in tobacco while ethylene stimulated its accumulation (Felix and Meins, 1986; Felix and Meins, 1987). The regulation was even more complex for chitinase and its mRNA which was down-regulated by a combination of cytokinin and auxin, but not by either hormone alone (Shinshi *et al.*, 1987). In tobacco cell cultures cytokinin or auxin reduced levels of β-gluycanase and its mRNA (Felix and Meins, 1986; Felix and Meins, 1987) and the combination of both growth regulators reduced the levels of chitinase (Shinshi *et al.*, 1987).

7.2.2.3 Cytokinin Response Mutants

There have been few reports in higher plants of variation in the cytokinin response (Skvirsky *et al.*, 1984) and only one published report of a higher plant mutant selected on the basis of its resistance to cytokinin, the *ckr1* mutant in *Nicotiana plumbaginifolia* (Blonstein *et al.*, 1991). The primary lesion in this mutant is not known, but it appears that the defect may be unrelated to cytokinin metabolism or perception. Instead, the mutant seems less able to mount a stress response in the presence of high levels of cytokinin. The mutant has a wilty phenotype, shows little abscisic acid (ABA) induction following stress, and is apparently blocked in the last step of ABA biosynthesis (Parry *et al.*, 1991). To avoid the problem of selecting for mutants that do not show general stress responses, Su and Howell (reported in Leader, 1991) have selected *Arabidopsis* mutants resistant to low concentrations of cytokinin. In doing so, they have found that a single gene locus (*ckr1*) in the *Arabidopsis* confers cytokinin resistance.

The analysis of mutants selected in tissue culture has been complicated. Meins and co-workers have studied the acquisition of cytokinin independence or autotrophy in tobacco tissue culture cells. The conversion of cells from a cytokinin-requiring state to a cytokinin-autotrophic state is known as cytokinin habituation. Habituation is heritable but appears to result most frequently from epigenetic changes (Meins and Binns, 1982; Meins and Lutz, 1980). Depending on their tissue of origin in regenerated plants, cells also vary in their tendency to habituate (Meins and Foster, 1985). Habituation may be a product of acquired capacity for cytokinin biosynthesis in cell culture. However, high cytokinin levels have not been observed in habituated cells (Hansen *et al.*, 1985). A genetic mutant has been obtained that yields cytokinin autotrophic cells from tissues that normally give rise to cytokinin-requiring cells (Meins and Foster, 1986). The mutation may confer constitutive cytokinin biosynthesis in tissues that are normally cytokinin requiring in tissue culture. The mutation can compensate for a defective *tmr* locus (see below) in *Agrobacterium*-induced tumor cells (Hansen and Meins, 1986).

7.3 Bacterial Genes

7.3.1 *AGROBACTERIUM* GENES

Much has been learned about plant growth hormone action and regulation from studies of virulent strains of *Agrobacterium, A. tumefaciens,* and *A. rhizogenes,* which produce crown gall and hairy root tumors, respectively. Crown gall tumors have different morphologies depending on the bacterial strain, the host plant, state of development, and position of inoculation. However, in general, crown gall tumors are composed or unorganized callus or loosely organized teratomatas containing abnormal leaf and stem structures. In contrast, hairy root tumors are tumors which produce abundant roots. Both *Agrobacterium* species owe their virulence to determinants on large plasmids, the Ti plasmid (tumor-inducing plasmid) in *A. tumefaciens* and the Ri plasmid (root-inducing plasmid) in *A. rhizogenes*. Both species transfer a segment of plasmid DNA, the T-DNA (transferred DNA), to inoculated plants. The tumor type is determined by the plasmid, and in *A. rhizogenes* the plasmid is called the Ri plasmid. *A. tumefaciens* and *A. rhizogenes* plasmids and bacterial strains can be further distinguished on the basis of the opines which they encode. In *A. tumefaciens* the octopine-type Ti plasmids transfer to plants two noncontiguous T-DNA segments, T_L and T_R, while the nopaline strains transfer only one T-DNA segment. Likewise in *A. rhizogenes,* the agropine-type Ri plasmids transfer two T-DNA segments, T_L and T_R, while the mannopine strains transfer only one T-DNA segment. In all cases, the T-DNA is integrated in the plant genome and expresses oncogenes which alter the production, levels, or perception of plant growth regulators.

The roles of the various *Agrobacterium* oncogenes in tumor formation were determined by very similar experimental procedures (Spanier *et al.*, 1989). First, the oncogenes were identified by transposon mutagenesis. From the phenotypes of the tumors incited by various *Agrobacterium* mutants inferences have been made about the functions of the inactivated oncogenes. Second, plasmids were constructed in which all other oncogenes were deleted except the one in question. These strains were used to generate tumors and the tumor phenotype was used to confirm oncogene functions deduced from the transposon insertion analysis. Third, individual oncogenes were linked to bacterial promoters and expressed in *Escherichia coli.* In *E. coli,* the presumed enzymatic function of the gene products could be tested. Finally, oncogenes linked to their own promoters or to other plant promoters were introduced and expressed in transgenic plants. The overproduction or ectopic production of the gene product often permitted greater insight into the function of the oncogene.

The structure of the T-DNA region of the Ti plasmid was first mapped by transposon analysis (Garfinkel *et al.*, 1981; Holsters *et al.*, 1980; Ooms *et al.*, 1981). In the Ti plasmids, three regions involved in tumor formation were identified: (1) a *tms* region which when disrupted produced "shooty" or "shooter" phenotype (tumors with abundant shoots); (2) a *tmr* region which when disrupted by transposons produced a "rooty" or "rooter" phenotype (tumors with abundant roots); and (3) a *tml* region which when similarly disrupted produced large tumors. The tumors bore the characteristics of plant growth hormone imbalances which can be observed in tissue culture at different cytokinin-to-auxin ratios. When the hormone content of the tumors was analyzed it was found that tumors produced by "rooter" mutants (inactivation of *tmr*) had low levels of cytokinins, relative to auxins, and tumors produced by the "shooter" mutants (inactivation of *tms*) had low levels of auxins relative to cytokinins (Akiyoshi *et al.*, 1983; Weiler and Spanier, 1981).

7.3.1.1 *tms* Locus (*iaaH* and *iaaM* Genes)

The *tms* region, which is inactivated in *A. tumefaciens* mutants that produce "shooty" tumors, was found to encode enzymes involved in auxin biosynthesis. The *tms* region was composed of two transcription units, genes 1 and 2. When gene 2 was cloned and expressed in *E. coli,* it encoded an activity that hydrolyzed indole-3-acetamide to IAA (Figure 7.1) (Schroder *et al.*, 1984), similar to an activity mediated by the *iaaH* gene in *Pseudomonas savastanoi,* another phytopathogenic bacterium. When gene 1 was expressed in *E. coli,* it encoded a monooxygenase activity that converted tryptophan to indole-3-acetamide (Thomashow *et al.*, 1986), similar to the action of the *iaaM* gene in *P. savastonaoi.* Hence, *tms* 1 and 2, which are equivalent to *iaaM* and *iaaH* in *P. savastanoi,* respectively, encode enzymes which are able to synthesize the auxin, IAA, in transformed plants via a prokaryotic biosynthetic pathway.

The two *tms* genes from *A. tumefaciens* have been linked to different plant promoters and expressed in transgenic petunia plants (Klee *et al.*, 1987). When plants were transformed with *tms* 2 (*iaaH*) linked to strong promoters, such as the cauliflower mosaic virus (CaMV) 35S RNA promoter or the *rbc ss* promoter, no demonstrable phenotype was observed. However, leaf disks explanted from these transgenic plants grew on medium containing low levels of naphthaleneacetamide (NAM) as the only auxin source. Leaf disks from nontransformed plants lacking hydrolase activity were unable to utilize the acetamide derivative as an auxin source. On the other hand, high concentrations of NAM, in the presence of an alternative auxin source, were toxic to transformed plants but not to untransformed plants. This is because the transformed plants converted NAM to high concentrations of the auxin naphthaleneacetic acid (NAA), which was toxic. Therefore, high-level expression of the *tms* 2 (*iaaH*) gene was of little consequence to transgenic petunia plants in the absence of the acetamide intermediate. However, at low concentration of intermediate, the plant can utilize NAM as an auxin source, but at high concentrations, transformed plants are killed by the intermediate. The lethal consequences of *tms* 2 (*iaaH*) gene expression in the presence of high concentrations of NAM has been put to good use in the development of a selectable marker in *Arabidopsis*. Karlin-Neumann and co-workers (Karlin-Neumann *et al.*, 1991) have linked the *tms* 2 (*iaaH*) gene to the chlorophyll *a*/*b*-binding protein (cab) 140 promoter and have introduced the construct into transgenic *Arabidopsis* plants in the effort to select for phytochrome-defective mutants that fail to turn on the cab gene in response to red light.

When *tms* 1 (*iaaM*) was similarly linked to a strong, constitutive promoter, such as the CaMV 19S promoter, and expressed in transgenic petunia plants, the plants displayed an unexpected, abnormal phenotype (Klee *et al.*, 1987). The plants were apically dominant, with little branching, and the stems were elongated and appeared woody. The leaves were smaller and narrower than the leaves from untransformed plants, and regenerated shoots were covered with adventitious roots. The plants accumulated high levels of indoleacetamide (IAM). The abnormal phenotype was unexpected despite the high levels of IAM accumulation because IAM is not thought to be a precursor of auxin biosynthesis. So, in the absence of an introduced *tms* 2 (*iaaH*) gene, IAM should not affect plant development. However, these plants also showed a 10-fold elevated level of IAA accumulation. So, clearly, chronic IAM accumulation had some direct or indirect effects on auxin levels in the transformed plants.

The *tms* 1 gene was also put under control of the soybean 7S storage protein gene promoter to drive seed-specific expression of the *tms* 1 gene (Klee *et al.*, 1987). Transgenic petunia material bearing this construct showed the above-described abnormalities in early stages of shoot regeneration, but the plants generally outgrew the abnormalities.

7.3.1.2 *tmr* Locus (*ipt* Gene)

The *tmr* locus that is inactivated in *A. tumefaciens* Ti plasmid mutants which produce "rooty" tumors was found to encode an enzyme involved in cytokinin biosynthesis. The *tmr* locus contained only a single transcription unit, gene 4. Gene 4 encodes isopentenyltransferase (*ipt* gene), which is thought to be similar to the activity catalyzing one of the first steps in plant cytokinin biosynthesis (Akiyoshi *et al.*, 1984; Barry *et al.*, 1984; Buchmann *et al.*, 1985). The effects of the *tmr* locus on plant development have been studied in the context of plant tumors and transgenic plants. *A. tumefaciens* with mutations at the *tmr* locus (gene 4) produces tumors with (1) low cytokinin-to-auxin ratios (Akiyoshi *et al.*, 1983); (2) a requirement for cytokinin to grow in culture (Inze *et al.*, 1984); and (3) a tendency to produce roots, not shoots. On the other hand, *A. tumefaciens* strains in which the *tmr* locus was "up-regulated," driven by a strong promoter, such as the CaMV 35S RNA promoter, produced tumors in tobacco and cucumber with highly elevated levels of zeatin (Z) and zeatin riboside (ZR) and somewhat depressed levels of IAA (Smigocki and Owens, 1988; Smigocki and Owens, 1989). Interestingly, these tumors exhibited both cytokinin and auxin autonomy in tissue culture. Galls on infected plants produced by these *A. tumefaciens* strains produced shoots at a higher frequency than that in strains where the *tmr* locus was driven by its own promoter.

Memelink *et al.* (1987) produced tobacco plants containing the *tmr* locus (referred to as T-cyt). These plants had elevated levels of cytokinin and exhibited characteristics of chronic cytokinin overproduction, that is, the plants had no roots and short adventitious shoots. Such plants accumulated RNA transcripts to high levels for a number of stress-related proteins including extensin, chitinase, and pathogenesis-related protein (PR-1). The same transcripts also accumulated in nontransformed tobacco shoots which had been cultured for 3 weeks on cytokinin. In an effort to regulate the *tmr* gene in transgenic plants, Medford *et al.* (1989) linked the *ipt* gene to an inducible heat-shock promoter and introduced the construct into tobacco and *Arabidopsis* plants. They found that although heat-shock elevated cytokinin levels, the background levels in the uninduced plants were high enough to generate a cytokinin overproduction syndrome, that is, release of axillary buds, reduced leaf area, and underdeveloped roots. Thus, transgenic plants expressing isopentenyl transferase show characteristics of cytokinin overproduction or exogenous cytokinin application.

Two other groups have placed the *ipt* gene under another inducible promoter, the *Drosophila hsp70* heat-shock promoter (Schmulling *et al.*, 1989; Smigocki, 1991). Both groups found an elevation in cytokinin levels (particularly zeatin and zeatin riboside) in response to heat, although like Medford *et al.* (1989) Smigocki found phenotypes associated with cytokinin overproduction even when the plants were not thermally induced. Recently, Smart *et al.* (1991) linked the *ipt* gene to yet another heat-shock promoter, the soybean heat-shock promoter HS6871. This promoter appeared to provide a lower uninduced level of cytokinin accumulation. Upon heat induction, isopentenyl adenine-type cytokinins accumulated quickly and were rapidly converted to zeatin-type cytokinins. Interestingly, cytokinin levels fell back down again within 24 hr of heat-shock, suggesting that isopentyl transferase and the cytokinins have a rapid rate of turnover. Under uninduced conditions the transgenic plants bearing the *ipt* gene linked to the soybean heat-shock promoter had larger side shoots and senesced more slowly. With multiple heat-shocks the transgenic plants produced more side branches and remained greener longer than nontransgenic plants (Smart *et al.*, 1991). The authors interpreted this to mean that cytokinin levels could be chronically elevated by multiple heat-shocks and that these elevated levels altered plant morphology and the rate of leaf senescence.

The *tmr* locus appeared to be influential in the transformation of some monocots (Prinsen *et al.*, 1990). *A. tumefaciens* infects *Asparagus officinalis* but transformed tissues selected for autotrophic growth did not show elevated levels of auxins. Surprisingly, the transformed tissues did not have detectable levels of tryptophan monooxygenase activity, and apparently the end product of the reaction, indole-3-acetamide, was toxic to the cells. It has been hypothesized that inactivation of the *tms* 1 gene is necessary for tumor growth and tissue survival, and the expression of *tmr* (gene 4) leading to elevated cytokinin levels is all that is required for tumor formation.

7.3.1.3 *tml* Locus

Another oncogenic region in the T-DNA defined by transposon mutagenesis is the *tml* (tumor-morphology large) locus. Mutations at this locus in the octopine Ti plasmid had a more subtle effect, leading to the formation of large tumors on *Kalanchoe daigremontiana*. Some confusion originally surrounded the analysis of the *tml* locus before it was recognized that the locus contained two genes, gene 6a and b. Because the transcripts for the two genes were nearly identical in size, Northern blot analysis obscured the fact that there were two adjacent transcription units. Gene 6a was found not to be an oncogene, but is involved in the secretion of opines (Messens *et al.*, 1985). On the other hand, gene 6b suppressed somewhat the shoot-producing phenotype of the "shooter" mutants. Gene 6b seemingly exerted an anticytokinin effect because T-DNA deletion mutants bearing gene 4 (the *ipt* gene) and genes 6a and b produced fewer shoots than deletions bearing *ipt* alone (Leemans *et al.*, 1983).

Because the promoter on gene 6b is very weak, Spanier and colleagues (Spanier *et al.*, 1989) linked the coding region of 6b (from the octopine Ti plasmid pTiAch5) to a strong promoter (T_R-1′-promoter) and expressed the construct in transformed tobacco leaf disks. Leaf disks in culture were used to assess the action of gene 6b, because leaf disks are more sensitive than intact plants to cytokinin effects. In leaf disk cultures, cytokinin concentrations were manipulated either by the inclusion of cytokinin in the culture medium or by action of the *tzs* gene. (The *tzs* gene located outside the T-DNA region sponsors cytokinin production in certain *Agrobacterium* strains bearing the gene.) They found that gene 6b reduced shoot formation at optimal or suboptimal cytokinin concentrations. At superoptimal cytokinin concentration conditions, that is, concentrations which reduce shoot formation, gene 6b actually stimulated shoot formation. Thus, in leaf disk explants gene 6b counteracted the action of cytokinin whether cytokinins were produced endogenously from the action of gene 4 or supplied exogenously from the medium or derived from cytokinin-producing *A. tumefaciens* strains. The authors suggested that the anticytokinin effect could be due to the action of enzymes encoded or activated by gene 6b that may conjugate, degrade, or otherwise inactivate cytokinins.

It has been argued that the action of gene 6b cannot be explained entirely on the basis of anticytokinin effects (Tinland *et al.*, 1989). The arguments derive from experiments in which *A. tumefaciens* T-DNA constructs expressing gene 6 (from the octopine/cucumopine Ti plasmid pTiTm4) were coinoculated with other *A. tumefaciens* strains on *Nicotiana rustica* plants. In these experiments, it was shown that when plants were coinoculated with constructs containing the *tms* genes and gene 6b that tumors were larger than when inoculated with *tms* genes alone. Thus, the gene 6b product appears to enhance the sensitivity of the transformed cells to auxin. However, because the system is complex, the authors could not rule out the possibility that the enhanced auxin effects were not due to a reduction in cytokinins which normally antagonize auxin responses.

7.3.1.4 Rol Genes

Agrobacterium rhizogenes is responsible for "hairy root disease." In hairy root disease, tumorous roots are formed at the site of *A. rhizogenes* infection. The disease results from the transfer of *A. rhizogenes* DNA to plant cells in the same manner as *A. tumefaciens*. Hairy root tumors are thought to differ from crown gall tumors in an important way. Crown gall tumors incited by *A. tumefaciens* are usually a mixture of transformed and untransformed cells. The cells transformed by *A. tumefaciens* elaborate growth hormones which diffuse from the transformed cells and promote the growth of neighboring untransformed cells. In contrast, hairy root tumors incited by *A. rhizogenes* are composed solely of transformed cells, and therefore, *A. rhizogenes*-induced tumors are thought to be cell autonomous. These observations were made before the functions of some of the *A. rhizogenes* oncogenes were known. The observations still seem valid, but the reasons for the differences are not clear. It has been reported recently that *A. rhizogenes,* like *A. tumefaciens,* produces tumors by promoting the overproduction of diffusible growth regulators, although the mechanism by which they do so differs from *A. tumefaciens*.

Agropine strains of *A. rhizogenes* carry auxin synthesis genes that are homologous to *tms* 1 and 2 in *A. tumefaciens* Ti plasmids (Huffman *et al.*, 1984; Jouanin, 1984). Unlike the situation in *A. tumefaciens,* these genes are quite weak because the hormone balance in transformed tissue is not substantially altered (Spano *et al.*, 1988). The auxin genes are encoded on T_R, but the T_R segment is only accessory to T_L in the formation of hairy root tumors (Cardarelli *et al.*, 1987a; Cardarelli *et al.*, 1985; Vilaine *et al.*, 1987). Furthermore, the mannopine strains do not have T_R and the mannopine plasmid T-DNAs do not have genes with homology to *tms* 1 and 2 (Lahners *et al.*, 1984). To account for these observations, it has been proposed that the agropine-type plasmids endow infected plants with two mechanisms of tumor formation—one which involves the modest overproduction of auxin through *de novo* biosynthesis and another which further enhances the auxin pool by other means.

The contrast between these two mechanisms can be observed in the polar induction of tumors in carrot root disks (for review see Gelvin, 1990). Virulent *A. tumefaciens* strains and agropine *A. rhizogenes* strains incite tumors when inoculated on either the apical or basal side of carrot root disks. *A. tumefaciens tms* mutants were only able to incite tumors when inoculated on the apical surface. This could be explained by the polar transport of auxin which concentrated auxin on the apical surface of the root disks and compensated for the auxin production deficiency of the *tms* mutants. The same phenomenon was observed with mannopine-type *A. rhizogenes* strains which did not have auxin genes. Interestingly, the infection properties of the mannopine type strains in the carrot root disk assays can be converted from polar to nonpolar by addition of the T_R region from the agropine plasmid to the mannopine plasmid (Cardarelli *et al.*, 1987b). Hence the mannopine strains were able to form tumors when adequate supplies of auxin were available as they were on the apical surface of the carrot disk. However, when auxin supplies were limiting as they were on the basal surface, tumors did not grow from the transformed tissues.

What the mannopine plasmid or the T_L region of the agropine *A. rhizogenes* plasmids appeared to do was to sensitize plant tissue to endogenous auxin concentrations or to enhance further the concentrations produced by the introduced *tms* genes. Shen and colleagues (Shen *et al.*, 1988) demonstrated that *Lotus corniculatus* root tips or protoplasts transformed with an agropine-type *A. rhizogenes* Ri plasmid showed greater auxin responses in three different assays—a short-term membrane potential response, an intermediate term (1 hr) proton secretion response, and a longer term (24 hr) root elongation response.

The genetic determinants in the T_L segment that are required for tumor formation in the agropine strains were identified by transposon and deletion mutagenesis of T_L-DNA. T_L-DNA contains 18 open reading frames (ORF) (Slighton *et al.*, 1986) and 4 of those, designated *rolA–D*, corresponding to ORFs 10, 11, 12, and 15, are required for tumor formation (White *et al.*, 1985). Inactivation of each *rol* loci gave a slightly different phenotype when assayed on *Kalanchoe* leaves. Under these conditions, *rolB* was found to be indispensable for hairy root tumor formation and insertions in *rolA, C,* or *D* gave an attenuated response. *rolA* mutants produced very straight roots, *rolC* mutants exhibited attenuated root growth and *rolD* mutants showed normal tumorous root initiation, but retarded root growth. In a tobacco root callus system, *rolA, B,* and *C* genes, in the absence of all other T_L genes, were able to induce hairy root formation (Spena *et al.*, 1987). In fact, each of these genes alone was able to stimulate root formation in the tobacco callus system although they differed in their capacity to do so. *rolB* was most effective but the combined action of the *rolA, B,* and *C* genes was cooperative. The effectiveness of *rolC* in stimulating root formation was enhanced by substituting the *rol* gene promoter with a stronger promoter, the CaMV 35S RNA promoter (Spena *et al.*, 1987). In transformed carrots, *rolA, B,* and *C* were effective in producing explantable hairy root tumors but only in combination with auxin synthesis genes on T_R or with ORFs 13 + 14. The functions of ORFs 13 + 14 are not known, however, the genes are cell autonomous and do not appear to involve the synthesis of diffusible growth regulators (Capone *et al.*, 1989).

Recently, some exciting results reported by Estruch *et al.* (1991a,b) have uncovered functions for the *rolB* and *C* gene products. These gene products can hydrolyze inactive hormone-glucoside conjugates and release biologically active, free hormone. The *rolC* gene product has been found to be a β-glucosidase that is capable of hydrolyzing cytokinin *N*-glucosides and releasing free cytokinins (Estruch *et al.*, 1991a). Likewise, the *rolB* gene product is another β-glucosidase that is capable of hydrolyzing indole β-glucosides (Estruch *et al.*, 1991b). Although these gene products have β-glucosidase activities when expressed in *E. coli* and in transgenic plant tissue, it remains to be seen whether the liberation of free hormone is, indeed, the mechanism by which tumor production occurs. It is difficult to reconcile these findings with earlier ones that hairy root tumors do not have elevated levels of free growth regulators. That data has been revisited by Estruch *et al.* (1991a) for *rolC* transformed plants, and it was observed that isopentenyl adenosine levels were elevated but dihydrozeatin levels were not. In any case, these findings provide a new perspective from which to view the production of hairy root tumors.

Unlike crown gall tumors, hairy root tumors from tobacco plants can be efficiently regenerated into plants. However, the regenerated plants exhibit a hairy root (*hr*) syndrome. The syndrome is characterized by extensive growth of roots in culture, the production of adventitious roots, reduced apical dominance, wrinkled leaves with increased width-to-length ratio, altered root morphology, root growth with altered geotropism, and reduced flower size, pollen, and seed production (Tepfer, 1984). The *hr* syndrome required the function of *rolA, B,* and *C* which appeared to act independently, but synergistically. Transgenic plants expressing individual *rol* genes produced distinct and specific growth abnormalities (Schmulling *et al.*, 1988). For example, transgenic tobacco plants expressing only *rolA* showed wrinkled leaves (Sinkar *et al.*, 1988) while those expressing only *rolC* were dwarfed (Oono *et al.*, 1987). The *rol* genes were overexpressed in transgenic plants by linking the genes to the CaMV 35S promoter. When *rolC* was overexpressed, the resulting plants were dwarf, bushy, pale green, with smaller leaves and flowers. When *rolB* was overexpressed, the resulting calli were largely necrotic and difficult to propagate or regenerate. The tobacco plants which could be regenerated from these calli (probably the low expressing plants) had necrotic and rounded leaves and displayed heterostyly. Thus, the overexpression of *rolB* and *C* had very different effects. Overexpression of *rolC* resulted in a juvenilization, while overexpression of *rolB* led to early necrosis.

The different *rol* gene promoters also showed different tissue-specific patterns of expression which undoubtedly conditioned the *hr* phenotype. The *rolA, B,* and *C* gene promoters have been linked to GUS reporter genes and demonstrated to be most active in phloem tissue (Schmulling *et al.*, 1989b; Sugaya *et al.*, 1989). However, the three promoters exhibited different patterns of expression in roots. The *rolB* promoter was most active in the root cap and apical region of cell division, but was not active along the main root except at regions corresponding to the sites of emergence of side roots. Expression of the *rolC* promoter was detectable on both the main and side roots while the *rolA* promoter was not detectable on side roots. The cell-autonomous behavior of the *rolC* effect was most dramatically demonstrated by introducing a construct into transgenic tobacco plants containing the transposable element *Ac* inserted into the 5′ untranslated region of the *rolC* gene (Spena *et al.*, 1989). Somatic excision of the *Ac* element produced mosaic pale green spots where *rolC* was locally expressed.

7.3.2 *PSEUDOMONAS SYRINGAE* SUBSPECIES *SAVASTANOI* GENES

Another bacterial pathogen that is well known for its production of tumors and its synthesis of plant hormones is *Pseudomonas syringae* subspecies *savastanoi*. As described above, this pathogen has *iaaH* and *iaaM* genes that are required for virulence and that are similar to those described above for the *tms* locus in *Agrobacterium*. In addition, the pathogen has an *iaaL* gene that encodes IAA-lysine synthetase which converts IAA to IAA-lysine or *N*ε-indole-3-acetyl-L-lysine (Figure 7.1) (Glass and Kosuge, 1986). The IAA-lysine conjugate is inactive, and it is thought that the expression of *iaaL* balances out the activity of *iaaH* and *iaaM* in tumor production. Only the *iaaL* gene will be described here because it has been used to perturb the hormone balance in transgenic plants into which the gene has been introduced.

iaaL Gene

Romano and co-workers (Romano *et al.*, 1991) examined the anti-auxin effects of the *iaaL* gene in transgenic plants to determine what were the biological consequences of lowering the concentration of active auxin. To do so, they linked the *iaaL* gene to the CaMV 35S promoter and introduced the chimeric construct into transgenic tobacco plants. They observed that endogenous IAA levels in the shoots of the juvenile transgenic plants dropped as much as 20 times lower than their nontransgenic counterparts. They found that lowering the auxin levels had a profound effect on the morphology of the transgenic plants. In these plants apical dominance was reduced, as was rooting, and vascular differentiation was inhibited. These experiments were unique among all the other transgenic plant experiments utilizing bacterial genes affecting plant hormones, because they demonstrate the consequences of reducing, not elevating, auxin levels. As might be expected, the morphology of the transgenic plants was similar to the phenotypes of certain auxin response mutants and mutants with reduced auxin accumulation due to lesions in the tryptophan biosynthetic pathway (Last *et al.*, 1991).

References

Ainley, W. M., Walker, J. C., Nagao, R. T., and Key, J. L. (1988). *J. Biol. Chem.* 263, 10658-10666.

Akiyoshi, D., Morris, R., Hinz, R., Mischke, B., Kosuge, T., Garfinkel, D., Gordon, M., and Nester, E. (1983). *Proc. Nat. Acad. Sci. U.S.A.* 80, 407–411.

Akiyoshi, D. E., Klee, H., Amasino, R. M., Nester, E. W., and Gordon, M. P. (1984). *Proc. Natl. Acad. Sci. U.S.A.* 81, 5994–5998.

Alliotte, T., Tire, C., Engler, G., Peleman, J., Caplan, A., Van Montagu, M., and Inze, D. (1989). *Plant Physiol.* 89, 743–752.

André, B., and G. F. E. Scherer. (1991). *Planta.* 185, 209–214.

Ashton, N. W., Grimsley, N. H., and Cove, D. J. (1979). *Planta.* 144, 427–435.

Axelos, M., Barbet, J., and Peaud-Lenoel, C. (1984). *Plant Sci. Lett.* 33, 201–212.

Barbier-Brygoo, H., Ephritikhine, G., Klambt, D., Ghislain, M., and Guern, J. (1989). *Proc. Nat. Acad. Sci. U.S.A.* 86, 891–895.

Barbier-Brygoo, H., Guern, J., Ephritikhine, G., Shen, W. H., Maurel, C., and Klambt, D. (1990). In *Plant Gene Transfer. UCLA Symposia on Molecular and Cellular Biology* (Ed. by C. Lamb and R. Beachy), pp. 165–173. New York: Alan R. Liss.

Barbier-Brygoo, H., Ephritikhine, G., Klambt, D., Maurel, C., Palme, K., Schell, J., and Guern, J. (1991). *Plant J.* 1, 83–93.

Barry, G. F., Rogers, S. G., Fraley, R. T., and Brand, L. (1984). *Proc. Natl. Acad. Sci. U.S.A.* 81, 4776–4780.

Bauw, G., DeLoose, M., Inze, D., van Montagu, M., and Vandekerckhove, J. (1987). *Proc. Natl. Acad. Sci. U.S.A.* 84, 4806–4810.

Berridge, M. J. (1987). *Ann. Rev. Biochem.* 56, 159–193.

Bitoun, R., Rousselin, P., and Caboche, M. (1990). *Mol. Gen. Genet.* 220, 234–239.

Blonstein, A. D., Parry, A. D., Horgan, R., and King, P. J. (1991). *Mol. Gen. Genet.* 220, 234–239.

Boss, W. F. (1989). In *Second Messengers in Plant Growth and Development* (Ed. by W. F. Boss, and D. J. Morre), pp. 29–56. New York: Alan R. Liss.

Bradford, K. J., and Yang, S. F. (1980). *Plant Physiol.* 65, 327–330.

Brinegar, A. C., Cooper, G., Stevens, A., Hauer, C. R., Shabanowitz, J., Hunt, D. F., and Fox, J. E. (1988). *Proc. Natl. Acad. Sci. U.S.A.* 85, 5927–5931.

Buchmann, I., Marner, F.-J., Schroder, G., Waffenschmidt, S., and Schroder, J. (1985). *EMBO J.* 4, 853–859.

Capone, I., Spano, L., Cardarelli, M., Bellincampi, D., Petit, A., and Costantino, P. (1989). *Plant Mol. Biol.* 13, 43–52.
Cardarelli, M., Spano, L., DePaolis, A., Mauro, M. L., Vitali, G., and Costanino, P. (1985). *Plant Mol. Biol.* 5, 385–391.
Cardarelli, M., Mariotti, D., Pomponi, M., Spano, L., Capone, I., and Costanino, P. (1987a). *Mol. Gen. Genet.* 209, 475–480.
Cardarelli, M., Spano, L., Mariotti, D., Mauro, M. L., Van Sluys, M. A., and Costantino, P. (1987b). *Mol. Gen. Genet.* 208, 457–463.
Chaleff, R. S., and Parsons, M. F. (1978). *Proc. Natl. Acad. Sci. U.S.A.* 75, 5104–5107.
Chatfield, J. M., and Armstrong, D. J. (1986). *Plant Physiol.* 80, 493–499.
Chen, C.-M. (1982). In *Plant Growth Substances* (Ed. by P. F. Wareing), pp. 155–164. New York: Academic Press, Inc.
Chen, C.-M., and Leisner, S. M. (1985). *Plant Physiol.* 77, 99–103.
Chen, C.-M., Melitz, D. K., and Eckert, R. L. (1980). *Eur. J. Biochem.* 108, 379–287.
Chen, C.-M., Ertl, J., Yang, M.-S., and Chang, C.-C. (1987). *Plant Sci.* 52, 169–174.
Davidonis, G. H., Hamilton, R. H., and Mumma, R. O. (1982). *Plant Physiol.* 70, 104–107.
Davies, P. J. (1987). In *Plant Hormones and Their Role in Plant Growth and Development* (Ed. by P. J. Davies), pp. 1–11. Dordrecht, The Netherlands: Martinus Nijhoff.
Dohrmann, U., Hertel, R., and Kowalik, H. (1978). *Planta.* 140, 97–106.
Epstein, E., Cohen, J. D., and Bandurski, R. S. (1980). *Plant Physiol.* 65, 415–421.
Erion, J. L., and Fox, J. E. (1981). *Plant Physiol.* 67, 156–162.
Estelle, M. A., and Somerville, C. (1987). *Mol. Gen. Genet.* 206, 200–206.
Estruch, J. J., Chriqui, D., Grossmann, K., Schell, J., and Spena, A. (1991a). *EMBO J.* 10, 2889–2895.
Estruch, J. J., Schell, J., and Spena, A. (1991b). *EMBO J.* 10, 3125–3128.
Ettlinger, C., and Lehle, L. (1988). *Nature (London)* 331, 176–178.
Felix, G., and Meins, F. (1986). *Planta.* 167, 206–211.
Felix, G., and Meins, F. (1987). *Planta.* 172, 386–392.
Flores, S., and Tobin, E. M. (1986). *Planta.* 168, 340–349.
Flores, S., and Tobin, E. M. (1988). *Plant Mol Biol.* 11, 409–416.
Fosket, D. E., and Tepfer, D. A. (1978). *In Vitro.* 14, 63–75.
Garfinkel, D. J., Simpson, R. B., Ream, L. W., White, F. F., Gordon, M. P., and Nester, E. W. (1981). *Cell* 27, 143–153.
Gelvin, S. B. (1990). *Plant Physiol.* 92, 281–285.
Glass, N. L., and Kosuge, T. (1986). *CRC Crit. Rev. Plant Sci.*, pp. 247–276. Boca Raton, Florida: CRC Press.
Guilfoyle, T. J. (1986). In *CRC Crit. Rev. Plant Sci.*, pp. 247–276. Boca Raton, Florida: CRC Press.
Hagen, G., and Guilfoyle, T. J. (1985). *Mol. Cell Biol.* 5, 1197–1203.
Hagen, G., Kleinschmidt, A., and Guilfoyle, T. (1984). *Planta.* 162, 147–153.
Hagen, G., Uhrhammer, N., and Guilfoyle, T. J. (1988). *J. Biol. Chem.* 263, 6442–6446.
Hansen, C. E., and Meins, F., Jr. (1986). *Proc. Natl. Acad. Sci. U.S.A.* 83, 2492–2495.
Hansen, C. E., Meins, F., Jr., and Milani, A. (1985). *Differentiation.* 29, 1–6.
Heim, S., and Wagner, K. G. (1986). *Biochem. Biophys. Res. Comm.* 134, 1175–1181.
Helsper, J. P. F. G., Heemskerk, J. W. M., and Veerkamp, J. H. (1987). *Physiol. Planta.* 71, 120–126.
Hertel, R., Thomson, K.-S., and Russo, V. (1972). *Planta.* 107, 325–340.
Hesse, T., Feldwisch, J., Balshusemann, D., Bauw, G., Puype, M., Vandekerckhove, J., Lobler, M., Klambt, D., Schell, J., and Palme, K. (1989). *EMBO J.* 8, 2453–2461.

Hicks, G. H., Rayle, D. L., and Lomax, T. L. (1989). *Science*. 245, 52–54.
Holsters, M., Silva, B., Van Vliet, F., Genetello, C., De Block, M., Dhaese, P., Depicker, A., Inze, D., Engler, G., Villarroel, R., Van Montagu, M., and Schell, J. (1980). *Plasmid*. 3, 212–230.
Huffman, G. A., White, F. F., Gordon, M. P., and Nester, E. W. (1984). *J. Bacteriol*. 157, 269–276.
Inohara, N., Shimomura, S., Fukui, T., and Futai, M. (1989). *Proc. Natl. Acad. Sci. U.S.A.* 86, 3564–3568.
Inze, D., Follin, A., Van Lijsebettens, M., Simoens, C., Genetello, C., Van Montagu, M., and Schell, J. (1984). *Mol. Gen. Genet*. 194, 265–274.
Irvine, R. F., Letcher, A. J., and Dawson, R. M. C. (1980). *Biochem. J.* 192, 279–283.
Jackson, M. B. (1979). *Physiol. Plant*. 46, 347–351.
Jayabaskaran, C. (1990). *Pl. Growth Regul*. 9, 9–17.
Jouanin, L. (1984). *Plasmid*. 12, 91–102.
Karlin-Neumann, G. A., Brusslan, J. A., and Tobin, E. M. (1991). *Plant Cell*. 3(6), 573–582.
Keim, P., Erion, J., and Fox, J. E. (1981). In *Metabolism and Molecular Activities of Cytokinins* (Ed. by J. Guern and C. Peaud-Lenoel), pp. 179–190. Berlin: Springer-Verlag.
Kelley, M. O., and Bradford, K. J. (1986). *Plant Physiol*. 82, 713–717.
Kende, H. (1965). *Proc. Natl. Acad. Sci. U.S.A.* 53, 1302–1307.
Klee, H. J., Horsch, R. B., Hinchee, M. A., Hein, M. B., and Hoffmann, N. L. (1987). *Genes Develop*. 1, 86–96.
Lahners, K., Byrne, M. C., and Chilton, M.-D. (1984). *Plasmid*. 11, 130–140.
Last, R. L., Bissinger, P. H., Mahoney, D. J., Radwanski, E. R., and Fink, G. R. (1991). *Plant Cell*. 3(4), 345–358.
Leader, W. M. (1991). *New Biologist* 3, 1161–1168.
Leemans, J., Hernalsteens, J. P., Deblaere, R., DeGreve, H., Thia-Toong, L., Van Montagu, M., and Schell, J. (1983). In *Molecular Genetics of Bacteria–Plant Interaction* (Ed. by A. Puehler), pp. 322–330. Berlin: Springer-Verlag.
Letham, D. S. (1963). *Life Sci*. 8, 569–573.
Letham, D. S., and Palni, L. M. S. (1983). *Annu. Rev. Plant Physiol*. 34, 163–197.
Letham, D. S., Palni, L. M. S., Tao, G. Q., Gollnow, B. I., and Bates, C. M. (1983). *J. Plant Growth Regul*. 2, 103–115.
Li, Y., Hagen, G., and Guilfoyle, T. J. (1991). *Plant Cell*. 3, 1167–1175.
Libbenga, K. R., Maan, A. C., V. D. Linde, P. C. G., and Mennes, A. M. (1986). In *Hormones, Receptors and Cellular Interactions in Plants* (Ed. by C. M. Chadwick and D. R. Garrod), pp. 1–68. Cambridge: Cambridge University Press.
Lincoln, C., Britton, J. H., and Estelle, M. (1990). *Plant Cell*. 2, 1071–1080.
Löbler, M., and Klämbt, D. (1985). *J. Biol. Chem*. 260, 9848–9853.
Maher, E. P., and Martindale, S. J. B. (1980). *Biochem. Gen*. 18, 1041–1053.
Martiny-Baron, G., and G. F. E. Scherer. (1989). *J. Biol. Chem*. 264, 18052–18059.
Matsubara, S. (1990). In *Critical Reviews in Plant Sciences* (Ed. by B. V. Conger), pp. 17–57. Boca Raton, Florida: CRC Press.
McClure, B. A., Hagen, G., Brown, C. S., Gee, M. A., and Guilfoyle, T. J. (1989). *The Plant Cell*. 1, 229–239.
McGaw, B. A. (1987). In *Plant Hormones and Their Role in Plant Growth and Development* (Ed. by P. J. Davies), pp. 76–93. Dordrecht, The Netherlands: Martinus Nijhoff.
McGaw, B. A., and Horgan, R. (1983a). *Phytochem*. 22, 1103–1105.
McGaw, B. A., and Horgan, R. (1983b). *Planta*. 159, 30–37.
McGaw, B. A., Scott, I. M., and Horgan, R. (1984). In *The Biosynthesis and Metabolism of Plant*

Hormones (Ed. by A. Crozier and J. R. Hillman), pp. 105–133. Cambridge: Cambridge University Press.
Medford, J. I., Horgan, R., El-Sawi, Z., and Klee, H. J. (1989). *The Plant Cell.* 1, 403–413.
Meins, F., Jr., and Binns, A. N. (1982). *Differentiation.* 23, 10–12.
Meins, F., Jr., and Foster, R. (1985). *Dev. Biol.* 108, 1–5.
Meins, F., Jr., and Foster, R. (1986). *Dev. Genet.* 7, 159–165.
Meins, F., Jr., and Lutz, J. (1980). *Planta.* 149, 402–407.
Melin, P. M., Sommarin, M., Sandelius, A. S., and Jergil, B. (1987). *Fed. Eur. Biochem. Soc. Lett.* 223, 87–91.
Memelink, J., Hoge, J. H. C., and Schilperoort, R. A. (1987). *EMBO J.* 6, 3579–3583.
Messens, E., Lenaerts, A., Van Montagu, M., and Hedges, R. W. (1985). *Mol. Gen. Genet.* 199, 344–348.
Mirza, J. I., Olsen, G. M., Iversen, T.-H., and Maher, E. P. (1984). *Physiol. Plant.* 60, 516–522.
Moore, F. H. (1979). *Plant Physiol.* 64, 594–599.
Morris, R. O. (1986). *Annu. Rev. Plant Physiol.* 37, 509–538.
Muller, J.-F., Goujaud, J., and Caboche, M. (1985). *Mol. Gen. Genet.* 199, 194–200.
Nakamura, C., Van Telgen, H.-J., Mennes, A. M., Ono, H., and Libbenga, K. R. (1988). *Plant Physiol.* 88, 845–849.
Napier, R. M., and Venis, M. A. (1990). *J. Plant Growth Regul.* 9, 113–126.
Ooms, G., Hooykaas, P., Moolenaar, G., and Schilperoort, R. (1981). *Gene.* 14, 33–50.
Oono, Y., Handa, T., Kanaya, K., and Uchimaya, H. (1987). *Jpn. J. Genet.* 62, 501–505.
Palmer, M. V., Horgan, R., and Wareing, P. F. (1981). *J. Exp. Bot.* 32, 1231–1241.
Palmgren, M. G., and M. Sommarin. (1989). *Plant Physiol.* 90, 1009–1014.
Palni, L. M. S., Horgan, R., Darrall, N. M., Stuckbury, T., and Wareing, P. F. (1983). *Planta.* 59, 50–59.
Parker, C. W., and Letham, D. S. (1973). *Planta.* 114, 199–218.
Parker, C. W., Letham, D. S., Gollnow, B. I., Summons, R. e., Duke, C. C., and MacLeod, J. K. (1978). *Planta.* 142, 239–251.
Parry, A. D., Blonstein, A. D., Babiano, M. J., King, P. J., and Horgan, R. (1991). *Planta.* 183, 237–243.
Pfaffmann, H., Hartmann, E., Brightman, A. O., and Moore, D. J. (1987). *Plant Physiol.* 85, 1151–1155.
Polya, G. M., and Davies, J. R. (1983). *Plant Physiol.* 71, 482–488.
Polya, G. M., and Davis, A. W. (1978). *Planta.* 139, 139–147.
Polya, J. A., and Bowman, J. A. (1979). *Plant Physiol.* 64, 387–392.
Prasad, P. V., and Jones, A. M. (1991). *Proc. Natl. Acad. Sci. U.S.A.* 88(13), 5479–5483.
Prinsen, E., Bytebier, B., Hernalsteens, J.-P., De Greef, J., and Van Onckelen, H. (1990). *Plant Cell Physiol.* 31, 69–75.
Reddy, A. S. N., Sopory, S. K., and Datta, A. (1983). *Biochem Int.* 6, 181–190.
Romano, C. P., Hein, M. B., and Klee, H. J. (1991). *Genes Develop.* 5(3), 438–446.
Romanov, G. A., Taran, V. Y., Chvojka, L., and Kulaeva, O. N. (1988). *J. Plant Growth Regul.* 7, 1–18.
Sandelius, A. S., and Sommarin, M. (1986). *Fed. Eur. Biochem. Soc. Lett.* 201, 282–286.
Sato, T., and Theologis, A. (1989). *Proc. Natl. Acad. Sci. U.S.A.* 86, 6621–6625.
Schaller, G. E., and M. R. Sussman. (1988). *Planta.* 173, 509–518.
Scherer, G. F. E., and B. André. (1989). *Biochem. Biosphos. Res. Comm.* 163, 111–117.
Schmulling, T., Schell, J., and Spena, A. (1988). *EMBO J.* 7, 2621–2629.
Schmulling, T., Beinsberger, S., DeGreef, J., Schell, J., Van Onckelen, H., and Spena, A. (1989a). *FEBS Lett.* 249, 401–406.

Schmulling, T., Schell, J., and Spena, A. (1989b). *Plant Cell.* 1, 665–670.

Schroder, G., Waffenschmidt, S., Weiler, E. W., and Schroder, J. (1984). *Eur. J. Biochem.* 138, 387–391.

Schumaker, K. S., and Sze, H. (1987). *J. Biol. Chem.* 262, 3944-3946.

Sembdner, G., Gross, D., Liebisch, H.-W., and Schneider, G. (1980). In *Hormonal Regulation of Development I* (Ed. By J. MacMillan), pp. 281–444. Berlin: Springer-Verlag.

Shen, W. H., Petit, A., Guern, J., and Tempe, J. (1988). *Proc. Natl. Acad. Sci. U.S.A.* 85, 3417–3421.

Shimomura, S., Sotobayashi, T., Futai, M., and Fukui, T. (1986). *J. Biochem.* 99, 1513–1524.

Shinshi, H., Mohnen, D., and Meins, F. (1987). *Proc. Natl. Acad. Sci. U.S.A.* 84, 89–93.

Short, K. C., and Torrey, J. G. (1972). *Plant Physiol.* 49, 155–160.

Sinkar, V. P., Pythoud, F., White, F. F., Nester, E. W., and Gordon, M. P. (1988). *Genes Dev.* 2, 688–697.

Skoog, F., and Miller, C. O. (1957). *Symp. Soc. Exp. Biol.* 11, 118–131.

Skvirsky, R. C., Hanson, M. R., and Ausubel, F. M. (1984). *Plant Sci. Lett.* 35, 237–246.

Slightom, J. L., Durand-Tardif, M., Jouanin, L., and Tepfer, D. (1986). *J. Biol. Chem.* 261, 108–121.

Smart, C. M., Scofield, S. R., Bevan, M. W., and Dyer, T. A. (1991). *Plant Cell.* 3(7), 647–656.

Smigocki, A. C. (1991). *Plant Mol. Biol.* 16, 105–115.

Smigocki, A. C., and Owens, L. D. (1988). *Proc. Natl. Acad. Sci. U.S.A.* 85: 5131–5135.

Smigocki, A. C., and Owens, L. D. (1989). *Plant Physiol.* 91, 808–811.

Sommarin, M., and Sandelius, A. S. (1988). *Biochim. Biophys. Acta* 958, 268–278.

Spanier, K., Schell, J., and Schreier, P. H. (1989). *Mol. Gen. Genet.* 219, 209–216.

Spano, L., Mariotti, D., Cardarelli, M., Branca, C., and Costantino, P. (1988). *Plant Physiol.* 87, 479–483.

Spena, A., Schmuelling, T., Koncz, C., and Schell, J. S. (1987). *EMBO J.* 6, 3891–3899.

Spena, A., Aalen, R. B., and Schulze, S. C. (1989). *Plant Cell.* 1, 1157–1164.

Stuchbury, T., Palni, L. M. S., Horgan, R., and Wareing, P. F. (1979). *Planta.* 147, 97–102.

Sugaya, S., Hayakawa, K., Handa, T., and Uchimiya, H. (1989). *Plant Cell Physiol.* 30, 649–653.

Sussman, M. R., and Kende, H. (1978). *Planta.* 140, 251–259.

Takahashi, Y., Kuroda, H., Tanaka, T., Machida, Y., Takebe, I., and Nagata, T. (1989). *Proc. Natl. Acad. Sci. U.S.A.* 86, 9279–9283.

Takegami, T., and Yoshida, K. (1975). *Biochem. Biophys. Res. Comm.* 67, 782–789.

Taya, T., Tanaka, Y., and Nishimura, S. (1978). *Nature (London)* 271, 545–547.

Tepfer, D. (1984). *Cell* 37, 959–967.

Teyssendier de la Serve, B., Axelos, M., and Peaud-Lenoel, C. (1985). *Plant Mol. Biol.* 5, 155–163.

Theologis, A. (1986). *Ann. Rev. Plant Physiol.* 37, 407–438.

Theologis, A., and Ray, P. M. (1982). *Proc. Natl. Acad. Sci. U.S.A.* 79, 418–421.

Theologis, A., Huynh, T. V., and Davis, R. W. (1985). *J. Mol. Biol.* 183, 53–68.

Thomashow, M. F., Hugly, S., Buchholz, W. G., and Thomashow, L. S. (1986). *Science* 231, 616–618.

Tinland, B., Huss, B., Paulus, F., Bonnard, G., and Otten, L. (1989). *Mol. Gen. Genet.* 219, 217–224.

Truelsen, T. A. (1973). *Physiol. Plant.* 28, 67–70.

Vanderhoef, L. N. (1980). In *Plant Growth Substances* (Ed. by F. Skoog,), pp. 90–96. Berlin: Springer-Verlag.

Venis, M. (1985). *Hormone Binding Sites in Plants,* pp. 114–128. New York: Longman.

Venis, M. A., Thomas, E. W., Barbier-Brygoo, H., Ephritikhine, G., and Guern, J. (1990). *Planta.* 182, 232–235.

Vilaine, F., Charbonnier, C., and Casse-Delbart, F. (1987). *Mol. Gen. Genet.* 210: 111–115.
Walker, J. C., and Key, J. L. (1982). *Proc. Natl. Acad. Sci. U.S.A.* 79, 7185–7189.
Weiler, E. W., and Spanier K. (1981). *Planta.* 153, 326–337.
White, E. F., Taylor, B. H., Huffman, G. A., Gordon, M. P., and Nester, E. W. (1985). *J. Bacteriol.* 164, 33–44.
Wightman, F. (1973). *Biochem. Soc. Symp.* 38, 247–275.
Wright, R., Mcclure, B., and Guilfoyle, T. (1989). *J. Plant Growth Regul.* 8, 283–292.
Zbell, B., and Walter-Beck, C. (1988). *J. Plant Physiol.* 133, 353–360.
Zobel, R. W. (1972). *J. Hered.* 63, 94–97.
Zobel, R. W. (1973). *Plant Physiol.* 52, 383–389.
Zobel, R. W. (1974). *Can. J. Bot.* 52, 735.
Zurfluh, L. L. and Guilfoyle, T. J. (1982a). *Planta.* 156, 525–527.
Zurfluh, L. L., and Guilfoyle, T. J. (1982b). *Plant Physiol.* 69, 332–337.

8

Gene Tagging in Transgenic Plants

Shirley A. Coomber[1] and Kenneth A. Feldmann[2]

E. I. Du Pont de Nemours and Company,
Wilmington, Delaware 19880

In the 1980s a large number of plant genes were isolated and studied at the molecular level for the first time. The most well-characterized genes are those involved in photosynthesis and chloroplast development. Despite the considerable progress made in the last decade there are many significant challenges facing plant molecular biologists excited by the prospect of unraveling the intricacies of plant development, disease and pest resistance, and plant response to environmental stimuli.

There are a number of approaches being taken to isolate and characterize the most interesting genes (phenotypes) from the tens of thousands of genes needed to produce a mature plant. A genetic approach, mutant isolation and characterization subsequent to gene isolation, is being pursued by most groups. The mutants currently being isolated fall into two categories, chemically induced mutants, predominantly ethylmethane sulfonate (EMS), and insertional mutants generated using transposons or T-DNAs. Insertional mutants have an advantage in that once a mutant has been identified it is relatively easy to isolate the tagged gene so that it can be used as a probe to recover the wild-type gene. This review deals with the variety of approaches that are being taken to "tag" plant genes using transposon, or T-DNA insertional, mutagenesis.

[1]Present address: Division of Biosphere Sciences, King's College, London, Campden Hill Road, London W8 7AH, UK.
[2]Present address: Department of Plant Sciences, University of Arizona, Tucson, Arizona 85721

8.1 Transposon Tagging in Transgenic Plants

Transposable elements were identified genetically in maize by McClintock (1948). The subsequent isolation of transposons from *Escherichia coli* and other prokaryotes (Kleckner, 1981) led to the realization that transposable elements and their immobile derivatives are natural components of most genomes. In prokaryotes, the use of transposon mutagenesis has led to the isolation of a large number of genes. Increasingly, transposable elements are being utilized to identify genes in higher eukaryotes including maize (Shepherd, 1988), *Caenorhabditis* (Moerman and Waterston, 1989) and *Drosophila* (Engels, 1989; Finnegan, 1989). In plants, the current number of genes isolated using transposons is relatively small. This is largely due to the nature of the two species, *Zea* and *Antirrhinum,* which contain active endogenous transposons. However, it is clear that with the molecular tools that are being developed transposons will play an important role in the isolation of many additional genes. As such, we will present a brief description of the most widely studied plant transposons.

8.1.1 GENES ISOLATED IN MAIZE USING NATIVE TRANSPOSONS

There are three major groups of endogenous maize transposons which have been isolated, the Activator-Dissociation (*Ac-Ds*), Suppressor-mutator (*Spm*) or Enhancer (*En*), and Mutator (*Mu*) elements (Fedoroff, 1989), and which are being used to isolate genes. One of the problems faced by molecular biologists interested in using native maize transposons to tag genes is their high copy number. This makes the task of identifying which insertion caused a specific mutant phenotype a fairly lengthy genetic exercise involving hundreds of individual plants over several generations. Another problem is that the transposition frequencies are quite low. Not surprisingly, the majority of genes isolated using the native transposon tagging approach in maize are those involved in easily scorable kernel phenotypes, such as color, which allow a rapid screen of thousands of individuals.

Among the genes isolated from corn are several that are involved in anthocyanin biosynthesis (e.g., Cone *et al.*, 1986; Wienard *et al.*, 1986; Fedoroff *et al.*, 1984; Theres *et al.*, 1987; O'Reilly *et al.*, 1985), and one each involved in carotenoid biosynthesis (Buckner *et al.*, 1990), chloroplast development (Martienssen *et al.*, 1989), and leaf morphology (Hake *et al.*, 1989). The number of new genes isolated in the last few years from native transposon tagged maize mutants, especially nonkernel genes, show that this approach, although time consuming, is valuable.

8.1.2 GENES ISOLATED IN *ANTIRRHINUM MAJUS* USING NATIVE TRANSPOSABLE ELEMENTS

Antirrhinum majus has three related endogenous transposons *Tam1, Tam2,* and *Tam3* that have been shown to be associated with a number of unstable flower morphologies. Two genes have been isolated from *A. majus* that were initially tagged with *Tam* insertions, the *pallida* (*pal*) locus involved in anthocyanin biosynthesis in flowers (Martin *et al.*, 1985) and the *deficiens* (*def*) locus involved in flower development (Sommer *et al.*, 1990). The *nivea* locus, which encodes chalcone synthase, a gene also involved in anthocyanin biosynthesis in flowers, although not cloned directly using the transposon, is also an example of a *Tam* tagged gene (Upadhyaya *et al.*, 1985).

8.1.3 USE OF MAIZE AND *ANTIRRHINUM MAJUS* TRANSPOSONS FOR TAGGING GENES IN HETEROLOGOUS PLANTS

Active transposons have been cloned only from *A. majus* and maize, although nonmobile transposon-like elements have been identified in a number of other plant species including *Arabidopsis thaliana* (Voytas and Ausubel, 1989) and tobacco (Grandbastien *et al.*, 1989). In order to develop transposon tagging systems which can be utilized in a large number of plant species, various groups are modifying the already well characterized maize and *A. majus* transposable elements.

The autonomous *Ac* element of maize, the simplest form of a transposon, is 4.65 kb in length, has an 11-bp imperfect terminal inverse repeat, and generates an 8-bp duplication of target sequences on insertion. The 3.5-kb major transcript encodes a single transposase that is required for the *Ac* element to undergo nonreplicative transposition, often to sites within 4 map units on the same chromosome (Finnegan *et al.*, 1988; Coupland *et al.*, 1989). Though all *Ac* elements are similar to each other, *Ds* elements are structurally heterogeneous and require the presence of the *Ac* transposase to transpose. Some of the characterized *Ds* elements are deletion derivatives of *Ac*, however, others only show homology to *Ac* terminal inverted repeats.

In transgenic plants, *Ac* was initially shown to transpose in tobacco cells after *Agrobacterium tumefaciens*-mediated transformation using a Ti plasmid vector (Baker *et al.*, 1986). The ability of the maize *Ac* element to transpose autonomously in a number of other plant species was quickly reported: carrot and *A. thaliana* (Van Sluys *et al.*, 1987), tomato (Yoder *et al.*, 1988), and potato (Knapp *et al.*, 1988). *Ds* transposition has been shown to require transactivation by *Ac,* or expressed *Ac* cDNA in tobacco, tomato, and *A. thaliana* (Hehl and Baker, 1989; Lassner *et al.*, 1989; Masterson *et al.*, 1989).

Having shown that maize *Ac/Ds* elements can transpose in heterologous plant species, the efforts are very focused on methods to increase transposition frequency in order to produce large numbers of tagged mutants. A number of groups have produced derivatives of *Ac* and *Ds* inserted within marker genes, such as neomycin phosphotransferase (Baker *et al.*, 1986, 1987; Coupland *et al.*, 1988), streptomycin phosphotransferase (Jones *et al.*, 1989), and hygromycin phosphotransferase II (Haring *et al.*, 1989), so that excision events can be monitored by restoration of the marker phenotype. The ability to select plants in which an excision event has occurred identifies plants which potentially have a new transposon insertion.

Ac/selectable marker constructs are initially being utilized to study the parameters influencing the use of *Ac* as a gene tagging system in transgenic plants. In maize, the *Ac* germinal excision rate has been reported to fall between 1 and 17% (Greenblatt, 1968). In transgenic tomato 30% of progeny have been shown to inherit a transposed *Ac* element (Belzile *et al.*, 1989). The frequency of germinal excision of *Ac* in tobacco has been variously reported to be as low as 1–5% (Jones *et al.*, 1990) and as high as 23–83% (Hehl and Baker, 1990). In addition, *Ac* has been shown to remain active and transpose in subsequent generations of tobacco, tomato, and *Arabidopsis* (Belzile *et al.*, 1989; Schmidt and Willmitzer, 1989; Taylor *et al.*, 1989; Hehl and Baker, 1990; Yoder, 1990). Similar to maize, the *Ac* element in transgenic tobacco has been shown to transpose predominantly to adjacent low copy or unique DNA (Chen *et al.*, 1987; Hehl and Baker, 1990; Jones *et al.*, 1990). However, unlike maize, where increasing *Ac* copy number leads to a decrease in germinal excision, an increase in *Ac* copy number results in an increase in excision frequency in transgenic tobacco (Hehl and Baker, 1990). Methylation has been shown to be responsible for *Ac* inactivation in maize and has been proposed to explain the lack of *Tam* transposition in transgenic tobacco (Martin *et al.*, 1989). Apparently *Ac* in transgenic tobacco is not susceptible to methylation.

One of the initial disappointments to geneticists studying *Arabidopsis* was the low rate of germinal excision of *Ac* (0.2–0.5%) in this species (Schmidt and Willmitzer, 1989). However, it was recently discovered that deleting the CpG-rich region upstream of the *Ac* transposase promoter, possible methylation region, led to a higher rate of germinal excision in *Arabidopsis* (4.8–45%). This modification should allow *Ac* to be utilized as an efficient gene tagging system (Dean *et al.*, 1990).

Other than simple saturation mutagenesis using a transposon system in *Arabidopsis,* it should be possible to do site-directed mutagenesis. A number of researchers have proposed establishing a population of mapped *Ds* elements in

Arabidopsis. When an EMS-induced mutant with a putative loss-of-function mutation has been identified and genetically mapped near several *Ds* sites of insertion, the identified *Ds* lines could be used as parents in a cross with a line containing an *Ac* element. The introduced *Ac* element should cause the transposition of the *Ds* element, hopefully to a nearby site, and should generate the desired mutation. By self-pollinating the progeny, the desired mutant can be identified.

In summary, endogenous plant transposons have been used to isolate a growing number of plant genes in a variety of metabolic and developmental pathways. The advantages of moving these transposons into a plant model system are that the transposons could be engineered and introduced rapidly into the plant, a broader set of mutants could be identified and characterized, and the transposon could be inactivated when desirable.

8.2 T-DNA Tagging of Genes in Transgenic Plants

Agrobacterium tumefaciens is capable of transferring defined segments of DNA to the plant nuclear genome (reviewed in Zambryski et al., 1989). The transferred DNA (T-DNA) is bounded by 25-bp repeats (called left and right borders) and normally carries genes which stimulate tumor formation when introduced into the plant. These tumor-inducing genes have been removed and in their place a variety of markers, both scorable and selectable, have been inserted. This phenomenon has been widely exploited to transform a variety of plant species with foreign DNA. The T-DNA appears to integrate into the plant genome randomly and a number of groups have attempted to utilize this as a method for generating insertional mutants. Thus far only in *A. thaliana* have there been any successes reported.

A model plant, *A. thaliana* has been utilized by a number of groups trying to develop effective T-DNA tagging methods. *A. thaliana* is favored because of its small size (30–40 cm), short life cycle (6–8 weeks), natural selfing, and high seed production (5000–10,000 seeds/plant). In addition to having the smallest known higher plant genome, *A. thaliana* has only 70,000 kb of DNA per haploid genome and genetic and physical restriction fragment length polymorphism (RFLP) maps are available (Koornneef *et al.*, 1983; Chang *et al.*, 1988; Nam *et al.*, 1989). As a result of having a small genome, *A. thaliana* is now the focus of an internationally coordinated program which has a goal of sequencing the entire genome by the year 2000. All these factors have contributed to the emergence of *A. thaliana* as the model system of choice for plant molecular biologists.

8.2.1 T-DNA TAGGING UTILIZING TISSUE CULTURE STRATEGIES

A number of protocols which utilize tissue culture approaches to transform *A. thaliana* with *A. tumefaciens* have been developed (Lloyd *et al.*, 1985; Valvekens *et al.*, 1988; Chaudhury and Signer, 1989). One of the drawbacks to utilizing tissue culture approaches to generate insertion mutants is that the tissue culture process itself is mutagenic. It thus becomes laborious to sort out mutants due to a T-DNA insert and those due simply to the tissue culture process. In spite of these obstacles there has been a gene cloned using a tissue culture approach (Koncz *et al.*, 1990) and it appears likely that additional genes will be cloned in the next couple of years.

Koncz *et al.*, (1989) introduced constructs carrying a promoterless reporter gene linked to the right border sequences to generate transformants of *Arabidopsis*. This strategy allowed the identification of transcriptional or translational plant gene fusions as well as insertion mutants. Initial studies showed that on average 30% of T-DNA insertions induced gene fusions in *A. thaliana* and tobacco (Koncz *et al.*, 1989). The tobacco genome is 1.5×10^9 bp (compared to *A. thaliana* 7×10^7 bp) and contains a larger proportion of redundant DNA. The high insertion rate into functional genes has led to the suggestion that T-DNA preferentially inserts into sequences that can be transcribed.

From a pool of 450 *Arabidopsis* T-DNA transformants produced by this approach, one recessive pale mutant was chosen for further study (Koncz *et al.*, 1990). The pale mutant, *chlorata-42*, was found to be caused by a T-DNA insert and the gene involved rescued and shown to be a novel chloroplast protein. Other mutants, for example, *apetala-2* (Jofuku *et al.*, 1991) and a leaf morphology mutant (Van Lijsebettens *et al.*, 1990), isolated from populations of transformants generated via tissue culture, appear to be due to an insertion event.

8.2.2 T-DNA TAGGING USING SEED INFECTION/TRANSFORMATION METHOD IN *ARABIDOPSIS*

To circumvent the problems associated with the high proportion of phenotypically unusual plants due to tissue culture, Feldmann and Marks (1987) developed a nontissue culture transformation protocol for *Arabidopsis*. This fairly simple protocol has been used to generate over 8000 independent T-DNA transformants in the ecotype Wassilewskija (WS; Feldmann, 1991).

This seed infection/transformation procedure was not developed as a transformation system per se but as a procedure to do insertion mutagenesis in *Arabidopsis*. It was believed that the *Agrobacterium* would move through the micropyle of the imbibed seed and by diffusion end up near the shoot meristem. It was also assumed that the *Agrobacterium* would penetrate the epidermal layer and potentially transform one of the genetically effective cells. Fortuitously this is not when transformation occurs. Instead, as we will describe, the data indicate that the transformation event occurs late in the development of the infected plant such that one of the meiotic products or a zygote is transformed. The procedure for seed infection/transformation is summarized below. This description is meant only to help explain when and how transformation happens.

8.2.2.1 *Arabidopsis* Seed Infection/ Transformation Method

Three thousand seeds are sterilized with bleach and placed in 50 ml of Murashige and Skoog (M & S) salts, 4% sucrose, and vitamins (see Feldmann and Marks, 1987) in a 125-ml Erlenmeyer flask. The seeds are imbibed for a period of 12–16 hr (predetermined for each seed lot) in an orbital incubator at 190 rpm, 22°C (~250 lux, fluorescent light). Following this imbibition, 3–5 ml of an overnight culture of *A. tumefaciens,* containing the cointegrate Ti plasmid p3850:1003 which confers resistance to kanamycin (Velten and Schell, 1985), is added. The mixture of *Agrobacterium* and seeds is placed in an orbital shaker for 24 hr at 28°C, 190 rpm (~250 lux). After cocultivation the seeds are dried. The seeds from a single flask are sown in a flat containing vermiculite subirrigated with Hoagland's solution (Feldmann and Marks, 1987). The plants are grown to maturity in a greenhouse with ambient temperature around 20–24°C. T_2 seed (numbering around 300,000) are collected from each flat and given a treatment number. Counting the plants in a number of flats indicates that only 1000 infected plants grow to maturity after this procedure. Under these dense growth conditions each plant produces approximately 300 seeds. Approximately 10% of the seeds from each treatment are tested for kanamycin-resistant transformants. This is accomplished by plating surfaced sterilized seeds at a density of 2000 seeds per plate on M & S media containing 1% sucrose and 50 mg/liter kanamycin. Treatments which are found to contain kanamycin-resistant transformants are plated out entirely on kanamycin-containing media. Each T_2 kanamycin-resistant seedling is transferred from agar to pots containing Metromix (WR Grace & Co.) at around 10 days of age and grown to maturity. The seeds are collected separately from each plant and given a number (1–8000+).

8.2.2.2 Genetic and Molecular Characterization of Transformants

We have collected considerable data to indicate that all of the transformants generated by this procedure are unique, that is, nonclonal. A number of analyses have led us to this conclusion. (1) Genetic analyses indicate that all of the transformants produce progeny which segregate on kanamycin. Hundreds of transformed lines have been tested on kanamycin and these lines contain 1–4 T-

DNAs, with an average of 1.5 functional inserts per transformant (Feldmann and Marks, 1987; Feldmann *et al.*, 1990). (2) When we have collected T_2 progeny from single T_1 plants we have observed only one transformant (T_2) per T_1 plant (K. Feldmann, unpublished results). (3) In addition, when the progeny (T_3) from the transformants that are derived from a single treatment are screened for visibly altered phenotypes, we do not find multiples of specific mutant phenotypes, again indicating that most or all of the transformants are unique.

We hypothesize that the transformational event occurs very late in the development of the infected sporophyte such that one of the gametophytes or even the newly forming sporophyte is transformed. The scenario is that the *Agrobacterium* enter the interstitial spaces near the growing shoot of the infected seed, that they are carried along with the plant as it develops, and that they are induced to attach to a specific cell type in the flower and to transfer the T-DNA.

Southern analyses of a number of transformants show that the bulk of the transformants contain concatamers of T-DNAs in direct and inverted repeats (Feldmann and Marks, 1987; Feldmann *et al.*, 1989, 1990; Marks and Feldmann, 1989; Yanofsky *et al.*, 1990).

To get some feel for the frequency of transformation it is important to realize that over 800 treatments have been tested for kanamycin-resistant seedlings. Each treatment generates approximately 300,000 seeds (50,000 seeds/gm). Approximately 215 of these 800 batches contained one or more kanamycin-resistant transformants. Therefore, from plating over 80 million seeds ($800 \times 30{,}000 + 215 \times 270{,}000$) on kanamycin media, 8000 kanamycin-resistant transformants have been isolated. The frequency of generating new transformants may become higher as parameters affecting the seed infection method are further refined.

8.2.2.3 Library of Transformants: Screening for Mutants with Visibly Altered Phenotypes

In the last two years the number of *A. thaliana* T-DNA transformants isolated using the seed infection/transformation protocol have increased from several hundred to over 8000. The goal is to generate a library of sufficient size so that virtually every gene has been hit by at least one T-DNA insertional event. It has been calculated that 75,000 independent transformants will have to be isolated to saturate the *A. thaliana* genome with T-DNA inserts using this method. This assumes a 95% probability of an insert every 2 kb, 1.4 inserts per transformant, and a genome size of 70,000 kb. This number will be modified as more data is collected.

The transformants generated have been screened for visibly altered phenotypes (Table 8.1). The T_3 seed families that are being tested are equivalent to an F_2 family. Thus, any line that possesses a mutation should be segregating for that phenotype; for example, a recessive mutation would segregate 3 wildtype:1 mutant. Screening has taken place under two environmental regimes. In the first

TABLE 8.1
Types and Frequencies of Mutants Observed in Screening of 8000 Transformants of *A. thaliana*

Phenotype	Loci (%)	Segregation ratio (wildtype : mutant)
Seedling lethals	3–5	3 : 1
Size variants	3–5	3 : 1
Embryo defective	2.3–3.5	1 : 2
Dramatic mutants	3–4	3 : 1
Pigment mutants	2–3	3 : 1
Reduced fertility	1–2	3 : 1
Physiological mutants	1–?	3 : 1
Total	17–28	

screen 30–40 seeds were sown in individual 4-inch pots containing Metromix and grown in the greenhouse to examine the hypocotyl, cotyledons, and shoot portion of the plant. The plants were screened 4–5 times during the 6- to 7-week life cycle. In the second screen 40–50 seeds from each transformed line were sterilized and sown on agar plates. These plates were transferred to a growth room (constant light, 24°C) and placed in a vertical position so that the roots would be readily visible. In this screen we were able to identify not only root and root hair mutants but also pigment mutants and other mutations that affect the hypocotyl, cotyledons, or young rosette (Mayer *et al.*, 1990). These plates were examined three times over a 2-week period. We have completed the screening of over 8000 lines. We have placed the observed mutants into several general classes (Table 8.1).

8.2.2.4 Classes of Mutants

The first two classes of mutants, seedling lethals and size variants, are those that are most subject to environmental influences and those in which the frequencies are somewhat tenuous. These putative mutants must be retested a second time to verify the phenotype.

The lines scored as seedling lethals are those in which 25% of the seedlings emerge but fail to grow past the cotyledon stage. Instead, the cotyledons turn brown after 2–3 weeks. We had previously estimated this frequency to be 10% but a more careful survey indicates that it is probably closer to 3–5%.

Size variants are lines in which 25% of the seedlings are small. These are not dwarfs but rather plants that are spindly and 25–50% of wild-type height. It is assumed that the disruption of any one of the thousands of genes involved in "housekeeping" functions could cause this phenotype or the seedling lethal phenotype.

Plants that appear to contain embryo lethal mutations can also be caused by environmental circumstances, however, this phenotype can usually be verified by counting the seeds in a number of siliques from a number of plants. This is a relatively easy, although laborious, phenotype to score. The siliques from a plant can be opened and screened for deformed seeds. The seeds will segregate 3:1 (wild type:mutant) in a plant that is heterozygous for the mutation. The mutants, of course, do not produce plants. As such, the T_3 family will segregate 2:1 (segregating for embryo lethals:wild type). Errampalli *et al.*, (1991) have performed cosegregation analyses on 18 of the embryo lethal lines observed in this population and have shown that at least 50% are due to a functional insert. Included in Errampalli's study are two mutants, a fusca (described below) and an albino, that we would put in the pigment class.

Reduced fertility mutants are those that make little or no seed such that the plants containing this mutation continue to flower later than their wild-type siblings. This is a particularly easy phenotype to identify when the plants are 6 weeks old and all of the wild-type plants have discontinued flowering. The mutants in this class are due to defects in pollen or ovule production, transmitting tissue organization or pollen recognition. A preliminary analysis of 42 of these mutants indicates that the bulk of the mutants have defects in pollen production or dehiscence (C. Makaroff, 1989, unpublished results).

Pigment mutants include, but are not limited to, those that are albino, yellow-green lethal, yellow-green viable, and purple lethal (fusca). Albinos account for 45% of the mutants in this class while yellow-greens and fuscas account for 51% and 4%, respectively. We have tested the cosegregation of the kanamycin resistance marker with the mutation in three of these lines and the results are consistent with a T-DNA disruption of a gene involved in conferring normal green color to the plants (S. A. Coomber, M. A. Mandel, and K. A. Feldmann, 1989, unpublished results).

Dramatic mutants are a large diverse class of mutants including all of those that affect the morphology of the plant in some way other than described above. These are, generally, mutants that we believe will help us to discover how a plant determines its shape. These include root hair and trichome mutants, root mutants, dwarfs, flower mutants, leaf and cotyledon mutants, stem mutants, and embryonic patterning mutants. A few of these mutants will be discussed below.

Physiological mutants are those that do not fit into any of the classes listed above. They include chlorate-resistant, ethylene-affected, long hypocotyl, early senescence, dark green, and wilty mutants. A few of these have been isolated from the pooled seeds as described in the section on pooled screens.

It is important to note that the frequency of mutants in the various classes is very similar to those reported for EMS-generated populations (Mayer *et al.*, 1990; Muller and Weiland, 1990).

8.2.3 USING T-DNA TAGGED MUTANTS TO UNRAVEL DEVELOPMENTAL PATHWAYS

Three of the developmental mutants have been investigated in considerable detail. These include a dwarf (Feldmann *et al.*, 1989), a homeotic floral mutant (*agamous,* Yanofsky *et al.*, 1990), and a trichome mutant (*glabrous 1-43,* Marks and Feldmann, 1989; Herman and Marks, 1989). All three mutants have been shown to be tagged by a T-DNA insertion using classical genetic analysis and complementation of the mutant plant with wild-type genomic DNA.

dwarf is around 8 cm tall at maturity, in contrast to 30 cm for wildtype height. In comparison to wildtype the hypocotyls of *dwarf* seedlings are reduced in length. The leaves are small, dark green and have an epinastic growth habit. In addition, the dwarf shows a reduction in apical dominance producing 40 or more inflorescences. The seed number is reduced in dwarf plants, in part due to a failure of the stamens to elongate. Although the gene interrupted by the *dwarf* T-DNA insertion has been sequenced, no homology to sequences in the EMBL database has been identified (T. Carlson, A. Wierzbicki, and K. A. Feldmann, 1990, unpublished results). The possible function of the *dwarf* gene product and its tissue-specific expression remain to be investigated. Eleven additional dwarf mutants have been identified. We are presently doing pairwise crosses with all 12 dwarfs to determine the number of complementation groups involved in dwarfism. Most of these dwarfs differ from each other in one or more morphological characters.

The first *glabrous* mutant identified from this population has normal leaf trichomes but completely lacks trichomes on the stem. Complementation analysis with other genetically characterized trichome mutants from six loci revealed that the T-DNA induced trichome mutant was allelic to *glabrous1* (*gl1*). The *gl1* gene has been sequenced and shows substantial homology to the DNA binding domain of the *myb* family of DNA binding proteins which participate in differentiation in animal cells. Preliminary Southern data using the *gl1* gene as a probe suggest that a family of *myb* homologues exists in *A. thaliana* (Oppenheimer *et al.*, 1991). We have now identified other trichome mutants including several that completely lack trichomes and mutants with spiky or distorted trichomes (M. David Marks, personal communication). These are presently being placed in complementation groups and being tested for cosegregation with the kanamycin resistance marker.

Mutants of the *agamous* gene produce flowers in which the third whorl of organs, normally stamens, is replaced by petals. In addition, the gynoecium is replaced by a repeating structure of sepals and petals. The *agm* gene has been cloned (Yanofsky *et al.*, 1990) and shows sequence homology to the DNA binding domain of a number of transcriptional factors, for example, human SRF (serum response factor, thought to be required for the transcriptional activation of the proto-oncogene c-*fos*) and yeast MCM1 (a transcriptional regulator of mating-type specific genes). Independently, *def*, a flower-specific gene isolated from a *Tam*-derived homeotic flower mutant of *A. majus* (*deficiens*) was also shown to share homology with the SRF and MCM1 transcription factors (Sommer *et al.*, 1990). Although *def* and *agm* share homology in the region thought to be the DNA binding region, they are distinct genes involved at different stages of flower development. This has raised the possibility that they are members of a related group of transcriptional factors which control flower development. The *agamous* gene has been used as a probe to isolate other *A. thaliana* flower-specific genes (Elliot Meyerowitz, personal communication). Sequence analysis of these genes together with the identification of the genes that they interact with provide a way to study the factors influencing flower development for the first time.

It is clear that the isolation of additional genes from other T-DNA tagged flower and *glabrous* mutants together with subsequent isolation of homologues will quickly lead to an explosion of information about genes involved in flower and trichome development. These are just two examples of the many genes, in a number of developmental pathways, that will be isolated and characterized over the next several years utilizing this approach.

8.2.4 POOLED SCREENS

One other general type of screen that is possible is a pooled screen. We take 12 T_3 seeds from each of 100 lines, for example, T_3-1 through T_3-100, and pool them. These 1200 seeds are planted fairly sparsely under greenhouse conditions to maximize the number of T_4 progeny that will be generated. Under our usual conditions we will collect 1.2 million seeds from these 1200 plants. An aliquot of seeds can be taken from each of the pools and tested under any variety of environmental conditions. This makes it efficient to test 8000 lines and ensures that the individual T_3 seed stocks will not be depleted. If a selected mutant is a lethal or sterile it is possible to rescreen the 100 families from which the mutant was derived to identify the specific line so that heterozygotes can be used to generate additional seed. As yet only minimal screening has been done with this procedure. We have, however, identified two chlorate-resistant mutants, a number of cold-sensitive mutants, and a number of ethylene-affected mutants (John Browse, Washington State University; Nigel Crawford, University of California San Diego; Joe Ecker, University of Pennsylvania, 1990, unpublished results). Many additional mutants will be identified as the population is subjected to various selective agents and conditions.

8.3 *in Planta* Transformation

At the Fourth International Conference on *Arabidopsis* Research in Vienna (June 2–5, 1990) Hong-Gil Nam described his initial results on transforming *A. thaliana* by a novel wounding method. The *in planta* transformation method starts with soil-grown *Arabidopsis* plants with primary bolts of around 3 cm in height. The inflorescence and many axillary buds are removed by cutting deeply into the rosette with a sharp-pointed scalpel. The wound is then inoculated with a fresh overnight culture of *A. tumefaciens* LBA4404 pBI121 (which has been previously washed to remove antibiotics). These plants are left for 7–10 days by which time a number of short secondary inflorescences have grown (~3 cm). These inflorescences and buds are cut out a second time and a fresh inoculum of *Agrobacterium* is applied. The plants are grown to maturity, and the seeds are collected and plated on selective media to select for transformed seedlings. Hong-Gil Nam has isolated over 100 transformants by this method (unpublished results).

This transformation method has been repeated successfully using *A. thaliana,* race WS (Robert Reiter, Shirley Coomber, Anna Wierzbicki, Roslyn Young, and Ken Feldmann, 1990, unpublished results). Two *A. tumefaciens* strains were used: C58C1Rif, containing the cointegrate Ti plasmid p3850:1003, and LBA4404 containing a pBin19 derivative. From ~200 plants that were used for each *Agrobacterium* strain we managed to isolate 20 independent transformants using C58C1Rif and three independent transformants from LBA4404-inoculated plants. The independent transformants suggest a transformation event that occurs late in the flower itself, which is similar to the model postulated to occur in the Feldmann and Marks (1987) seed infection procedure. It should be pointed out that we have never successfully used LBA4404 strains of *Agrobacterium* to generate transformants of *Arabidopsis* with the seed infection/transformation technique.

Interestingly, three plants inoculated with C58C1 resulted in hundreds of transformants, suggesting that an entire shoot had been transformed. This would mean that a transformed cell had to participate in the organization of a new shoot meristem. The *in planta* method of transformation of *A. thaliana* has the potential to hasten the production of transgenic plants, as it is a fairly quick way of introducing T-DNA into plants, circumventing the more tedious and labor-intensive tissue culture protocols.

This procedure may also be useful in generating insertion mutants. However, given that it is possible to generate clonal transformants it becomes necessary to collect seed from one or only a few plants for testing on selective media. This approach would make it too labor intensive to generate thousands of unique transformants.

8.4 Summary

The last few years have seen the emergence of two systems to tag genes in plants; transposons and T-DNAs. Tens of genes have already been isolated from corn, and two from snapdragon, using their endogenous transposon systems. The transfer of these transposons to heterologous plant model systems, where they can be manipulated, will make transposons even more attractive. T-DNA insertion mutagenesis has been developed only in the last three years, but already several genes have been isolated and characterized; dozens more are presently under study. It is likely that these two systems will facilitate the isolation of thousands of new genes involved in all aspects of plant development in the next decade. The exciting era of unraveling and understanding complex developmental pathways that are specific to plants has now truly begun.

Acknowledgments

We would like to thank numerous academic researchers for contributing their time to screening this population of *Arabidopsis* transformants at Du Pont. We would also like to thank Karen Oishi for her comments on the manuscript.

References

Baker, B., Schell, J., Lorz, H., and Fedoroff, N. V. (1986). *Proc. Natl. Acad. Sci. U.S.A.* 83, 4844–4848.

Baker, B., Coupland, G., Fedoroff, N. V., Starlinger, P., and Schell, J. (1987). *EMBO J.* 6, 1547–1554.

Belzile, F., Lassner, M. V., Tong, Y., Khush, R., and Yoder, J. I. (1989). *Genetics* 123, 181–189.

Buckner, B., Kelson, T. L., and Robertson, D. S. (1990). *Plant Cell* 2, 867–876.

Chang, C., Bowman, J., DeJohn, A., Lander, E. S., and Meyerowitz, E. M. (1988). *Proc. Natl. Acad. Sci. U.S.A.* 85, 6856–6860.

Chaudhury, A. M., and Signer, E. R. (1989). *Plant Mol. Biol. Rep.* 7, 258–265.

Chen, J., Greenblatt, I. M., and Dellaporta, S. L. (1987). *Genetics* 117, 109–116.

Cone, K. C., Burr, F. A., and Burr, B. (1986). *Proc. Natl. Acad. Sci. U.S.A.* 83, 9631–9635.

Coupland, G., Baker, B., Schell, J., and Starlinger, P. (1988). *EMBO J.* 7, 3653–3659.

Coupland, G., Plum, C., Chatterjee, S., Post, A., and Starlinger, P. (1989). *Proc. Natl. Acad. Sci. U.S.A.* 86, 9385–9388.

Dean, C., Sjodin, C., Lawson, E., Lister, C., and Bancroft, I. (1990). Fourth International Conference on Arabidopsis Research Abstracts, Vienna, June 2–5.

Engels, W. R. (1989). In *Mobile DNA* (Ed. by D. E. Berg and M. Howe), pp. 437–484. American Society for Microbiology Press.

Errampalli, D., Patton, D., Castle, L., Mickelson, L., Hansen, K., Schnall, J., Feldmann, K., and Meinke, D. (1991). *Plant Cell* 3, 149–157.

Fedoroff, N. V. (1989). *Cell* 56, 181–191.

Fedoroff, N. V., Furtek, D. B., and Nelson, O. E. (1984). *Proc. Natl. Acad. Sci. U.S.A.* 81, 3825–3829.

Feldmann, K. A. (1991). *Plant J.* 1, 71–82.

Feldmann, K. A., and Marks, M. D. (1987). *Mol. Gen. Genet.* 208, 1–9.

Feldmann, K. A., Marks, M. D., Christianson, M. L., and Quatrano, R. S. (1989). *Science* 243, 1351–1354.

Feldmann, K. A., Carlson, T. J., Coomber, S. A., Farrance, C. E., Mandel, M. A., and Wierzbicki, A. (1990). In *Horticultural Biotechnology* Vol. 11 (Ed. by A. B. Bennett and S. D. O'Neill), pp. 109–120. New York: Alan R. Liss.

Finnegan, D. J., Taylor, B. H., Dennis, E. S., and Peacock, W. J. (1988). *Mol. Gen. Genet.* 212, 505–509.

Finnegan, D. J. (1989). In *Mobile DNA* (Ed. by D. E. Berg and M. Howe), pp. 503–518. American Society for Microbiology Press.

Grandbastien, M-A., Spielmann, A., and Caboche, M. (1989). *Nature (London)* 337, 376–380.

Greenblatt, I. M. (1968). *Genetics* 58, 585–597.

Hake, S., Vollbrecht, E., and Freeling, M. (1989). *EMBO J.* 8, 15–22.

Haring, M. A., Gao, J., Volbeda, T., Rommens, C. M. T., Nijkamp, H. J. J., and Hille, J. (1989). *Plant Mol. Biol.* 13, 189–201.

Hehl, R., and Baker, B. (1989). *Mol. Gen. Genet.* 217, 53–59.

Hehl, R., and Baker, B. (1990). *Plant Cell* 2, 709–721.

Herman, P. L., and Marks, M. D. (1989). *Plant Cell* 1, 1051–1055.

Jofuku, K. D., Van Montagu, M., and Okamuro, J. K. (1991). *J. Cell. Biochem.* 15A, 34.

Jones, J. D. G., Carland, F. M., Harper, L., Lim, E., and Dooner, H. K. (1989). *Science* 244, 204–207.

Jones, J. D. G., Carland, F., Lim, E., Ralston, E., and Dooner, H. K. (1990). *Plant Cell* 2, 701–707.

Kleckner, N. (1981). *Ann. Rev. Genet.* 15, 341–404.

Knapp, S., Coupland, G., Uhrig, H., Starlinger, P., and Salamini, F. (1988). *Mol. Gen. Genet.* 213, 285–290.

Koncz, C., Martini, N., Mayerhofer, R., Koncz-Kalman, Z., Korber, H., Redei, G. P., and Schell, J. (1989). *Proc. Natl. Acad. Sci. U.S.A.* 86, 8467–8471.

Koncz, C., Mayerhofer, R., Koncz-Kalman, Nawrath, C., Reiss, B., Redei, G. P., and Schell, J. (1990). *EMBO J.* 9, 1337–1346.

Koornneef, M., van Eden, J., Hanhart, C. J., Stam, P., Braaksma, F. J., and Feenstra, W. J. (1983). *J. Hered.* 74, 265–272.

Lassner, M. W., Palys, J. M., and Yoder, J. I. (1989). *Mol. Gen. Genet.* 218, 25–32.

Lloyd, A. M., Barnason, A. R., Rogers, S. G., Byrne, M. C., Fraley, R. T., and Horsch, R. B. (1985). *Science* 234, 464–466.

Marks, M. D., and Feldmann, K. A. (1989). *Plant Cell* 1, 1043–1050.

Martienssen, R. A., Barken, A., Freeling, M., and Taylor, W. C. (1989). *EMBO J.* 8, 1633–1639.

Martin, C., Carpenter, R., Sommer, H., Saedler, H., and Coen, E. S. (1985). *EMBO J.* 4, 1625–1630.

Martin, C., Prescott, A., Lister, C., and MacKay, S. (1989). *EMBO J.* 8, 997–1004.

Masterson, R. V., Furtek, D. B., Grevelding, C., and Schell, J. (1989). *Mol. Gen. Genet.* 219, 461–466.

Mayer, U., Torres Ruiz, R. A., Berleth, T., Misera, S., and Jurgens, G. (1990). *Fourth International Conference on Arabidopsis Research,* p. 113, University of Vienna, Vienna, Austria.

McClintock, B. (1948). *Carniegie Inst. Washington Year Book* 46, 155–169.

Moerman, D. G., and Waterston, R. H. (1989). In *Mobile DNA* (Ed. by D. E. Berg and M. Howe), pp. 537–556. American Society for Microbiology Press.

Muller, A. J., and Weiland, U. (1990). *Fourth International Conference on Arabidopsis Research,* p. 119, University of Vienna, Vienna, Austria.

Nam, H-G., Giraudat, J., den Boer, B., Moonan, F., Loos, W. D. B., Hauge, B. M., and Goodman, H. M. (1989). *Plant Cell* 1, 699–705.

Oppenheimer, D. G., Herman, P. L., Siva Kumaran, S., Esch, J., and Marks, M. D. (1991). *Cell* 67, 483–493.

O'Reilly, C., Shepherd, N. S., Pereira, A., Shwartz-Sommer, Z., Bertram, I., Robertson, D. S., Peterson, P., and Saedler, H. (1985). *EMBO J.* 41, 877–882.

Schmidt, R., and Willmitzer, L. (1989). *Mol. Gen. Genet.* 220, 17–24.

Shepherd, N. (1988). In *Plant Molecular Biology: A Practical Approach* (Ed. by C. Shaw), pp. 187–220. IRL Press.

Sommer, H., Beltran, J-P., Huijser, P., Pape, H., Lonnig, W-E., Saedler, H., and Schwartz-Sommer, Z. (1990). *EMBO J.* 9, 605–613.

Taylor, B. H., Finnegan, D. J., Dennis, E. S., and Peacock, W. J. (1989). *Plant Mol. Biol.* 13, 109–118.

Theres, N., Scheele, T., and Starlinger, P. (1987). *Mol. Gen. Genet.* 209, 193–197.

Upadhyaya, K., Sommer, H., Krebbers, E., and Saedler, H. (1985). *Mol. Gen. Genet.* 199, 201–206.

Valvekens, D., Van Montagu, M., and Van Lijsebettens, M. (1988). *Proc. Natl. Acad. Sci. U.S.A.* 85, 5536–5540.

Van Lijsebettens, M., Vanderhaeghen, R., Scheres, B., and Van Montagu, M. (1990). *Fourth International Conference on Arabidopsis Research,* p. 48, University of Vienna, Vienna, Austria.

Van Sluys, M. A., Tempe, J., and Fedoroff, N. V. (1987). *EMBO J.* 6, 3881–3889.

Velten, J., and Schell, J. (1985). *Nuc. Acid Res.* 13, 6981–6998.

Voytas, D. F., and Ausubel, F. M. (1989). *Nature (London)* 242–244.

Wienand, U., Weydemann, U., Niesbach-Kloesgen, U., Peterson, P. A., and Saedler, H. (1986). *Mol. Gen. Genet.* 203, 202–207.

Yanofsky, M. F., Ma, H., Bowman, J. L., Drews, G. N., Feldmann, K. A., and Meyerowitz, E. M. (1990). *Nature (London)* 346, 35–39.

Yoder, J. I. (1990). *Plant Cell* 2, 723–730.

Yoder, J. I., Palys, J., Alpert, K., and Lassner, M. (1988). *Mol. Gen. Genet.* 213, 219–296.

Zambryski, P., Tempe, J., and Schell, J. (1989). *Cell* 56, 193–201.

Part III

Transgenic Plants and Crop Improvement

9

Herbicide-Tolerant Crops

Maud A. W. Hinchee, Stephen R. Padgette, Ganesh M. Kishore, Xavier Delannay, and Robert T. Fraley

Monsanto, Co., St. Louis, Missouri 63198

Weed control in crops is critical for high farm productivity. Weeds compete with crops for moisture, nutrients, and light. In addition, the presence of weeds and weed seeds in the final crop product significantly reduces the quality of the crop. The use of herbicides to control weeds allows the grower more efficient crop management together with improved yield. The monetary losses due to weeds in recent years have averaged about 10–20% of the crop value. In the United States alone, on an annual basis, this can easily amount to 3–4 billion dollars. The principal method of weed control in the 1990s will most likely be based on chemicals. The technological breakthroughs of the 1980s will pave the way for the use of new and existing herbicides which are both extremely effective and environmentally sound.

Many herbicides are currently available for weed management. Herbicide choice is based on the effectiveness of the herbicide against problem weeds, the crop sensitivity to the herbicide, the cost of the herbicide treatment, and in many cases the environmental properties of the herbicide. Before commercialization, herbicides are evaluated for their broad-spectrum activity against weeds and nontoxicity to target crops, mammals, and invertebrates. Several classes of herbicides are quite effective for broad-spectrum weed control (Figure 9.1), but they are either nonselective and kill crop plants or they significantly injure some crops at the application rates required for sufficient weed control. Examples of such herbicides are the imidazolinones, the sulfonylureas, and glyphosate.

Chemical and mechanical methods for protecting crop plants from herbicide damage are possible. Chemical safeners can be used to reduce herbicide toxicity

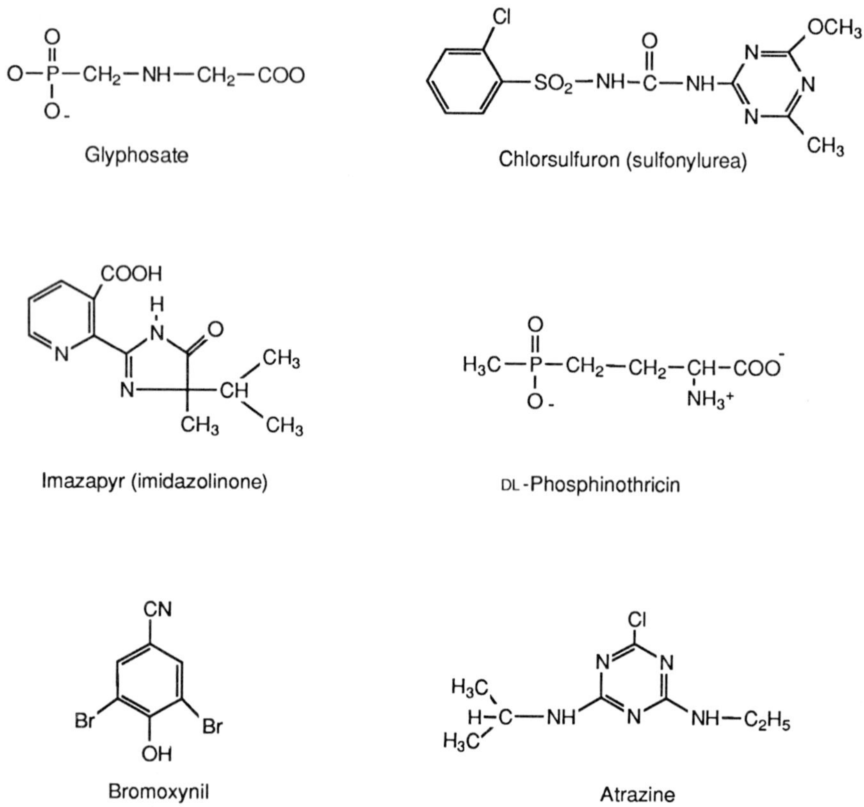

Figure 9.1 Chemical structures of different classes of herbicides.

to a crop. An example of a safener is 1,8-naphthalic anhydride, which can be used as a seed treatment to protect corn seedlings from injury during germination in soil treated with thiocarbamate herbicides. Special herbicide application equipment can be used to avoid herbicide contact with a crop. The herbicide Roundup can be applied using a special wick applicator which reaches only weeds in between crop rows. These two approaches for protecting crops are limited in use since they are restricted to certain herbicides and/or application methods, and can be cost prohibitive.

The development of herbicide-tolerant crops is another way in which crops can be protected from herbicide damage. Two general methods have been used to generate herbicide-tolerant crops: selection and direct herbicide-tolerant gene insertion using genetic engineering techniques. An overview of results using each of these methods is provided below.

9.1 Selection of Herbicide-Tolerant Crops

Several classes of herbicides, such as the triazines, imidazolinones, and sulfonylureas, are more active on weeds than they are on target crops. Herbicide selectivity is a result of the interaction between crop physiology and the physicochemical properties of the herbicide. Crop plant selectivity can result from metabolic detoxification of the herbicide, internal compartmentalization and sequestering of the herbicide within crop plant cells, and physical or physiological barriers in the crop plant which interfere with herbicide uptake, translocation, or target-site interaction.

Some herbicides are not as selective as is desired on particular crops. However, careful germplasm screening within a crop can potentially identify a genotype with herbicide tolerance. Germplasm screening of *Glycine max* (soybean) germplasm has resulted in the identification of cultivars with resistance to the herbicide metribuzin (Barratine *et al.*, 1975). Inbred lines of corn have been shown to differ in their sensitivity to some sulfonylurea herbicides, with one line showing complete tolerance to the sulfonylurea herbicide, primisulfuron, while other lines were completely killed (Harms *et al.*, 1990). A limitation using this approach is that not all crops have genes within their gene pool which allow complete tolerance to field dose rates of the herbicides most efficacious in weed control. Spontaneous mutants for herbicide tolerance have been used for breeding herbicide tolerance into crop plants. For example, atrazine tolerance was introduced into *Brassica napus* (canola) (Beversdorf *et al.*, 1989) by genetic crosses with a spontaneous atrazine-tolerant *Brassica rapa* mutant (Maltais and Bouchard, 1987). Genetic crosses with atrazine-tolerant *B. napus* resulted in the production of atrazine-tolerant *B. oleracea* (broccoli) (Ayotte *et al.*, 1989). Cybrids (cytoplasmic hybrids), resulting from protoplast fusion of *B. oleracea* (cauliflower) and atrazine-resistant *B. napus,* were regenerated into plants whose selfed progeny were atrazine tolerant (Jourdan *et al.*, 1989).

Another genetic approach is to induce genetic resistance by the use of mutagens and/or herbicide selection pressure during *in vitro* cell culture or seed germination. Herbicide selective pressure during plant cell culture has been used successfully to produce imidazolinone and/or sulfonylurea-tolerant cell lines and plants. *Zea mays* (corn) cell lines, selected for growth on the imidazolinone imazapyr, were mutants in acetolactate synthase (ALS) with more than 100-fold resistance to the herbicide (Anderson and Georgeson, 1989). These corn lines were also resistant to sulfonylurea herbicides. Imidazolinone-resistant corn plants, regenerated from one of these lines, passed on herbicide tolerance to their progeny as a single, dominant nuclear trait (Shaner and Anderson, 1985). These plants have been used to introduce imidazolinone tolerance into elite corn cultivars used for hybrid seed production (Fincher, 1989). Several sulfonylurea-resistant mutant cell cultures of *Nicotiana tabacum* (tobacco) have been generated after herbicide selection pressure, and genetic crosses of regenerated plants from these cell lines indicated that the resistance was due to either a single dominant or semidominant mutation (Chaleff and Ray, 1984). Linkage analysis of six mutants identified two unlinked genetic loci. Homozygous plants for one of the mutants were at least 100-fold more resistant to chlorsulfuron than the control plants, and contained a mutant ALS with a high I_{50} for chlorsulfuron (Chaleff and Mauvais, 1984). Not all herbicide-tolerant cell lines produced after herbicide selection pressure are the result of a mutation in a biochemical herbicide target. Roundup-tolerant tobacco plants, regenerated from a glyphosate selected cell line, overproduced the enzyme which is inhibited by the herbicide (Dyer *et al.*, 1988).

Mutagenesis, prior to seedling growth or plant regeneration, has also yielded herbicide-tolerant mutants. Canola plants with field-level tolerance to the imidazolinones have been generated using mutagenesis of isolated microspores prior to plant regeneration under herbicide selective pressure (Swanson *et al.*, 1989). *Arabidopsis thaliana* mutants, isolated by mutagenizing seeds and germinating them in the presence of chlorsulfuron, contained a gene (*csr1*) for a mutant ALS with I_{50} values for chlorsulfuron over 1000-fold that of the control (Haughn and Somerville, 1986). The plants containing this mutant gene were 3000 times more tolerant to chlorsulfuron than the wild-type plants. Recently, another locus (*imr*) was identified in *Arabidopsis* which confers imidazolinone tolerance, but not chlorsulfuron resistance (Haughn and Somerville, 1990). Resistance to the herbicide isoxaben, which biochemically acts as a cell wall inhibitor, was obtained from two unlinked mutations generated after *Arabidopsis* seed mutagenesis (Heim *et al.*, 1988, 1989). A similar approach was taken for producing sulfonylurea-tolerant soybean plants. Soybean seeds, which were mutagenized and then germinated under continuous chlorsulfuron selection pressure, produced a soybean mutant with a high degree of resistance to both postemergence and preemergence applications of a variety of sulfonylurea herbicides (Sebastian *et al.*, 1989).

It is important to note that mutagenesis and genetic selection techniques may not allow identification of herbicide-tolerant crop genotypes for certain herbicides. For example, resistance at the whole plant level to the herbicide Roundup has never been achieved by selection for spontaneous, mutagen-induced or tissue culture-induced genetic variation. This may be due to differences in the way a particular herbicide interacts with its biochemical target. For example, when an enzyme's active site and herbicide binding site coincide, the frequency of generating a mutation which results in an herbicide-tolerant, yet still enzymatically efficient, target is extremely low. In these cases, direct insertion of herbicide tolerance genes can be performed using genetic engineering techniques.

9.2 Genetic Engineering for Herbicide-Tolerant Crops

To date, genetic engineering of herbicide tolerance has been based on inserting genes to express proteins which confer herbicide tolerance by three mechanisms: (1) target-site modification; (2) target-site overproduction; and (3) metabolic inactivation of the herbicide (for reviews, see Quinn, 1990; Oxtoby and Hughes, 1990; Mazur and Falco, 1989; Padgette *et al.*, 1989). Target-site modification, resulting from the introduction of an herbicide-tolerant protein, has been used in developing tolerance to the sulfonylureas, imidazolinones, glyphosate, and atrazine. The target-site overproduction approach has been applied for glyphosate and phosphinothricin tolerance. Genetic insertion of an herbicide inactivation enzyme has been a successful approach for producing plants which are resistant to phosphinothricin, bromoxynil, or 2,4-dichlorophenoxyacetic acid. These approaches will be discussed in more detail below.

9.2.1 PHOSPHINOTHRICIN RESISTANCE

DL-Phosphinothricin is the active ingredient in a nonselective, broad-spectrum, postemergent herbicide, sold by the Hoechst Co. (Germany) as Basta (ammonium glufosinate). The D-isomer is herbicidally inactive. L-Phosphinothricin (PPT) was initially identified as the active moiety of the herbidical tripeptide bialaphos (L-phosphinotricinyl-L-alanyl-L-alanine) produced by *Streptomyces hygroscopicus* (Kondo *et al.*, 1973). PPT is a potent inhibitor of the enzyme glutamine synthetase (GS; glutamate–ammonia ligase) (Bayer *et al.*, 1972). In plants, GS is critical for the assimilation of ammonia as well as for general nitrogen metabolism. It is the only enzyme in plants that can detoxify ammonia released by nitrate reduction, amino acid catabolism, and respiration. Multiple isozymic forms of GS have been detected in different plant organs (McNally *et al.*, 1983). In addition, leaves contain two isozymic forms of GS; one of the isozymes is cytosol localized while the other is associated with chloroplasts.

Two mechanisms have been explored for engineering tolerance to PPT. The first approach was the overproduction of the target enzyme, GS. Since there are a number of GS isozymes in plants, there is no obvious choice as to whether it is the cytosolic and/or plastidic enzymes which have to be overproduced to achieve tolerance. A cytosolic form was shown to be responsible for the PPT resistance of PPT-selected alfalfa suspension cells (Donn *et al.*, 1984; Tischer *et al.*, 1986). Tolerance to PPT has been achieved in transgenic tobacco plants which overexpressed an alfalfa cytosolic GS gene regulated by either the cauliflower mosaic virus (CaMV) 35S promoter or the potato ST/LS1 promoter. These plants regenerated at twenty times the lethal concentration of PPT, but showed severe damage when sprayed at four times the normal field application rate (Eckes *et al.*, 1989). Tissues from the GS overproducing plants contained a 7-fold lower level of free NH_3 relative to control tissue, but grew and reproduced normally.

The second approach for achieving tolerance is metabolic inactivation of the herbicide. An inactivating enzyme has been identified which catalyzes the conversion of L-phosphinothricin to *N*-acetyl-L-phosphinothricin in the presence of acetyl-CoA as a cosubstrate. The gene (*bar*) for the phosphinotricin acetyl transferase (PAT) was isolated from *Streptomyces hygroscopicus*. Since GS of *S. hygroscopicus* is sensitive to L-phosphinothricin, presumably the bacterium protects itself by metabolism of the herbicide. The PAT enzyme shows relatively higher specificity for L-phosphinothricin as a substrate (Thompson *et al.*, 1987), although some acetylation of L-glutamate can be detected. Transgenic tobacco, tomato, and potato plants, expressing the *bar* gene under the regulation of the CaMV 35S promoter, were tolerant to 4–10 times the field rate of application of the herbicide (DeBlock *et al.*, 1987). Even plants expressing PAT at a level of 0.001% of the total leaf protein could survive herbicide application. The ammonia level of transgenic plants was unaltered following herbicide treatment, indicating complete tolerance. Transgenic *Brassica napus* and *B. oleracea* have both been produced which are Basta resistant (DeBlock *et al.*, 1989). Several research groups have produced PPT-resistant corn plants. These herbicide tolerant corn plants have been regenerated from protoplasts after polyethylene glycol-mediated DNA uptake (Donn *et al.*, 1990) and from DNA-coated microprojectile bombardment of suspension cultures (Fromm *et al.*, 1990; Gordon-Kamm *et al.*, 1990). The plants were completely tolerant to over 2 times the application rate of PPT (Gordon-Kamm *et al.*, 1990).

Field tests of phosphinothricin-resistant alfalfa, tobacco, and potato have been conducted. Alfalfa plants, which expressed the *bar* gene regulated by the CaMV 35S promoter, were fully resistant to two applications of Basta at 6 liter/ha. Whole plant tolerance was on the average much lower when the *bar* gene was regulated with the T-DNA TR-2 promoter (d'Halluin *et al.*, 1990). Field tests were conducted with two different lines of transgenic tobacco plants. These two lines differed approximately 100 times in their PAT expression although both contained the same CaMV 35S-*bar* gene. Despite their differences in PAT expression, both grew well at four times the normal field application rate (4000 g a.i./ha of phosphinothricin). However, tobacco plants which expressed the PAT gene regulated by the T-DNA-TR-2 promoter were resistant to only 1000 g a.i./ha of phosphinothricin (DeGreef *et al.*, 1989). Promoter choice for the regulating *bar* gene expression appeared critical for achieving maximum PPT tolerance in field-grown plants.

Phosphinotricin resistance has proven to be a useful selectable marker in the transformation of both monocotyledonous and dicotyledonous plants. The *bar* gene was used successfully as a selectable marker for the transformation of rice protoplasts after electroporation (DeKeyser *et al.*, 1989). Transgenic Black Mexican Sweet corn suspension cells were selected on bialaphos after transformation using DNA-coated microprojectile bombardment (Spencer *et al.*, 1990). Selection of embryogenic suspension culture cells of corn with bialaphos (Gordon-Kamm *et al.*, 1990) following microprojectile bombardment has also been used to produce fertile, transgenic corn plants. PPT selection of cell colonies, resulting after protoplast transformation, has resulted in transgenic corn plant production (Donn *et al.*, 1990). Direct selection of hybrid aspen (*Populus alba* × *P. tremula*) and hybrid poplar (*P. trichocarpa* × *P. deltoides*) stem explants following *Agrobacterium*-mediated transformation produced the same number of transgenic shoots as for kanamycin selection (DeBlock, 1990). Direct selection of tobacco and potato leaf discs with PPT after *Agrobacterium*-mediated transformation has resulted in transgenic plant production (DeBlock *et al.*, 1987), but a similar approach on stem and petiole explants was not successful in alfalfa (d'Halluin *et al.*, 1990).

9.2.2 BROMOXYNIL TOLERANCE

Bromoxynil, the active ingredient in the herbicide Buctril (Rhone-Poulenc, France), is used for selective, postemergence control of seedling broad leaf weeds in wheat, barley, and seedling turf grasses. The primary mode of action of bromoxynil appears to be inhibition of photosynthetic electron transport (Friend and Moreland, 1969), and it acts by binding a component of the quinone-binding complex of Photosystem II.

A mechanism for the detoxification of bromoxynil was discovered. A nitrilase activity was identified in the extracts of a soil bacterium, *Klebsiella pneumoniae* subsp. *ozaenae,* which specifically hydrolyzed bromoxynil to 3,5-dibromo-4-hydroxybenzoic acid (McBride *et al.*, 1986). This microorganism completely transformed the nitrile moiety of bromoxynil to the carboxylic acid with liberation of ammonia. The carboxylic acid was inactive as an herbicide. The nitrilase gene was shown to be plasmid encoded in *K. p. ozaenae* and was expressed constitutively in *E. coli* by direct transfer of the 82-kb naturally occurring plasmid (Stalker and McBride, 1987). A 2.6-kb subclone of this plasmid was obtained which encoded for the bromoxynil nitrilase gene (*bxn*). Transgenic tobacco plants, which contained the nitrilase in their cytosol as a result of expression of the *bxn* gene driven by the light-regulated ribulose-bisphosphate carboxylase (RuBPC) small subunit promoter, demonstrated excellent resistance to bromoxynil (Stalker *et al.*, 1988). Different levels of nitrilase activity in different transgenic plants correlated with different degrees of bromoxynil tolerance. The low expressors (0.0007% of total leaf protein) were tolerant up to 1.5–2.0 lb/acre and high expressors (0.01% of total protein) were tolerant at greater than 4 lb/acre. Wild-type plants were killed at 0.5 lb/acre application rate. These results indicated that these tobacco plants were tolerant to bromoxynil at 8 times the normal field application rate.

9.2.3 2,4-D RESISTANCE

2,4-Dichlorophenoxyacetic acid (2,4-D) is an herbicide which mimics the actions of auxins, a class of plant growth regulators. 2,4-D is highly active against many species of broad-leaved weeds and is widely used as a weed control agent on cereal crops. Broad-leaved crops, such as cotton, can be severely damaged by accidental drift from the aerial spraying of nearby cereal crops. Soil microorganisms which can degrade 2,4-D have been identified. One of the most well understood is the bacterium *Alcaligenes eutrophus* which has a 2,4-D degradative pathway comprised of 6 enzymes that convert 2,4-D to chloromalylacetic acid, a compound subsequently metabolized to succinic acid (Don *et al.*, 1985). The first gene in the 2,4-D degradation pathway (*tfdA*) encodes a 2,4-D monoxygenase which converts 2,4-D into 2,4-dichlorophenol by cleavage of the aliphatic side chain (Streber *et al.*, 1987). The *A. eutrophus tfdA* gene was cloned and placed in a CaMV 35S promoter expression vector (Lyon *et al.*, 1989). Transgenic tobacco plants were produced via *Agrobacterium*-mediated transformation. The tobacco plants expressing the highest levels of the monooxygenase enzyme exhibited increased tolerance to 2,4-D in leaf disk and seed germination assays. Transgenic progeny plants survived up to 8 times the usual field application rate. Transgenic seed was slightly resistant to the herbicide MCPA (4-chloro-2-methylphenoxyacetic acid), a compound structurally similar to 2,4-D; but there was not cross-resistance to the closely related herbicide 2,4,5-T (2,4,5-trichlorophenoxyacetic acid). Similar results were obtained with transgenic tobacco plants which contained the *tfdA* gene under the regulation of the CAMV 35S promoter (Streber and Willmitzer, 1989). These plants had equivalent tolerance to 4-CPA (4-chlorophenoxyacetic acid), lesser tolerance to MCPA, and no tolerance to the trisubstituted phenoxyacetic acids or the phenoxypropionic acids.

9.2.4 SULFONYLUREA AND IMIDAZOLINONE TOLERANCE

The sulfonylurea herbicides are potent, low-rate herbicides (Levitt, 1983). These herbicides inhibit acetolactate synthase (ALS) or acetohydroxy acid synthase (AHAS), the first enzyme in the biosynthetic pathway for the branched-chain amino acids valine, leucine, and isoleucine (Chaleff and Mauvais, 1984). This enzyme is also inhibited by the imidazolinone herbicides (Shaner *et al.*, 1984). Bacterial, plant, and yeast ALS enzymes have been identified which are tolerant to sulfonylureas. A yeast ALS gene was used as a heterologous probe for the isolation of herbicide-sensitive and herbicide-insensitive ALS genes from plants (Mazur *et al.*, 1987). The herbicide-insensitive ALS gene was obtained from tobacco plants which had been generated from sulfonylurea-resistant mutant cell cultures (Chaleff and Ray, 1984), while the herbicide-sensitive gene was obtained from wild-type *Arabidopsis* plants. The plant ALS genes did not contain introns. The plant enzymes contained additional amino acids at the amino terminus compared to an *E. coli* ALS enzyme. These additional amino acids have been established as being a transit peptide which targets the ALS precursor into the chloroplast for processing into the mature 70 kDa polypeptide (Bascomb *et al.*, 1987). Mature tobacco and *Arabidopsis* ALS proteins were 84% homologous, while their transit peptides were only 27% homologous. The two plant ALS enzymes showed significant homology to the yeast and bacterial ALS enzymes.

Two mutant tobacco ALS (Hra and C3 lines) genes have been isolated (Smith *et al.*, 1988). The C3 line mutant gene had a single amino acid substitution (proline to glutamine at amino acid 196), while the Hra line had two amino acid substitutions (proline to alanine substitution at amino acid 196 and a tryptophan to leucine substitution at amino acid 573) (Lee *et al.*, 1988). The resistance resulting from the single amino acid change at position 196 is enhanced by the additional mutation at codon 573 (Lee *et al.*, 1988; Creason and Chaleff, 1988). The double tobacco mutant with the Pro-196 to Ala and the Trp-573 to Leu was also cross-resistant to both classes of herbicides (Mazur *et al.*, 1987). A mutant *Arabidopsis* ALS gene (*csr1-1*) had a single mutation of proline to serine at position 197 (Haughn *et al.*, 1988) and this change was identical to that found in the herbicide-tolerant yeast ALS mutant (Yadav *et al.*, 1986).

The sulfonylurea-resistant *Arabidopsis csr1* gene, under the regulation of its own promoter, has been expressed in transgenic tobacco plants via *Agrobacterium*-mediated transformation (Haughn *et al.*, 1988; Gabard *et al.*, 1989). Progeny of these transgenic plants grew well on germination medium containing 1000 n*M* chlorsulfuron and were over 100-fold more resistant than wild-type plants (Haughn *et al.*, 1988). Transgenic tobacco plants demonstrated cross-resistance to 3 other short residual sulfonylurea herbicides in addition to chlorsulfuron, but not to any of the 4 imidazolinone herbicides tested (Gabard *et al.*, 1989). The tobacco Hra mutant ALS gene was also introduced into tobacco plants, and field tests of these tobacco plants showed that the transgenic plants were unaffected by herbicide application at 32 g/ha (4 times the application rate) (Smith *et al.*, 1988). The ALS sequences responsible for resistance were in domains which are conserved between plants, yeast, and bacteria, and identical molecular substitutions in ALS have conferred herbicide resistance in both yeast and several species of crop plants (Hartnett *et al.*, 1990). Additional ALS mutations were made by oligonucleotide-directed *in vitro* mutagenesis of plant ALS genes. These were introduced into transgenic plants, and the expression of herbicide tolerance has provided evidence on the critical sequences for sulfonylurea tolerance in both plants and yeast (Hartnett *et al.*, 1990).

Agrobacterium-mediated transformation has been used to produce sulfonylurea-resistant *Brassica napus* (canola) plants (Hartnett *et al.*, 1990; Wiersma *et al.*, 1989). An *Arabidopsis* transit peptide sequence for the ALS gene was used to isolate a *B. napus* ALS genomic clone. The genomic clone was mutated *in vitro* at codon 173 to replace a proline residue with serine in one of the conserved regions responsible for herbicide sensitivity (Wiersma *et al.*, 1989). Leaf pieces from transgenic canola plants were able to produce callus on concentrations of chlorsulfuron which killed wild-type leaf pieces.

Sulfonylurea herbicide resistance can be useful in the process of producing transgenic plants. Recently it has been reported that transgenic corn cells and plants can be selected on chlorsulfuron after particle gun-mediated transformation (Fromm *et al.*, 1990). Embryogenic corn cell suspensions were bombarded with DNA containing a mutant corn ALS gene driven by the CaMV 35S promoter. Mutant ALS genes have not yet proved advantageous as selectable markers in dicotyledonous plants. Although selection of transgenic tobacco shoots was possible after direct selection of leaf disks on medium containing chlorsulfuron, the selected shoots developed slowly (Gabard *et al.*, 1989) and, in some cases, abnormally (Charest *et al.*, 1990).

9.2.5 ATRAZINE TOLERANCE

The primary action of the triazine herbicides, such as atrazine, is to inhibit photosynthesis in plant chloroplasts. Specifically, they inhibit photosynthetic electron transfer from reduced plastoquinone Q to the protein-bound plastoquinone B in PS II (Arntzen *et al.*, 1982). A 32-kD protein (herbicide binding protein, Q_b protein, D1 protein) was specifically labeled with [^{14}C]azidoatrazine in thylakoid membranes from triazine-sensitive *Amaranthus hybridus*, while no labeling occurred in the triazine-resistant lines (Steinback *et al.*, 1981). Further studies of the triazine-resistant *A. hybridus* line showed that herbicide resistance was due to a single amino acid substitution at position 228 (serine to glycine) in the 32-kDa protein (Hirschberg and MacIntosh, 1983). A mutation of Ser-264 to Gly in the 32-kDa protein has also been shown to result in triazine resistance in other plant species (Bettini *et al.*, 1987; Golubinoff *et al.*, 1985). Attempts have been made to engineer atrazine tolerance into transgenic tobacco plants via introduction of a mutant 32-kDa herbicide-binding protein (Bogorad, 1986; Cheung *et al.*, 1988). The 32-kDa protein was targeted to chloroplasts using the transit peptide sequence of the small subunit of RuBPC. Transgenic calli which expressed the mutant 32-kDa protein demonstrated atrazine tolerance. However, when plants regenerated from these calli were challenged with atrazine, it appeared that their tolerance was marginal. More recently, transgenic soybean have been reported to demonstrate atrazine tolerance (Yue *et al.*, 1990).

9.2.6 ROUNDUP TOLERANCE

Roundup is a foliar-applied, broad-spectrum, nonselective, postemergent herbicide (Baird *et al.*, 1971). It is highly effective against the majority of annual and perennial grasses and broad-leaved weeds. Since it is nonselective, Roundup also kills crop plants. Glyphosate, the active ingredient in Roundup, has favorable environmental features, such as rapid soil binding and biodegradation, and extremely low mammalian toxicity (Franz, 1985).

Plants which have been sprayed with glyphosate rapidly transport the herbicide to their growing points, and the morphological result is the cessation of growth in the shoot and root apical meristems several days postapplication (Baird *et al.*, 1971). This is most likely due to the loss of essential aromatic amino acids required for meristem function. Glyphosate blocks aromatic amino acid biosynthesis in both plants and bacteria, by blocking the incorporation of shikimic acid into aromatic amino acids (Hollander and Amrhein, 1980; Amrhein *et al.*, 1980).

The enzyme targeted by glyphosate is the shikimate pathway enzyme 5-enolpyruvylshikimate-3-phosphate synthase (EPSP synthase, EPSPS) (Steinrucken and Amrhein, 1980). EPSPS catalyzes the reversible condensation of shikimate 3-phosphate (S3P) and phosphoenolpyruvate (PEP) to give EPSP and inorganic phosphate (P_i). EPSP is then converted to tyrosine, tryptophan, and phenylalanine, as well as a variety of essential vitamins and secondary plant metabolites. The shikimate pathway in plants is localized in the plastids. Isolated *Spinacia oleracea* chloroplasts incorporated ^{14}C-labeled bicarbonate and shikimic acid into aromatic amino acids *in vitro* (Bickel *et al.*, 1978; Schultz *et al.*, 1981). However, there is evidence that at least three enzymes in the pathway have distinct plastidic and extraplastidic isozymes. These are DAHP synthase (Ganson *et al.*, 1986), shikimate oxidoreductase (Rothe *et al.*, 1983), and chorismate mutase (d'Amato *et al.*, 1984). Subcellular fractionation in *P. sativum* (Mousdale and Coggins, 1985) and *Petunia hybrida* (petunia) (della-Cioppa *et al.*, 1986a) showed that EPSPS is located in the plastids.

Two approaches have been evaluated for the genetic engineering of glyphosate tolerance. The first mechanism is based on overproduction of wild-type plant EPSPS. Glyphosate tolerance in this case results from unhibited EPSP synthase activity in the presence of glyphosate due to overproduction of EPSPS relative to glyphosate concentration in the affected plant tissues. The second mechanism relies on the expression of genes encoding glyphosate-tolerant EPSP synthases in plants.

9.2.6.1 Overproduction of Wild-Type EPSPS

It was show that *E. coli,* transformed with a high copy number plasmid containing the *aroA* gene encoding EPSPS, produced 16-fold the normal amount of EPSPS and conferred an 8-fold increase in glyphosate tolerance (Rogers *et al.*, 1983). Subsequently, a petunia cell line was developed which was tolerant to high levels of glyphosate by stepwise selection on increasing concentrations of glyphosate. This cell line had significantly enhanced EPSPS specific activity compared to its parent cell line (Steinrucken *et al.*, 1986); however the EPSPS was glyphosate sensitive. The glyphosate-resistant petunia cell line was used to isolate EPSPS cDNA and to purify a plant EPSPS enzyme. The amino-terminal sequence of the petunia enzyme facilitated the design of oligonucleotide probes for the isolation of full-length EPSPS cDNA from a λgt10 cDNA library of the glyphosate-tolerant petunia cell line. Using the cDNA as a probe in genomic blot analysis, it was demonstrated that the glyphosate tolerance in the petunia cell line was due to gene amplification (Shah *et al.*, 1986).

The petunia EPSPS cDNA clone encoded for a precursor protein (pre-EPSPS) with an additional 72 amino acids at the amino-terminal end of the EPSPS amino acid sequence previously determined for the purified protein. EPSPS produced by the tolerant petunia cell line was found localized in the plastids (della-Cioppa *et al.*, 1986b) and the 72-amino acid extension of the preEPSPS proved to be a transit peptide regulating translocation of the preEPSPS into the chloroplast. Inside the chloroplast, the transit peptide was removed by a metalloendoprotease releasing the mature 444-amino acid EPSPS. However, the preEPSPS was able to catalyze the EPSPS reaction without maturation. The petunia preEPSPS was shown to be imported and processed in chloroplasts of other plant species, indicating conservation in plant EPSPS import and maturation mechanisms. The transit peptide also targeted a bacterial EPSPS into plant chloroplasts (della-Cioppa *et al.*, 1987).

In addition to the petunia EPSPS cDNA, a full-length cDNA clone of tomato EPSPS and the genes for petunia and *Arabidopsis* EPSPS have been isolated. These genes are interrupted by seven introns in identical positions and there is a high degree of conservation (84–93%) in the amino acid sequences of the mature EPSPS of these plants (Gasser *et al.*, 1988). Corn and soybean EPSPS genes have also been isolated (Gasser *et al.*, unpublished results).

Following its isolation, the petunia EPSPS cDNA was engineered into an *Agrobacterium* Ti plasmid vector system and placed under the regulation of the CaMV 35S promoter. Transformation of petunia leaf disks resulted in the production of callus and plants which were tolerant to glyphosate. Transgenic callus, which overproduced EPSPS 40 to 80-fold, survived on normally lethal levels of glyphosate (0.5 and 1.0 m*M* glyphosate). Petunia plants were sprayed with 0.8 lb/acre of Roundup, and the transgenic plants survived while control plants were killed (Shah *et al.*, 1986). However, the growing tips and newly emerged leaves of the transgenic plants were chlorotic and exhibited growth retardation. In similar experiments, the *Arabidopsis* EPSPS coding sequence, including introns, was fused to the CaMV 35S promoter and reintroduced into *Arabidopsis* plants. These plants were also Roundup tolerant based on seedling germination assays (Klee *et al.*, 1987), but were not completely tolerant at the whole plant level (Klee, personal communication). The above results indicated that overproduction of EPSPS was sufficient to allow tolerance to Roundup, but that the extent of overproduction was not sufficient to confer complete tolerance in the tissues in which glyphosate accumulated.

9.2.6.2 Glyphosate-Tolerant EPSPS

The alternative approach of introducing a gene encoding a glyphosate-tolerant EPSPS should significantly reduce the level of enzyme overexpression required for whole plant Roundup tolerance. Not all EPSP synthases are equally sensitive to glyphosate. EPSPS enzymes from different sources vary in their glyphosate inhibition (Schulz *et al.*, 1985). EPSP synthases from *Klebsiella pneumoniae* and several strains of pseudomonads were shown to be glyphosate tolerant. However, the highly glyphosate-tolerant EPSPS mutants displayed increases in the K_m for PEP (Sost *et al.*, 1984; Schulz *et al.*, 1985). This suggests that glyphosate tolerance is mediated through a site on the enzyme which overlaps with the PEP binding site. Indeed, glyphosate is a competitive inhibitor of EPSPS versus PEP.

There are several instances in which glyphosate-tolerant EPSPS mutants have been identified. Glyphosate-tolerant mutants of *Salmonella typhimurium* were isolated, and one of the mutations mapped in the *aroA* (EPSPS) locus (Comai *et al.*, 1983). This mutant enzyme demonstrated glyphosate-tolerant EPSPS activity and 2-fold increase in crude extract specific activity. When this *aroA* gene was cloned into *E. coli*, the transformed *E. coli* were able to grow in an inhibitor concentration of glyphosate (2000 μg/ml). The glyphosate-tolerant *S. typhimurium* EPSPS contained a single amino acid change of proline to serine at position 101 of the polypeptide sequence (Stalker *et al.*, 1985), and the mutant enzyme had a glyphosate I_{50} of 1 mM compated to 10 μM for the wild-type enzyme.

Using *Agrobacterium* vectors, transgenic tobacco plants were produced which contained the mutant *aroA* gene (without a chloroplast transit peptide) driven by either of two constitutive promoters, octopine synthase (octopine dehydrogenase) and mannopine synthase (mannopine dehydrogenase) (Comai *et al.*, 1985). These plants were sprayed with 0.5 kg/ha (1 kg/ha = 1.12 lb/acre) of Roundup and were partially protected from glyphosate. After 20 days, the most tolerant tobacco plant grew to 70% of the weight of the unsprayed control plants while the sprayed control plant weighed only 10% of the unsprayed control. Transgenic tomato plants containing the same mutant EPSPS gene also demonstrated a similar level of Roundup tolerance (Fillatti *et al.*, 1987).

Improved Roundup tolerance in transgenic plants has been obtained by using other mutant EPSPS genes with chloroplast transit peptide sequences. A highly glyphosate-insensitive mutant EPSPS was isolated from *E. coli* B after selection for growth in the presence of glyphosate (Kishore *et al.*, 1986). The *aroA* gene which encodes this mutant enzyme (SM-1) was cloned and contained a single amino acid substitution. The mutant enzyme had an 8000-fold decreased sensitivity to glyphosate based on kinetic studies (K_i for glyphosate is 4 mM compared to 0.5 μM for the wild-type *E. coli* enzyme). The K_m for S3P was approximately 4-fold greater than wild-type, while the K_m for PEP increased approximately 13-fold over wild-type. This increase in K_m for PEP was expected since glyphosate is a competitive inhibitor with respect to PEP. The K_i/K_m ratio is much greater than the wild-type enzyme, indicating that glyphosate binding was affected to a greater extent than was PEP binding.

Transgenic tobacco plants have been produced which express glyphosate-tolerant bacterial and plant mutant EPSPS genes. The expression of these genes was regulated by the CaMV 35S promoter which causes constitutively high-level expression. *In vitro* studies using the SM-1 mutant fused to the petunia EPSPS transit peptide showed that the hybrid enzyme was imported into isolated chloroplasts with similar efficiency as the petunia pre-EPSPS (della-Cioppa *et al.*, 1987). The SM-1 enzyme without the transit peptide was not imported. Transgenic tobacco calli expressing the SM-1 enzyme without the transit peptide were killed at 0.5 m*M* glyphosate, while those transgenic calli containing the gene for the petunia transit peptide/SM-1 fusion grew normally. Transgenic tobacco plants demonstrated the same differential response to Roundup treatment. The tobacco plants expressing the SM-1 enzyme fused to the transit peptide grew normally after Roundup spray at 0.4 lb/acre, while those expressing the SM-1 enzyme without the transit peptide showed morphological abnormalities and chlorosis (Padgette *et al.*, 1989). Enzyme analysis following subfractionation verified that the highly tolerant plants contained the hybrid enzyme in their chloroplasts, while the weakly tolerant plants contained the enzyme in the cytosol fraction. Chloroplast targeting of a mutant enzyme clearly improved the glyphosate tolerance of the plant.

Soybean and canola have been transformed to contain glyphosate-tolerant plant or bacterial EPSP synthases. Improvements in the enzymatic efficiency of glyphosate-tolerant EPSPS genes has allowed for significant improvements in greenhouse and field tolerance after glyphosate treatment. By using successive gene improvements, the glyphosate tolerance demonstrated by transgenic soybeans under greenhouse conditions was increased from 0.8 kg (acid equivalent, a.e., of glyphosate)/ha in the earliest tested transgenics to 3.3 kg a.e./ha for the latest lines tested (Hinchee *et al.*, 1988; X. Delannay, unpublished results). Field tests in 1989 and 1990 demonstrated that the first transgenic soybeans had significant glyphosate tolerance, but this tolerance was not adequate for commercial use. Transgenic soybean plants, which contain the improved glyphosate tolerance genes (Figure 9.2), attained the glyphosate tolerance levels necessary for adequate commercial weed control (0.6–0.8 kg a.e./ha) in 1991 summer field tests in the United States. Similar improvements in the development of glyphosate-tolerant canola have also been achieved. Increased glyphosate tolerance, due to EPSPS gene improvements, has been demonstrated in successive field tests of glyphosate-tolerant canola (Figure 9.3). In 1990 winter field tests, the best transgenic canola lines showed tolerance to 1 kg/ha a.e. of glyphosate (X. Delannay, unpublished results). This is more than 2 times greater than the rates sufficient to control weeds (0.4–0.5 kg a.e./ha) in canola fields in Canada.

Figure 9.2 Glyphosate tolerance in transgenic soybean field tested in Puerto Rico in the winter of 1990. The left rows show nontransgenic soybeans sprayed with 1 kg a.e./ha of glyphosate. The rows on the right are planted with transgenic soybean which have been sprayed at the same rate and are demonstrating tolerance.

This commercial level performance was confirmed in Canadian field tests in 1991.

9.3 Conclusions and Future Prospects

Based on the number of different herbicide tolerance systems described in this chapter, it is evident that it is possible to introduce herbicide tolerance traits for a diverse range of herbicides into a wide number of crop species. Some critics of herbicide tolerance research feel that the knowledge gained from this research is not widely applicable. On the contrary, the results described in this chapter indicate that herbicide tolerance is, and will be, an extremely useful research tool for plant genetics in a manner analogous to the use of antibiotic tolerance in bacterial genetics. The science of herbicide tolerance research has made a tremendous impact on several areas of plant enzymology, gene expression, and physiology. Further research in this area is expected to not only impact herbicide and herbicide tolerance technology, but also to facilitate the development of new

Figure 9.3 Glyphosate tolerance in transgenic canola field tested in Canada in the summer of 1989. The row on the right shows nontransgenic canola sprayed at 0.56 kg a.e./ha of glyphosate. The center and left rows show two different levels of tolerance to the same rate, demonstrating improvements in the design of the glyphosate tolerance gene.

applications for crop products such as food and fiber production and feed materials for chemical industries.

From a technical perspective, primarily two mechanisms have been chosen for introduction of herbicide tolerance: herbicide target site tolerance and herbicide degradation. Genetically engineered tolerance does not result in tolerance to all herbicides, but only to the single herbicide or class of herbicides against which the mechanism is designed. The fact that target site alteration can result in tolerance to particular herbicides, such as glyphosate, the sulfonylureas, or the imidazolinones, has shown convincingly that these herbicides act as highly specific metabolic inhibitors. This is contrary to a widely held opinion that these herbicides inhibit several plant-specific processes. Interestingly, natural crop selectivity to herbicides also has a similar molecular basis. New mechanisms for achieving herbicide tolerance in the future may be derived from an understanding of other processes involved in natural selectivity, such as restrictive herbicide uptake and translocation.

Research in the area of herbicide tolerance has vastly enhanced our understanding of plant biochemistry, gene expression, and physiology. Studies on herbicide mode of action and the genetic engineering of herbicide tolerance has contributed significantly to our knowledge of amino acid biosynthetic enzymes in plants (Kishore and Shah, 1988). Particular tolerance phenotypes can be correlated to patterns of herbicide tolerance gene expression and protein localization. This correlation can provide information on herbicide translocation and accumulation as well as on concentrations of critical metabolites required in the targeted biochemical pathways of plant organs, cells, and organelles. Coding and regulatory sequences have been isolated and characterized for the amino acid biosynthetic genes EPSPS (Benfey *et al.*, 1990) and ALS (Mazur *et al.*, 1987). Regulation of petal and root specific expression of the EPSPS gene was shown to be relatively complex and to require extensive stretches of sequence which apparently contained several widely separated *cis*-elements. The understanding of how a complex plant promoter, such as EPSPS, is controlled at different stages of development within different tissues should help to further understand the regulation of a large class of genes that are expressed at more than one time in plant development. This type of information on the regulation of plant genes and their products should facilitate the genetic engineering of other traits such as disease resistance and environmental stress tolerance.

Although theoretically it should be possible to introduce tolerance to any herbicide in any crop, commercialization of an herbicide tolerance trait for the purposes of weed control should be governed by a number of stringent criteria. Herbicides used for this application must have desirable features, such as broad-spectrum and high unit activity, rapid soil inactivation, minimum volatility and soil mobility, selective toxicity to plants, and minimum or no toxicity to aquatic, avian, and human/animal species. The potential for weeds acquiring resistance to the herbicide should also be low. If these criteria are judiciously followed, commercial availability of an herbicide for use with an herbicide-tolerant crop can potentially improve weed control systems as well as environmental quality.

Such weed control benefits are critical for the 1990s because several countries desire to reduce the amount of chemicals introduced into the environment. While critics of herbicide tolerance research argue that commercialization of herbicide-tolerant crops will only increase herbicide use, in reality it should be possible with this technology to reduce the extent of herbicide use and, at the same time, promote the use of environmentally sound herbicides. Postemergent herbicides, such as glyphosate, offer other advantages as well. Since glyphosate kills both mature and immature weeds of many diverse species, a single postemergent application of this herbicide could potentially be used to control several weeds normally controlled by preemergent herbicides. The grower can then use the herbicide only when it is needed, and not in anticipation of the need for weed control. This flexibility in application timing, rather than strict adherence to a prescribed herbicide application regime, may also result in lower chemical introduction into the environment.

One of the major goals of the agricultural chemist has been to discover herbicides with crop selectivity. The advent of plant transformation technology has allowed selectivity to be engineered into the crop instead of designed into the herbicide. Herbicide tolerance has the potential to enhance significantly the agronomic utility of broad-spectrum herbicides. It will impact other agronomic practices and potentially result in lower total herbicide use rates, the use of inherently safer herbicides, and provide cost-effective broad-spectrum weed control. In addition, our understanding of plant metabolism, enzymes, and genes is being advanced by the basic research involved in developing herbicide-tolerant plants. This is the type of knowledge which is the basis for the design and discovery of more potent, herbicidally active, and environmentally acceptable organochemicals.

Acknowledgments

We would like to thank C. Gasser, E. Levine, C. Hironaka, J. Elmer, R. Deaton, S. Metz, B. LaVallee, and G. Parker for sharing their unpublished results, as well as our other colleagues at Monsanto for their important contributions to the work discussed in this chapter.

References

Amrhein, N., Deus, B., Gehrke, P., and Steinrucken, H. C. (1980). *Plant Physiol.* 66, 830–834.

Anderson, P. C., and Georgeson, M. (1989). *Genome* 31, 994–999.

Arntzen, C. J., Pfister, K., and Steinback, K. E. (1982). In *Herbicide Resistance in Plants* (Ed. by H. M. LeBaron and J. Gresse), pp. 185–214. New York: Wiley.

Ayotte, R., Harney, P. M., and Sousa Machado, V. (1989). *Euphytica* 40, 15–19.

Baird, D. D., Upchurch, R. P., Homesley, W. B., and Franz, J. E. (1971). *Proc. North Cent. Weed Control Conf.* 26, 64–68.

Barrantine, W. L., Edwards, C. J., Jr., and Hartwid, E. E. (1975). *Abstr. Weed Sci. Soc. Amer.*, p. 8.

Bascomb, N. F., Gutteridge, S., Smith, J. K., and Leto, K. J. (1987). *J. Cell Biochem.* 11B, 88.

Bayer, E., Guge, K. H., Hagele, K., Hogenmajer, H., Jessipow, S., Konig, W. A., and Zaher, H. (1972). *Helv. Chim. Acta* 55, 224–239.

Benfey, P. N., Takatsuji, H., Ren, L., Shah, D., and Chua, N-H. (1990). *Plant Cell* 2, 849–856.

Bettini, P., McNally, S., Sevignac, M., Darmency, H., Gasquez, J., and Dron, M. (1987). *Plant Physiol.* 84, 1442–1446.

Beversdorf, W. D., Weiss-Lerman, J., Erickson, L. R., and Souza-Machado, V. (1989). *Can. J. Genet. Cytol.* 22, 167–172.

Bickel, H., Palme, L., and Schultz, G. (1978). *Phytochemistry* 17, 199–124.

Bogorad, L. (1986). (oral presentation), *Tailoring Genes for Crop Improvement. University of California at Davis, Aug. 24–28, 1986.*

Chaleff, R. S., and Mauvais, C. J. (1984). *Science* 224, 1443–1445.

Chaleff, R. S., and Ray, T. B. (1984). *Science* 223, 1148–1151.

Charest, P. J., Hattori, J., DeMoor, J., Iyer, V. N., and Miki, B. (1990). *Plant Cell Rep.* 8, 643–646.
Cheung, A. Y., Bogorad, L., Van Montagu, M., and Schell, J. (1988). *Proc. Natl. Acad. Sci. U.S.A.* 85, 391–395.
Comai, L., Sen, L., and Stalker, D. M. (1983). *Science* 221, 370–371.
Comai, L., Facciotti, D., Hiatt, W. R., Thompson, G., Rose, R. E., and Stalker, D. M. (1985). *Nature (London)* 317, 741–744.
Creason, G. L., and Chaleff, R. S. (1988). *Theor. Appl. Genet.* 76, 177–182.
d'Amato, T. A., Ganson, R. J., Gaines, C. G., and Jensen, R. A. (1984). *Planta* 162, 104–108.
DeBlock, M. (1990). *Plant Physiol.* 93, 1110–1116.
DeBlock, M., De Brower, D., and Tenning, P. (1989). *Plant Physiol.* 91, 694–701.
DeBlock, M., Botterman, J., Vandewiele, M., Dockx, J., Thoen, C., Gossele, V., Movva, N. R., Thompson, C., Van Monatagu, M., and Leemans, J. (1987). *EMBO J.* 6, 1242–1244.
DeGreef, W., Delon, R., DeBlock, M., Leemans, J., and Botterman, J. (1989). *Bio/Technology* 7, 61–64.
DeKeyser, R., Claes, B., Marichal, M., Van Monatagu, M., and Caplan, A. (1989). *Plant Physiol.* 90, 217–223.
della-Cioppa, G., Bauer, S. C., Klein, B. K., Shah, D. M., Fraley, R. T., and Kishore, G. M. (1986a). *Proc. Natl. Acad. Sci. U.S.A.* 83, 6873–6877.
della-Cioppa, G., Hauptman, R. M., Fraley, R. T., and Kishore, G. M. (1986b). *Curr. Top. Plant Biochem. Physiol.* 5, 194.
della-Cioppa, G., Bauer, S. C., Taylor, M. L., Rochester, D. E., Klein, B. K., Shah, M., Fraley, R. T., and Kishore, G. M. (1987). *Bio/Technology* 5, 579–584.
d'Halluin, K., Botterman, J., and DeGreef, W. (1990). *Crop Science* 30, 866–871.
Don, R. H., Weightman, A. J., Knackmuss, H-J., and Timmis, K. N. (1985). *J. Bact.* 161, 85–90.
Donn, G., Tischer, E., Smith, J. A., and Goodman, H. M. (1984). *J. Mol. Appl. Genet.* 2, 621–635.
Donn, G., Nilges, M., and Morocz, S. (1990). In *Abstracts VIIth International Congress on Plant Tissue and Cell Culture,* pg. 53, Amsterdam, the Netherlands.
Dyer, W. E., Weller, S. C., Bressan, R. A., and Herrmann, K. M. (1988). *Plant Physiol.* 88, 661–666.
Eckes, P., Schmitt, P., Daub, W., and Wengenmayer, F. (1989). *Mol. Gen. Genet.* 217, 263–268.
Fillatti, J. J., Kiser, J., Rose, R., and Comai, L. (1987). *Bio/Technology* 5, 726–730.
Fincher, R. R. (1989). *BCPC Monogr.* 42, 69–73.
Franz, J. E. (1985). In *The Herbicide Glyphosate* (Ed. by E. Grossman and D. Atkinson), pp. 3–17. London: Butterworths.
Friend, J., and Moreland, D. E. (1969). *Nature (London)* 214, 942–943.
Fromm, M. E., Morrish, F., Armstrong, C., Williams, R., Thomas, J., and Klein, T. M. (1990). *Bio/Technology* 8, 833–839.
Gabard, J. M., Charest, P. J., Iyer, V. N., and Miki, B. L. (1989). *Plant Physiol* 91, 574–580.
Ganson, R. J., d'Amato, T. A., and Jensen, R. A. (1986). *Plant Physiol.* 82, 203–210.
Gasser, C. S., Winter, J. A., Hironaka, C. M., and Shah, D. M. (1988). *J. Biol. Chem.* 263, 4280–4289.
Golubinoff, P., Edleman, M., and Hallick, R. B. (1985). *Nucleic Acids Res.* 12, 9489–9496.
Gordon-Kamm, W. J., Spencer, T. M., Mangano, M. L., Adams, T. R., Daines, R. J., Start, W. G., O'Brien, J. V., Chambers, S. A., Adams, W. R., Jr., Willetts, N. G., Rice, T. B., Mackey, C. J., Krueger, R. W., Kausch, A. P., and Lemaux, P. G. (1990). *Plant Cell* 2, 603–618.
Harms, C. T., Montoya, A. L., Privalle, L. S., and Briggs, R. W. (1990). *Theor. Appl. Genet.* 80, 353–358.
Hartnett, M. E., Chuk, C-F., Mauvais, C. J., McDevitt, R. E., Knowlton, S., Smith, J. K., Falco, S. C., and Mazur, B. J. (1990). *J. Cell Biochem. Suppl.* 0 (14 Part E), 301.
Haughn, G. W., and Somerville, C. (1986). *Mol. Gen. Genet.* 204, 430–434.

Haughn, G. W., and Somerville, C. (1990). *Plant Physiol.* 92, 1081–1085.

Haughn, G. W., Smith, J., Mazur, B., and Somerville, C. (1988). *Mol. Gen. Genet.* 211, 266–271.

Heim, D. R., Roberts, J. L., Pike, P. D., and Larrinua, I. M. (1988). *Plant Physiol.* 90, 146–150.

Heim, D. R., Roberts, J. L., Pike, P. D., and Larrinua, I. M. (1989). *Plant Physiol.* 92, 858–861.

Hinchee, M. A. W., Connor-Ward, D. V., Newell, C. A., McDonnell, R. E., Sato, S. J., Gasser, C. S., Fischhoff, D. A., Re, D. B., Fraley, R. T., and Horsch, R. B. (1988). *Bio/Technology* 6, 915–922.

Hirschberg, J., and MacIntosh, L. (1983). *Science* 222, 1346–1349.

Hollander, H., and Amrhein, N. (1980). *Plant Physiol.* 66, 823–829.

Jourdan, P. S., Earle, E. D., and Mutschler, M. A. (1989). *Theor. Appl. Genet.* 78, 271–279.

Kishore, G. M., and Shah, D. M. (1988). *Ann. Rev. Biochem.* 57, 627–663.

Kishore, G. M., Brundage, L., Kolk, K., Padgette, S. R., Rochester, D., Huynh, Q. K., and della-Cioppa, G. (1986). *Fed. Proc. Fed. Am. Soc. Exp. Biol.* 45, 1506.

Klee, H. J., Muskopf, Y. M., and Gasser, C. S. (1987). *Mol. Gen. Genet.* 210, 437–442.

Kondo, Y., Somura, T., Ogawa, Y., Tsuruoka, T., Watanabe, A., Totukawa, K., Suzuki, T., Moriyama, C., Yoshida, J., Inouye, S., and Niida, T. (1973). *Sci. Rep. Meij. Seika* 13, 34–41.

Lee, K. Y., Townsend, J., Tepperman, J., Black, M., Chui, C. F., Mazure, B., Dunsmuir, P., and Bedbrook, J. (1988). *EMBO J.* 7, 1241–1248.

Levitt, G. (1983). In *Pesticide Chemistry. Human Welfare and the Environment. Vol. 1 Synthesis and Structure-Activity Relationships* (Ed. by J. Miyamoto and P. C. Kearney), pp. 243–250. Oxford: Pergamon Press.

Lyon, B. R., Llewellyn, D. J., Huppatz, J. L., Dennis, E. S., and Peacock, W. J. (1989). *Plant Mol. Biol.* 13, 533–540.

Maltais, B., and Bouchard, C. J. (1987). *Phytoprotection* 59, 117–119.

Mazur, B. J., and Falco, S. C. (1989). *Ann. Rev. Plant Physiol. Plant Mol. Biol.* 40, 441–470.

Mazur, B. J., Chui, C-F., and Smith, J. K. (1987). *Plant Physiol.* 85, 1110–1117.

McBride, K. E., Kenny, J. W., and Stalker, D. M. (1986). *Appl. Env. Microbiol.* 52, 325–330.

McNally, S. F., Hirel, B., Gadal, P., Mann, A. F., and Stewart, G. R. (1983). *Plant Physiol.* 72, 22–25.

Mousdale, D. M., and Coggins, J. R. (1985). *Planta* 163, 241–249.

Oxtoby, E., and Hughes, M. A. (1990). *Trends Biotech.* 8, 61–65.

Padgette, S. R., and Kishore, G. M. (1987). *Fed. Proc., Fed. Am. Soc. Exp. Biol.* 45, 1612.

Padgette, S. R., della-Cioppa, G., Shah, D. M., Fraley, R. T., and Kishore, G. M. (1989). In *Cell Culture and Somatic Cell Genetics* (Ed. by J. Schell and I. K. Vasil), pp. 441–475. New York: Academic Press, Inc.

Quinn, J. P. (1990). *Biotech. Adv.* 8, 321–333.

Rogers, S. G., Brand, L. A., Holder, S. B., Sharps, E. S., and Brackin, M. J. (1983). *Appl. Environ. Microbiol.* 46, 37–43.

Rothe, G. M., Hangst, G., Mildenberger, I., Scharer, H., and Utesch, D. (1983). *Planta* 157, 358–366.

Schultz, G., Bickel, H., Buchholz, B., and Soll, J. (1981). In *Photosynthesis* (Ed. by G. Akoyunoglou), pp. 311–318. Philadelphia, Pennsylvania: Balaban Int. Sci. Serv.

Schulz, A., Kruper, A., and Anrhein, N. (1985). *FEMS Microbiol. Lett.* 28, 297–301.

Sebastian, S. A., Fader, G. M., Ulrich, J. F., Forney, D. R., and Chaleff, R. S. (1989). *Crop Sci.* 29, 1403–1408.

Shah, D. M., Horsch, R. B., Klee, H. J., Kishore, G. M., Winter, J. A., Tumer, N. E., Hironaka, C. M., Sanders, P. R., Gasser, C. S., Aykent, S., Siegel, N. R., Rogers, S. G., and Fraley, R. T. (1986). *Science* 233, 478–481.

Shaner, D. L., and Anderson, P. C. (1985). In *Biotechnology in Plant Science* (Ed. by M. Zaitlin, P. Day, and A. Hollaender), pp. 287–299. Orlando, Florida: Academic Press, Inc.

Shaner, D. L., Anderson, P. C., and Stidham, M. A. (1984). *Plant Physiol.* 76, 545–546.
Smith, J. K., Mauvais, C. J., Knowlton, S., and Mazur, B. J. (1988). In *Proc. ACS Symp. Biotechnol Crop Protect.,* pp. 25–36, Washington, D.C.
Sost, D., Schulz, A., and Amrhein, N. (1984). *FEBS Lett.* 173, 238–242.
Spencer, T. M., Gordon-Kamm, W. J., Daines, R. J., Start, W. G., and Lemaux, P. G. (1990). *Theor. Appl. Genet.* 79, 625–631.
Stalker, D. M., and McBride, K. E. (1987). *J. Bacteriol.* 169, 955–960.
Stalker, D. M., Hiatt, W. R., and Comai, L. (1985). *J. Biol. Chem.* 260, 4724–4728.
Stalker, C. M., McBride, K. E., and Malyj, L. D. (1988). *Science* 242, 419–423.
Steinback, K. E., McIntosh, L., Bogorad, L., and Arntzen, C. J. (1981). *Proc. Natl. Acad. Sci. U.S.A.* 78, 1207–1212.
Steinrucken, H. C., and Amrhein, N. (1980). *Biochem. Biophys. Res. Commun.* 94, 1207–1212.
Steinrucken, H. C., Schulz, A., Amrhein, N., Porter, C. A., and Fraley, R. T. (1986). *Arch. Biochem. Biophys.* 244, 169–178.
Streber, W. G., and Willmitzer, L. (1989). *Bio/Technology* 7, 811–816.
Streber, W. S., Timmis, K. N., and Zenk, M. H. (1987). *J. Bacteriol.* 169, 2950–2955.
Swanson, E. B., Herrgesell, M. J., Arnoldo, M., Sippell, D. W., and Wong, R. S. C. (1989). *Theor. Appl. Genet.* 78, 525–530.
Thompson, C. J., Movva, N. R., Tizard, R., Crameri, R., Davies, J. E., Lauwereys, M., and Botterman, J. (1987). *EMBO J.* 6, 2519–2523.
Tischer, E., DasSarma, S., and Goodman, H. M. (1986). *Mol. Gen. Genet.* 203, 221–229.
Wiersma, P. A., Schmiemann, M. G., Condie, J. A., Crosby, W. L., and Moloney, M. M. (1989). *Mol. Gen. Genet.* 219, 413–420.
Yadav, N., McDevitt, R. E., Benard, S., and Falco, S. C. (1986). *Proc. Natl. Acad. Sci. U.S.A.* 83, 4418–4422.
Yue, S-X., Liu, B-L., Mao, D-Z., Hu, N-B., Fu, J-H., Li, L-C., and Zuh, L-H. (1990). *Acta Bot. Sin.* 32, 343–349.

10

Production of Transgenic Plants with Enhanced Resistance to Microbial Pathogens

Richard Broglie, Karen Broglie, Dominique Roby,[1] and Ilan Chet[2]

Agricultural Products Department
E. I. Du Pont de Nemours and Company
Wilmington, Delaware 19880

Plants have evolved a variety of defense mechanisms to protect themselves from invading microorganisms such as bacteria, viruses, and fungi. One example is the hypersensitive response in which a localized region of necrotic cells is produced in the vicinity of the primary infection site in order to limit the spread of the invading pathogen. During activation of the defense response a number of biochemical and physical changes take place in the host plant (see Figure 10.1). These include: (1) cell wall modifications, such as the accumulation of hydroxyproline-rich glycoproteins (Esquerre-Tugaye *et al.*, 1979; Showalter *et al.*, 1985), lignification (Vance *et al.*, 1980; Espelie *et al.*, 1986), or the deposition of callose (Ride, 1983); (2) the induction of enzymes involved in the biosynthesis of phytoalexins, secondary metabolites that are toxic to bacteria and fungi (Hahlbrock and Grisebach, 1979; Darvill and Albersheim, 1984); (3) the synthesis of hydrolytic enzymes, such as chitinase (Boller, 1987) and β-1,3-glucanase (Ferraris *et al.*, 1987; Mauch *et al.*, 1988; Meins and Ahl, 1989), which are capable of degrading the cell walls of invading pathogens; and (4) the synthesis of

[1]Present address: Laboratoire de Biologie Moleculaire des Relations Plantes-Microorganismes, CNRS/INRA, Chemin de Borderouge. BP 27, 31326 Castanet Tolosan Cedex, France
[2]Present address: Department of Plant Pathology and Microbiology, Faculty of Agriculture, The Hebrew University, Rehovot, 76100 Israel

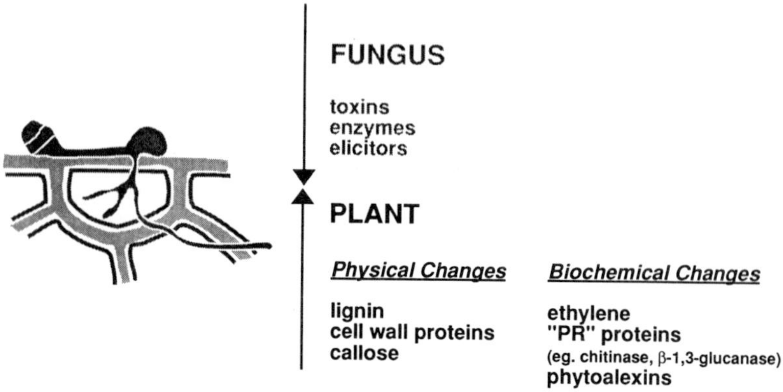

Figure 10.1 Schematic diagram of the physical and biochemical changes that occur in plants during pathogen attack. See text for details.

inhibitors of various proteinases (Ryan, 1978; Esquerre-Tugaye *et al.*, 1985). Increasing evidence suggests that the success of the plant in warding off a pathogen invasion is dependent on the coordination between the different defense strategies and the rapidity of the overall response (Bell *et al.*, 1986; Bonhoff *et al.*, 1987; Joosten and De Wit, 1989; Benhamou *et al.*, 1989; Benhamou *et al.*, 1990). In this chapter, we will concentrate on the role of chitinase in the defense response of plants, with special emphasis on experiments designed to modify chitinase gene expression in order to produce transgenic plants with enhanced resistance to fungal attack.

10.1 Chitinases and Plant Defense

Chitinase is a lytic enzyme found in most higher plants that catalyzes the hydrolysis of chitin, a β-1,4-linked homopolymer of *N*-acetyl-D-glucosamine. Endo-type chitinase activities have been observed in many plant species including bean, pea, soybean, tomato, sunflower, melon, cotton, corn, wheat, barley, and tobacco (Molano *et al.*, 1979; Boller *et al.*, 1983; Shinshi *et al.*, 1987; Leah *et al.*, 1987; Nasser *et al.*, 1988). In healthy, uninfected plants, chitinase levels are usually low or undetectable. However, in many cases, the chitinase levels are increased on treatment with the phytohormone ethylene (Boller *et al.*, 1983) or with oligosaccharide elicitors (Kurosaki *et al.*, 1986; Roby *et al.*, 1986; Herget *et al.*, 1990) or on infection by fungal pathogens (Pegg and Vessey, 1973; Roby and Esquerre-Tugaye, 1987; Joosten and Dewit, 1989). In some tissues, such as cereal aleurone or endosperm tissues or in lactifers, chitinases are found at high, constitutive levels (Molano *et al.*, 1979; Leah *et al.*, 1987; Swegle *et al.*, 1989).

While no endogenous substrate for this enzyme is known to exist in plants, chitin is a ubiquitous component of the cell walls of most fungi (Wessels and Sietsma, 1981). For this reason, it has been suggested that chitinase functions as a defense against chitin-containing pathogens (Abeles *et al.*, 1970). Indeed, several of the pathogenesis-related proteins known to be induced in dicots following pathogen infection have been identified as chitinases and β-1,3-glucanases (Legrand *et al.*, 1987; Kombrink *et al.*, 1988; Awade *et al.*, 1989; Joosten and De Wit, 1989).

Chitinases have been purified from several plant species (Molano *et al.*, 1979; Boller *et al.*, 1983; Leah *et al.*, 1987; Legrand *et al.*, 1987; Metraux *et al.*, 1988; Kragh *et al.*, 1990); the majority of the enzymes isolated to date have been found to correspond to basic proteins of approximately 30 kDa, although acidic isoforms have also been identified (Awade *et al.*, 1989; Payne *et al.*, 1990; Majeau *et al.*, 1990; Samac *et al.*, 1990). Subcellular localization studies indicate that the basic isoforms are found in the vacuole (Boller and Vogeli, 1984; Mauch and Staehelin, 1989), while the acidic forms accumulate in the intercellular spaces (Legrand *et al.*, 1987; Boller and Metraux, 1988; Metraux *et al.*, 1988; Payne *et al.*, 1990). The purified enzymes from tomato and bean have been shown to hydrolyze isolated fungal cell walls (Young and Pegg, 1982; Boller *et al.*, 1983; Schlumbaum *et al.*, 1986). Additionally, purified pea chitinase, alone and in combination with β-1,3-glucanase, has been shown to inhibit the growth of certain phytopathogenic fungi in culture. (Mauch *et al.*, 1988).

In bean, chitinase is encoded by a small multigene family consisting of at least three members (Broglie *et al.*, 1986). Using cloned cDNAs as hybridization probes, the level of chitinase mRNA in bean seedlings was found to increase 50 to 75-fold following ethylene or elicitor treatment (Broglie *et al.*, 1986; Roby *et al.*, unpublished data). Nucleotide sequence analysis of isolated cDNA clones indicated that at least two of the chitinase genes are transcribed in response to ethylene treatment. The expression of these genes, designated CH5A and CH5B, in transgenic tobacco plants was found to be dependent on the application of exogenous ethylene or oligosaccharide elicitors (Broglie *et al.*, 1989; Roby *et al.*, unpublished data). DNA sequences that are responsible for mediating this response have been shown to be situated upstream of the transcription initiation site in the CH5B gene. Analysis of CH5B promoter deletions suggests that a 43 base pair (bp) sequence that is precisely conserved between the CH5A and CH5B genes may play a role in transcriptional regulation of chitinase gene expression by the phytohormone ethylene (Broglie *et al.*, 1989). Consistent with this finding, we have identified a nuclear protein factor by gel mobility shift and DNase I protection assays that binds to this DNA sequence element (Broglie *et al.*, in preparation).

In addition to analyzing the effects of ethylene and oligosaccharride elicitors on chitinase gene expression, we have also studied the induction of the CH5B gene in response to attack by several pathogenic fungi. In these experiments, the temporal and spatial pattern of CH5B expression was assessed during fungal infection by using β-glucuronidase as a reporter gene. A chimeric gene was constructed by fusing a 1.7-kb fragment bearing CH5B upstream sequences to the coding region of β-glucuronidase followed by the termination signals of the nopaline synthase gene (Broglie *et al.*, 1989). As shown in Figure 10.2, when transgenic tobacco plants containing this chimeric gene were infected with the

Figure 10.2 Histochemical staining of β-glucuronidase activity in transgenic tobacco plants infected with *Botrytis cinerea*. Leaves of transgenic tobacco plants containing a chimeric gene consisting of the chitinase 5B gene promoter fused to the reporter gene β-glucuronidase were inoculated by placing droplets of a conidial suspension (10^6 conidia /ml) on the surface of the leaf (*left side*). Plants were incubated in a dew chamber for 24 hours and removed to a growth chamber. Leaves were stained for β-glucuronidase activity 48 hours postinfection as described by Jefferson *et al.* (1987).

foliar pathogen *Botrytis cinerea,* a high-level, localized induction of the chitinase promoter was observed. Similar results were obtained following infection by the soil-borne pathogens *Rhizoctonia solani* and *Sclerotium rolfsii* (Roby *et al.*, 1990). The pattern of the CH5B promoter induction was found to parallel the activation of endogenous tobacco chitinase enzyme activity. These experiments demonstrate that CH5B gene expression is strongly induced in response to fungal attack. Moreover, activation of the bean chitinase promoter was found to occur preferentially at the site of fungal infection.

10.2 Modification of Chitinase Gene Expression

To evaluate critically the role of chitinase induction in the plant defense response and to determine whether modification of chitinase gene expression would be a useful strategy for the production of fungal resistant plants, a novel chitinase gene was constructed by replacing the native, inducible promoter with the high-level, constitutive 35S promoter of cauliflower mosaic virus (Figure 10.3). Transgenic plants containing this chimeric gene were first assayed for expression of the bean chitinase polypeptide by Western blot analysis. The results of this

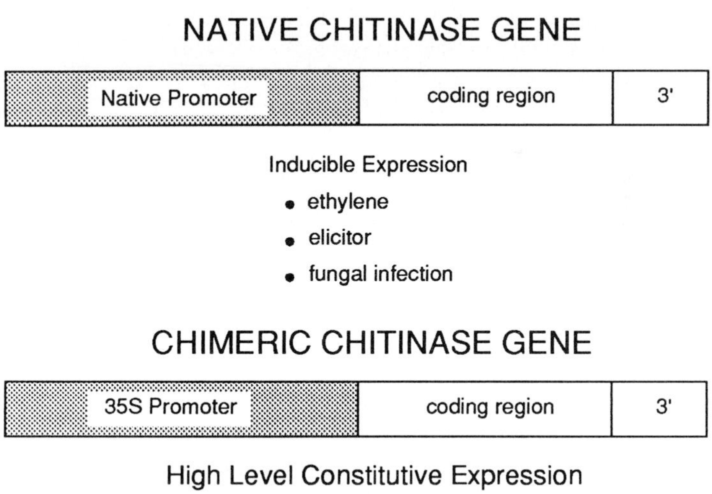

Figure 10.3 Strategy for producing high-level, constitutive expression of chitinase in transgenic tobacco. The native promoter of the bean CH5B chitinase gene, which allows inducible expression in response to pathogen attack, was replaced by the 35S promoter of cauliflower mosaic virus. This chimeric gene was introduced into tobacco plants by *Agrobacterium*-based transformation using Ti plasmid-based vectors.

experiment showed that high levels of the bean protein are produced in healthy, unstressed plants (Broglie *et al.*, 1991). The size of the polypeptide detected in the transgenic plants was identical to that of the mature form of bean chitinase, suggesting that the precursor polypeptide was correctly processed in the heterologous system and localized in the correct subcellular compartment. Analysis of several independent transformants revealed an approximately 2- to 4-fold increase in the level of chitinase enzyme activity in roots, a 10- to 20-fold increase in stems, and a 20- to 40-fold increase in leaves of these plants.

Previous studies have shown that the growth of certain phytopathogenic fungi can be inhibited by the addition of chitinase to the culture media (Mauch *et al.*, 1988). Our studies indicate that the growth of the soil-borne fungus *Rhizoctonia solani* is particularly sensitive to inhibition by bean chitinase and that this inhibition presumably results from the enzyme-catalyzed disruption of fungal hyphal tips (Broglie *et al.*, 1991; Benhamou *et al.*, submitted for publication). We were therefore interested in determining whether transgenic tobacco plants exhibiting high-level constitutive chitinase expression would be more resistant to attack by this fungus than are control tobacco plants. To test this, young tobacco seedlings were transplanted into soil that had been artificially infested with a level of *R. solani* empirically determined to yield approximately 50% disease incidence on control or wild-type plants. All of the transgenic tobacco plants that contained the modified chitinase gene showed increased survival relative to control plants when grown for extended periods in *R. solani*-infested soil. In preliminary experiments, progeny of several independently isolated transformants displayed an average 80% survival after 16 days growth in inoculated soil. In contrast, control plants showed only 40% survival under the same conditions (Broglie *et al.*, 1991). The rate of survival of the transgenic plants was found to be dependent upon the concentration of fungal inoculum applied to the soil.

In addition to a difference in mortality, control and 35S-chitinase tobacco plants also exhibited differences in morphology following growth in fungal-infested soil. Figure 10.4 shows a comparison between surviving control plants (#548) which lack the 35S-chitinase gene and a representative 35S-chitinase-containing transformant (#373) fourteen days after transplantation into *R. solani*-infested soil. Control plants were visibly stunted and showed poor root development. In contrast, all of the 35S-chitinase transformants tested were hardier and displayed near-normal root growth and morphology. In order to quantitatively compare control and 35S-chitinase-containing plants, 12 to 14-day-old seedlings were transplanted into soil containing increasing amounts of *R. solani* inoculum. The levels of inoculum chosen were such that death of the plants did not occur; instead the course of the infection was followed by measuring the effect of the fungus on the root fresh weight of the plants. At the highest level of fungal inoculum tested (4 ml/pint of soil), control plants suffered an average 50% loss in root fresh weight after growth in infected soil (Table 10.1).

Figure 10.4 Effect of *R. solani* infection on plant growth and development in transgenic tobacco plants. Control (*upper row*) and 35S-chitinase containing seedlings (*lower row*) were transplanted into soil infested with *R. solani*. At 15 days postinfection, the plants were gently uprooted and photographed.

TABLE 10.1
Effect of Increased *Rhizoctonia solani* Concentration on Root Fresh Weight of Transgenic Tobacco Plants

	Root fresh weight (%)[a]		
	Inoculum level (ml/pint soil)		
Plant number	0	1	4
548 (control)	100	73.8	54.2
230	100	106	84.7
238	100	96.9	97.7
329	100	108	95.2
373	100	96	86.4

[a]Values represent a comparison between the root fresh weights of inoculated plants and uninoculated plants. Data points represent the mean root fresh weight values of 10 plants.

Under the same conditions, five independently isolated transgenic plants containing the 35S-chitinase gene showed an average 10% reduction in root fresh weight.

The previous results indicate that transgenic tobacco plants with high, constitutive levels of bean endochitinase exhibit increased resistance to infection by the soil-borne fungus *Rhizoctonia solani*. Diseases attributable to *Rhizoctonia,* such as root and stem rot, seedling blight, and damping off, have been reported for a number of important crop species including oilseed rape and soybean. When environmental conditions are optimal for fungal growth, disease severity can be directly related to inoculum potential (Sinclair, 1982). This value represents an estimate of the potential for infection at the host surface and is a function of the quantity and the quality of fungal hyphae rather than a measure of the total fungal biomass in the soil (Cook and Weller, 1987). To determine what effect, if any, 35S-chitinase plants have on the inoculum potential, control and transgenic plants harboring the modified chitinase gene were first grown in *Rhizoctonia solani*-infested sand. After thirteen days, the plants were uprooted and new seedlings were sown in the same soil. The newly planted seedlings were allowed to grow for an additional nine days. After the first cycle, only 40% of the control seedlings survived. In contrast, disease development in the 35S-chitinase plants was delayed when compared to control plants, resulting in the survival of 70–80% of the seedlings (Figure 10.5, Cycle 1). When new seedlings were replanted in soil originally containing control plants, symptoms of *Rhizoctonia* infection appeared immediately with only 60% of the plants surviving after two days (Figure 10.5, Cycle 2). However, when seedlings were replanted in soil originally containing the 35S-chitinase transformants, disease development remained slower with greater than 85% of the plants surviving after two days. The results of this preliminary study suggest that growth of transgenic plants with high, constitutive levels of chitinase may be beneficial in reducing the indigenous levels of *Rhizoctonia* in the soil by interfering with their ability to produce successful infections.

10.3 Future Prospects

The production of transgenic plants that are resistant to attack by fungal and bacterial pathogens has lagged behind the development of virus- and insect-resistant varieties (see, for example, this volume, Chapters 11–14). This is due, in part, to the complexity and diversity of the interactions that exist among the many microbial pathogens and their hosts and to the lack of understanding of the molecular basis of disease resistance in plants. In the experiments summarized here, we have shown that transgenic tobacco plants that contain high, constitutive levels of bean endochitinase are more resistant to attack by the soil pathogen *Rhizoctonia solani* than is wild-type tobacco. The resistance of these

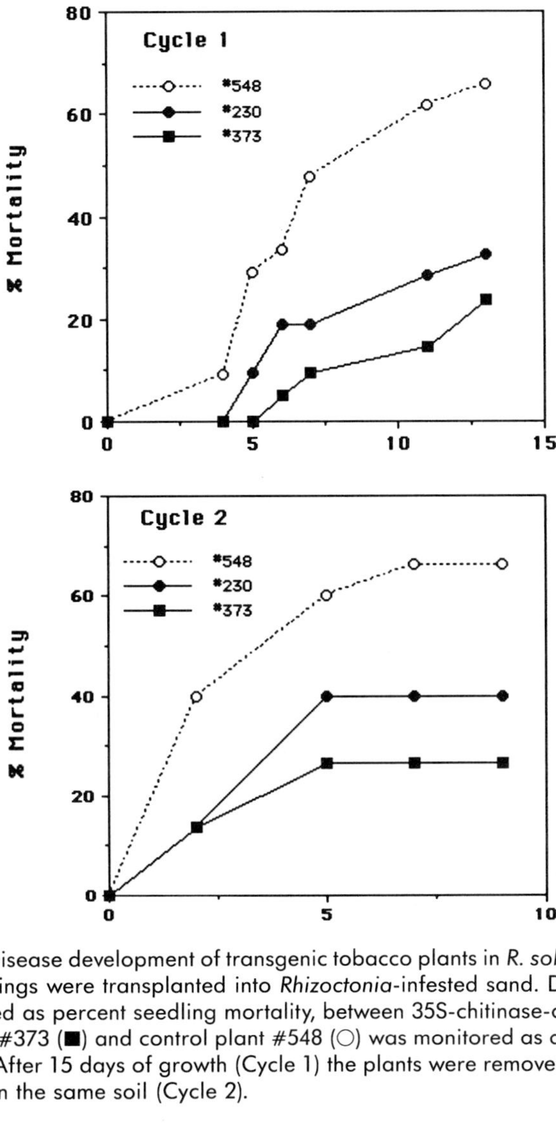

Figure 10.5 Disease development of transgenic tobacco plants in *R. solani*-infested soil. Tobacco seedlings were transplanted into *Rhizoctonia*-infested sand. Disease development, expressed as percent seedling mortality, between 35S-chitinase-containing plants #230 (●) and #373 (■) and control plant #548 (○) was monitored as a function of time postinfection. After 15 days of growth (Cycle 1) the plants were removed and new seedlings planted in the same soil (Cycle 2).

plants to infection was found to vary with the level of fungal inocoulum and is indicative of a quantitative type of resistance. Both the delay in the appearance of symptoms as well as the lower incidence of disease seen on these plants may have important practical implications. In field situations, a major consequence of *Rhizoctonia* infection is seedling damping-off disease which results in a reduction in stand establishment. The presence of high, constitutive levels of chitinase in transgenic plants may enable young seedlings to survive the critical period during stand establishment in the field when they are most susceptible to attack by soil-borne pathogens. These plants may also have a competitive advantage when adverse environmental conditions exist.

In vitro studies of the effects of purified bean chitinase on the growth of *R. solani* mycelium suggest that the fungal-resistant phenotype displayed by the transgenic plants may result from the inhibition of fungal growth due to the chitinase-catalyzed disruption of hyphal tips. A similar mode of action for chitinase has been suggested from studies employing purified enzymes to hydrolyze isolated fungal cell walls and to inhibit the growth of certain pathogenic fungi in culture (Young and Pegg, 1982; Boller *et al.*, 1983; Schlumbaum *et al.*, 1986; Mauch *et al.*, 1988; Roberts and Selitrennikoff, 1988). To date, data concerning the role of chitinase in the defense response of the plant have been obtained *ex planta*. In the studies summarized here we have demonstrated that chitinase plays an important role in protecting plants against potentially pathogenic organisms. Furthermore, we have shown that the manipulation of natural host defense mechanisms may be used to produce fungal-resistant plants. At present, it is not known whether chitinase alone is sufficient to provide protection against a wide range of fungal pathogens. Recently, Neuhaus *et al.*, (1991) introduced a gene for class I (basic) tobacco chitinase regulated by the cauliflower mosaic virus 35S promoter into *Nicotiana sylvestris*. Although chitinase levels were increased 10-fold on infection of these transgenic tobacco plants with *Cercospora nicotiani,* only a slight decrease in disease symptoms was observed. These results indicated that basic chitinase may not be the limiting factor in the resistance of tobacco plants to this specific pathogen. However, it may be possible to extend and/or enhance the resistance response by integrating constitutive chitinase expression with the expression of other available antifungal polypeptides. For example, in some instances the antifungal action of chitinase *in vitro* is evident only when applied in combination with β-1,3-glucanase (Mauch *et al.*, 1988) or is enhanced synergistically in the presence of chitin-binding lectins (Broekaert *et al.*, 1989; Chrispeels and Raikhel, 1991). The effects of transgenic plants containing such combinations of polypeptides would be limited to defense against chitin-containing pathogens.

Finally, another approach that has proven successful for producing pathogen-resistant transgenic plants is through the introduction of enzymes capable of detoxifying a pathogenic toxin (Anzai *et al.*, 1989). Infection of tobacco plants by the bacterial pathogen *Pseudomonas syringae* pv. *tabaci* results in the production of chlorotic halos that typify wildfire disease. The presence of a toxin, tabtoxin, produced by *P. syringae* has been shown to be associated with the development of these disease symptoms (Sinden and Durbin, 1968). Anzai *et al.*, (1989) isolated a tabtoxin resistance gene (*ttr*) from *P. syringae* that encoded an acetyltransferase enzyme. The coding sequence of this gene was fused to the 35S promoter of cauliflower mosaic virus and introduced into tobacco cells by *Agrobacterium*-mediated transformation. When transgenic plants were inoculated with *P. syringae* none of the tobacco plants expressing the *ttr* gene produced any symptoms of wildfire disease. This strategy of expressing detoxification genes in transgenic plants could, in theory, be applied to other diseases caused by pathogenesis-related toxins.

References

Abeles, F. B., Bonhart, R. P., Forrence, L. E., and Habig, W. H. (1970). *Plant Physiol.* 47, 129–134.

Anzai, H., Yoneyama, K., and Yamaguchi, I. (1989). *Mol. Gen. Genet.* 219, 492–494.

Awade, A., DeTapia, M., Didierjean, L., and Burkard, G. (1989). *Plant Sci.* 63, 121–130.

Bell, J. N., Ryder, T. B., Wingate, V. P. M., Bailey, J. A., and Lamb, C. J. (1986). *Mol. Cell. Biol.* 6, 1615–1623.

Benhamou, N., Grenier, J., Asselin, A., and LeGrand, M. (1989). *The Plant Cell* 1, 1209–1214.

Benhamou, N., Joosten, M. H. A. J., and De Wit, P. J. G. M. (1990). *Plant Physiol.* 92, 1108–1120.

Bonhoff, A., Reith, B., Golecki, J., and Grisebach, H. (1987). *Planta* 172, 101–105.

Boller, T. (1987). In *Plant-Microbe Interactions, Molecular and Genetic Perspectives* (Ed. by T. Kosuge and E. W. Nester), pp. 385–414. New York: Macmillan.

Boller, T., and Metraux, J. P. (1988). *Physiol. Mol. Plant Pathol.* 33, 11–16.

Boller, T., and Vogeli, U. (1984). *Plant Physiol.* 74, 442–444.

Boller, T., Gehri, A., Mauch, F., and Vögeli, U. (1983). *Planta* 157, 22–31.

Broekaert, W. F., Parijs, J. V., Leyns, F., Joos, H., and Peumans, W. J. (1989). *Science* 245, 1100–1102.

Broglie, K. E., Gaynor, J. J., and Broglie, R. M. (1986). *Proc. Natl. Acad. Sci. U.S.A.* 83, 6820–6824.

Broglie, K. E., Biddle, P., Cressman, R., and Broglie, R. (1989). *The Plant Cell* 1, 599–607.

Broglie, K. E., Chet, I., Holliday, M., Cressman, R., Biddle, P., Knowlton, S., Mauvais, J. and Broglie, R. (1981). *Science,* **254,** 1194–1197.

Chrispeels, M. J., and Raikhel, N. (1991). *The Plant Cell* 3, 1–9.

Cook, R. J., and Weller, D. M. (1987). In *Innovative Approaches to Plant Disease Control* (Ed. by I. Chet), pp. 41–76. New York: Wiley.

Darvill, A. G., and Albersheim, P. (1984). *Ann. Rev. Plant Physiol.* 35, 243–275.

Espelie, K. E., Francheschi, V. R., and Kolattukudy, P. E. (1986). *Plant Physiol.* 81, 487–492.

Esquerre-Tugaye, M. T., Lafitte, C., Mazau, D., Toppan, A., and Touze, A. (1979). *Plant Physiol.* 64, 320–326.

Esquerre-Tugaye, M. T., Mazau, D., Pelissier, B., Roby, D., Rumeau, D., and Toppan, A. (1985). In *Cellular and Molecular Biology of Plant Stress* (Ed. by J. L. Key and T. Kosuge), pp. 459–473. New York: Alan R. Liss.

Ferraris, L., Abbattista Gentile, I., and Matta, A. (1987). *J. Phytopathol.* 118, 317–321.

Hahlbrock, K., and Grisebach, H. (1979). *Ann. Rev. Plant Physiol.* 30, 105–130.

Herget, T., Schell, J., and Schreier, P. (1990). *Mol. Gen. Genet.* 224, 469–476.

Jefferson, R. A., Kavanagh, T. A., and Bevan, M. W. (1987). *EMBO J.* 6, 3901–3907.

Joosten, M. H. A. J., and De Wit, P. J. G. M. (1989). *Plant Physiol.* 89, 945–951.

Kombrink, E., Schroeder, M., and Hahlbrock, K. (1988). *Proc. Natl. Acad. Sci. U.S.A.* 85, 782–786.
Kurosaki, F., Tashirs, N., and Nishi, A. (1986). *Plant Cell Physiol.* 27, 1587–1591.
Kragh, K. M., Jacobsen, S., and Mikkelsen, J. D. (1990). *Plant Science* 71, 55–68.
Leah, R., Mikkelsen, D., Mundy, J., and Svendsen, I. (1987). *Carlsberg Res. Commun.* 52, 31–37.
Legrand, M., Kauffman, S., Geoffroy, P., and Fritig, B. (1987). *Proc. Natl. Acad. Sci. U.S.A.* 84, 6750–6754.
Majeau, N., Trudel, J., and Asselin, A. (1990). *Plant Science* 68, 9–16.
Mauch, F., and Staehelin, L. A. (1989). *The Plant Cell* 1, 447–457.
Mauch, F., Hadwiger, L. A., and Boller, T. (1988). *Plant Physiol.* 87, 325–333.
Meins, F., and Ahl, P. (1989). *Plant Science* 61, 155–161.
Metraux, J. P., Streit, L., and Staub, Th. (1988). *Physiol. Mol. Plant Pathol.* 33, 11–16.
Molano, J., Polacheck, I., Duran, A., and Cabib, E. (1979). *J. Biol. Chem.* 254, 4901–4907.
Nasser, W., Tapia, M., Kauffmann, S., Montasser-Kouhsari, and Burkard, G. (1988). *Plant Mol. Biol.* 11, 529–538.
Neuhaus, J-M., Ahl-Goy, P., Hinz, U., Flores, S. and Meins, F. (1991). *Plant Mol. Biol.* 16, 141–152.
Payne, G., Ahl, P., Moyer, M., Harper, A., Beck, J., Meins, F., and Ryals, J. (1990). *Proc. Natl. Acad. Sci. U.S.A.* 87, 98–102.
Pegg, G. F., and Vessey, J. C. (1973). *Physiol. Plant Pathol.* 3, 207–222.
Ride, J. P. (1983). In *Biochemical Plant Pathology* (Ed. by J. A. Callow), pp. 215–236. New York: Wiley.
Roberts, W. K., and Selitrennikoff, C. P. (1988). *J. Gen. Microbiol.* 134, 169–176.
Roby, D., and Esquerre-Tugaye, M. T. (1987). *Plant Sci.* 52, 175–185.
Roby, D., Toppan, A., and Esquerre-Tugaye, M. T. (1986). *Plant Physiol.* 81, 228–233.
Roby, D., Broglie, K. E., Cressman, R., Biddle, P., Chet, I., and Broglie, R. (1990). *The Plant Cell* 2, 999–1007.
Ryan, C. A. (1978). *Trends in Biochem. Sci.* 5, 148–151.
Samac, D., Hironaka, C. M., Yallaly, P. E., and Shah, D. M. (1990). *Plant Physiol.* 93, 907–914.
Schlumbaum, A., Mauch, F., Vögeli, U., and Boller, T. (1986). *Nature (London)* 324, 365–367.
Shinshi, H., Mohnen, D., and Meins, F. (1987). *Proc. Natl. Acad. Sci. U.S.A.* 84, 89–93.
Showalter, A. M., Bell, J. N., Cramer, C. L., Bailey, J. A., Varner, J. E., and Lamb, C. J. (1985). *Proc. Natl. Acad. Sci. U.S.A.* 82, 6551–6555.
Sinclair, J. B. (1982). *Compendium of Soybean Diseases.* The American Phytopathological Society, St. Paul, Minnesota.
Sinden, S. L., and Durbin, R. D. (1968). *Nature (London)* 219, 3790380.
Swegle, M., Huang, J-K., Lee, G., and Muthukrishnan, S. (1989). *Plant Molecular Biology* 12, 403–412.
Vance, C. P., Sherwood, R. T., and Kirk, T. K. (1980). *Annu. Rev. Phytopathol.* 81, 259–288.
Wessels, J. G. H., and Sietsma, J. H. (1981). In *Encyclopedia of Plant Physiology New Ser.* (Ed. by W. Tanner and F. A. Loewus), pp. 352–394. New York: Springer-Verlag.
Young, D. H., and Pegg, G. F. (1982). *Physiol. Plant Pathol.* 21, 411–417.

11

Transgenic Plants for Conferring Virus Tolerance: Satellite Approach

Peter B. McGarvey and Jacobus M. Kaper

Microbiology and Plant Pathology Laboratory
Plant Sciences Institute, ARS, USDA
Beltsville Agricultural Research Center—West
Beltsville, Maryland 20705

Virus-induced plant diseases are ubiquitous. In some areas of the world it is nearly impossible to grow a virus-free crop in the field. The severity of viral plant diseases varies from mild symptoms to death. The economic damage caused by plant viruses may go unnoticed but in extreme cases, such as the large losses inflicted to the Italian tomato crop of 1988, it attracts worldwide attention (Kaper *et al.*, 1990a). Traditional breeding methods to produce virus-resistant cultivars have been useful for some virus–host combinations but not for all; other methods commonly used to protect plants are often labor intensive or involve the use of environmentally hazardous chemicals to control the virus transmission through insect vectors. For general reviews of plant virology see Bos (1983), Gibbs and Harrison (1980), and Matthews (1991). Advances in the molecular biology of plants and plant viruses coupled with the development of techniques for the genetic transformation of plants have recently made it possible to produce transgenic plants that are resistant to specific plant viruses by expressing genes that interfere with some stage in the viral life cycle. The successful approaches investigated so far involve the expression of sequences derived from either the viral genome (i.e., coat protein) or from virus-associated satellite RNAs, although other approaches based on plant genes, ribozymes, and defective interfering RNAs hold promise for the future. This chapter reviews the use of viral satellite RNA to produce transgenic plants resistant to virus-induced crop diseases. Other approaches for producing virus resistance in plants have been reviewed elsewhere (van den Elzen *et al.*, 1989; Wilson, 1989; Golemboski *et al.*, 1990) and will be discussed only briefly here.

11.1 Viral Satellites

Viral satellites are small RNA molecules that (1) require the presence of a specific "helper" virus for their replication but themselves are not required for the replication of the helper virus; and (2) have no appreciable sequence similarity with the genome of their helper virus. This second criterion distinguishes viral satellites from defective interfering RNAs (DI-RNAs), other defective viral mutants, and subgenomic RNAs, all of which have significant sequence similarity to their helper virus genome. Viral satellites can be further categorized as either satellite RNAs which are encapsidated in the coat protein of their helper virus or satellite viruses which encode their own coat protein. Although these definitions are useful for classification of the various types of secondary RNA agents found in association with plant viruses they are, like most biological definitions, imperfect. For instance, the nucleotide sequence of a satellite RNA of turnip crinkle virus (TCV) has revealed that it is a hybrid molecule containing sequences derived from another satellite RNA and a portion of the helper virus genome (Simon and Howell, 1986). For more complete reviews of viral satellites and other secondary RNA agents see Collmer and Howell (1992), Kaper (1992), Kaper and Collmer (1988), Fritsch and Mayo (1988), Simon (1988), and Francki (1985).

First described in 1962 by Kassanis, viral satellites have received increasing attention in recent years because of their ability to modify the disease symptoms of their helper viruses and because of autocatalytic cleavage and ligation reactions that the satellites of tobacco ringspot virus (STobRV) (Buzayan *et al.*, 1986), arabis mosaic virus (SArMV) (Kaper *et al.*, 1988), and certain viroid-like satellite RNAs called virusoids can undergo (Symons, 1989). To date one or more viral satellites or satellite-like replicating RNAs have been found in association with some 30 plant viruses from 6 virus groups. These represent a list that is being rapidly expanded.

Most viral satellites reduce the severity of viral infections, presumably through interference with viral replication (see below), or have no obvious biological effect. In a few cases certain satellite RNAs in combination with specific helper viruses and host plants can aggravate disease symptoms (discussed below). For the purpose of this chapter we will focus on satellite RNAs and in particular the satellite RNAs of cucumber mosaic virus (CMV). We take this approach because the majority of research using satellite RNAs to protect plants from viral diseases has been done using CMV satellites. However, other viral satellites and secondary RNA agents such as DI-RNAs have similar properties that make them suitable subjects for protection strategies similar to those we will describe here and are the subject of ongoing research in a number of laboratories including our own (Morris and Hillman, 1989; Morch *et al.*, 1987).

11.2 Cucumber Mosaic Virus Satellite RNA and Disease Modulation

Cucumber mosaic virus is a widespread plant virus that can infect over 800 plant species (Douine *et al.*, 1979) and is one of the most damaging viruses to field-grown vegetable crops worldwide (Tomlinson, 1987). The virus has a tripartite (+) strand RNA genome and produces at least one subgenomic RNA that serves as messenger RNA for the viral coat protein. In addition many isolates of CMV contain one or more linear satellite RNAs of varying length (334–386 nucleotides) that we refer to as *C*MV *A*ssociated *RNA 5* or CARNA 5.

In 1972 an outbreak of lethal tomato necrosis disease in the Alsace region of France (Putz *et al.*, 1974; Marrou and Duteil, 1974), followed by the demonstration that a CMV satellite and not CMV per se was the causal agent of this disease (Kaper and Waterworth, 1977), stimulated interest in the disease modulating properties of these molecules that continues today. In the years since 1972 numerous studies on the structural and biological properties of CMV and its satellites have been reported [reviewed in Collmer and Howell (1992), Kaper and Collmer (1988), Fritsch and Mayo (1988)]. In addition to the disease exacerbation caused by some CARNA 5 sequence variants, many (if not most) natural CARNA 5 variants have the opposite effect in that the severity of the disease symptoms are greatly reduced (see Figure 11.1) concurrent with a reduction in

Figure 11.1 Modulation of viral plant disease symptoms by CARNA 5. Pots contain tomato (*Lycopersicon esculentum* Mill cv Rutgers) plants infected with (*left* to *right*): CMV-R76B, showing mild fern leaf symptoms; CMV-R76B + T-CARNA 5, showing lethal tomato necrosis; CMV-R76B + S-CARNA 5, showing attenuation of viral disease symptoms; uninfected healthy control.

the amount of CMV genomic RNA and intact virions that can be purified from infected plants (Kaper *et al.*, 1976; Kaper and·Tousignant, 1977; Mossop and Francki, 1979; Takanami, 1981; Jacquemond and Leroux, 1982). Indeed even combinations of CMV and necrotic CARNA 5 that are lethal to tomato are relatively harmless in other host plants such as pepper (see Figure 11.2) and protect the plant from the more severe symptoms caused by CMV alone (Waterworth *et al.*, 1979).

The biochemical mechanisms by which CARNA 5 modulates symptom expression are not understood. Experiments using infectious *in vitro* RNA transcripts of either chimeric CARNA 5 molecules or site-specific mutants engineered from cDNA clones have determined sequence domains that distinguish a necrosis- or chlorosis-causing CARNA 5 variant from those that attenuate the infection (Collmer and Kaper, 1986; Masuta and Takanami, 1989; Devic *et al.*, 1989; Kurath and Palukaitis, 1989; Jaegle *et al.*, 1990; Devic *et al.*, 1990; Sleat and Palukaitis, 1990). It appears that in some cases a change of one or two nucleotides in specific domains of the molecule may be sufficient to dramatically

Figure 11.2 Host plant-induced variation in the type of modulation of cucumber mosaic virus disease symptoms by its necrogenic satellite CARNA 5. Upper row of plants was infected with virus alone (CMV-WT); lower row was infected with virus plus CARNA 5 (CMV-WT + WT-CARNA 5). In tabasco pepper (*Capsicum frutescens* L.) (plants on left) CARNA 5 attenuates disease symptoms; in tomato (*Lycopersicon esculentum* Mill cv Rutgers) (plants on right) CARNA 5 induces lethal tomato necrosis.

alter the biological properties of these small RNAs. Though open reading frames are present on some CARNA 5 variants, and while they can be translated *in vitro* (Owens and Kaper, 1977; Avila-Rincon *et al.*, 1986a; Avila-Rincon *et al.*, 1986b; Steen *et al.*, 1990; Hidaka *et al.*, 1990), thus far they have not been found to affect the biological properties of these satellites (Collmer and Kaper, 1988). Since the attenuation of viral infections by CARNA 5 are correlated with a reduction in both viral RNA and infectious particles, and since both genomic RNA and satellite RNA are replicated in the same particulate fraction (Jaspars *et al.*, 1985; Diaz-Ruiz *et al.*, 1987), it seems likely that some aspect of the replicative machinery mediates these effects. Recently sequence analysis of the genomic RNAs of several helper viruses suggests that some of the variations in the disease modulation seen with certain virus/satellite combinations can be attributed to viral encoded components of the putative viral replicase (Collmer *et al.*, 1992; P. McGarvey, unpublished, 1992). Of several possible mechanisms proposed, one is consistent with most of the biological data recorded so far. Its basic premise, the direct competition by the satellite for all or part of the viral replicative machinery (Kaper, 1982; Kaper, 1984), has been validated by the recent demonstration that CMV satellites are replicated *in vitro* by a highly purified CMV RNA-dependent RNA polymerase (Wu *et al.*, 1991). This mechanism is based on accepted knowledge concerning (+) strand RNA bacteriophage replication and well-established biochemical principals of enzymatic competition and feedback regulation. It accounts fully for the accumulation of large amounts of the double-stranded form (dsCARNA 5) of the satellite, which is a hallmark for most or all CMV satellite infections studied thus far (Diaz-Ruiz and Kaper, 1977; Habili and Kaper, 1982; Piazzolla *et al.*, 1982; White and Kaper, 1989; Kaper *et al.*, 1990b), even in mature plants from the field (Rosner *et al.*, 1983; Valverde and Dodds, 1986; Montasser *et al.*, 1991). Other mechanisms suggest a direct interaction between the satellite and one or more genomic RNAs, although the specificity of such interactions has come into question after further experimentation [reviewed in Collmer and Howell (1992), Kaper and Collmer (1988)].

11.3 Biological Control of Cucumber Mosaic Virus

The observed reduction in disease symptoms mediated by satellite RNA has become the basis for experimental and practical field strategies for the biological control of CMV infections. The first report of successful protection described tomato plants preinoculated with a CMV strain containing a nonnecrogenic variant of CARNA 5, and demonstrated protection against challenge infections with a necrosis-inducing CMV/CARNA 5 combination and also against a severe satellite-free CMV isolate (Jacquemond, 1982). Later work in Japan (Yoshida *et al.*, 1985), China (Wu *et al.*, 1989), and most recently our laboratory (Montasser *et al.*, 1991), has shown that the conventional cross-protection effects (mediated in part by coat protein) seen in plants preinoculated with a satellite-free strain of CMV were enhanced when satellite RNA was added to the preinoculation mixture. These techniques have been successfully applied in field tests to tomato and pepper plants (Montasser *et al.*, 1991; Gallitelli *et al.*, 1991; Tien *et al.*, 1987). In China where CMV infestations are endemic, satellite protection strategies are being applied on a large scale for tomato, a number of pepper varieties, cucurbits, eggplant, cabbage, and tobacco plants in over 10,000 hectares in at least 16 localities with outstanding results (Tien and Wu, 1991).

The practice of preinoculation of virus and satellite as a biological control method has been justifiably criticized as presenting potential dangers [reviewed in Fulton (1986)]. The high degree of variability inherent in multicomponent RNA genomes could lead to a breakdown in protection. There is the danger that a small amount of a disease-aggravating CARNA 5 either present in the inoculum or arising from mutation could come to dominate the satellite population after passage through certain hosts (Garcia-Luque *et al.*, 1984; Kaper *et al.*, 1986). The methods may also be harmful if large amounts of the virus used for preinoculation would build up in the environment, as unknown synergistic effects with other pathogens may occur. In order to circumvent these dangers, rigorous screening and testing of the inoculum and plants surrounding the protected fields must be utilized, a process that can be labor intensive and requires a relatively high degree of expertise. In addition even the mild CMV/satellite combinations used to date can reduce the yields of tomato crops 10–20% from that of virus-free crops (Montasser *et al.*, 1991; P. Tien, personal communication, 1990), although other crops such as pepper are not seriously affected (Tien *et al.*, 1987). One way to address some of these concerns is to make the satellite RNA a constitutive part of the plant cell and thus eliminate the need for a helper virus to maintain the satellite.

11.4 Expression of Satellite RNA in Transgenic Tobacco Plants

The initial studies using satellite RNA to produce virus-tolerant transgenic plants were performed using tobacco, which is a susceptible host for CMV and other plant viruses and is among the easier plants to transform using *Agrobacterium tumefaciens*-based transformation systems. The first report on the expression of CMV-satellite sequences in transgenic tobacco (*Nicotiana tabacum* Samsun NN) described the insertion of head-to-tail concatemeric constructs containing 1.3 and 2.3 cDNA copies of the CMV satellite I17N under the control of the 35S promoter from cauliflower mosaic virus (CaMV) (Baulcombe *et al.*, 1986). The

I17N satellite of CMV ameliorates viral infections in tobacco plants but causes lethal necrosis in tomato plants (Jacquemond and Lot, 1981). The transgenic plants grew normally and expressed polyadenylated transcripts of approximately 700 and 1000 nucleotides (nt) consisting of the satellite sequence plus extra sequences on the 5′ and 3′ ends derived from the cloning vectors and including portions of the promoter and terminator sequences. Following challenge infections with a satellite-free CMV isolate R76b, unit length (335 nt) satellite RNA appeared in plant tissue and increased in concentration over time. Unit length single-stranded satellite was encapsulated in virus particles and could infect *N. clevelandii* plants where it clearly exhibited the ameliorative properties expected from CMV satellites. In later experiments other plants of the same recombinant genotype were challenged with a satellite-free isolate of CMV-K which, unlike CMV-R76b, causes visible symptoms on tobacco (Harrison *et al.*, 1987). Here the amounts of satellite RNA, viral RNA, and infectious virus were monitored in inoculated leaves and leaves infected systemically. In the inoculated leaves and the first 2–3 systemically infected leaves there was an increase in viral RNA and unit length satellite RNA plus the appearance of symptoms characteristic of CMV-K infection. However, younger leaves showed a reduction of at least 80% in viral RNA and the infectivity of leaf extracts and exhibited little or no symptoms, resulting in much healthier plants than the nontransformed controls. Plants were also challenged with tomato aspermy virus (TAV), a cucumovirus closely related to CMV (Kaper and Waterworth, 1981) and known to support replication of some CMV satellite RNAs (Mossop and Francki, 1979). In these plants a similar amplification of satellite RNA and reduction in symptoms occurred in systemically infected leaves but was not associated with a reduction in the amounts of TAV RNA or infectivity, both of which remained relatively high. This suggested that here the protective effects of satellite RNA may not come solely from the inhibition of viral replication. In both challenge experiments the degree of protection appeared independent of the strength of the virus inoculum used.

Protection from systemic infection was also seen using the satellite of tobacco ringspot virus (STobRV) (Gerlach *et al.*, 1987). In this work an *A. tumefaciens* system was used to produce transgenic tobacco plants expressing head-to-tail trimers of STobRV in either the (+) strand or (−) strand orientation under control of the 35S promoter of CaMV. In both cases the long polyadenylated transcripts self-cleaved to monomeric size prior to infection. On infection with TobRV the satellite concentration increased and virus symptoms appeared on infected leaves. However, both the amount of virus and symptoms were greatly reduced or completely absent in systemically infected leaves. The protection was strongest in the plants expressing (+) strand RNA. That the plants expressing (−) strand RNA showed satellite replication at all was surprising since *in vitro* transcripts of (−) strand STobRV had proved biologically inactive when inoculated with TobRV on plants (Gerlach *et al.*, 1986). A third construct expressing a (+) strand monomer of STobRV (along with similar extra 5′ and 3′ sequences) showed no unit length satellite before or after infection and no protection was observed. It appears that for STobRV correct ends of the linear form of the satellite are essential to initiate the replication cycle of the satellite. The self-cleavage properties of trimers or at least permuted dimers of STobRV (absent in monomers) can provide these ends. Although with CARNA 5 and other cucumoviral satellite infections multimers are also detected, they have neither convincingly been shown to self-cleave nor to coexist with circular forms (Linthorst and Kaper, 1984, 1985). Cucumoviral satellites in all likelihood replicate by a mechanism similar to that of the helper virus RNA (Collmer and Kaper, 1985) and not the rolling circle model proposed for STobRV (Kiefer *et al.*, 1982).

Tobacco (*N. tabacum* Xanthi nc) has also been transformed with a monomer construct of the I17N satellite of CMV using an *Agrobacterium rhizogenes* transformation system and the CaMV 19S promoter (Jacquemond *et al.*, 1988). On challenge infection of these plants with the satellite-free I17F strain of CMV unit length satellite appeared, the amount of infectious virus one week postinoculation was reduced 3- to 4-fold, and a corresponding reduction in viral symptoms was observed in systemically infected leaves. Individual transgenic plants that expressed different amounts of RNA from the inserted constructs, as determined by gel electrophoretic analysis of total RNA, showed similar amounts of tolerance to challenge infections indicating that the mechanism of tolerance was independent of the amount of RNA expressed. In addition, protection from viral infection by aphids (the natural vector of CMV) was demonstrated. Recently similar plants expressing the (−) strand of the same satellite RNA constructs have shown similar levels of tolerance to challenge infections as the plants expressing (+) strand RNA (M. Tepfer, personal communication, 1990). As was the case with STobRV experiments using (−) strand *in vitro* transcripts (see above), the (−) strand transcripts of cloned CMV satellite RNA have also failed to show satellite replication when inoculated on plants, yet seem to be replicated when the (−) strand RNA is derived from a transgenic insert (M. Tepfer, personal communication, 1990).

All the satellites discussed above attenuate the symptoms of viral infection in tobacco. Sometimes this is not immediately obvious, however, as with the satellite of CMV-Y which ameliorates the symptoms of CMV-Y in tobacco but superimposes a new symptom of strong chlorosis resulting in yellow mosaic leaves (Takanami, 1981). When this satellite was expressed in transgenic tobacco plants as either a monomer or dimer the plants appeared normal until infection with a satellite-free isolate of CMV-Y. Plants containing either recombinant genotype quickly showed replication of unit length satellite and within 7 days developed yellow mosaic leaves characteristic of the satellite (Masuta *et al.*, 1989). Long-term symptoms were not reported.

Field tests of transgenic tobacco plants expressing satellite RNA sequences have taken place in China (Tien and Wu, 1991). Transgenic tobacco plants containing a monomer of the attenuating satellite 1-CARNA 5 were grown in fields where natural CMV infections are common, and compared to non-transgenic tobacco. Results similar to those described in the laboratory were observed, with the early leaves of the transgenic plants showing symptoms and tolerance developing in the later leaves after 2–3 weeks. Unfortunately, although the transgenic plants were healthier than the nonprotected controls the most valuable part of tobacco plants are the large early leaves that were not protected. Thus from a growers point of view the crop still had unacceptable economic losses. In China satellite preinoculated tobacco plants have been better protected against early infection than the transgenic tobacco plants probably due to the presence of larger amounts of unit length CARNA 5 in these plants prior to infection with CMV strains in the field. Other crops such as peppers or tomatoes, in which later developing tissues such as fruits are of economical importance, are more likely to benefit from late developing tolerance.

11.5 Expression of Satellite RNA in Transgenic Tomato Plants

Many severe outbreaks of virus-induced plant diseases occur in vegetable crops. This is especially true of diseases caused by CMV because of its broad host range and the ability of over 60 species of aphids to transmit the virus (Kaper and Waterworth, 1981). In addition to the well-documented crop losses in France, Italy, and China mentioned in the previous sections we know of recurring problems with CMV-induced crop disease in Japan, Spain, Indonesia, Taiwan, Egypt, and several African and European countries. A common problem in several areas of the world has been outbreaks of CMV-induced diseases in tomato crops. Tomato plants have been especially interesting to plant virologists because in this host CMV diseases can be characteristically modified by CARNA 5, making it a useful plant with which to study the relationship of symptom expression to the trilateral interactions between the viral, satellite, and host plant genomes. However, since it is also one of the two plants for which certain CARNA 5 variants are known to clearly exacerbate disease symptoms while others attenuate them, it is a somewhat risky plant to use in satellite protection strategies.

In order to better assess the benefits and risks that could arise from using transgenic plants expressing CMV satellite RNA *in planta* and in order to study the trilateral interactions that presumably ensue upon CMV infection and that result in viral symptom modulation, transgenic tomato plants expressing D-CARNA 5, a necrogenic satellite, and S-CARNA 5, a nonnecrogenic attenuating satellite, were generated using an *A. tumefaciens* transformation system (McGarvey *et al.*, 1990; McGarvey and Kaper, unpublished data, 1991). Of particular interest was the question what effect (if any) necrogenic sequence expression *in planta* would have on the individual transformants.

In a collaborative investigation with a group of Spanish investigators we have inserted the three DNA constructs shown in Figure 11.3 into tomato plants

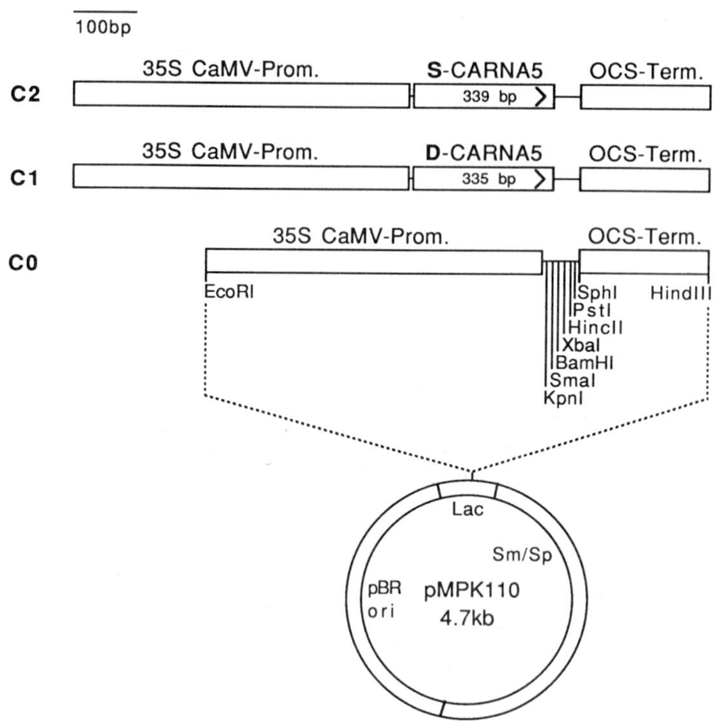

Figure 11.3 Construction of chimeric genes C0, C1, and C2. 35S CaMV-Prom represents 535 bp of DNA from the cauliflower mosaic virus 35S promoter. CARNA 5 represents either a D-CARNA 5 or an S-CARNA 5 cDNA oriented so that transcripts from the 35S promoter contain (+) strand RNA. OCS-Term refers to 190 bp from the termination region of the octopine synthase (octopine dehydrogenase) gene. Transcripts from the C1 and C2 genes should contain 12 nucleotides (35S promoter and polylinker) upstream of and 200–350 nucleotides [polylinker, ocs terminator, and poly(A) tail] down stream of the CARNA 5 insert. pBR ori designates the origin of replication from pBR322, sm/sp refers to the streptomycin/spectinomycin resistance gene from R702, and *lac* refers to the *E. coli* lactose operon from pUC-18. For details on construction and plant transformation see McGarvey *et al.* (1990) and references therein.

(*L. esculentum* cv UC82b). The construct designated: C0, lacked a CARNA 5 insert and was used as a control; C1, contained a monomer of D-CARNA 5; and C2, contained a monomer of S-CARNA 5. Subsequently, the F_1 offspring of self-crossed primary transformants that expressed polyadenylated RNA with the inserted sequences were challenge infected with genomic RNA from a satellite-free isolate of CMV-1. This CMV strain produced chlorosis, shoestring leaves, and stunting in tomato plants. Nontransformed and C0 transgenic tomato plants challenge infected with the combinations CMV-1 RNA/D-CARNA 5 or CMV-1 RNA/S-CARNA 5 served as positive controls to verify satellite-induced symptom modulation. The C0, C1, and C2 transgenic plants all germinated and developed normally. The symptoms of the challenged C1 and C2 transgenic plants and their nontransformed controls can be seen in Figures 11.4 and 11.5. Most of the C1 plants expressing D-CARNA 5 sequences developed lethal tomato necrosis 2–3 weeks after challenge (see Figure 11.4) (McGarvey *et al.*, 1990; McGarvey and Kaper, unpublished data, 1991). The nontransformed control plants challenged with CMV-1/D-CARNA 5 had responses identical to those of the C1 transgenic plants challenged with CMV-1 lacking any CARNA 5 (Figure 11.4). The C0 transgenic plants had responses identical to the nontransformed control plants with all challenge inoculums (not shown). Northern blot hybridizations of selected plants after challenge are shown in Figure 11.6. Extracts of the C1 transgenic plants showed evidence of replication of unit length (335 nt)

Figure 11.4 Transgenic C1 tomato (*Lycopersicon esculentum* cv UC82b) plants with a necrogenic D-CARNA 5 insert (white pots) and nontransgenic control tomato plants (colored pots) 19 days after challenge infection. The C1 plant on the extreme right and the control plant on the extreme left are normal looking uninfected controls. The plant in the middle was challenge infected with the combination CMV-1 RNA/D-CARNA 5 and served as positive control for lethal necrosis. The plants second from left and second from right were challenge infected with satellite free CMV-1 RNA. The control plant on the left has normal CMV-1 symptoms; the C1 plant on the right has lethal necrosis. For experimental details see McGarvey *et al.* (1990) and references therein.

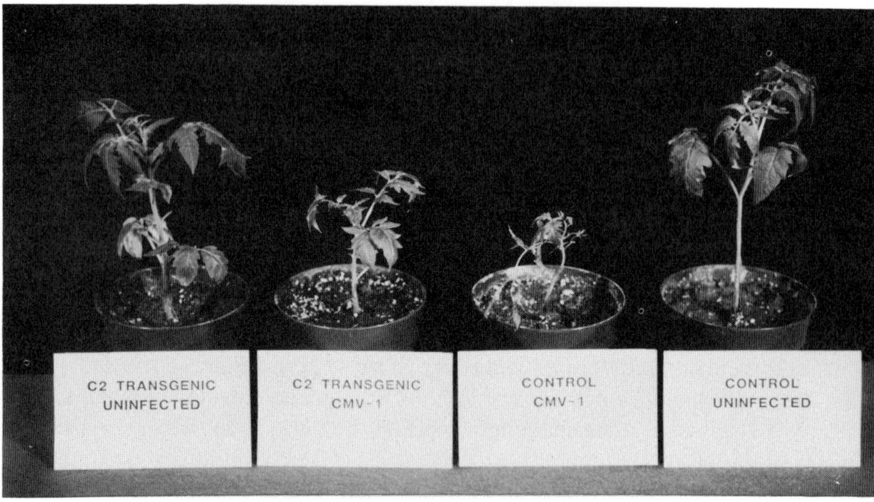

Figure 11.5 Transgenic C2 tomato (*Lycopersicon esculentum* cv UC82b) plants with an attenuating S-CARNA 5 insert and normal control tomato plants 14 days after challenge infection. The C2 plant on the extreme left and the control plant on the extreme right are normal looking uninfected controls. The C2 plant second from the left and the control plant second from the right were challenged with satellite-free CMV-1 RNA. The C2 plant shows only stunting with no chlorosis or shoestring leaves, while the control plant shows normal CMV-1 symptoms of shoestring leaves, chlorosis, and stunting. For experimental details see McGarvey *et al.* (1990) and references therein.

single- and double-stranded D-CARNA 5, as anticipated from the previous work on CMV satellite in transgenic tobacco (see Figure 11.6, C1 lane, middle). The nontransformed and C0 controls challenged with CMV-1/D-CARNA 5 showed similar results (Figure 11.6, right-hand side). Characteristic satellite symptoms were not observed in the tomato plants containing the C1 (necrogenic D-CARNA 5) construct until after challenge infection with CMV-1, as reported with the yellow chlorosis-producing satellite of CMV-Y in transgenic tobacco (Masuta *et al.*, 1989). The C2 plants expressing S-CARNA 5 sequences did not show any chlorosis or shoestring leaves and were only slightly stunted 3 weeks after challenge (see Figure 11.5; McGarvey and Kaper, unpublished data, 1991). Northern blot hybridizations of extracts of the C2 transgenic tomato plants also showed evidence of replication of unit length (339 nt) single- and double-stranded S-CARNA 5 (see Figure 11.6, C2 lane, middle), as did the nontransformed and C0 controls challenged with CMV-1 plus S-CARNA 5 (Figure 11.6, right-hand side). Our preliminary work with the C2 transgenic plants expressing an attenuating S-CARNA 5 thus far has shown that two of the three symptoms characteristic of CMV-1 are completely absent in these plants after challenge infection. The initial stunting seen in these plants has also been observed in some non-

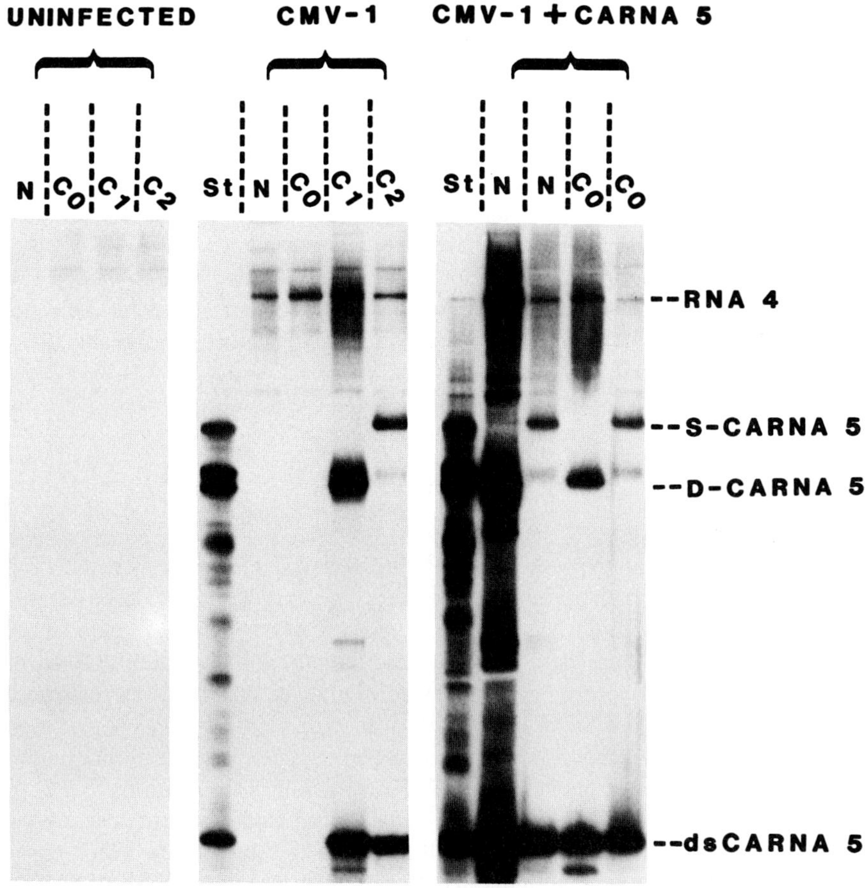

Figure 11.6 Northern blot hybridizations of total nucleic acids from transgenic and control tomato plants 14 days after challenge infection. RNA was electrophoresed through 9% semidenaturing polyacrylamide gels and probed with CARNA 5 and CMV RNA-4 riboprobes. Lanes marked: St., contain single- and double-stranded CARNA 5 markers (some of whose positions are shown in the right-hand margin); N, contain RNA from normal nontransgenic tomato plants; C0, C1, C2, contain RNA from the respective C0 (no CARNA 5 insert), C1 (D-CARNA 5 insert), and C2 (S-CARNA 5 insert) transgenic tomato plants. The left-hand side represents uninfected plants; the middle, plants challenged with satellite-free CMV-1 RNA; and the right-hand side, pairs of nontransformed or C0 control plants challenged with CMV-1 RNA plus either D-CARNA 5 (left lane) or S-CARNA 5 (right lane). Single- and double-stranded CARNA 5s appear only in the C1 and C2 transgenic plants challenged with satellite-free CMV-1 (*middle*) or in normal and C0-positive controls in which CARNA 5 was added to the inoculum (*right*). For detailed methods see White and Kaper (1989) and McGarvey *et al.* (1990) and references therein.

transformed control plants inoculated with CMV and S-CARNA 5 (not shown). This stunting is overcome as the plants mature.

Other groups have obtained results similar to those described above in independent studies with transgenic tomato plants using different necrogenic and nonnecrogenic CARNA 5 sequence variants, different vector constructions, and different tomato cultivars (M. Tepfer, personal communication, 1991). In addition, transgenic tomato plants containing sequences from another nonnecrogenic CMV satellite named 1-CARNA 5 have also been developed and field tested in China. the F_1 generation of these plants (including offspring that contained and those that lacked the insert) were tested under conditions of high natural disease pressure and yielded 50% more tomatoes than nontransgenic control plants (Tien and Wu, 1991).

11.6 Discussion

Transgenic plants that express viral satellite RNA sequences can resist crop diseases induced by their specific helper viruses. This has been demonstrated with two different satellite RNA systems (those of CMV and TobRV) in tobacco, and in two species of host plants (tobacco and tomato) with the satellite RNA of CMV. In the latter case preliminary field studies indicate that the CMV resistance observed in the laboratory with transgenic tobacco and tomato plants is also seen under conditions of natural disease pressure. This suggests that the technique can be adapted to other host plants and other virus/satellite combinations.

Tolerance is probably the most appropriate word with which to describe the transgenic plant's response to viral infection since some of the plant tissue becomes infected with virus initially and may show disease symptoms. However, following this initial infection the rapid replication of satellite RNA and the concomitant suppression of viral replication prevent the systemic spread of viral disease and can result in a plant substantially free of any disease symptoms. The biological expression *in planta* of "harmful" satellite RNA sequences [i.e., the tobacco yellowing of the satellite of CMV-Y (Masuta *et al.*, 1989) and the tomato necrosis of D-CARNA 5 (McGarvey *et al.*, 1990)] in a susceptible host plant has proved not to occur until the plant is challenged with helper virus. In all likelihood this means that the unit length single- or double-stranded satellite RNA "rescued" and amplified from longer RNA transcripts by the CMV replicative machinery is the agent responsible for the "pathogenic" response, much as in a conventional CMV/satellite infection. An interaction of the satellite sequence with virus-encoded factors to elicit a direct pathogenic response seems unlikely although such an interaction cannot be excluded. To address this question with transgenic plants would require expressing CARNA 5 at higher levels with and without any extra sequences on the 5′ and 3′ ends. One experimental strategy would be the insertion of self-cleavage domains at the ends of the CARNA 5 sequence to produce a transcript that would cut itself into unit length RNAs similar to that seen with the naturally self-cleaving satellite RNA of TobRV. Double-stranded RNA might be produced by expression of both (+) and (−) strand satellite RNA in the same genome. Similar experiments with an attenuating CARNA 5 variant might enhance tolerance if the unit length satellite RNA could initiate replication more quickly in response to a challenge infection than the larger RNA transcripts used to date.

The apparent advantages of satellite-mediated tolerance are that it seems to be independent of both the strength of the challenge inoculum (Harrison *et al.*, 1987) and the amount of RNA expressed from the inserted gene (Jacquemond *et al.*, 1988). This is in contrast to some other forms of engineered viral resistance, such as that provided by antisense RNA (Rezaian *et al.*, 1988; Cuozzo *et al.*, 1988) or coat protein (Abel *et al.*, 1986; Nelson *et al.*, 1987; Loesch-Fries *et al.*, 1987; Tumer *et al.*, 1987; Nelson *et al.*, 1988; Cuozzo *et al.*, 1988), which can both be overcome by high concentrations of virus in the inoculum and where the level of tolerance varies with the amount of expressed RNA or coat protein present in the plants cells. It is possible that this lower metabolic commitment required by satellite-mediated tolerance could be advantageous to a plant under environmental stress. Disadvantages of the satellite-mediated approach are that some plant tissue is infected and tolerance is delayed. This is in contrast to the coat protein approach which seems to completely prevent virus infection at low inoculum concentrations and often delays the appearance and extent of virus symptoms at higher concentrations. Comparisons between these two strategies are somewhat speculative since no direct comparisons using the same virus, host plant, and experimental protocols have been made. It would be useful to do such a comparison and to combine the two techniques in the same plant since cross-protection experiments have shown that preinoculation with virus plus satellite provides enhanced protection from subsequent challenge infection than does preinoculation with virus alone (Yoshida *et al.*, 1985; Wu *et al.*, 1989; Montasser *et al.*, 1991). Combining the two techniques in transgenic plants could provide two lines of defense against viral diseases. Coat protein, which can prevent or delay the onset of viral disease, would provide the first line of defense and satellite RNA, which provides a late developing tolerance independent of the strength of the inoculum, would provide a second line of defense. This has recently been done in tobacco (Yie *et al.*, 1992) and is in progress in tomato (McGarvey and Kaper, unpublished, 1992). However, a note of caution is in order with the case of CMV since the virus seems to lack specificity as to which RNAs it encapsulates, at least *in vitro* (Kaper and Geelen, 1971; Kaper, 1975). Thus, plants expressing large amounts of both CMV coat protein and CARNA 5 might encapsidate the expressed satellite sequences or genomic RNA from other infecting viruses. Such RNA, if not usually aphid transmissible, could be transmitted via aphids without prior infection by CMV (Chen and Francki, 1990). Indeed, this is a general concern in using transgenic plants expressing viral coat protein since the transcapsidation of viral and viroid RNAs is known to occur in some mixed virus infections (Francki *et al.*, 1986; Waterhouse and Murant, 1983; Rochow, 1977). The use of mutant or altered coat proteins may prevent this from occurring.

Concerns have been expressed about the safety of using transgenic plants expressing satellite RNA sequences (Baulcombe *et al.*, 1989). The main danger being the possibility that if the plant is infected by helper virus the virus/satellite combination could then be transmitted to other nonprotected plants and thus introduce a potential pathogenic molecule into the surrounding environment. Naturally any candidate satellite RNA should first be tested by preinoculation with helper virus into a variety of crops and native plant species to assure that no harmful effects occur before being used for expression in transgenic plants. This has already been done for several CARNA 5s in a combined total of over 100 species (Tien *et al.*, 1987; Tien and Wu, 1991). However, it remains possible that an attenuating satellite variant could undergo mutations converting it into a harmful variant. These mutations are more likely to occur in an "escaped" satellite replicated by virus than one transcribed from the DNA of the plant due to the much higher mutation rate of RNA genomes as compared to DNA genomes (Domingo and Holland, 1988). Ongoing research into determining the sequence domains responsible for satellite replication and symptom attenuation or aggravation may allow the design of satellites that can attenuate viral diseases but cannot mutate into a harmful variant. In addition, specific sequences may be required in certain virus/satellite combinations for a packaging of satellite RNA in viral proteins, as is the case for viruses such as TMV (Fukuda *et al.*, 1980). If so, it may be possible to engineer variants that cannot be packaged by an invading virus. However, as mentioned above this approach may not work with CMV. In addition, due to the compact size of satellite RNAs some of these functional domains may overlap to some degree and involve interactions at the level of secondary and tertiary structure which would make it difficult to delete one function without impinging on others.

In the light of extensive experience gathered in China using the preinoculation of CMV/CARNA 5 combinations as a means of controlling viral disease (Tien and Wu, 1991), it seems that the risks of satellite-based protection strategies may have been overstated. The preinoculation method of biological control is potentially more dangerous since both virus and satellite are present in all the protected plants, making the risk of satellite mutation and subsequent transmission much greater. The technology has been applied in China on an increasingly large scale for over nine years in some areas, yet to date no harmful CARNA 5 variants or other virus diseases have appeared in protected or nearby unprotected crops. This is especially relevant since two of the protected crops have been tomato and tobacco, which are the only plants for which harmful CARNA 5 variants have been discovered.

In future greenhouse and field tests of transgenic plants expressing CARNA 5 sequences the risk of satellite spread to the surrounding environment will have to be assessed. Plants expressing necrosis- or chlorosis-inducing CARNA 5s should be useful in these studies since they provide an easily recognizable marker for the spread of virus plus satellite. It will also be important to study the tolerance of the transgenic plant to infection with CMV strains that carry a virulent satellite such as D-CARNA 5. Studies with preinoculated plants have shown protection against challenge with a virulent CMV/necrogenic CARNA 5 combination (Jacquemond, 1982; Montasser *et al.*, 1991; Gallitelli *et al.*, 1991). Transgenic tobacco plants expressing satellite RNA upon challenge with CMV-Y/Y-satellite combination showed tolerance, but satellite-induced yellow chlorosis symptoms did appear and only slowly regressed as the endogenous satellite outcompeted the exogenous Y-satellite (Baulcombe *et al.*, 1989). Basic research on the competitive replication of benign and virulent CARNA 5s (Smith *et al.*, 1992) would help improve resistance to virulent infections.

In conclusion, we feel the future of the satellite approach for producing viral tolerance in transgenic plants looks promising. More laboratory and field research is required to assess the practicality and safety of this technique. Studies on the origins, replication, and biological activities of viral satellites should lead to important insights into the molecular biology of plant resistance to viruses as well as provide practical means to enhance the effectiveness of these small molecular parasites as biological control agents.

References

Abel, P. P., Nelson, R. S., De, B., Hoffmann, N., Rogers, S. G., Fraley, R. T., and Beachy, R. N. (1986). *Science* 232, 738–743.

Avila-Rincon, M J., Collmer C. W., and Kaper, J. M. (1986a). *Virology* 152, 455–458.

Avila-Rincon, M. J., Collmer, C. W., and Kaper, J. M. (1986b). *Virology* 152, 446–454.

Baulcombe, D. C., Saunders, G. R., Bevan, M. W., Mayo, M. A., and Harrison, B. D. (1986). *Nature (London)* 321, 446–449.

Baulcombe, D. C., Devic, M., Jaegle, J., and Harrison, B. (1989). In *Molecular Biology of Plant-Pathogen Interactions. UCLA Symposia on Molecular and Cellular Biology, New Series, Vol. 101* (Ed. by B. Staskawicz, P. Ahlquist, and O. Yoder), pp. 257–268. New York: Alan R. Liss.

Bos, L. (1983). *Introduction to Plant Virology*, pp. 1–160. New York: Longman.

Buzayan, J. M., Gerlach, W. L., Bruening, G., Keese, P., and Gould, A. R. (1986). *Nature (London)* 323, 349–353.

Chen, B., and Francki, R. I. B. (1990). *J. Gen. Virol.* 71, 939–944.

Collmer, C. W., and Howell, S. H. (1992). *Ann. Rev. Phytopathol* (In Press).

Collmer, C. W., and Kaper, J. M. (1985). *Virology* 145, 249–259.

Collmer C. W., and Kaper, J. M. 1986). *Biochem. Biophys. Res. Commun.* 135, 290–296.
Collmer, C. W., and Kaper, J. M. (1988). *Virology* 163, 293–298.
Collmer, C. W., Stenzler, L., Chen, X., Fay, N., Hacker, D., and Howell, S. H. (1992). *Proc. Natl. Acad. Sci. U.S.A.* 89, 309–313.
Cuozzo, M., O'Connell, K. M., Kaniewski, W., Fang, R. X., Chua, N., and Tumer, N. E. (1988). *Bio/Technology* 6, 549–557.
Devic, M., Jaegle, M., and Baulcombe, D. (1989). *J. Gen. Virol.* 70, 2765–2774.
Devic, M., Jaegle, M., and Baulcombe, D. (1990). *J. Gen. Virol.* 71, 1443–1449.
Diaz-Ruiz, J. R., and Kaper, J. M. (1977). *Virology* 80, 204–213.
Diaz-Ruiz, J. R., Avila-Rincon, M. J., and Garcia-Luque, I. (1987). *Plant Science* 50, 239–248.
Domingo, E., and Holland, J. J. (1988). In *RNA genetics. Vol. III, Variability of RNA genomes* (Ed. by E. Domingo, J. J. Holland, and P. Ahlquist), pp. 3–36. Boca Raton, Florida: CRC Press.
Douine, L., Quiot, J. B., Marchoux, G., and Archange, P. (1979). *Ann. Phytopathol.* 11, 439.
Francki, R. I. B. (1985). *Ann. Rev. Microbiol.* 39, 151–174.
Francki, R. I. B., Zaitlin, M., and Palukaitis, P. (1986). *Virology* 155, 469–473.
Fritsch, C., and Mayo, M. A. (1988). In *Plant Viruses. Vol. I, Structure and Replication* (Ed. by C. L. Mandahar), pp. 289–321. Boca Raton, Florida: CRC Press.
Fukuda, M., Okada, Y., Otsuki, Y., and Takebe, I. (1980). *Virology* 101, 493–502.
Fulton, R. W. (1986). *Ann. Rev. Phytopathol.* 24, 67–81.
Gallitelli, D., Vovlas, C., Martelli, G. P., Montasser, M. S., Tousignant, M. E., and Kaper, J. M. (1991). *Plant Dis.* 75, 93–95.
Garcia-Luque, I., Kaper, J. M., Diaz-Ruiz, J. R., and Rubio-Huertos, M. (1984). *J. Gen. Virol.* 65, 539–547.
Gerlach, W. L., Buzayan, J. M., Schneider, I. R., and Bruening, G. (1986). *Virology* 151, 172–185.
Gerlach, W. L., Llewellyn, D., and Haseloff, J. (1987). *Nature (London)* 328, 802–806.
Gibbs, A., and Harrison, B. (1980). *Plant Virology, The Principles,* pp. 1–292. London: Edward Arnold, Ltd.
Golemboski, D. B., Lomonossoff, G. P., and Zaitlin, M. (1990). *Proc. Natl. Acad. Sci. U.S.A.* 87, 6311–6315.
Habili, N., and Kaper, J. M. (1981). *Virology* 112, 250–261.
Harrison, B. D., Mayo, M. A., and Baulcombe, D. C. (1987). *Nature (London)* 328, 799–802.
Hidaka, S., Hanada, K., and Ishikawa, K. (1990). *J. Gen. Virol.* 71, 439–442.
Jacquemond, M. (1982). *C. R. Acad. Sci. Paris* 290, 991–994.
Jacquemond, M., and Leroux, J. (1982). *Agronomie* 2, 55–62.
Jacquemond, M., and Lot, H. (1981). *Agronomie* 1, 927–932.
Jacquemond, M., Amselem, J., and Tepfer, M. (1988). *Molecular Plant-Microbe Interactions* 1, 311–316.
Jaegle, M., Devic, M., Longstaff, M., and Baulcombe, D. (1990). *J. Gen. Virol.* 71, 1905–1912.
Jaspars, E. M. J., Gill, D. S., and Symons, R. H. (1985). *Virology* 144, 410–425.
Kaper, J. M. (1975). *The Chemical Basis of Virus Structure, Dissociation and Reassembly,* pp. 1–485. Amsterdam: North-Holland Publishing Company.
Kaper, J. M. (1982). *Biochem. Biophys. Res. Commun.* 105, 1014–1022.
Kaper, J. M. (1984). In *Control of Virus Diseases* (Ed. by E. Kurstak), pp. 317–343. New York and Basel: Marcel Dekker, Inc.
Kaper, J. M. (1992). *Res. Virol.* 143, 5–10.
Kaper, J. M., and Collmer, C. W. (1988). In *RNA Genetics, Vol. III: Variability of RNA Genomes* (Ed. by E. Domingo, J. Holland, and P. Ahlquist), pp. 171–194. Boca Raton, Florida: CRC Press.
Kaper, J. M., and Geelen, J. L. M. C. (1971). *J. Mol. Biol.* 56, 277–294.
Kaper, J. M., and Tousignant, M. E. (1977). *Virology* 80, 186–195.

Kaper, J. M., and Waterworth, H. E. (1977). *Science* 196, 429–4431.

Kaper, J. M., and Waterworth, H. E. (1981). In *Handbook of Plant Virus Infections and Comparative Diagnosis* (Ed. by E. Kurstak), pp. 257–332. Amsterdam: Elsevier/North Holland.

Kaper, J. M., Tousignant, M. E., and Lot, H. (1976). *Biochem. Biophys. Res. Commun.* 72, 1237–1243.

Kaper, J. M., Duriat, A. S., and Tousignant, M. E. (1986). *J. Gen. Virol.* 67, 2241–2246.

Kaper, J. M., Tousignant, M. E., and Steger, G. (1988). *Biochem. Biophys. Res. Commun.* 154, 318–325.

Kaper, J. M., Gallitelli, D., and Tousignant, M. E. (1990a). *Res. Virol.* 141, 81–95.

Kaper, J. M., Tousignant, M. E., and Geletka, L. M. (1990b). *Res. Virol.* 141, 487–503.

Kassanis, B. (1962). *J. Gen. Microbiol.* 27, 477–488.

Kiefer, M. C., Daubert, S. D., Schneider, I. R., and Bruening, G. (1982). *Virology* 121, 262–273.

Kurath, G., and Palukaitis, P. (1989). *Molecular Plant–Microbe Interactions* 2, 91–96.

Linthorst, H. J. M., and Kaper, J. M. (1984). *Virology* 139, 317–329.

Linthorst, H. J. M., and Kaper, J. M. (1985). *J. Gen. Virol.* 66, 1839–1842.

Loesch-Fries, L. S., Merlo, D., Zinnen, T., Burhop, L., Hill, K., Krahn, K., Jarvis, N., Nelson, S., and Halk, E. (1987). *EMBO J.* 6, 1845–1851.

Marrou, J., and Duteil, H. (1974). *Ann. Phytopathol.* 6, 155–171.

Masuta, C., and Takanami, Y. (1989). *Plant Cell* 1, 1165–1173.

Masuta, C., Komari, T., and Takanami, Y. (1989). *Ann. Phytopath. Soc. Japan* 55, 49–55.

Matthews, R. E. F. (1991). *Plant Virology,* pp. 1–835. San Diego: Academic Press, Inc.

McGarvey, P. B., Kaper, J. M., Avila-Rincon, M. J., Pena, L., and Diaz-Ruiz, J. R. (1990). *Biochem. Biophys. Res. Commun.* 170, 548–555.

Montasser, M. S., Tousignant, M. E., and Kaper, J. M. (1991). *Plant Dis.* 75, 86–92.

Morch, M. D., Joshi, R. L., Manifacier, S. A., and Haenni, A. L. (1987). *Nucleic Acids Research* 15, 4123–4130.

Morris, T. J., and Hillman, B. I. (1989). In *Molecular Biology of Plant-Pathogen Interactions, UCLA Symposia on Molecular and Cellular Biology, New Series, Vol. 101* (Ed. by B. Staskawicz, P. Ahlquist, and O. Yoder), pp. 185–197. New York: Alan R. Liss.

Mossop, D. W., and Francki, R. I. B. (1979). *Virology* 95, 395–404.

Nelson, R. S., Abel, P. P., and Beachy, R. N. (1987). *Virology* 158, 126–132.

Nelson, R. S., McCormick, S. M., Delannay, X., Dube, P., Layton, J., Anderson, E. J., Kaniewska, M., Proksch, R. K., Horsch, R. B., Rogers, S. G., Fraley, R. T., and Beachy, R. N. (1988). *Bio/Technology* 6, 403–409.

Owens, R. A., and Kaper, J. M. (1977). *Virology* 80, 196–803.

Piazzolla, P., Tousignant, M. E., and Kaper, J. M. (1982). *Virology* 122, 147–157.

Putz, C., Kuszala, J., Kuszala, M., and Spindler, C. 1974). *Ann. Phytopathol.* 6, 139–154.

Rezaian, M. A., Skene, K. G. M., and Ellis, J. G. (1988). *Pl. Mol. Biol.* 11, 463–471.

Rochow, W. F. (1977). In *Aphids as Virus Vectors* (Ed. by K. F. Harris and K. Maramorosch), pp. 253–273. New York: Academic Press, Inc.

Rosner, A., Bar-Joseph, M., Moscovitz, M., and Mevarech, M. (1983). *Phytopathology* 73, 699–702.

Simon, A. E. (1988). *Plant Molecular Biology Reporter* 6, 240–252.

Simon, A. E., and Howell, S. H. (1986). *EMBO J.* 5, 3423–3428.

Sleat, D. E., and Palukaitis, P. (1990). *Proc. Natl. Acad. Sci. U.S.A.* 87, 2946–2950.

Smith, C. R., Tousignant, M. E., Geletka, L. M., and Kaper, J. M. (1992). *Plant Dis.* (in Press).

Steen, M. T., Kaper, J. M., Pleij, C. W. A., and Hansen, J. N. (1990). *Virus Genes* 4, 41–52.

Symons, R. H. (1989). *TIBS* 14, 445–449.

Takanami, Y. (1981). *Virology* 109, 120–126.

Tien, P., and Wu G-S. (1991). In *Advances in Virus Research, Vol. 39* (Ed. by K. Maramorosch, F. Murphy, and A. Shatkin), pp. 321–338. New York: Academic Press, Inc.

Tien, P., Zhang, X-H., Qiu, B-S., Qin, B., and Wu, G-S. (1987). *Ann. Appl.Biol.* 111, 143–152.

Tomlinson, J. A. (1987). *Ann. Appl. Biol.* 110, 661–681.

Tumer, N. E., O'Connell, K. M., Nelson, R. S., Sanders, P R., Beachy, R. N., Fraley, R. T., and Shah, D. M. (1987). *EMBO J.* 6, 1181–1188.

Valverde, R. A., and Dodds, J. A. (1986). *J. Gen. Virol.* 67, 1875–1884.

van den Elzen, P. J. M., Huisman, M. J., Willink, D. P-L., Jongedijk, E., Hoekema, A., and Cornelissen, B. J. C. (1989). *Pl. Mol. Biol.* 13, 337–346.

Waterhouse, P. M., and Murant, A. F. (1983). *Ann. Appl. Biol.* 103, 455–464.

Waterworth, H. E., Kaper, J. M., and Tousignant, M. E. (1979). *Science* 204, 845–847.

White, J. L., and Kaper, J. M. (1989). *J. Virol. Meth.* 23, 83–94.

Wilson, T. M. A. (1989). *BioEssays* 10, 179–186.

Wu, G-S., Kang, L-Y., and Tien, P. (1989). *Ann. Appl. Biol.* 114, 489–496.

Wu, G-S., Kaper, J. M., and Jaspars, E. M. J. (1991). *FEBS Letters* 292, 213–216.

Yie, Y., Zhao, F., Zhao, S. Z., Liu, Y. Z., and Tien, P. (1992). *Molecular Plant-Microbe Interactions* (in Press).

Yoshida, K., Goto, T., and Iizuka, M. (1985). *Ann. Phytopath. Soc. Japan* 51, 238–242.

12

Production of Bacillus thuringiensis Insecticidal Proteins in Plants

Kenneth A. Barton and Michael J. Miller

Agracetus, Inc.
Middleton, Wisconsin 53562

Intensive efforts to exploit classical breeding and molecular technologies in the protection of crops from insects make it easy to forget that the majority of plants in the world are effectively protected from most significant insect pests. Natural resistance is a complex blend of inherent genetic and ecological resistance factors that are derived from many years of coincidental evolution of plant and pest species. Intensive agricultural practices, however, now generally displace the thoroughbred crops from natural settings, creating artificial environments that are quite suitable for pest infestations. The ecological safeguards developed through years of natural selection have been supplanted by modern, highly intensive growth practices which in some ways actually foster pest proliferation. For example, an increasing dependence on chemical control of insect pests has created a strong selection pressure for the development of resistant insect populations, while coincidentally reducing the populations of insect predators of the pest species. New forms of genetic resistance in plants or improvements in the use of externally applied pesticides are needed to improve on the current susceptibility of major crops to insect pests.

Unfortunately, our present understanding of the natural mechanisms for plant resistance to insects is not well advanced. In those cases where a significant understanding is in hand, we have generally not yet developed the necessary technologies to exploit the mechanisms in other crops of agronomic importance. Development of molecular tools to assist in the design of new insect resistance mechanisms for plants is therefore sorely needed.

Current molecular efforts to develop insect-resistant crops are following the simplistic theme of transforming plants with single genes encoding insecticidal enzymes or toxins. The most widely researched genes of this type are the δ-endotoxins of *Bacillus thuringiensis*. The characterization of the bacterial hosts and their toxic components, the understanding of toxin mechanisms, the development of plant gene expression and transformation capabilities, and the movement toward commercialization of crops that demonstrate field resistance to insect pests have each become highly competitive activities pursued by many research groups throughout the world. Technical feasibility, commercial strategies to enable displacement or complementation of classical chemical pesticides, and acquisition of both regulatory approvals and society acceptance are all issues now being addressed.

12.1 *Bacillus thuringiensis*

Bacillus thuringiensis is a free-living soil bacterium that accumulates high levels of insecticidal proteins during sporulation (Hofte and Whiteley, 1989). The fermented bacteria have been utilized for several decades in insecticidal formulations, which have shown high specificity toward lepidopteran pests and are generally recognized to be environmentally safe (Andrews *et al.*, 1987). Practical limitations of the present microbial products include a relatively high cost of production compared to chemical alternatives, and poor persistence under field conditions resulting in a need for multiple applications. However, the high specificity of *B. thuringiensis* toxins for target pests presents a potential ecological advantage over many chemical insecticides, which are lethal to beneficial insects as well as pathogens, and may also be subject to persistence well after application. The acquisition of our current level of understanding of the *B. thuringiensis* organisms, and application of *B. thuringiensis* in development of commercial pesticides have been long processes. However, new microbial products and also transgenic plants may soon be moving toward commercialization.

12.1.1 HISTORICAL PERSPECTIVE OF *BACILLUS THURINGIENSIS*

The first reported isolation of a bacterium later determined to be *B. thuringiensis* occurred in 1902, with the characterization of the etiological agent of a disorder of silkworms (Aoki and Chegasaki, 1915). At that time, Ishiwata noted that young, actively growing cultures of the bacterium were not pathogenic to insect larvae, in contrast to the older cultures undergoing sporulation which were quite toxic. Some years later, Berliner isolated another bacterial strain responsible for a disorder of Mediterranean flour moth larvae (Berliner, 1915). He noted that a crystal inclusion body was present within the spores of this bacterium, and he was first to designate the isolates as *B. thuringiensis*. He did not then correlate the presence of the inclusion directly with pathogenicity. However, as might be expected, the identification of these microorganisms as the causative agent of insect larvae disease quickly led to activities to develop them for practical use as biological insecticidal agents. Through the first half of the twentieth century various attempts were made, with limited success, to achieve control of lepidopteran pests using microbial formulations of *B. thuringiensis*. Very little significant understanding of the toxic mechanisms was yet available.

In 1951, a crystal inclusion body was again noted to be present in a *B. thuringiensis* isolate that was responsible for pathogenic symptoms in still another lepidopteran insect species (Steinhaus, 1951). At about the same time, Hannay (1953) proposed that the crystals themselves, which he found to be primarily proteinaceous, were directly responsible for the insecticidal activity of *B. thuringiensis*. It was then not long until partial purifications of the peptides from the crystals were accomplished, and sufficient cytological evidence was gathered to show conclusively that the purified crystal proteins imparted the insecticidal activity (Angus, 1956). This understanding provoked renewed interest in exploiting the toxic proteins to develop highly selective, environmentally safe biopesticides, with such product development activity continuing to this time. Various *B. thuringiensis* isolates have now been empirically shown to be selectively lethal to many lepidopteran, coleopteran, and dipteran insects, where the two former orders encompass many of the approximately 10,000 recognized insect pathogens of agricultural crops.

Bacillus thuringiensis has been utilized as a commercial insecticide in the United States since 1958, primarily in formulations of spores generated by fermentation (Aronson *et al.*, 1986; Andrews *et al.*, 1987). Annual production is still increasing in the United States and throughout the world, and currently amounts to greater than 2 million kg (Rowe and Margaritis, 1987). Practical limitations of the products (high production costs and poor persistence) continue to limit product acceptance. However, the high specificity of *B. thuringiensis* toxins presents a significant ecological advantage over broad-spectrum chemical insecticides, which are lethal to beneficial insects as well as to pathogens. The prospect of retaining or further improving upon the high specificity of this insecticide toward phytophagous pest species, while minimizing the limitations inherent in present microbial formulations is the motivation for efforts to produce the bacterial peptides directly in those crops in need of protection.

12.1.2 *BACILLUS THURINGIENSIS* GENE CHARACTERIZATION

Since the first *B. thuringiensis* toxin gene DNA sequence was published in 1985 (Schnepf *et al.*, 1985; Schnepf and Whiteley, 1985), approximately fifty additional gene sequences have been reported. Many of the genes have been found to share extensive homologies based on DNA sequence, immunological cross reactivity, or derived protein sequence information. The homology exists despite isolation of the genes from widely gathered organisms that are characterized by many distinct physiological attributes. It now appears that at least four unique classes of specifically insecticidal genes can be identified, based on similarity between *cry* (crystal protein) gene sequence homologies (Hofte and Whiteley, 1989). The currently accepted classifications are lepidopteran-specific, *cryI;* lepidopteran- and dipteran-specific, *cryII;* coleopteran-specific, *cryIII;* and those active against dipteran species, *cryIV*. Many *B. thuringiensis* strains contain additional genes encoding alternative toxin peptides, or genes encoding enzymatic pathways for biosynthesis of other metabolites such as nucleoside analogues that have broad cytotoxic activities against organisms other than insects. However, due to the broad target spectrum, such gene products have limited utility for expression in plants in the near future. An exception may be nematode-active *B. thuringiensis* genes, if it is found that proteins from such genes have a similar mode of action to the insect-active δ-endotoxin genes characterized to date (Edwards *et al.*, 1990).

The lepidopteran-active *cryI* genes are the most thoroughly studied of all δ-endotoxin genes. The *cryI* genes encode proteinaceous protoxins of 130–140 kDa, which are localized within bacterial spores as bipyramidal crystal inclusions. The δ-endotoxins, and associated peptides present in lesser quantities, can amount to greater than 20% of the weight of the spore. The proteins of these crystals will disperse when exposed to an alkaline medium (pH 9–12), such as that found in the gut of lepidopteran insects. Proteolytic activity from various sources, including both bacterial residue and insecticidal gut juices, is then capable of degrading the high molecular weight protoxins to smaller, insecticidally active peptides of 65–70 kDa (Whiteley and Schnepf, 1986).

The toxic domain of the lepidopteran-active endotoxins is the amino-terminal portion of the protoxin, contained within approximately the first 600 amino acids. The immediate terminus, approximately amino acids 1–29, and the carboxyl terminus, approximately amino acids 603–1176 on *cryI* gene products, are not essential for toxin activity. Both deletions and fusions of additional peptides to the amino terminus are feasible (Schnepf and Whiteley, 1985). The carboxyl terminus of the protoxin is dispensable by either truncation of the protoxin at various points following amino acid 602, or fusion of the cleaved carboxyl terminus to various other peptides (Schnepf and Whiteley, 1985; Vaeck *et al.*, 1987). The natural function of the carboxyl terminus of the protoxins is not known, beyond presumed contributions of these sequences to crystallization within the bacterial spores. Surprisingly, the carboxyl termini of various sequenced protoxin genes appear to be more highly conserved than the amino-terminal sequences (Hofte and Whiteley, 1989). While the reasons for such high conservation are unclear, the majority of cysteine residues are located in the carboxyl terminus, further implying that this region provides structural character to the protoxin. The high degree of sequence conservation and the role of the carboxy region in crystal formation presumably facilitates coassembly of multiple δ-endotoxin species into the same crystals in sporulating bacteria, in those hosts where different δ-endotoxin genes are present. Because amino acid composition of the amino- and carboxyl-terminal halves vary considerably and are generally complementary, the protoxin may function in the germinating spore as a stored nutrient reserve. However, no contributions of the carboxyl-terminal portion of the protoxin have yet been ascribed directly to insect toxicity.

Among the lepidopteran-active δ-endotoxins, sequence differences can be found throughout much of the toxin region, although various domains of strong sequence homology are evident. The individual *cryI* genes generally display broad toxicity toward a number of different lepidopteran species. However, variations in the *cryI* protein sequences contribute to differing levels of toxic activities toward the different susceptible insects.

One problem that surfaces frequently in published literature describing *B. thuringiensis* toxin analyses is the lack of recognition that many *B. thuringiensis* spore formulations or even partially purified protein preparations are actually complex mixtures of various δ-endotoxins. Occasionally, such preparations are additionally contaminated with alternative toxic activities such as the β-exotoxin (Lecadet and de Barjac, 1981). Moreover, the potential for septicemia from germination of bacterial spores within infected larvae, a factor which plays a further role in toxicity against various hosts, is also infrequently considered. As a result, interpretation of published literature discussing the toxic spectrum of individual *B. thuringiensis* genes must be tempered by a consideration of the purity of the individual toxin proteins under experimentation. The importance of appropriate interpretation of published results from toxins produced in microbes relates to a significant aspect of plant engineering: only a single, or very few, well-characterized toxin-encoding genes will be engineered into a resistant plant species. In those reports where detailed work has been conducted utilizing well purified *cryI* gene products, overlapping lepidopteran toxicities have been observed, and varying specific activities have been noted between the different *cryI* δ-endotoxins.

A second class of δ-endotoxins of significant agricultural import includes those with coleopteran toxicity, the *cryIII* genes (Hofte and Whiteley, 1989). These toxins form rhomboidal crystal inclusions in spores, and have been found to date in three morphologically different *B. thuringiensis* isolates (Herrnstadt *et al.*, 1986; Krieg *et al.*, 1983; Donovan *et al.*, 1988). Surprisingly, the DNA sequences of the protoxin genes from each of the three strains have been found to be identical. However, the precise character of the crystallized protoxin or the final insecticidal fragments may differ, due to the potential for alternative proteolytic processing occurring within the several hosts. The genes each encode a

72-kDa peptide that is processed to an approximately 66-kDa toxin by the proteolytic cleavage of 57 amino-terminal residues (McPherson *et al.*, 1988). The toxin-active fragment bears noticeable homology to the other *cry* classes of toxins, although the entire carboxyl-terminal domain is absent.

The *cryIII* gene products have shown toxicity toward various *Tenebrio* species (Krieg *et al.*, 1983) as well as to the Colorado potato beetle (McPherson *et al.*, 1988), which is a significant agricultural pest. Reports of toxicity toward alternative coleopteran pests, including various rootworms and cotton boll weevils, have not been confirmed. The activity of *cryIII* peptides toward potato beetles has been shown to be greatest during the early larval stages, although some residual activity remains against the adults.

Two additional *B. thuringiensis* δ-endotoxin classes have been identified, including the *cryII* genes, with complex activity spectra including various lepidopteran, dipteran, and possibly coleopteran insects, and the *cryIV* genes which are primarily dipteran active (Hofte and Whiteley, 1989). The former class, the *cryII* genes, may prove to be agriculturally useful following more extensive characterization. The latter class, *cryIV*, has already been subject to considerable scrutiny due to the disease-carrying properties of susceptible mosquitoes and flies, but relatively few dipterans are considered to be major agronomic pest species. An additional proteinaceous toxin activity of some *B. thuringiensis* isolates is a 27-kDa peptide (encoded by *cytA*) with little or no sequence homology to the δ-endotoxins. This toxin is widely cytolytic to cells of both vertebrates and invertebrates and has questionable insecticidal activity (Federici *et al.*, 1990). Even though the *cytA* gene product has not been reported to be toxic to plant cells, like the β-exotoxins, it is of doubtful utility in the design of insect-resistant plants due to the broad spectrum of target organisms.

Despite extensive divergence between the four *cry* gene classes, regions of both protein and DNA homology located throughout the toxin domain have been identified. As mentioned in discussions of *cryI* genes, the carboxyl terminus of each *cry* classification shows there is strong homology to the others (Hofte and Whiteley, 1989). The conservation of strategic regions of the toxin domain of the δ-endotoxins implies a conservation of secondary structural characteristics that are presumably involved with insecticidal character of the toxins in general. As this aspect of the δ-endotoxins becomes better understood, it may be possible to design chimeric peptides from the various genes of differing specific activity, to derive new genes with broader and more efficacious toxicity spectra.

12.1.3 MODE OF TOXIN ACTION

In nature, the parasporal crystal proteins are ingested by insect larvae, with the crystal then being solubilized in the highly alkaline environment of the gut (Brousseau and Masson, 1988). Peptides from the crystal which have already been partially processed by bacterial enzymes are further proteolyzed by activity of larvae gut enzymes. Both the solubilization and proteolytic processing of protoxins can be affected specifically by the insect larvae, due to differences in gut protease activities and other genotype-specific chemistry (Jaquet *et al.*, 1987; Haider and Ellar, 1987). The resulting protease-resistant core of the protoxin forms a toxin-active fragment (Aronson *et al.*, 1986; Lecadet and Martouret, 1967) which can display differing pesticidal activities depending upon the specific path of processing toward toxin activation (Haider and Ellar, 1987). Both amino- and carboxyl-terminal endpoints have been empirically determined to define the final toxin fragment of some toxins, which is defined by approximately amino acids 29–602 for lepidopteran-active genes (Adang *et al.*, 1985; Hofte *et al.*, 1986; Schnepf and Whiteley, 1985). The toxic fragment of the coleopteran-active *cryIII* genes appears to correspond quite closely to that of the lepidopteran genes when aligned based upon primary sequence homology, with structurally similar amino and carboxyl endpoints.

Once in active form within the gut of a susceptible insect, the toxin acts on the apical membrane of the midgut epithelial cells (Fast and Donaghue, 1971; Harvey and Wolfersberger, 1979). Proposed mechanisms of toxicity most likely involve specific interactions between the toxins and individual receptors located in the midgut. Perhaps the most enlightening work in this area has come from the use of monoclonal antibodies targeted toward specific epitopes found on various of the δ-endotoxin species (Huber-Lukac *et al.*, 1986). The binding of various monoclonals to newly identified δ-endotoxins has been correlated with the spectrum of insecticidal specificities of the toxins, providing a correlative screen for toxin activity based on sequence variation at the primary amino acid level. Since the initial work demonstrating the useful potential for monoclonal screening of δ-endotoxins, antibodies have been used to distinguish not only the broad classifications of new toxins (Hofte *et al.*, 1988) but also to predict the target spectrum among more closely related insect species. While subtle variations in primary sequence of the various δ-endotoxins might elude an immunological screen, the ease of such screening provides a significant advantage over alternative methods in identifying and characterizing new toxins at present. In addition, the use of monoclonals has enabled identification of specific toxin receptors within susceptible insects, and as a result has contributed toward our understanding of toxic mechanisms.

Receptor sites for the toxins vary between insects, both in number of potential sites and affinities for individual peptide toxins, thus imparting another level of control over the activity spectra of individual toxins (Hofmann *et al.*, 1988). Although the precise mode of toxic action of the *B. thuringiensis* toxins has not yet been revealed, various other bacterial toxins have the ability to recognize surface receptors prior to insertion into or translocation across membranes. It seems quite likely that this is the case with δ-endotoxins. The toxic mechanism may involve a receptor-binding activity on the cellular surface, followed by cytolysis due to either membrane disruption or alternative gross effects on translocation processes. The two events would likely have individual elements of specificity; that is, cell binding may occur without cytolysis if the two different specificities are not met.

A variety of physiological and behavioral effects of the ingested toxins have been reported, providing additional clues to the toxic mode of action. Shortly after ingestion of toxin, larvae appear to experience paralysis of both the gut and mouth parts, which leads to inhibition in feeding (Cooksey, 1971; Heimpel and Angus, 1959). Cells within the midgut epithelium have been shown to alternatively swell or collapse, with disintegration of the microvilli and associated tissues (Luthy and Ebersold, 1981). Some of these events can be observed *in vitro* with cultured insect cells (English and Cantley, 1986). At this level, cells exposed to active toxins have been reported to show stimulated glucose uptake, altered midgut and hemolymph pH, reduced midgut ATP concentrations, and altered potassium transport. Through various possible mechanisms, the result is larval death. A series of recent reviews will more completely indicate findings to date relating to the precise mechanism of toxicity, which appears to involve alteration of membrane transport mechanisms (Andrews *et al.*, 1987; Himeno, 1987). Host specificity for toxin activity, as mentioned above, can be imparted at various levels. However, results of many years of experimentation with a variety of toxins strongly conclude that the *cryI* δ-endotoxins under study up to this time appear to be quite insect specific.

12.2 *Bacillus thuringiensis* Toxin in Plants

In the years following cloning and characterization of the first *B. thuringiensis* δ-endotoxin gene in 1985 (Schnepf and Whiteley, 1985), more than fifty additional sequences have been reported. With concurrent development of plant transformation and gene expression capabilities, it is understandable that these genes have become a focal point for use in plant variety improvement.

Although considerable research effort has centered on production of these insecticidal toxins in plants, the strong commercial focus of such work has limited the amount of experimental detail which has appeared to date in reviewed journals. Those publications which have appeared cite comparable difficulties and varying degrees of success in achieving the research goals (Barton *et al.*, 1987; Fischoff *et al.*, 1987; Vaeck *et al.*, 1987). A problem encountered by all research groups pursuing this activity has been the difficulty in obtaining high-level expression in plants of the chimeric *B. thuringiensis* toxin genes, at least relative to alternative experiences with various other prokaryotic or eukaryotic genes. The identification of problems relating to *B. thuringiensis* gene expression in plants, and overcoming these experimental difficulties in order to achieve efficient gene expression, have served as valuable experience for broad development of plant gene expression capabilities.

12.2.1 TARGET CROPS

Economic incentives dictate that the initial commercial targets for expression of *B. thuringiensis* toxin genes will be crops with the most significant problems. In such crops, seeds with the engineered and heritable resistance will compete for grower acceptance against the current practices of intensive and costly pesticide application. Candidate crops with major commercial potential include corn and cotton, which in the United States suffer massive annual crop losses despite extensive use of insecticides. Of these two crops, both technical feasibility and tremendous market potential are converging to make it quite likely that insect-resistant cotton will become the first "blockbuster" product to emerge from the agricultural biotechnology industry.

Cotton is grown on greater than 10 million acres in the United States alone and generates an estimated $4 billion dollars in annual crop value worldwide (Suguiyama and Osteen, 1988). Insect damage to cotton bolls is the problem of major concern, with solutions made increasingly complex by the development of chemically resistant insect pest species and an increasing public awareness of broad chemical effects on the environment. The major lepidopteran insect pests to be controlled in cotton include the *Heliothis* species, cotton bollworm (*H. zea*) and tobacco budworm (*H. virescens*), and in the southwestern United States the pink bollworm, *Pectinophora gossypiella*. Occasional localized invasions of armyworms (*Spodoptera exigua*) occur, and the coleopteran species, *Anthonomus grandis* or cotton boll weevil, continues to plague regions of the southern United States. Current yield losses and costs of insect control measures for these pests of cotton amount to an estimated $645 million dollars per year in the United States (Perlak *et al.*, 1990).

Corn is subject to economic concerns of similar magnitude, but technical progress relative to that in cotton has been delayed. With the recent development of transformation methodology for corn (Gordon-Kamm *et al.*, 1990), efforts to develop resistance to the European corn borer (*Ostrinia nubilalis*) and the corn rootworms (*Diabrotica longicornis barberi, D. virgifera virgifera, D. undecimpunctata howardi*) will intensify. Soybeans, wheat, sorghum, peanuts, tobacco, and an assortment of vegetable crops are each subject to lesser but still significant total insect pressures. These crops and many others may eventually benefit from technology now under development, but cotton is the distinct early target where projected revenues can most reasonably be expected to offset the substantial research, field testing, and registration expenditures.

12.2.2 CROP TRANSFORMATION TECHNOLOGY

Over the past decade, various techniques have been developed for moving genes into plants. With few exceptions, the utility of each new technology has been short term and limited in scope. Typically, transformation techniques necessitate the manipulation of plant cells in tissue culture, frequently to the point of protoplasting, followed by the subsequent regeneration of transformed cells into fertile plants. A hurdle inherent to this approach is the generally poor efficiency of most major crops in movement into cell culture and return to intact, healthy plants. Even for those crops where this culturing process is feasible, genotype specificity, extensive manpower, and prolonged time frames are frequently required to generate each transformant.

Of all the more widely explored transformation techniques, many described in other chapters of this volume, systems derived from the plant pathogen *Agrobacterium tumefaciens* have met with the broadest success. *Agrobacterium* vectors have been integrally responsible for initial development and further refinement of the selectable and screenable markers used to optimize most alternative transformation methods, and *Agrobacterium* vectors remain useful today for the many dicotyledonous crops receptive to *Agrobacterium* infection.

Crops that served as useful models for early development of *Agrobacterium* transformation technology, tobacco and tomato, are coincidentally crops with high susceptibility to the major cotton insect pests. It is therefore not surprising that initial reports of the expression of *B. thuringiensis* δ-endotoxins in plants were conducted in these two model plant systems (Barton *et al.*, 1987; Fischoff *et al.*, 1987; Vaeck *et al.*, 1987). With the simultaneous development of cotton transformation techniques utilizing *Agrobacterium* vectors (Umbeck *et al.*, 1987), it was a logical progression to extend the *B. thuringiensis* studies from model systems into commercial cotton (Umbeck *et al.*, unpublished; Perlak *et al.*, 1990). Although a number of laboratories have been simultaneously pursuing research toward insect-resistant crops, with cotton as a common target there has been extensive overlap between each of the program activities. Coincidental solutions were therefore frequently derived for many of the difficulties that were encountered by each laboratory. We will discuss below our own development of the technology required for *B. thuringiensis* expression in cotton, and note unique aspects of other programs where this has been evident.

12.2.3 "FIRST GENERATION" CHIMERIC GENES

Utilizing the first cloned and characterized δ-endotoxin gene (Schnepf and Whiteley, 1985), our laboratory initiated construction of chimeric genes to enable transcriptional and translational expression of the coding regions in plant cells. We constructed chimeras encoding the complete protoxin sequence, a peptide of

1176 amino acids (Schnepf *et al.*, 1985), as well as genes encoding an truncated portion of the protoxin, the amino-terminal 645 amino acids (Schnepf and Whiteley, 1985). We anticipated that the entire coding region might be essential for appropriate folding of the peptide into the stable secondary structure found in parasporal crystals. However, we were uncertain whether that or the smaller of the two toxin forms would prove optimal for both initial expression and structural stability in plant cells. Various early constructions included nopaline synthase (nopaline dehydrogenase) or CaMV 35S transcriptional promoters, with polyadenylation signal sequences from various *Agrobacterium* or plant-derived genes (Barton *et al.*, 1987).

Our attempts to express the entire 1176-amino-acid protoxin in plant cells presented unexpected problems. Following *Agrobacterium* transformations and selection for plant cells exhibiting kanamycin resistance, cell cultures developed that were typically slow growing, with considerable necrotic tissue present. Immunoblots demonstrated expression of the *B. thuringiensis* chimeras in the cells, with calli typically containing 10–50 ng of protoxin per mg plant protein. However, all calli expressing significant levels of protoxin gradually became necrotic and died. Frequently, regions of green and rapidly growing tissue would outgrow the remaining calli, but the healthy calli no longer showed positive immunoresponses to *B. thuringiensis* toxin antisera. No plants regenerated from calli transformed with this protoxin gene were resistant to insect larvae. A high proportion of the regenerated plants were found to have broken or entirely deleted *B. thuringiensis* genes. And finally, no plants with intact protoxin genes showed measurable levels of either protoxin mRNA or protein. We concluded that expression of the intact protoxin gene was deleterious to plant cell viability, resulting in a negative selection against highly expressing gene insertion events. We did not know at that time whether the mechanism of toxicity to plant cells was related to the insecticidal mode of action.

Schnepf and Whiteley reported that a truncated form of the gene, encoding only the amino-terminal portion of the protoxin, retained essentially all the insecticidal activity of the intact gene (Schnepf and Whiteley, 1985). Based on this report, we truncated our protoxin gene to encode 645 amino acids. Expression of such a gene in *E. coli* enabled production of an insecticidally active peptide that did not form the typical crystal inclusion (Schnepf and Whiteley, 1985). Plant expression cassettes encoding such a peptide were transformed into tobacco. In this case, transgenic calli were obtained that appeared morphologically indistinguishable from control transformants without *B. thuringiensis* genes. Immunoblot screening of the transformed calli showed a broad range of expression, with absolute levels of toxin generally at 0.001% of the total cell protein or lower. The range of expression of these genes was typical of chimeric gene expression, with variation in levels of expression due to insertion into differing chromosomal locations. Plants were easily regenerated from the transformed calli, and the plants appeared to be as healthy as controls. Insect feeding trials were conducted to assess resistance of the transformed plants to tobacco hornworm larvae (*Manduca sexta*), and Northern, Southern and Western blots were conducted to characterize the inserted gene and the levels of gene expression. Following these analyses, we noted good correlations between the observed levels of mRNA, toxin peptides, and insect resistance in each transformant (Barton *et al.*, 1987).

Insect bioassays proved to be our most sensitive monitor of gene expression. The effects of the toxin in reduction of both insect feeding and growth rates were evident even when levels of gene expression were beneath the sensitivity of our molecular assays. Furthermore, the behavioral effects on larvae during feeding on the transgenic plants and the pathological symptoms we observed in diseased larvae were similar to those noted in larvae that were fed a diet of plant tissues "spiked" with toxin purified from bacterial spores (Barton and Miller, unpublished). It therefore appeared that truncation of the protoxin enabled production of the *B. thuringiensis* peptides in plant cells without deleterious effects to the plant cells, and that ingestion of the plant-produced toxins had similar effects upon insect larvae as the natural toxin from bacterial spores. One very encouraging finding was that ingestion of plant-produced toxin also caused a reduction in feeding activity even prior to mortality. However, we were concerned that the absolute levels of *B. thuringiensis* gene expression from the population of plants generated in these experiments were far below those observed for other gene chimeras utilized in our laboratories up to that time, for reasons we could not explain.

Other laboratories were reporting similar research programs in 1987, with similar experimental results (Fischoff *et al.*, 1987; Vaeck *et al.*, 1987). Vaeck and co-workers developed transgenic tobacco plants following transfer of several gene constructions including intact protoxin genes, truncated genes, and truncated genes with carboxyl peptide fusions. The plants generated with intact protoxin genes showed no significant insect resistance or *B. thuringiensis* gene expression; they attributed the lack of resistance to inefficient translation of the protoxin gene, and did not report analysis of gene expression in cell culture. Genes encoding truncated or fusion peptides provided measurable levels of gene expression in their study, as well as significant insect resistance. However, they also noted that levels of gene expression were considerably lower than anticipated, and too low to provide practical levels of resistance toward less sensitive insect pests such as *Spodoptera* or *Heliothis*.

12.2.4 SYNTHETIC *BACILLUS THURINGIENSIS*

The level of expression found with truncated gene constructs from our experiments was in many transformants sufficient to provide complete resistance to tobacco hornworms, characterized by minimal plant damage due to insect feeding and complete mortality of test larvae. However, tobacco hornworms are relatively sensitive to the *B. thuringiensis* toxin, while more significant agronomic pest species such as *Heliothis* (cotton boll worms or tobacco budworms) or *Spodoptera* (various armyworms) require significantly higher doses of toxin to cause death (MacIntosh *et al.*, 1990). In our most resistant tobacco plants with the "first generation" truncated gene constructions we did observe reduced feeding by these species, reduced larvae growth rates, abnormal pupae development, and increased mortality. However, significant plant tissue destruction was apparent (Barton *et al.*, 1987).

The reason for low levels of accumulation of the *B. thuringiensis* peptides was not clear. The problem was noted by other researchers despite their use of various *B. thuringiensis* genes differing significantly in both DNA and peptide sequences, and despite design differences in the resulting chimeric genes. We therefore reasoned that the cause of poor expression was inherent in *B. thuringiensis* toxin-coding regions in general, rather than a result of our specific experimental design. While this assumption helped to limit the potential courses of action we might consider to obtain increased gene expression, many possibilities still remained. Because we observed no apparent morphological differences between transgenic plants containing the highest levels of *B. thuringiensis* peptides and wild-type plants containing no toxin, we assumed that the truncated toxin was not deleterious to plant cell growth and that no negative selection was in effect against high-level gene expression. The problem appeared to be a result of either inefficiency in the synthesis of gene products, or rapid degradation of either the *B. thuringiensis* mRNA or protein following synthesis. Following analyses of results from some additional experiments involving targeting of toxin peptides to various cellular locations (unpublished), we concluded that protein instability leading to rapid turnover was not the primary block to achieving high-level expression. We then began to focus our interest on *B. thuringiensis* mRNA synthesis and stability.

Northern blots of mRNA from tobacco transformants expressing our truncated *B. thuringiensis* genes showed very low steady-state levels of mRNA, with evidence of varying degrees of degradation (Barton *et al.*, 1987). Other groups reported similar observations (Fischoff *et al.*, 1987; Vaeck *et al.*, 1987). Following indications that rapid turnover of *B. thuringiensis* mRNA might be occurring, we conducted a computer analysis of the primary sequences of various δ-endotoxin genes. We then compared *B. thuringiensis* gene sequences to plant gene sequences, to identify common features of the *B. thuringiensis* gene family that might be responsible for the problems we observed. We noted that the *B. thuringiensis* genes are characterized by an exceptionally high (A + T) content relative to most previously characterized genes from higher plants. Our computer analyses revealed additional sequence differences between *B. thuringiensis* and plant genes, where most variations (secondary structure predictions, codon usage, melting temperatures) were correlated with the differences in base composition. Scanning of *B. thuringiensis* gene sequences for potential nuclease sensitivity, polyadenylation sites, or splice junctions showed candidate sites, but the accuracy of such predictions is questionable.

Following our analysis of the *B. thuringiensis* gene mRNA primary structure, which identified general differences from plant gene structure, we set out to construct a series of synthetic genes. We sequentially replaced the native DNA sequence from *B. thuringiensis* gene *cryI(a)* with synthetic DNA. Utilizing degeneracy in the genetic code to conserve the original peptide sequence, we followed rules we derived from a comprehensive survey of plant genes to generate a coding region more resembling that of plant gene mRNAs. Our synthetic alterations were made in blocks of approximately 50 codons, in order to monitor the effect on gene expression of each synthetic block. This was done because of our concern that a comprehensive redesign of the native gene sequence might impart deleterious effects on gene expression as well as potentially positive effects. In analysis of these constructions, we observed dramatic elevations in gene expression following synthetic changes near the amino terminus of the coding region, with a continued increase in expression as larger blocks of synthetic DNA were inserted. Surprisingly, with less than 10% of the peptide coding region redesigned, peptide accumulations of greater than 0.01% of the cellular protein were found in transformed plants, representing an increase in average expression of greater than 100-fold relative to the levels seen with transformants expressing the wild-type DNA coding sequence. The most significant obstacle to *B. thuringiensis* gene expression in plants therefore appears to be due to gross primary sequence differences from typical plant genes. We are continuing to explore the more dramatic enhancement we observed following sequence alterations near the amino terminus of the gene.

A recent report by Perlak *et al.*, (1990) describes a gene reconstruction similar to our own, in which they generated synthetic genes encoding two different *B. thuringiensis* toxins, *cryI(b)* and *cryI(c)*, that showed less than 80% homology with the original wild-type DNA sequences. Based on a combination of effects from DNA synthesis, promoter enhancement, and vector design, they reported toxin peptide concentrations of 0.1% of the cell protein. These levels of expression are quite sufficient to provide complete mortality to susceptible insects. Even those insect species resistant to *B. thuringiensis* toxins might be subject to increased control under field conditions, as a result of predator insect populations which would be expected to increase with the reduced insecticidal applications projected for such crops. There now is little doubt that it is quite feasible to generate new crop varieties expressing very high levels of insect resistance.

12.2.5 FIELD TRIALS OF TRANSGENIC PLANTS

Research to overcome poor efficiency of *B. thuringiensis* gene expression in plants has only recently been successful. However, movement toward commercialization of this technology has proceeded, beginning with field trials of plants containing early versions of chimeric *B. thuringiensis* genes. Through the end of 1990, at least 7 different research groups had conducted field trials of plants containing *B. thuringiensis* genes (Goldburg and Tjaden, 1990). Plant species undergoing trials up to that time include tobacco, tomato, potato, and cotton. Because synthetic gene technology is quite new, no published results of field trials utilizing such genes are yet available. However, any lack of adequate resistance reported in trials prior to synthetic gene technology is less of an important concern than it might have been, if improvements in expression of *B. thuringiensis* genes in field crops were not to be readily anticipated. In spite of the comparatively low expression levels, even "first generation" genes have shown some success in insect control (Delannay *et al.*, 1989). Additional knowledge derived from trials to date include the opportunity to resolve concerns of society, reveal the expectations of regulatory agencies, and correlate laboratory data with field trial results.

12.3 Future Issues

The transfer of *B. thuringiensis* genes from the genomes of bacteria to those of transgenic commercial crops has been a long process, one which is not yet complete. While molecular technology has been developed to design genes that will efficiently express in plant cells, the effectiveness of the resulting transgenic plants under field conditions has not yet been fully explored. Relatively little information has been published relating to field trials of plants containing "first generation" *B. thuringiensis* genes, and no reports have yet appeared of trials with plants exhibiting the high-level expression made possible from synthetic coding sequences. In a report by Delannay *et al.* (1989) describing the field performance of transgenic tomato expressing a first generation *B. thuringiensis* gene, a series of tests showed promise for the control of both tomato pinworm (*Keiferia lycopersicella*) and fruitworm (*Heliothis zea*). With the higher concentrations of toxin peptide in plant tissues now possible through new gene design, it is apparent that effective control of susceptible insect species is possible in tests of this limited scope.

By approximately 1995, transgenic plants producing levels of δ-endotoxins that confer commercially acceptable levels of insect resistance may be reaching the marketplace. To meet this target date, a great deal of additional information must be gathered through significantly expanded field trials. Commercial advantages proposed to justify the high development cost of *B. thuringiensis*-expressing plants are derived from financial savings at the crop production level. Through a reduction in pesticide utilization, and potentially through increased crop yields in situations where current practices still leave crops subject to damage, the farmer will receive financial incentives from the use of an insect-resistant crop. However, integrated pest management practices in high-value crops such as cotton currently involve a complex blend of genetics, cultivation, and pesticide usage (Rude, 1984). Field trials of the new insect-resistant varieties must therefore be designed to determine the most cost-efficient way to manage such a crop. In current practice with cotton, the frequent pesticide applications targeted toward *Heliothis* coincidentally control many other pest and beneficial insect populations. However, it is unlikely that all pest species will be eliminated by *B. thuringiensis* gene expression. Some alternative pests are not susceptible to *B. thuringiensis* toxins, and the feeding habits of others may effectively limit their ingestion of toxin. Whether such species will become of greater economic concern once chemical pesticide applications are reduced can only be determined through field observations. In fact, because current high doses of broad-spectrum insecticides also eliminate beneficial predator insect species, it is equally conceivable that an ecological synergy might provide a broader pest protection than is expected based on *B. thuringiensis* toxin spectrum alone. The production in plants of a toxin as selective toward target species as is *B. thuringiensis* will provide dramatic changes in cultivation practices, but precisely what those changes will be cannot be accurately predicted prior to extension of field use.

Other ecological effects of *B. thuringiensis* toxins presented in transgenic plants will also need to be addressed experimentally. Historically, strong selection pressures from intensive chemical application targeted at high numbers of insects have quickly fostered the development of resistant pest species. While *B. thuringiensis* microbial formulations have been relatively spared from the appearance of broad-spectrum resistance to date (McGaughey, 1985), the composition and application characteristics of microbial formulations are quite different from the presentation of toxin to insects directly from transgenic plants. Resistance management must therefore be explored, in order to establish appropriate protocols that will prolong the utility of resistant plant varieties. While resistance management practices are already an increasing emphasis for chemical insecticides, this is perhaps more critical an issue for transgenic crops due to the long development time frames leading to release of new plant varieties. Various theoretical models have been forwarded to propose how best to avoid development of insect resistance. Included are suggestions that resistant and wild-type plants might be mixed in the field to minimize selection pressures, or that multiple toxin activities presented either simultaneously or sequentially would prevent dispersal of new resistance genes acquired through mutation (Gould,

1988). Now that effectively resistant plants are in hand, such models can be experimentally tested.

Bacillus thuringiensis in microbial formulations is considered to be sufficiently safe that use of the pesticides on edible vegetables is permitted to day of harvest. Despite the absence of any evidence that *B. thuringiensis* toxins are harmful when ingested by mammals, many questions are now being raised pertaining to the safety of this new form of production of the peptides in plants (Goldberg and Tjaden, 1990). Relatively little experimental evidence has yet been compiled to address such questions. Although there is no reason to suspect increased hazards due to production of these insecticidal peptides in plants, some assessment is undoubtedly warranted. Two different considerations are apparent: Is there any new and real hazard generated by production of *B. thuringiensis* toxins in plants? Also, how much and what types of evidence of safety will be sufficient to warrant regulatory approval for commercialization? Whether answers to these questions will delay the release of plant varieties is not yet clear. While cotton derives the vast majority of commercial value from fiber, which at present provokes little apparent safety concern, the protein and oil of cotton is routinely destined for various feed purposes. The human and animal consumption of plant products containing *B. thuringiensis* peptides will be scrutinized in the United States by the Food and Drug Administration (FDA) and it is not yet clear what the response of the FDA will be toward these new applications of *B. thuringiensis*. High value of cotton fiber coupled with savings from the reduction in extremely high pesticide costs for the crop could result in a situation where the cotton oil and protein were discarded if not approved for consumption, but this is an unlikely occurrence due to the necessary segregation of transgenic and wild-type crops. The valuable components of many other crops, such as the tomato fruit, are exclusively targeted toward human or animal consumption, so crop introduction absolutely depends upon an approval for human consumption.

Although recognized to be safe based on decades of practical use in microbial pesticide formulations, the δ-endotoxins are pesticides and as such are subject to United States Environmental Protection Agency (EPA) oversight. Involvement of the U.S. Department of Agriculture (USDA) for field trials, FDA for food and health concerns, and EPA for environmental oversight in development of *B. thuringiensis*-producing crops imposes a high level of regulatory complexity. Gradual progress over the past decade has resulted in a "coordinated framework" for movement of transgenic plant applications through the various federal agencies toward commercial release. However, the path is still in flux due to the novelty of plants under consideration. Discussions of how plants with δ-endotoxin genes may proceed toward regulatory approval is detailed elsewhere (Purchase and MacKenzie, 1990). How readily approvals will be achieved is still a significant question.

Beyond considerations of scientific feasibility and product efficacy, after gaining regulatory approvals for commercialization, and further assuming that cost advantages will encourage farmer acceptance, there remains an issue of public perception toward *B. thuringiensis* toxin expression in plants. Will society view an insect-resistant plant as a solution or as another problem? The weight of scientific arguments from decades of *B. thuringiensis* use indicate that these toxins are not hazardous to nontarget organisms. In fact, *B. thuringiensis* pesticides are currently favored over chemicals for control of insects in many home gardens due to the perception of safety of these products. While our experience with *B. thuringiensis* production in plants is currently limited, there is at this time no evidence of increased hazard resulting from this technology. In contrast, some of the commonly applied insecticides which may be displaced following introduction of resistant plants are now under increasing scrutiny due to undesirable environmental side effects. It therefore seems reasonable that society might welcome such crops, and that regulatory agencies might expedite the progress toward crop introductions now that we are technically able to produce such plant varieties. Whether this proves to be the case will become more clear as we move *B. thuringiensis* expressing plants toward commercialization over the next several years.

References

Adang, M. J., Staver, M. J., and Rocheleau, T. A., *et al.*, (1985). *Gene* 36, 289–300.

Andrews, R. E., Faust, R. M., Wabiko, H., Raymond, K. C., and Bulla, L. A., Jr. (1987). *CRC Crit. Rev. Biotechnol.* 6, 163–230.

Angus, T. A. (1956). *Can. J. Microbiol.* 2, 122.

Aoki, K., and Chegasaki, Y. (1915). *Mitt. Med. Fak. Kais.* 13, 419.

Aronson, A. I., Beckman, W., and Dunn, P. (1986). *Microbiol. Rev.* 50, 1–24.

Barton, K. A., Whiteley, H. R., and Yang, N.-S. (1987). *Plant Physiol.* 85, 1103–1109.

Berliner, E. (1915). *Z. Angew. Entomol.* 3, 29.

Brousseau, R., and Masson, L. (1988). *Biotech. Adv.* 6, 697–724.

Cooksey, K. E. (1971). In *Microbial Control of Pests and Mites* (Ed. by H. D. Burges and N. W. Hussey), pp. 249–274. London, New York: Academic Press, Inc.

Delannay, X., LaVallee, B. J., Proksch, R. K., *et al.*, (1989). *Bio/Technology* 7, 1265–1269.

Donovan, W. P., Gonzalez, J. M., Gilbert, M. P., and Dankosik, C. (1988). *Mol. Gen. Genet.* 214, 365–372.

Edwards, D. L., Jewel, P., and Soares, G. G. (1990). United States Patent Application #4,948,734.

English, L. H., and Cantley, L. C. (1986). *J. Biol. Chem.* 261, 1170–1173.

Fast, P. G., and Donaghue, T. P. (1971). *J. Invert. Pathol.* 18, 135–138.

Federici, B. A., Luthy, P., and Ibarra, J. E. (1990). In *Bacterial Control of Mosquitoes and Blackflies: Biochemistry, Genetics and Applications of Bacillus thuringiensis and Bacillus sphaericus* (Ed. by H. de Barjac and S. Sutherland). New Brunswick, New Jersey: Rutgers University Press.

Fischhoff, D. A., Bowdish, K. S., Perlak, F. J., *et al.* (1987). *Bio/Technology* 5, 807–813.

Goldburg, R. J., and Tjaden, G. (1990). *Bio/Technology* 8, 1011–1015.

Gordon-Kamm, W. J., Spencer, T. M., Mangano, M. L., *et al.* (1990). *Plant Cell* 2, 603–618.

Gould, F. (1988). *BioScience* 38, 26–33.
Haider, M. Z., and Ellar, D. J. (1987). *Mol. Microbiol.* 1, 59–66.
Hannay, C. L. (1953). *Nature (London)* 172, 1004.
Harvey, W. R., and Wolfersberger, M. G. (1979). *J. Exp. Biol.* 83, 293–304.
Heimpel, A. M., and Angus, T. A. (1959). *J. Insect Pathol.* 1, 152–170.
Hernstadt, C., Soares, G. G., Wilcox, E. R., and Edwards, D. L. (1986). *Bio/Technology* 4, 305–308.
Himeno, M. (1987). *J. Toxicol.-Toxin Rev.* 6, 45–71.
Hofmann, C., Luthy, P., Hutter, R., and Pliska, V. (1988). *Eur. J. Biochem.* 173, 85–91.
Hofte, H., and Whiteley, H. R. (1989). *Microbiol. Rev.* 53(2), 242–255.
Hofte, H., De Greve, H., Seurinck, J., *et al.* (1986). *Eur. J. Biochem.* 161, 273–280.
Hofte, H., Van Rie, J., Jansens, S., *et al.* (1988). *Appl.Environ. Microbiol.* 54, 2010–2017.
Huber-Lukac, M., Jaquet, F., Luthy, P., Huetter, R., and Braun, D. G. (1986). *Infect. Immun.* 54, 228–232.
Jaquet, F., Hutter, R., and Luthy, P. (1987). *Appl. Environ. Microbiol.* 53, 500–504.
Krieg, A., Huger, A., Langenbruch, G., and Schnetter, W. (1983). *J. Appl. Entomol.* 96, 500–508.
Lecadet, M.-M., and de Barjac, H. (1981). In *Pathogenesis of Invertebrate Microbial Diseases* (Ed. by E. W. Davidson), pp. 293–321. Totowa, New Jersey: Allenheld, Osmun, and Co.
Lecadet, M.-M., and Martouret, D. (1967). *J. Invert. Pathol.* 9, 322–330.
Luthy, P., and Ebersold, H. R. (1982). In *Pathogenesis of Invertebrate Microbial Diseases* (Ed. by E. W. Davidson), pp. 235–267. Totowa, New Jersey: Allanheld, Osmun, and Co.
MacIntosh, S. C., Stone, T. B., Sims, S. R., *et al.* (1990). *J. Invert. Pathol.* 56, 258–266.
McGaughey, W. H. (1985). *Science* 299, 193–195.
McPherson, S. A., Perlak, F. J., Fuchs, R. L., *et al.* (1988). *Bio/Technology* 6, 61–66.
Perlak, F. J., Deaton, R. W., Armstrong, T. A., *et al.* (1990). *Bio/Technology* 8, 939–943.
Purchase, H. G., and MacKenzie, D. R. (1990). *Agriculture Biotechnology: Introduction to Field Testing* (Ed. by H. G. Purchase and D. R. MacKenzie). Office of Agricultural Biotechnology, U.S. Dept. of Agriculture.
Rowe, G. E., and Margaritis, A. (1987). *Crit. Rev. Biotechnol.* 6, 87–127.
Rude, P. A. (1984). *Integrated Pest Management for Cotton in the Western Region of the United States.* University of California, Division of Agriculture and Natural Resources, Publication 3305.
Schnepf, H. E., and Whiteley, H. R. (1985). *J. Biol. Chem.* 260, 6273–6280.
Schnepf, H. E., Wong, H. C., and Whiteley, H. R. (1985). *J. Biol. Chem.* 260, 6264–6272.
Steinhaus, E. A. (1951). *Hilgardia* 20, 259.
Suguiyama, L., and Osteen, C., (1988). *The Economic Importance of Cotton Insects and Mites.* U.S. Dept. of Agriculture Economic Research Service, Agricultural Economic Report Number 599.
Umbeck, P., Johnson, G., Barton, K., and Swain, W. (1987). *Bio/Technology* 5, 263–266.
Vaeck, M., Reynaerts, A., Hofte, H., *et al.* (1987). *Nature (London)* 328, 33–37.
Whiteley, H. R., and Schnepf, H. E. (1986). *Annu. Rev. Microbiol.* 40, 549–576.

13

Transgenic Plants Conferring Insect Tolerance: Protease Inhibitor Approach

Vaughan A. Hilder, Angharad M. R. Gatehouse, and Donald Boulter

Department of Biological Sciences
University of Durham
Durham DH1 3LI, United Kingdom

Insect predation is a major factor affecting the reduction in the worldwide crop yield. The advantages of using crop varieties which are inherently resistant to, or tolerant of, insect pests, as opposed to using exogenously applied controls or simply accepting insect damage, have already been outlined (e.g., Meeusen, 1986; Hilder *et al.*, 1990). Genetic engineering of crop plants can make a major contribution to the production of such inherently insect-tolerant varieties, since it opens up a virtually limitless source of germplasm variability from which to select insect control genes for introduction into elite crop varieties. As such it should complement, though not replace, conventional plant breeding.

As an increasing number of the crops having major worldwide importance become amenable to genetic transformation, the question of where to find genes encoding desirable characteristics becomes increasingly important. A logical source of such genes, described in Chapter 12 in this volume, is the entomicidal bacteria, such as *Bacillus thuringiensis*. Formulations of *B. thuringiensis* have been used for insect control long before genetic engineering became possible (Dulmage, 1981). We have adopted an alternative approach, focusing on those genes which have evolved in plants themselves for protection against herbivorous insects—the plant's own solution to the problem. In particular, we have initially studied inhibitors of insect digestive proteases.

13.1 Natural Defense Mechanisms of Plants

Ever since they first emerged, perhaps some 400 million years ago (Labandiera *et al.*, 1988), insects have fed on plants and various mechanisms have evolved in plants to limit the damage caused. In their turn certain insects have evolved specialized mechanisms of overcoming particular plant defenses. Thus we have the coevolution of insect–host plant–non-host plant complexes. In the artificial ecosystems created by agriculture, however, plants become exposed to insects which may not have been agents in their natural selection. Agricultural practices also may favor the rapid buildup of certain insect populations due to interference with the mechanisms which limit these populations naturally. Further, the threshold of damage to plants which can be tolerated in an agricultural system may be less than that which plants can sustain in the wild. The presence of pests is an indication of the disruption of the coevolutionary process.

Plants are unable to escape from insect predation by running away, but they have evolved an impressive array of alternative defenses. These plant-based defense mechanisms generally depend on three factors.

(1) *Temporal avoidance.* In some plants the life cycle is such that it avoids producing sensitive tissues at times when pest activity is maximal. Such a strategy has been mimicked in some integrated pest management agricultural schemes where the use of "early" or "late" varieties may allow the crop to escape serious predation.

(2) *Physical defenses.* Many mechanisms have evolved which render sensitive plant tissues physically difficult to "get at." Prime examples are the hard, lignified coats which enclose many seeds. Vegetative tissues may be made inaccessible to many types of insect by surface hairs, spines, or thickened, waxy cuticles. There has been an interesting demonstration of how the overall morphology of the plant can affect the relative performance of herbivores and their predators: ladybirds (*Coccinella septempunctata* and *Hippodamia variegata*) are less likely to fall off leafless peas and are, therefore, more effective at predating the pea aphid (*Acyrthosiphon pisum*) on leafless varieties (Kareiva and Sahakian, 1990). This example is cited because it highlights the complexity of real biological systems: It is seldom likely to be sufficient to consider just the short-term interaction of a particular plant phenotype with a single species of herbivore. Most crops are subject to a whole complex of pests, which in turn are susceptible to a complex of predators and pathogens, and our science should now be sufficiently advanced to at least recognize this complexity.

(3) *Chemical defenses.* One of the most important classes of natural defense mechanisms is chemical defense. Plants have a vast metabolic capability and produce many secondary chemicals, not necessary for their own vital processes, which are toxic, antimetabolic, or aversive to species which might other-

wise be potential predators (Norris and Kogan, 1980). For the most part these secondary compounds are relatively small organic molecules which are produced by complex metabolic pathways involving the controlled interaction of many different gene products. Familiar examples include the pyrethrins from chrysanthemums and alkaloids like nicotine from tobacco whose insecticidal effectiveness is such that they were amongst the earliest compounds to be used as exogenously applied pesticides. Other classes of plant secondary compounds which have been implicated in protection from insects include the terpenoids, steroids, flavanoids, phenolics, glucosinolates, cyanogenic glycosides, rotenoids, saponins, and nonprotein amino acids (see Gatehouse *et al.*, 1991). Production of some of these compounds imposes a demonstrable metabolic cost on the plant, indicated by a reduced fitness in the absence of predation, which suggests that their production in the plant is a selective response to insect feeding (e.g., Kakes, 1989; Baldwin *et al.*, 1990).

As they are the products of multienzyme pathways which involve the interaction of many gene products, such defense systems are in most cases too complex for the state of the art of plant genetic engineering, although work is in hand toward exploiting them (e.g., Dawson *et al.*, 1989). A few plant defense mechanisms, however, are based on proteins themselves, the product of a single gene. Because proteins are nonvolatile, they must be ingested to have an effect and the target site of most of them is the insect digestive system. For example, resistance of certain wild accessions of the common bean (*Phaseolus vulgaris*) to the Mexican bean beetle (*Zabrotes subfasciatus*) is due to a protein, arcelin-4, which largely replaces the conventional bean storage protein, phaseolin. Arcelin-4 is not digestible by this insect, and the larvae therefore starve to death on the resistant beans (Minney *et al.*, 1990). Proteinaceous inhibitors of insect digestive enzymes have a similar effect (Gatehouse *et al.*, 1979) and certain lectins bind to, and disrupt, cells of the insect gut epithelium (Gatehouse *et al.*, 1984). Ribosome-inactivating proteins (RIPs) such as the highly toxic ricin from castor beans have also been shown to be highly toxic to certain insects (Gatehouse *et al.*, 1990).

These types of single gene characters are eminently suitable for gene transfer using current technology. Being of plant origin, they have the added advantage that they are likely to have a high degree of compatibility with a transgenic host plant's metabolic system.

13.1.1 PROTEASE INHIBITORS AS INSECT CONTROL COMPOUNDS

Protease inhibitors are widely distributed within the plant kingdom. They are particularly abundant in seeds and storage organs where they may accumulate to around 1–10% of the total protein content (Ryan, 1981), or in exceptional cases to much higher levels (Pearce *et al.,* 1987). Although some of these inhibitors may have a role in endogenous protein metabolism (e.g., Ryan and Walker-Simmons, 1981; Poerio *et al.,* 1989), most of the protease inhibitors that have been characterized from plants do not inhibit endogenous plant proteases but have specificities for animal or microbial enzymes (Laskowski and Sealock, 1977). This led to the view that they might be involved in the protection of these vulnerable plant tissues from pest and pathogen attack by virtue of an antinutritional interaction with the pest's digestive enzymes (e.g., Applebaum, 1964; Ryan, 1973).

Support for a protective role for protease inhibitors came from studies of insects raised on artificial diets into which inhibitors had been incorporated. The effects on susceptible insects are generally seen as an increase in mortality, decrease in growth rate, and prolongation of developmental period of the larvae. Thus Lipke *et al.* (1954) demonstrated a protease inhibitor in soybeans which was toxic to *Tribolium confusum;* lima bean protease inhibitor and both the Kunitz and Bowman-Birk inhibitors from soybeans are effective against corn borer (*Ostrinia nubilalis*) larvae (Steffens *et al.,* 1978); soybean Kunitz trypsin inhibitor is effective against *Manduca sexta* (Shukle and Murdock, 1983); cowpea trypsin inhibitor is effective against *Callosobruchus maculatus* (Gatehouse *et al.,* 1979) and a range of other Coleoptera, Lepidoptera, and Orthoptera (see Hilder *et al.,* 1990; Gatehouse *et al.,* 1990). Pest species of the plant from which inhibitors are prepared are generally not susceptible, although they may be susceptible to related inhibitors from other sources (Steffens *et al.,* 1978).

Some of the most convincing evidence for a direct role of protease inhibitors in plant protection comes from studies of the wound-induced synthesis of inhibitors in potato and tomato plants. Wounding of leaves by the feeding behavior of adult or larval Colorado potato beetles (*Leptinotarsa decemlineata*) or the mechanical simulation of such wounding induced a rapid accumulation of protease inhibitors throughout the aerial tissues of the plant (Green and Ryan, 1972). The systemic response, whereby inhibitors are synthesized in leaves distant from the wound site, is mediated by a hormonelike factor, protease inhibitor-inducing factor (PIIF), which is released from the damaged leaves. The inhibitors accumulate rapidly to over 1% of the leaf-soluble proteins and remain at these elevated levels for a considerable time (Shumway *et al.,* 1976; Walker-Simmons and Ryan, 1977). The nature of this response is highly suggestive of a defense reaction to limit further damage (Ryan, 1983).

The primary effect of protease inhibitors is assumed to be to cause a critical shortage of essential amino acids; their effectiveness is, therefore, heavily dependent on the nutritional background in which they are presented, and can be overcome by amino acid supplementation in artificial diets (Gatehouse and Boulter, 1983; Broadway and Duffey, 1986b). Within the plant, their effectiveness is likely to depend on a complex interaction between the level of inhibitor and the overall nutritional quality of the plant tissue (Broadway and Duffey, 1986a).

The enhanced resistance to insect attack of transgenic plants expressing foreign protease inhibitors finally provided direct evidence for a protective role of these compounds.

13.2 Transgenic Plants Expressing Introduced Protease Inhibitors

13.2.1 COWPEA TRYPSIN INHIBITOR

The cowpea, or black-eyed bean (*Vigna unguiculata* L. [Walp.]) is an important grain legume crop, particularly in West Africa and parts of South America (FAO, 1970). The crop is subject to huge losses in storage due to the depredations of the cowpea bruchid beetle (*Callosobruchus maculatus* F). An extensive screening program at the International Institute of Tropical Agriculture in Nigeria has identified one out of several thousand accessions which had good resistance to the larvae of *C. maculatus*. This variety, designated TVu2027, provided a potentially useful source of insect resistance genes, although it was devoid of almost any other agronomically desirable characters.

Bruchid resistance in this accession did not have a physical basis and of the range of toxic or antimetabolic secondary compounds screened for the only association found was with elevated levels of trypsin inhibitors. The differences in trypsin inhibitor content between TVu2027 and susceptible varieties were purely quantitative, rather than qualitative, being some 2- to 4-fold higher in the former (Gatehouse *et al.*, 1979).

The antimetabolic properties of cowpea trypsin inhibitor (CpTI) against *C. maculatus* were confirmed in bioassays on artificial seeds. Those fractions of cowpea meal which contained active CpTI were effectively insecticidal to the bruchid (Gatehouse *et al.*, 1979). The trypsin inhibitors extracted from cowpeas were more effective antimetabolites than those extracted from a number of other legumes (Gatehouse and Boulter, 1983).

Of prime interest in terms of a wider use of CpTI was the observation that it was antimetabolic to a wide range of insects in feeding trials on artificial diets. The range of susceptible insects has now covered pest species from the orders Lepidoptera (caterpillars in the feeding stage) including *Heliothis* species (tobacco budworms and corn silkworms) and *Spodoptera* (armyworms) species; Coleoptera (beetles) including *C. maculatus* (bruchid beetle), *Diabrotica* species (corn rootworms), and *Anthonomus grandis* (cotton boll weevil); and an orthopteran (locust) *Locusta migratoria*. This broad range of effectiveness is typical of many plant-based protection mechanisms.

Biochemical studies showed that the trypsin inhibitors in cowpea seeds belonged to the Bowman-Birk family of double-headed protease inhibitors (Gatehouse *et al.*, 1980). They constitute a small family of four major isoinhibitors. They are encoded by a somewhat larger gene family, although the number of active genes is unknown. Three of the isoinhibitors are specific for trypsin at each active center, the fourth is a trypsin/chymotrypsin bifunctional inhibitor. The primary sequence of one isoinhibitor was established by the manual DABITC protein sequencing, confirming their identity as Bowman-Birk-type inhibitors (Hilder *et al.*, 1989).

The determined protein sequence provided the information necessary to allow CpTI encoding clones to be selected from a cowpea cotyledon cDNA library. DNA sequencing has established that these included representatives of trypsin–trypsin and trypsin–chymotrypsin inhibitors (Hilder *et al.*, 1989).

A full-length cDNA clone encoding a trypsin–trypsin inhibitor [pUSSRc3/2] was chosen for transfer into tobacco plants. The coding sequence was transferred to the plant expression vector pRok2, where it was brought under the control of the CaMV35S gene promoter. These constructs were mobilized into *Agrobacterium tumefaciens* LBA4404 and used to transform leaf discs of two species of tobacco plants—*Nicotiana tabacum* and *N. plumbaginifolia*. Transgenic plants were regenerated from these by standard procedures (Hilder *et al.*, 1987).

Molecular biological characterizations of the transgenic tobacco plants revealed that all of them contained multiple, unrearranged copies of the construct inserted at a single chromosomal locus in most plants, at two independent loci in the others (Hilder *et al.*, 1990). Expression of CpTI was measured by dot immunobinding assays using rabbit anti-CpTI antiserum binding to protein extracts from young leaves. Immunochemical assay of CpTI is particularly difficult due to the poor antigenicity of the inhibitor which results from the rigidity of the molecule imposed by the high degree of disulfide cross-linking. However, detection at the levels expressed in these plants is quite unambiguous using the appropriate techniques. A wide range, from below the limit of detection to about 0.9% of total soluble protein, was detected in independently transformed plants. Western blots demonstrated that correct processing of the CpTI occurred (Hilder *et al.*, 1987). Plants which expressed high levels of CpTI in the leaves also expressed the protein in roots, stems, and floral tissues at levels which should be insecticidally effective.

Before any other characterization was carried out, the original transformed plant population was tested for enhanced resistance to insect pests. Whole plant bioassays were carried out with first instar larvae of *H. virescens,* the tobacco

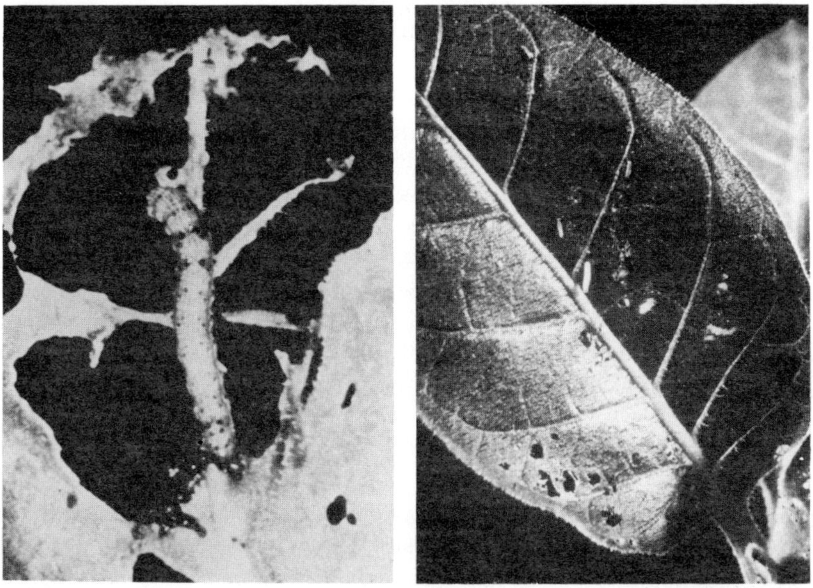

Figure 13.1 Tobacco budworm (*H. virescens*) larvae on control (*left*) and CpTI-expressing transgenic (*right*) tobacco plants.

budworm, which is a serious pest that readily eats tobacco plants (Figure 13.1). Plants which survived the first round were subjected to further rounds of infestation with *H. virescens*. Plants which still appeared to have enhanced resistance, plus some controls, were clonally replicated as stem cuttings to allow statistically meaningful trials to be performed. It was not until this stage that levels of CpTI expression were actually measured. The plants which had been selected as "potentially resistant" were those which were expressing CpTI to the highest levels. The best plant [+5/5] was the highest CpTI expresser at about 0.9% of total soluble protein (Hilder *et al.*, 1987).

Subsequent insect bioassays were carried out on clonal replicates of +5/5 or on seed-grown plants from the homozygous line +5/51-A20, which was produced by selfing +5/5. These latter express almost twice as much CpTI as the original hemizygous transformants. Resistance was measured from the point of view of the insects, as survival and insect biomass, and of the plants, as percent leaf area destroyed, estimated by image analysis of the leaves. These bioassays have conclusively established that CpTI-expressing plants have significantly enhanced resistance to *Heliothis* and to a broad range of other lepidopteran pests which are capable of eating tobacco (Hilder *et al.*, 1987; Gatehouse and Hilder, 1988; Hilder *et al.*, 1990).

13.2.2 WOUND-INDUCED PROTEASE INHIBITORS

The inhibitors PI-I and PI-II constitute two distinct families of protease inhibitors which are unrelated either to one another or to CpTI. They accumulate in the tubers of potatoes, but are also produced in the leaves of potato and tomato plants in response to wounding (Green and Ryan, 1972). Systemic signals, called PIIF are produced at the sites of wounding and induce high-level expression of these inhibitors not only in the wounded leaf, but also in more distant, unattacked leaves (Ryan, 1978; Bishop *et al.*, 1984). Sequences encoding cDNAs for PI-I and PI-II proteins have been isolated (Graham *et al.*, 1985a,b). These have been used to select genomic clones for both inhibitors from potato and tomato λ libraries (Lee *et al.*, 1986; Keil *et al.*, 1986; Cleveland *et al.*, 1987).

13.2.2.1 Regulation of Wound-Induced Expression in Transgenic Plants

The entire potato PI-II sequence has been introduced into transgenic tobacco by *Agrobacterium*-mediated leaf disk transformation. These plants showed tissue-specific, wound-induced expression of potato PI-II mRNA in the transgenic leaves (Sanchez-Serrano *et al.*, 1987). The 5′ upstream sequence of the PI-II gene and the 3′ region of either the PI-II gene or the 6b gene of octopine T-DNA (An, 1986) have been fused to a reporter gene—the coding sequence of the chloramphenicol acetyltransferase (CAT) gene (Herrera-Estrella *et al.*, 1983)—and these chimeric constructs transferred to transgenic tobacco plants. Transgenic plants containing the 5′ and 3′ sequences from the PI-II gene showed wound-induced CAT activity in wounded leaves and also in upper, unwounded leaves (Thornberg *et al.*, 1987).

These studies were not primarily concerned with insect control but rather with the characterization of factors involved in the wound-induced response. They have established that wound-induced regulation of gene expression can be transferred between species and have narrowed down the regulatory sequences responsible. It has been suggested that wound induction of introduced insect control genes might be desirable in transgenic plants, since then the "foreign" gene would only be active when the plant was under attack from insects.

13.2.2.2 Insect Control with PI-I and PI-II Genes in Transgenic Plants

Chimeric gene constructs for plant expression have been made using the coding sequence of PI-II gene coding sequences from tomato and potato. These were fused to the CaMV35S gene constitutive promoter and introduced into transgenic tobacco. Leaves from these plants were used in bioassay against tobacco hornworm (*Manduca sexta*) caterpillars. Growth and development of the hornworm caterpillars were shown to be significantly reduced on plants expressing PI-II protein (Johnson *et al.*, 1989). The severity of the effect on the caterpillars was related to the level of trypsin inhibitor which the transgenic plants produced (Figure 13.2). Leaf damage was also significantly reduced in the PI-II-expressing transgenics. These results are very similar to those obtained with transgenic tobacco plants expressing CpTI. However, similar constructs containing the coding sequence for the PI-I inhibitor gene from tomato, which inhibits chymotrypsin rather than trypsin-like digestive enzymes, were ineffective in controlling this particular species of insect when introduced into tobacco plants (Johnson *et al.*, 1989). This demonstrates that any protease inhibitor will not be effective against a specific pest insect; care must be taken over the selection of which inhibitor to transfer in any particular situation.

13.3 Selection of Protease Inhibitor Genes for Transfer

CpTI genes were selected for transfer into transgenic plants on the basis of a demonstrated association with field resistance to an insect pest and a broad

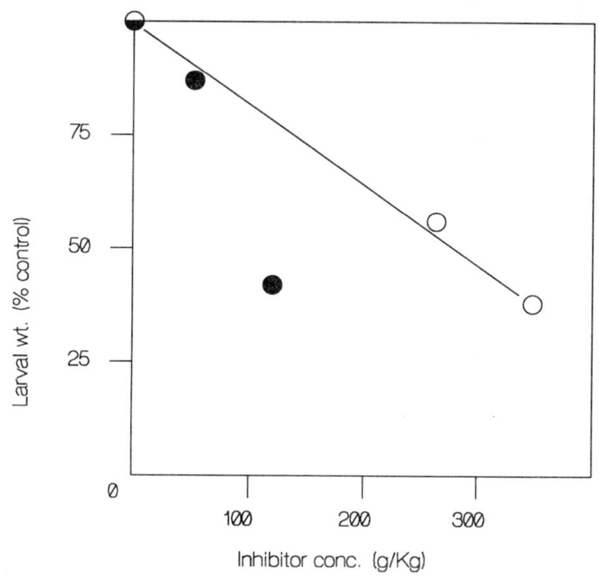

Figure 13.2 Effect of expression of tomato proteinase inhibitor II (●) or potato proteinase inhibitor II (○) in transgenic tobacco plants in bioassay against *Manduca sexta* larvae. Data from Johnson *et al.* (1989).

spectrum of activity in artificial diets to a range of economically important pest species. The TI-I and TI-II genes are naturally expressed in response to insect attack. There are in plants a variety of other, unrelated types of protease inhibitors, as well as many different variants within a particular class of inhibitor. Because the results of transgenic plants expressing the TI-I gene demonstrate that any protease inhibitor will not confer tolerance of any particular insect species, it is important to have some means of rationally selecting from among the alternatives.

13.3.1 PLANT PROTEASE INHIBITOR FAMILIES

Within the plant kingdom are found inhibitors of three of the four major classes of proteases: serine proteases, thiol proteases, and metalloproteases; proteinaceous inhibitors of aspartyl proteases such as pepsin have not been reported in plants. Inhibitors have clearly arisen more than once in evolution, giving rise to functionally related but nonhomologous families of inhibitors (Table 13.1). Some of these, particularly the inhibitors of serine proteases, have been studied in considerable detail. Thus there is extensive information on their distribution, primary sequences, structure, mode of action, evolution, and nutritional properties. They have been the subject of numerous excellent reviews to which the

TABLE 13.1
Major Families of Protease Inhibitors in Plants

Class of enzymes inhibited	Monomer M_r	Disulfide bonds	Protein gene sequence	
Serine proteases (trypsin, chymotrypsin elastase, subtilisin, kallikrein)				
Kunitz family	20,000	2	+	+
Bowman-Birk family	8400	7	+	+
PI-I family	8100	1	+	+
PI-II family	12,000	5	+	+
Cereal trypsin–amylase inhibitors	14,000	5	+	–
Squash inhibitors	3600	3	+	–
Thiol proteases				
Papain, bromelain, ficin (various nonhomologous families)		+	+	+
Metallocarboxypeptidases				
Carboxypeptidase A	4200	3	+	–

reader is referred (Ryan, 1973; Richardson, 1981, 1990; Garcia-Olmedo *et al.*, 1987).

13.3.2 PROTEASE INHIBITOR SPECIFICITIES

The gross specificity of an inhibitor is established by its ability to reduce the rate of proteolysis catalyzed by the archetypal enzymes: generally bacterial subtilisin and mammalian trypsin, chymotrypsin, elastase, and kallikrein for serine protease inhibitors; plant-derived papain, bromelain, or ficin for thiol protease inhibitors; and the mammalian enzymes for carboxypeptidases. Within these gross categories, however, different inhibitors display a wide range of activity. In a survey of 26 different serine protease inhibitors a 1700-fold range in equilibrium dissociation constants (K_d) with bovine trypsin was found. Even within a single family, for example, the plant Kunitz inhibitors or the Bowman-Birk inhibitors, a several hundredfold range in K_d was observed (Christeller and Shaw, 1989).

There is tremendous variability in the primary structure of homologous proteins, not only between species but also between different isoinhibitors within a species. This is particularly evident in the Bowman-Birk family. Residues in the active centers appear to be hypervariable; they have certainly evolved at a greater rate than has the leader sequence of the preproteins (Hilder *et al.*, 1989) and this must have a significant effect on the binding specificity to target proteases. An interesting parallel is seen in the accelerated evolution of the active site residues of serpins (plasma serine protease inhibitors of mammals) which are also believed to fulfill a protective role against exogenous proteases, in this case those of blood parasites and pathogens (Hill and Hastie, 1987).

In the case of plant inhibitors we have no real knowledge of what target protease drives the evolution of any particular inhibitor. It is fairly certain, however, that in no case is it the enzymes commonly used to assay them. It is also clear that the efficacy of inhibition of mammalian proteases is not a good predictor of effectiveness against a specific insect protease of the same gross specificity. The range of K_d for serine protease inhibitors with a specific insect digestive trypsin is even greater (ca. 5000-fold) than with the bovine enzyme. The ratios of K_d for the bovine and insect enzymes showed a similar range of values (Christeller and Shaw, 1989) (Table 13.2).

Christeller and Shaw (1989) have proposed that the K_d might be used as a determinant of the effectiveness of particular protease inhibitors for gene transfer once a specific target enzyme has been identified. It should be noted that a "good" inhibitor of one insect protease may not be a "good" inhibitor of others; there is a great, and unpredictable, variability in the relative strength of inhibition of bovine and human trypsin and chymotrypsin by particular inhibitors (Belitz *et al.*, 1982; Feeney *et al.*, (1969).

TABLE 13.2
Dissociation Constants (K_d) for Complexes between Various Bowman-Birk-Type Trypsin Inhibitors and Mammalian (*Bos taurus*) and Insect (*Costelytra zealandica*) Trypsins[a]

	Dissociation constant ($\times 10^{-9} M$)		
Inhibitor[b]	*Bos*	*Costelytra*	*B* : *C*
SBBI	0.96	0.28	3.43
WGI-I	0.54	0.57	0.95
WGI-II	2.1	0.021	100
LbTI	0.03	0.61	0.05
CpTI	8.7	0.58	15
MbTI	3.2	0.72	4.44

[a]Data from Christeller and Shaw (1989).
[b]SBBI, Soybean Bowman-Birk inhibitor; WGI-I, wheat germ inhibitor I; WGI-II, wheat germ inhibitor II; LbTI, lima bean trypsin inhibitor; CpTI, cowpea trypsin inhibitor; MbTI, mung bean trypsin inhibitor.

13.3.3 DIGESTIVE PROTEASES OF INSECTS

Insecta is the most species-diverse class of animals and this is reflected in the diversity of digestive physiology encountered within them (see Terra, 1990). Until recently there had been relatively few detailed studies to characterize biochemically insect digestive enzymes, but the practical importance of such information is now becoming increasingly widely recognized. Proteases with activities related to (though not always identical with) vertebrate trypsin, chymotrypsin, elastase, cathepsins, and carboxypeptidases have been described (Table 13.3). Note that the absence of a particular activity in Table 13.3 does not necessarily indicate its complete absence in the insect, only that it was not detected in the particular study.

In general, we may divide phytophagous insects into three broad groups on the basis of their major protein digestive mechanisms. Many insects have neutral or alkaline guts and depend predominantly on serine proteases. Others have acid guts and depend on thiol proteases. Finally, many hemipterans which feed on sap appear to lack digestive proteases entirely. Some knowledge of the digestive physiology would obviously be useful if a particular insect pest is being targeted with protease inhibitors. Thus, the predominance of trypsin-like, as opposed to chymotrypsin-like enzymes in lepidopterans was a factor in determining the

TABLE 13.3
Major Proteolytic Activities Identified in the Guts of Agronomically Interesting Species of Insects

Insect	Serine proteases[a]	Thiol proteases[b]	Carboxy-peptidases[c]	Ref.[d]
Coleoptera				
Acanthoscelides obtectus (Bean bruchid)		B		(*13*)
Callosobruchus maculatus (Cowpea seed weevil)		B		(*7, 14*)
Costelytra zealandica (Grass grub)	T (E)		A,B	(*5*)
Cylas formicarius (Sweet potato weevil)	T,C,E		A,B	(*3*)
Diabrotica sps. (Corn rootworms)		+		(*8*)
Leptinotarsa decemlineata (Colorado potato beetle)		B,D,H		(*12*)
Sitophilus sps. (Rice, maize weevils)	T,C,E		A,B	(*2*)
Tenebrio molitor (Yellow mealworm)	T,(C)		B	(*1*)
Diptera				
Drosophila sps. (Fruit flies)	T,C			(*1*)
Mayetiola destructor (Hessian fly)	T			(*11*)
Hymenoptera				
Apis melifera (adult)	T,C			(*1*)
(Honey bee) (larva)	C			
Lepidoptera				
Bombyx mori (Silkworm)	T			(*1*)
Erynnyis ello (Cassava hornworm)	T		A	(*10*)
Heliothis sps. (Bollworm, silkworm)	T			(*4, 8*)
Manduca sexta (Tobacco hornworm)	T			(*1*)
Ostrinia nubilalis (European corn borer)	T,C,(E)		+	(*9*)
Spodoptera sps. (Armyworms)	T(C)			(*4, 8*)
Trichoplusia ni (Cabbage semi-looper)	T,C			(*1*)
Orthoptera				
Locusta migratoria (Migratory locust)	T,C			(*1*)
Teleogryllus commodus (Black field cricket)	T,E		A,B	(*6*)

[a]T, Trypsin-like; C, chymotrypsin-like; E, elastase-like.

[b]Letters refer to cathepsin B-, D-, or H-like activity.

[c]Letters refer to carboxypeptidase A- or B-like activity.

[d]Key to references: (*1*) see Applebaum, 1985; (*2*) Baker, 1982; (*3*) Baker *et al.*, 1984; (*4*) Broadway and Duffey, 1986a; (*5*) Christeller *et al.*, 1989; (*6*) Christeller *et al.*, 1990; (*7*) Gatehouse *et al.*, 1985; (*8*) Gatehouse *et al.*, unpublished; (*9*) Houseman *et al.*, 1989; (*10*) Santos *et al.*, 1983; (*11*) Shukle *et al.*, 1984; (*12*) Thie and Houseman, 1990); (*13*) Wieman and Nielsen, 1988; (*14*) Kitch and Murdock, 1986.

transfer of the cowpea trypsin–trypsin inhibitor, rather than the trypsin–chymotrypsin inhibitor.

13.3.4 OTHER SYSTEMS DEPENDENT ON PROTEASES

It is assumed that the primary site of action of these protease inhibitors is the digestive proteases in the insect gut, leading to a disruption in essential amino acid metabolism. However, there are a number of other systems within insects which are critically dependent on proteolytic activity. If the insect gut is at all "leaky" to small proteins, then ingested inhibitors could enter the hemolymph where they could seriously affect a number of vital processes.

Trypsin-like proteases are involved in the molting process in lepidopterans (Brookhart and Kramer, 1990). Insects are generally particularly vulnerable during molting and any disruption of this process is likely to have severe consequences. It has been proposed that the prophenol oxidase system is involved in cellular immunity in arthropods (Sodderhall, 1982). Prophenol oxidase activation in the hemolymph is under cascade regulation by serine proteases (Yoshida and Ashida, 1986) and these in turn are probably regulated by endogenous protease inhibitors of the Kunitz or serpin type (Aspan *et al.*, 1990; Kanost, 1990). Interference with this regulation by exogenous inhibitors might render the insect more susceptible to pathogens. Various insect hormones are neuropeptides which are produced by specific proteolytic cleavage of larger precursors (see Holman *et al.*, 1990). It is easy to see that disruption of such proteases would seriously affect insect behavior and development.

13.4 Possible Limitations of Protease Inhibitor Approach

13.4.1 TOXICITY TO NONTARGET SPECIES

One of the major advantages of inherent resistance, irrespective of whether it is introduced into crops by conventional breeding or genetic engineering, is that only those organisms which actually eat the plant are exposed. However, in most cases crops are grown specifically to be consumed either by humans or animals. Since most of the protease inhibitors are effective inhibitors of mammalian proteases, the question of whether they might be at least as harmful to the intended users as to the pests has to be addressed.

The criticism that plants expressing trypsin inhibitors ought to be harmful to mammals is largely based on a misunderstanding of the biology of the system. Food is consumed by organisms, not by enzymes, and the organization of the mammalian and insect digestive systems are quite different (Figure 13.3). Thus, for example, before CpTI encounters trypsin in the mammalian gut it spends some time in the very acid environment of the stomach where it is exposed to pepsin. CpTI is very sensitive to pepsin digestion. This is not true of all trypsin inhibitors—the Kunitz inhibitors appear to be much more resistant to digestion in mammals. Thus the relative toxicity of any particular inhibitor needs to be assessed *in vivo*.

It may be noted that those inhibitors which have been transferred to transgenic plants to date are naturally expressed in the consumed parts of food crops. They are already consumed in large quantities (although usually processed by cooking) with no ill effects.

Another concern is the possible effects of protease inhibitors on beneficial insects, specifically on pollinating species. There is clearly no problem with cereals and other wind-pollinated crops, but it would be highly undesirable to kill bees and other pollinators on crops such as oilseed rape or alfalfa. The potential toxicity of protease inhibitors to pollinating insects has not yet been adequately studied; this is an area which requires further investigation. However, the tissues on which these insects feed (pollen and nectar) are so specific that the use of promoters which do not function in them should be relatively straightforward.

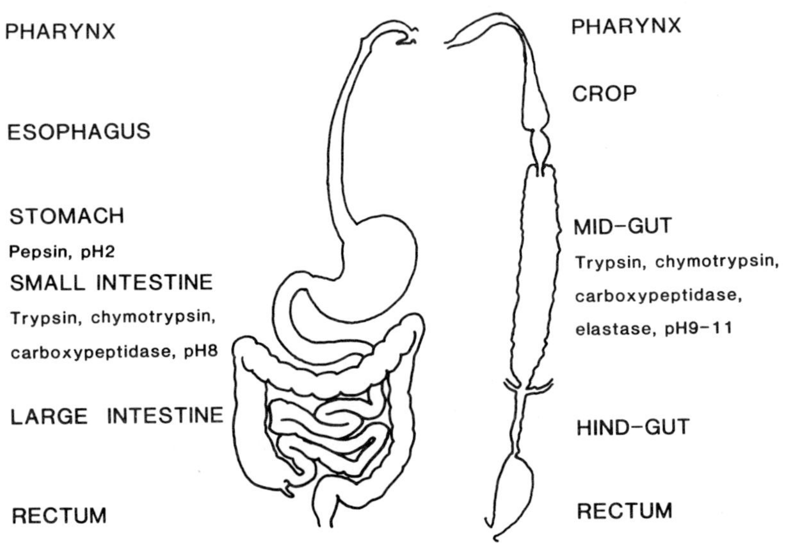

Figure 13.3 Diagrammatic representation of a mammalian (*left*) and an insect (*right*) digestive system, indicating the sites of action of the major digestive proteases.

Indeed, the CaMV35S promoter used to express CpTI in transgenic plants is said to be inactive in pollen.

13.4.2 SAP-SUCKING INSECTS

Sap-sucking insects of the order Homoptera are a group of serious insect pests, particularly on certain cereal and vegetable crops. Not only are they responsible for causing intrinsic damage, but they serve as important vectors of various plant diseases. The phloem sap on which they feed may be viewed as essentially a dilute solution of simple sugars, amino acids, and minerals. As such, the insects do not need, and are usually reported not to have, hydrolytic enzymes in their guts. Thus, it is highly unlikely that protease inhibitors would be effective against them.

13.4.3 PENALTIES TO PLANTS

Protease inhibitors have relatively low absolute toxicities to insects when compared to chemical pesticides or to *B. thuringiensis* toxins. Expression of CpTI at levels approaching 1% of total soluble protein is required for an observable protective effect in transgenic plants. The question of what effect this level of expression of a foreign protein has on the yield of the plant is of considerable importance.

We have addressed this question in a series of experiments in the controlled environment growth room. Various quantifiable morphological, vegetative, and reproductive characteristics have been measured on transgenic tobacco plants which express CpTI at levels from 0 to 1% of total soluble protein and on untransformed controls. This was done under strictly controlled environmental conditions in the absence of any insects, that is, conditions under which CpTI is not expected to confer any advantage. The results of these studies may be summarized. (1) Small but significant differences (with large enough sample sizes) were observed in a number of yield-related parameters between control and transformed plants. (2) These differences were generally within the normal range of variability of the untransformed controls. (3) The extent to which transformed plants deviated from the controls was absolutely not related to the level of CpTI expression. The deviation is not due to expression of the inhibitor but to some other aspect of the transformation process. (4) The deviation of high CpTI-expressing plants remained marginal even under conditions of severe water or nutrient stress (under which conditions the relative level of CpTI expression was constant).

Thus we could detect no yield penalty associated with high levels of expression of this foreign protein, although there was some marginal penalty associated with some aspect of the genetic engineering process (Hilder and

Gatehouse, 1991). This surprising observation indicates the "spare" synthetic capacity which plants possess. The conclusions are supported by yield data from field trials of transgenic tomato and potato plants expressing virus coat proteins (Nelson *et al.*, 1988; Kaniewski *et al.*, 1990) and transgenic tobacco and potato plants expressing a herbicide resistance gene (De Greef *et al.*, 1989).

The absolute toxicity of these compounds is, therefore, largely irrelevant. What matters is that they *can* be expressed at effective levels in transgenic plants *without* this expression imposing any serious yield penalty.

13.4.4 EVOLUTION OF RESISTANCE IN INSECTS

The rapidity with which insects that are resistant to chemical pesticides emerge is a major problem (see Metcalf, 1986). Laboratory resistance to *B. thuringiensis* toxins has already been observed (McGaughey, 1985; Stone *et al.*, 1989). To some extent this is due to the high selective pressure imposed upon the insect population by the "wipe out" approach to pest control. Protease inhibitors are likely to function less by eliminating the insects (although mortality is increased) and more by slowing down the buildup of the pest population, thus preventing it from reaching a catastrophe level. The selective pressure to evolve resistance under these circumstances should be lower.

The target of protease inhibitors is the active site of proteolytic enzymes. This could not readily change by simple mutations so as to exclude the inhibitor yet still retain proteolytic activity. Insects as a group, however, do contain genes for a wide range of proteases and the possibility of activating the gene for an unused protease or of increasing the use of a currently minor protease exists. This, or the evolution of some mechanism for inactivating the inhibitor before it reached its target enzyme, would call for a major reorganization of the digestive physiology of the insect. Nature is, however, wonderful, and we should be mindful of the dangers of relying on any type of single factor resistance (see Section 13.5.2). The extent of variability within insect populations and the potential to develop resistance to specific protease inhibitors is another area which requires further investigation.

13.5 Future Developments

13.5.1 EFFECTIVENESS IN FIELD

Ultimately, the only place where this application of science and technology matters is in the field. In order to evaluate its likely effect there it will be necessary to carry out realistic field trials of transgenic crop plants expressing protease inhibitors. Since it is not anticipated that such plants would wipe out the insect pests, but rather limit their population growth, such trials would need to be carried out over more than one growing season and the effects on endemic pest populations monitored.

To date, a single, small-scale, short-term field trial of CpTI-expressing transgenic tobacco plants has been carried out. This trial involved artificial infestation of the plants with caterpillars. The results were similar to those obtained in bioassays conducted in the growth room (Hoffmann, Zalom, Smilanick, Malyj, Kiser, Wilson, Hilder, and Barnes, submitted). The degree of protection afforded by CpTI alone is probably insufficient to be commercially acceptable in developed, high-input agricultural systems. It is not, however, expected that CpTI would be used alone; rather it would be one part of an integrated pest management scheme—perhaps combined with other foreign genes, perhaps with the judicious use of some exogenous pesticides.

13.5.2 PYRAMIDING GENES

In order to increase the protective efficacy, spectrum of activity, and durability of resistance it is envisaged that "packages" of different genes will be introduced into crops. The components of such packages should each act on quite different targets within the insect. In this way we would seek to mimic the type of broad-range, multimechanistic resistance which is typical of plants (e.g., Janzen *et al.*, 1986). Protease inhibitors would be particularly valuable components of such "packages" since, apart from their inherent insecticidal effect, they would protect other introduced gene products from premature digestion in the insect gut.

The first demonstration of this approach has been the introduction of both CpTI and pea lectin into transgenic tobacco plants. The two gene products together showed an additive protective effect against tobacco budworm caterpillars (Boulter *et al.*, 1990a). It is also interesting to note that trypsin inhibitors have a marked potentiating effect on *B. thuringiensis* toxins (MacIntosh *et al.*, 1990).

13.5.3 REGULATORY ISSUES AND TECHNOLOGY TRANSFER

The speed with which transgenic plants with agronomically useful traits will be moved from the laboratory into general use is now limited more by uncertainties over regulatory approval, proprietary protection, commercial potential, and public acceptability than by any scientific or technological constraints. The present state and future prospects of these complex issues have been reviewed by Connett and Barfoot (1991). This is a new and unproven technology and it is entirely correct that a careful, case-by-case and step-by-step approach is being adopted in these matters. Careful consideration of the ecological consequences of using such material is needed (see Tiedje *et al.*, 1989). Perhaps most importantly, public confidence in what ought to be an environmentally friendly technology needs to be encouraged (see Boulter *et al.*, 1990b). Such considerations apply to all applications of transgenic crops. Insect-tolerant crops based on introduced protease inhibitors ought to be particularly acceptable since essentially all that has been done is to move a natural defense compound from one (crop) plant to another.

Much of the research and development in this area has been supported by the commercial sector and the transfer of this technology from the laboratory to the field is a commercial goal. The strategy for the commercialization of CpTI-expressing transgenic crops has been reviewed by Barfoot and Connett (1989).

There has been considerable comment on the possible consequences of such commercial sector involvement for the transfer of this technology, particularly to less agronomically developed countries of the world (e.g., Crouch, 1990; Panos Institute, 1990). Insect-tolerant transgenic plants are one product of plant biotechnology which could be of immense benefit to agricultural development in these countries which suffer the worst problems of crop losses due to insects and where misuse of pesticides is widespread. There appears to be a widespread misunderstanding of the purposes and application of proprietary protection. Certainly the Agricultural Genetics Company, Cambridge, U. K., which has proprietary rights to CpTI, recognizes not only the inevitability that this technology will be, but also the desirability that it should be, transferred to these countries, and have made efforts to facilitate this. If agricultural biotechnology does eventually fulfill its promise, then it is likely to do so to the benefit of all humanity.

Acknowledgements

We are grateful to our many colleagues for helpful discussions and to the Agricultural Genetics Company for financial support.

References

An, G. (1986). *Plant Physiol.* 81, 86–91.

Applebaum, S. W. (1964). *J. Insect Physiol.* 10, 783–788.

Applebaum, S. W. (1985). In *Comparative Insect Physiology, Biochemistry and Pharmacology Vol. 4*. (Ed. by G. A. Kerkut and L. I. Gilbert), pp. 278–312. London: Pergamon Press.

Aspan, A., Hall, M., and Sodderhall, K. (1990). *Insect Biochem.* 20, 485–492.

Baker, J. E. (1982). *Can. J. Zool.* 60, 3206–3214.

Baker, J. E., Woo, S. M., and Mullen, M. A. (1984). *Entomol. Exp. Appl.* 36, 97–105.

Baldwin, I. T., Sims, C. L., and Kean, S. E. (1990). *Ecology* 71, 252–262.

Barfoot, P. D., and Connett, R. A. (1989). *AgBiotech. News and Information* 1, 177–182.

Belitz, H., Lynen, F., and Weder, J. K. P. (1982). *Z. Lebens. Unters Forsch.* 174, 442–444.

Bishop, P., Pearce, G., Bryant, J. E., and Ryan, C. A. (1984). *J. Biol. Chem.* 259, 13172–13177.

Boulter, D., Edwards, G. A., Gatehouse, A. M. R., Gatehouse, J. A., and Hilder, V.A. (1990a). *Crop Protection* 9, 351–354.

Boulter, D., Gatehouse, J. A., Gatehouse, A. M. R., and Hilder, V. A. (1990b). *Endeavour,* New Series 14, 185–190.

Broadway, R. M., and Duffey, S. S. (1986a). *Insect Physiol.* 32, 673–680.

Broadway, R. M., and Duffey, S. S. (1986b). *Insect Physiol.* 32, 827–833.

Broadway, R. M., Duffey, S. S., Pearce, G., and Ryan, C. A. (1986). *Entomol. Exp. Appl.* 41, 33–38.

Brookhart, G. L., and Kramer, K. J. (1990). *Insect Biochem.* 20, 467–477.

Christeller, J. R., and Shaw, B. D. (1989). *Insect Biochem.* 19, 233–241.

Christeller, J. T., Shaw, B. D., Gardiner, S. E., and Dymock, J. (1989). *Insect Biochem.* 19, 221–231.

Christeller, J. T., Laing, W. A., Shaw, B. D., and Burgess, E. P. J. (1990). *Insect Biochem.* 20, 1157–164.

Cleveland, T. E., Thornberg, R. W., and Ryan, C. A. (1987). *Plant Mol. Biol.* 8, 199–207.

Connett, R. A., and Barfoot, P. D. (1991). In *Plant Genetic Manipulation for Crop Protection.* (Ed. by A. M. R. Gatehouse, V. A. Hilder, and D. Boulter). pp. 45–74. Wallingford, England: C. A. B. International.

Crouch, M. (1990). *Plant Cell* 2, 275–276.

Dawson, G. W., Hallahan, D. L., Mudd, A., *et al.* (1989). *Pestic. Sci.* 27, 191–201.

De Greef, W., Delon, R., De Block, M., Leemans, J., and Botterman, J. (1989). *Bio/Technology* 7, 61–64.

Dulmage, H. T. (1981). In *Microbial Control of Pests and Plant Diseases 1970–1980.* (Ed. by H. D. Burges), pp. 193–222. New York: Academic Press, Inc.

FAO (1970). The State of Food and Agriculture. p. 274. Rome.

Feeney, R. E., Means, G. E., and Bigler, C. (1969). *J. Biol. Chem.* 244, 1957–1960.

Garcia-Olmedo, F., Salcedo, G., Sanchez-Monge, R., Gomez, L., Royo, J. and Carbonero, P. (1987). In *Oxford Suveys of Plant Molecular and Cell Biology, Vol. 4.* (Ed. by B. J. Miflin), pp. 275–334. Oxford: Oxford University Press.

Gatehouse, A. M. R., and Boulter, D. (1983). *J. Sci. Food Agric.* 34, 345–350.

Gatehouse, A. M. R., and Hilder, V. A. (1988). In *Brighton Crop Protection Conference—Pests and Diseases, 1988, Vol. 3,* pp. 1245–1254. Lavenham, England: Lavenham Press.

Gatehouse, A. M. R., Gatehouse, J. A., Dobie, P., Kilminster, A. M., and Boulter, D. (1979). *J. Sci. Food Agric.* 30, 948–958.

Gatehouse, A. M. R., Gatehouse, J. A., and Boulter, D. (1980). *Phytochem.* 19, 751–756.

Gatehouse, A. M. R., Dewey, F. M., Dove, J., Fenton, K. A., and Pusztai, A. (1984). *J. Sci. Food Agric.* 35, 373–380.

Gatehouse, A. M. R., Butler, K. J., Fenton, K. A., and Gatehouse, J. A. (1985). *Entomol. Exp. Appl.* 39, 279–286.

Gatehouse, A. M. R., Barbieri, L., Stirpe, F., and Croy, R. R. D. (1990). *Entomol. Exp. Appl.* 54, 43–51.

Gatehouse, J. A., Hilder, V. A., and Gatehouse, A. M. R. (1991). In *Plant Genetic Engineering.* (Ed. by D. Grierson), pp. 105–135. Glasgow, Scotland: Blackie & Son.

Graham, J. S., Pearce, G., Merryweather, J., Titani, K., Ericsson, L. and Ryan, C. A. (1985a). *J. Biol. Chem.* 260, 6555–6560.

Graham, J. S., Pearce, G., Merryweather, J., Titani, K., Ericsson, L. and Ryan, C. A. (1985b). *J. Biol. Chem.* 260, 6561–6564.

Green, T., and Ryan, C. A. (1972). *Science* 175, 776–777.

Herrera-Estrella, L., Deblock, M., Messens, M., Hernalsteens, J., van Montagu, M. and Schell, J. (1983). *EMBO J.* 2, 987–995.

Hilder, V. A., Gatehouse, A. M. R., Sheerman, S. E., Barker, R. F., and Boulter, D. (1987). *Nature (London)* 330, 160–163.

Hilder, V. A., Barker, R. F., Samour, R. A., Gatehouse, A. M. R., Gatehouse, J. A., and Boulter, D. (1989). *Plant Mol. Biol.* 13, 701–710.

Hilder, V. A., Gatehouse, A. M. R., and Boulter, D. (1990). In *Genetic Engineering of Crop Plants.* (Ed. by G. Lycett and D. Grierson), pp. 51–66. London: Butterworths.

Hilder, V. A. and Gatehouse, A. M. R. (1991). *Transgenic Research* 1, 54–60.

Hill, R. E., and Hastie, N. D. (1987). *Nature (London)* 326, 96–99.

Holman, G. M., Nachman, R. J., and Wright, M. S. (1990). *Ann. Rev. Entomol.* 35, 201–217.

Houseman, J. G., Philogene, B. J. R., and Downe, A. E. R. (1989). *Can. J. Zool.* 67, 864–868.

Janzen, D. H., Ryan, C. A., Liener, I. E., and Pearce, G. (1986). *J. Chem. Ecol.* 12, 1469–1480.

Johnson, R., Narvaez, J., An, G., and Ryan, C. (1989). *Proc. Natl. Acad. Sci. U.S.A.* 86, 9871–9875.

Kakes, P. (1989). *Theor. Appl. Genet.* 77, 111–118.

Kaniewski, W., Lawson, C., Sammons, B., Haley, L., Hart, J., Delannay, X., and Tumer, N. (1990). *Bio/Technology* 8, 750–754.

Kanost, M. R. (1990). *Insect Biochem.* 20, 141–148.

Kareiva, P., and Sahakian, R. (1990). *Nature (London)* 345, 433–434.

Keil, M., Sanchez-Serrano, J., Schell, J., and Willmitzer, L. (1986). *Nucl. Acid Res.* 14, 5641–5650.

Kitch, L. W., and Murdock, L. L. (1986). *Arch. Ins. Biochem. Physiol.* 3, 561–575.

Labandiera, C. C., Beall, B. S., and Hueber, F. M. (1988). *Science* 242, 913–916.

Laskowski, M., and Sealock, R. W. (1977). In *The Enzymes Vol. 3.,* 3rd ed. (Ed. by P. Boyer), pp. 375–473. New York: Academic Press, Inc.

Lee, J. S., Brown, W. E., Graham, J. S., Pearce, G., Fox, E. A., Dreher, T. W., Ahern, K. G., Pearson, G. D. and Ryan, C. A. (1986). *Proc. Natl. Acad. Sci U.S.A.* 83, 7277–7281.

Lipke, H., Fraenkel, G. S., and Liener, I E. (1954). *Ag. Food Chem.* 2, 410–414.

MacIntosh, S. C., Kishore, G. M., Perlak, F. J., Marrone, P. G., Stone, T. B., Sims, S. R. and Fuchs, R. L. (1990). *J. Agric. Food Chem.* 38, 1145–1152.

McGaughey, W. H. (1985). *Science* 229, 193–195.

Meeusen, R. L. (1986). Reported in *Agrow* 29, 9.

Metcalf, R. L. (1986). In *Ecological Theory and Integrated Pest Management Practice.* (Ed. by M. Kogan), pp. 251–297. New York: Wiley.

Minney, B. H. P., Gatehouse, A. M. R., Dobie, P., Dendy, J., Cardona, C. and Gatehouse, J. A. (1990). *J. Insect Physiol.* 36, 757–767.

Nelson, R. S., McCormick, S. M., Delannay, X., Dube, P., Layton, J., Anderson, E. J., Kaniewska, M., Proksch, R. K., Horsch, R. B., Rogers, S. G., Fraley, R. T. and Beachy, R. N. (1988). *Bio/Technology* 6, 403–409.

Norris, D. M., and Kogan, M. (1980). In *Breeding Plants Resistant to Insects.* (Ed. by F. G. Maxwell and P. R. Jennings), pp. 23–60. London: Wiley.

Pannos Institute (1990). *Miracle or Menace? Biotechnology and the Third World.* Pannos Institute, London.

Pearce, G., Liljgren, D., and Ryan, C. A. (1987). In *Tomato Biotechnology.* (Ed. by D. J. Nevins and R. A. Jones), pp. 139–144. New York: Alan R. Liss.

Poerio, E., Carrano, L., Garzillo, A. M., and Buoncore, V. (1989). *Phytochemistry* 28, 1307–1311.

Richardson, M. (1981). *Food Chem.* 6, 235–253.

Richardson, M. (1990). In *Methods in Plant Biochemistry Vol. 5*. (Ed. by M. Richardson), pp. 261–307. New York: Academic Press, Inc.
Ryan, C. A. (1973). *Ann. Rev. Plant Physiol.* 24, 173–196.
Ryan, C. A. (1978). *Trends in Biochemical Sciences* 3, 148–150.
Ryan, C. A. (1981). In *The Biochemistry of Plants, Vol. 6*. (Ed. by A. Marcus), pp. 351–370. New York: Academic Press, Inc.
Ryan, C. A. (1983). In *Variable Plants and Herbivores in Natural and Managed Systems*. (Ed. by R. F. Denno and M. S. McClure), pp. 43–60. New York: Academic Press, Inc.
Ryan, C. A., and Walker-Simmons, M. (1981). In *The Biochemistry of Plants, Vol. 6*. (Ed. by A. Marcus), pp. 321–347. New York: Academic Press, Inc.
Sanchez-Serrano, J. J., Keil, M., O'Connor, A., Schell, J., and Willmitzer, L. (1987). *EMBO J.* 6, 303–306.
Santos, C. D., Ferreira, C., and Terra, W. R. (1983). *J. Insect Physiol.* 29, 707–714.
Shukle, R. H., Murdock, L. L., and Gallun, R. L. (1984). *Insect Biochem.* 15, 93–101.
Shumway, L. K., Yang, V. V., and Ryan, C. A. (1976). *Planta* 129, 161–165.
Sodderhall, K. (1982). *Devl. Comp. Immunol.* 6, 601–611.
Steffens, R., Fox, B., and Kassell, B. (1978). *J. Agric. Food Chem.* 26, 170–175.
Stone, T. B., Sims, S. R., and Marrone, P. G. (1989). *J. Invertebr. Pathol.* 53, 228–234.
Terra, N. M. R., and Houseman, J. G. (1990). *Insect Biochem.* 20, 313–318.
Thornberg, R. W., Cleveland, T. E., and Ryan, C. A. (1987). *Proc. Natl. Acad. Sci. U.S.A.* 84, 744–748.
Tiedje, J. M., Colwell, R. K., Grossman, Y. L., Hodson, R. E., Lenski, R. E., Mack, R. N., and Regal, P. J. (1989). *Ecology* 70, 298–315.
Walker-Simmons, M., and Ryan, C. A. (1977). *Plant Physiol.* 59, 437–439.
Wieman, K. F., and Nielsen, S. S. (1988). *Comp. Biochem. Physiol.* 89B, 419–426.
Yoshida, H., and Ashida, M. (1986). *Insect Biochem.* 16, 539–545.

14

Transgenic Plants for Improving Seed Storage Proteins

Samuel S. M. Sun

Department of Plant Molecular Physiology
University of Hawaii
Honolulu, Hawaii 96822

Brian A. Larkins

Department of Plant Sciences
University of Arizona
Tucson, Arizona 85721

Plants are the primary source of all protein consumed by humans and livestock. Seeds, especially those of legumes and cereals, contain large quantities of protein and are a major source of dietary protein. In comparison to meat, seed protein is far more economic to produce, store, and transport. However, it is deficient in several essential amino acids. In general, cereal seed proteins are low in lysine and tryptophan while legume seed proteins are deficient in the sulfur amino acids, methionine and cysteine (FAO, 1970; Yamaguchi, 1980). Consequently, seed proteins are nutritionally incomplete.

Plant geneticists and breeders have made significant efforts in the past to improve the quality of seed proteins. Thus, mutations resulting in high-lysine corn and barley have been identified and developed in elite germplasm (Nelson, 1968; Bright and Shewry, 1983). Unfortunately, there are undesirable traits associated with these mutations, such as lower yields and greater susceptibility to pests and diseases, and these have prevented their agronomic utilization (Nelson, 1980; Bright and Shewry, 1983). Efforts in breeding legumes containing proteins with increased levels of sulfur amino acids have not met with significant success (Payne, 1983).

With recent developments in recombinant DNA technology in plant molecular biology research, including plant transformation and regeneration, alternative molecular approaches can now be implemented to alter the amino acid composition of seed proteins for improved nutritional quality. This chapter summarizes recent research and comments on future prospects of using molecular

TRANSGENIC PLANTS, VOLUME 1
Engineering and Utilization

approaches for seed protein improvement. A brief and generalized summary of the biochemistry and molecular biology of seed proteins, especially the expression of seed protein genes in transgenic plants, will first be presented for background information. Additional detailed information on these aspects of seed protein research can be obtained from several recent reviews (Heidecker and Messing, 1986; Casey and Domoney, 1987; Kuhlemeier *et al.*, 1987; Shotwell and Larkins, 1989; Goldberg, *et al.*, 1989).

14.1 Biochemistry and Molecular Biology of Seed Storage Proteins

14.1.1 COMMON PROPERTIES AND CLASSIFICATION

Seeds accumulate large quantities of storage protein during development which serves as a source of nitrogen, sulfur, and carbon for germinating seedlings. Storage proteins have six common characteristics: (1) they are the most abundant proteins in the seed; (2) they have no enzymatic activity; (3) their synthesis is under developmental control; (4) they are stored in membrane-enclosed structures (protein bodies); (5) they are typically composed of a group of structurally related polypeptides; and (6) they are rapidly degraded during seed germination.

Seed storage proteins are classified by solubility as: (1) globulins, salt soluble; (2) prolamines, alcohol soluble; (3) glutelins, acid or alkaline soluble; and (4) albumins, water soluble. Globulins are the major storage proteins in the cotyledons of most dicot plants. These are frequently classified as 11S, 7S, and 2S based on their sedimentation coefficient. Prolamines are the predominant proteins in the endosperm of monocot species. They are named according to the species from which they are isolated, for example, *Zea mays,* zein, and *Avena sativa,* avenin.

Oats and rice are unusual cereals because they contain only small amounts of prolamine proteins and their major storage protein is an 11S-type globulin. Although the albumin fraction in most seeds was thought to be largely enzymatic, some albumins have been identified as storage proteins (Youle and Huang, 1981).

14.1.2 AMINO ACID COMPOSITION

Seed storage proteins are rich in the amide amino acids, glutamine and asparagine. These two amino acids often account for 40% of the total. The seed proteins of legumes generally contain low levels (1–2%) of the sulfur amino acids, methionine and cysteine. Methionine is an essential amino acid for humans and monogastric animals; cysteine can partially replace the methionine requirement. Cereal storage proteins are poor in the essential amino acids lysine (1.5–4.5%), tryptophan (0.8–2.0%), and threonine (2.7–3.9%) (Bright and Shewry, 1983). The ideal levels of these essential amino acids as recommended by FAO are: methionine 3.5%, lysine 5.5%, tryptophan 1.0%, and threonine 4.0% (FAO, 1973).

Some seed proteins have been shown to contain unusually large amounts of sulfur amino acids. Thus the 15-kDa maize prolamine protein contains 18 Met and 8 Cys residues, accounting for 16% of the total amino acids (Pedersen *et al.*, 1986), while the 10-kDa maize prolamine is even richer in the sulfur amino acids with 22.5% Met and 4% Cys (Phillips and McClure, 1985; Kirihara *et al.*, 1988a). Two low molecular weight pea seed albumins, PA1a (6 kDa) and PA1b (4 kDa), contain 7.5% and 16.2% cysteine, respectively (Higgins *et al.*, 1986). The 2S albumin protein of the Brazil nut is also known for its high content of Met (18%) and Cys (8%) (Ampe *et al.*, 1986; Sun *et al.*, 1987a; Altenbach *et al.*, 1987). Recently, a 2S albumin in sunflower seeds has been shown to contain 16% Met and 7% Cys (Lilley *et al.*, 1989).

14.1.3 PROTEIN STRUCTURE

The 11S globulins from diverse dicot species are structurally very similar (Plietz and Damaschun, 1986). The protein is a hexamer composed of two 8S trimers stacked one atop of the other, but offset by 60 degrees, forming a trigonal antiprism. Hydrophobic bonds and electrostatic interactions contribute to the joining of the two trimers. The trimer is made up of three spherical subunits each of which contains an acidic and a basic polypeptide (Reichelt *et al.*, 1980; Plietz and Damaschun, 1986; Plietz *et al.*, 1988). The C terminus of the basic chain is hydrophobic and is located in the center of the molecule. Hydrophobic interactions between the C termini of these basic polypeptides contribute to the stability of the hexamer. The C terminus of the acidic chain, in contrast, is hydrophilic (Lycett *et al.*, 1984; Nielsen *et al.*, 1989) and is located at the surface of the molecule, suggesting that it would have little effect on the stability of the oligomer. This portion of the acidic chain also contains peptide inserts of different lengths and amino acid compositions and is thus referred to as the hypervariable region (Argos *et al.*, 1985). It appears that the C terminus of the acidic polypeptide is an attractive target for amino acid modifications.

The 7S globulin is composed of three Y-shaped subunits arranged around a threefold symmetry axis (Plietz *et al.*, 1983a,b), but the intermolecular interactions involved in stabilizing this structure are unknown. The C-terminal region of the 7S globulin subunit also contains small inserts rich in glutamine and aspartate residues (Argos *et al.*, 1985), and appears to be located at the surface of the molecule. These insertion regions, like those in the 11S globulins, are attractive sites for sequence modifications.

The polypeptides of monocot prolamines can be divided into different domains or modules according to their amino acid composition (Kasarda *et al.*, 1984; Reeck and Hedgcoth, 1985). A central region containing tandemly repeated peptides and regions that are particularly poor or rich in certain amino acids, such as proline and glutamine have been identified (Okita *et al.*, 1985). Some of these regions contain sequences that are conserved, while others are variable. For example, the A, B, and C regions of a proline-poor domain are strongly conserved among all the S-rich wheat prolamines, while the I1-I4 regions are variable (Kreis *et al.*, 1985).

Information on the secondary and tertiary structures of prolamines comes mostly from circular dichroism (CD) measurements or from computer-generated model studies (Shotwell and Larkins, 1989). Based on analyses of this type, a structural model has been proposed for the α-zein of maize (Argos *et al.*, 1982). In this model, the nine repeated peptides in the central region of the molecule form α-helices that wind up and down to fold the protein into a rod-shaped structure. The turn regions of the helices are glutamine rich and could form side-chain interactions that contribute to the aggregation of the α-zein in the protein bodies.

14.1.4 BIOSYNTHESIS, PROCESSING, TRANSPORT, AND DEPOSITION

Seed storage proteins are synthesized in the endosperm, the cotyledons, or other tissues of the embryo at defined stages during seed development. Globulin proteins are synthesized on rough endoplasmic reticulum (RER), cotranslationally glycosylated (for those containing sugar moieties), and translocated into the lumen of the ER. The signal peptide of the nascent polypeptide directs translocation and is removed during this process. Disulfide bond formation and assembly into oligomers occur in the ER lumen and the oligomers are then transported via the Golgi body to the storage vacuoles (protein bodies) (Barton *et al.*, 1982; Chrispeels *et al.*, 1982; Herman *et al.*, 1986). Glycosylation is primarily restricted to the 7S globulins (Hurkman and Beevers, 1982). Results from experiments using glycosylation inhibitors suggest that the sugar moiety of the protein is not necessary for targeting of the protein to the vacuole (Badenoch-Jones *et al.*, 1981). Proteolytic processing of the precursor polypeptide into mature subunits occurs in the storage vacuole; this may be followed by a second assembly of the trimers into hexamers, as in the case of 11S globulin (Dickinson *et al.*, 1989).

Prolamines are also synthesized on RER but are transported into storage vacuoles through two different pathways. The prolamines of maize (Larkins and Hurkman, 1978), sorghum (Taylor *et al.*, 1985), and rice (Krishnan *et al.*, 1986) form protein bodies directly in the RER. In wheat (Kim *et al.*, 1988) and barley (Cameron-Mills and von Wettstein, 1980), the prolamines are transported from ER via Golgi vesicles into the vacuole.

The 2S albumins in pea (Higgins *et al.*, 1986) and the Brazil nut (Sun *et al.*, 1987b) are synthesized at specific developmental stages during seed maturation and are deposited in storage vacuoles, but their subcellar transport has not been characterized. The pea albumins PA1a and PA1b are synthesized as a precursor, a 13-kDa preproprotein, which undergoes cotranslational removal of the signal peptide (2 kDa) and posttranslational cleavage of the proprotein into the two polypeptides (Higgins *et al.*, 1986). In the Brazil nut, the sulfur-rich 2S albumin is synthesized as an 18-kDa precursor. After cotranslational removal of a 3-kDa signal peptide, the 15-kDa intermediate precursor undergoes two additional proteolytic cleavages to form the mature 9-kDa and 3-kDa subunits (Sun *et al.*, 1987b).

14.1.5 DEGRADATION

Storage proteins are hydrolyzed during seed germination to provide nitrogen, carbon, and sulfur for the young seedlings. Specific proteinases and carboxy- and aminopeptidases are involved in the degradation. Although these proteolytic enzymes act sequentially, it appears that they are not sequence specific (Shutov and Vaintraub, 1987).

14.1.6 STORAGE PROTEIN GENE STRUCTURE AND ORGANIZATION

Because seed storage proteins are abundant and tissue specific, isolation and study of the genes encoding them have progressed rapidly. Since the isolation of the β-phaseolin gene in 1981 (Sun *et al.*, 1981), over 100 genomic and cDNA clones for seed storage proteins have been reported (Elliston *et al.*, 1988).

DNA sequence analysis of cloned seed storage protein genes has revealed the presence of canonical eukaryotic promoter elements such as TATA and CAAT boxes in the 5′ noncoding regions and polyadenylation signal sequence (AATAAA) in the 3′ flanking regions. In many cereal genes, a CATC rather than a CAAT sequence is a common structure (Kreis *et al.*, 1986). In addition to these putative regulatory sequences, other consensus sequences have been identified in the 5′ regions of many legume seed protein genes. These include the RY repeat element (5′-CATGCAT-3′) (Hoffman and Donaldson, 1985; Dickinson *et al.*, 1988; Nielsen *et al.*, 1989), the CACA sequence

$$\left(\begin{array}{c} \text{A} \qquad\qquad \text{AA} \\ 5'\text{-TAACACACT-}3' \\ \text{C} \qquad\qquad \text{C} \end{array}\right)$$

(Goldberg, 1986), the 28-bp "legumin box" (5′-TCCATAGCCATGCATG-CTGAAGAATGTC-3′) (Baumlein *et al.*, 1986), and the "vicilin box" (5′-GCCACCTCAATTT-3′) (Gatehouse *et al.*, 1986). The RY repeats also occur in many cereal seed protein genes (Forde *et al.*, 1985) and a "−300 element"

(5′-ACA . . TGTAAAGGTGAATGNAGATGAGTCATG-3′)
TG G TG

is unique to the prolamine genes (Kreis *et al.*, 1986). Although these conserved sequences have been implicated to play a role in regulating the expression of these genes, their function is unknown.

Some seed storage protein genes contain intervening sequences (Sun *et al.*, 1981). Most 11S globulin genes contain three introns of different lengths (Shotwell and Larkins, 1989) while the 7S globulin genes analyzed so far have five introns (Doyle *et al.*, 1986). Genes encoding cereal prolamines apparently lack introns (Shotwell and Larkins, 1989; Casey and Domoney, 1987).

Nearly all seed storage protein genes investigated to date are members of large multigene families. These gene families range in size from a few members, for example, one or two genes each for the β-, γ-, and δ-zeins (Marks *et al.*, 1984; Gallardo *et al.*, 1988; Kirihara *et al.*, 1988b), and five genes for the soybean glycinin (Nielsen *et al.*, 1989), to perhaps a hundred members, as for the maize α-zein genes (Wilson and Larkins, 1984) and the α/β-gliadin genes in wheat (Okita *et al.*, 1985). Thus, the storage protein in a mature seed usually represents the combined expression of many genes.

The molecular organization of seed storage protein genes varies among plant species. Some genes are tightly clustered, while others are scattered throughout the genome. For example, two of the three subfamilies of the pea 11S globulin genes are located on chromosomes 7 and 1 and these genes are tightly clustered; a third subfamily is also on chromosome 1, but these genes are less closely spaced (Domoney *et al.*, 1986). The pea 7S globulin genes map to six different loci. The convicilin genes have a single locus on chromosome 2 (Matta and Gatehouse, 1982), while the vicilin genes are at five loci, three of which map to chromosome 7 (Ellis *et al.*, 1986).

In wheat and barley, little is currently known with regard to the molecular organization of the prolamine genes, but studies on zein genes in maize reveal that the genes of subfamily Z1A are located on chromosomes 4, 7, and 10, while those of subfamily Z1B are on chromosomes 4 and 7. Genes of subfamily Z1C are only on chromosome 4 (Valentini *et al.*, 1979; Soave *et al.*, 1981, 1982). Genomic linkage experiments indicate that these genes are closely spaced along the chromosome.

14.1.7 STORAGE PROTEIN GENE EXPRESSION

Seed storage protein gene expression is regulated both temporally and spatially (Higgins, 1984; Goldberg *et al.*, 1989). The time of expression during seed maturation differs among gene families and may even vary between members of the same family (Higgins, 1984). In most seeds, however, maximum transcriptional activity, mRNA concentration, and rate of protein synthesis are around mid-to-late maturation stages. The expression of storage protein genes is tissue specific. Primary synthesis of these proteins occurs in the cotyledons and endosperm and only lower amounts of these proteins are present in the embryonic axis (Meinke *et al.*, 1981). Storage protein mRNAs have been localized in specific cells within the cotyledons and axis (Goldberg *et al.*, 1989). Seed storage protein genes are primarily regulated at the transcriptional level, but posttranscriptional mechanisms have also been suggested to play an important role in the synthesis of these proteins (Chappell and Chrispeels, 1986; Goldberg *et al.*, 1989).

Several types of mutations affecting the expression of seed storage protein genes have been described. These include DNA insertions (Vodkin *et al.*, 1983; Goldberg *et al.*, 1983) and deletions (Kreis *et al.*, 1983) in the coding sequences, and mutations in the regulatory loci (Larkins *et al.*, 1984). Environmental factors such as nutrients and temperature have also been shown to regulate the expression of seed storage protein genes (Higgins, 1984).

14.1.8 TRANS-ACTING PROTEIN FACTORS

There have been some studies on the interaction between trans-acting factors and cis-acting regulatory elements involved in controlling seed storage protein gene expression. These results are largely based on the characterization of embryo proteins that bind to specific DNA sequences in the 5′ flanking regions of storage protein genes. In soybean, a 60-kDa nuclear protein was found to bind to two regions (I: 12 bp, −184 to −173; II: 40 bp, −165 to −126) in the 5′ flanking sequence of a lectin gene (Jofuku *et al.*, 1987). Both of these regions share a 7-bp core motif,

5′-ATTAAAT-3′.
T

This nuclear protein is not detected in other plant tissues and its accumulation in the seed parallels the lectin transcription rate during seed development.

A specific nuclear protein in the endosperm of maize was reported to bind to a 22-bp sequence in the 5′ flanking region of a 19-kDa zein gene (Maier *et al.*, 1987). This protein binding site includes 14 bp of a 15-bp sequence (5′-CACATGTGAAAGGT-3′, −317 to −331) which is conserved in all zein genes.

In a further search for nuclear protein binding sites in the 1.5-kb sequence preceding the 19-kDa maize zein gene, at least four protein-binding sites were identified and two of these show tissue specificity (Maier *et al.*, 1988). However, there is no direct evidence that these proteins or the sequences with which they interact are important in regulating the transcription of the genes.

Recently, Bustos *et al.* (1989) identified an upstream region (−628 to −682) of the bean β-phaseolin gene that binds to nuclear proteins from immature bean cotyledons. This region contains two inverted A/T-rich motifs, 5′-TTAATTTTAAG-3′, at −650 to −640 and 5′-AATATTTTAAT-3′, at −657 to −667. When a 103-bp fragment containing these two A/T-rich motifs was inserted 5′ to a truncated 35S CaMV promoter/GUS chimeric gene and its activity assayed in tobacco, strong GUS expression was detected in several tissues, indicating that these A/T-rich sequences can activate transcription *in vivo*.

14.1.9 EXPRESSION OF SEED STORAGE PROTEIN GENES IN TRANSGENIC PLANTS

Recent advances in transformation and regeneration procedures allow the transfer of cloned genes into plants. In this way, the expression of specific genes can be analyzed. The ability to create transgenic plants containing seed storage protein genes is particularly attractive for (1) studies on the developmental and spatial aspects of gene regulation and mechanism controlling the level of expression; (2) functional analysis of promoter sequence elements; and (3) examination of the synthesis, transport, processing/modification, deposition, and stability of these protein products in seeds. The introduced genes can be heterologous, homologous, natural, modified, or synthetic. Information from these studies not only contributes to our understanding of plant gene expression but also to the improvement of seed protein quality through genetic engineering.

14.1.9.1 Transfer and Expression of Storage Protein Genes

As summarized in Table 14.1, many seed storage protein genes have been studied in transgenic plants. These include globulins, prolamines, albumins, and lectins from diverse plant species. *Agrobacterium tumefaciens* Ti plasmid vectors have been used to incorporate these genes into host plants, frequently solanaceous species such as tobacco and petunia. In general, the expression of the introduced genes shows normal developmental and spatial regulation. In most cases, the protein products have normal processing in the developing seeds, including the removal of the signal peptide, proteolytic cleavage, glycosylation, transport, assembly, and deposition in protein bodies. This suggests that diverse plant species have similar mechanisms of gene regulation and protein processing. In transgenic tobacco seeds, the introduced storage protein was found in the protein bodies of the embryo while a small portion was in the endosperm tissue (Greenwood and Chrispeels, 1985; Hoffman *et al.*, 1987; Sturm *et al.*, 1988). Protein products of the expected size were generally detected, although in several instances smaller polypeptides were observed (Sengupta-Gopalan *et al.*, 1985; Beachy *et al.*, 1985).

The amount of introduced protein in transgenic plants is generally low, ranging from 0.001 to 8% of the total seed protein. There are many possible explanations for the small accumulation of these proteins. The promoters of the introduced seed protein genes may function less efficiently in the heterologous seeds. The proteins may not be processed or transported properly following their synthesis, or they may be degraded by proteases. In addition, these proteins result from a single or few gene copies, which would also limit protein accumulation as compared to the normal multigene family situation.

It has been found that the copy number of the introduced storage protein genes can vary from one to more than 10 copies per haploid genome (Refs. in Table 14.1). However, a general conclusion cannot be drawn as to whether there is a correlation between gene copy number and amount of protein. Although there was a good correlation between gene number and protein accumulation in tobacco containing a field bean legumin B gene (Baumlein *et al.*, 1988), a soybean β-subunit gene of β-conglycinin (Bray *et al.*, 1987), and an *Arabidopsis* 2S albumin gene (De Clercq *et al.*, 1990), there was significant variation in the gene number and expression of maize α-zein genes in petunia (Ueng *et al.*, 1988) and Brazil nut 2S albumin genes in rapeseed (Guerche *et al.*, 1990). This variation in expression and gene copy number may be partially due to position effects resulting from the site of integration of the transferred genes in the chromosomes (Jones *et al.*, 1985).

The 5′ flanking sequence of the genes used in these studies has varied from 0.3 to 3.0 kb. It has been found that 1 kb of 5′ flanking sequence generally contains the essential signals for developmental and spatial regulation. However, maize α-zein genes do not fit this pattern, at least in tobacco and petunia (Ueng *et al.*, 1988; Schernthaner *et al.*, 1988). Transcription of these genes was found to initiate and terminate correctly in the seeds of the transgenic plants, but only small amounts of zein transcripts were synthesized and no protein was detected at any stage of development. In addition, low levels of the mRNA were also detected in leaves, stems, and flowers of transformed petunia plants. It was concluded that the DNA sequences responsible for developmental regulation of this zein gene are not readily recognized in these dicot plants, or the genes tested had weak promoters or are pseudogenes.

TABLE 14.1
Expression of Seed Storage Protein Genes in Transgenic Plants

Protein	Gene	Source plant	Host plant	Expression		Level (%)[b]	Processing[c]	Stability[d]	Flanking sequence (kb)		Ref.
				Temporal	Spatial[a]				5′	3′	
β-Subunit (48 kDa)	Phaseolin	French bean	Tobacco	+	+	1.0	+	Up to 70% degradation	0.86	1.23	Sengupta-Gopalan *et al.*, 1985
α-Subunit (76 kDa)	β-Conglycinin	Soybean	Petunia	+	+	0.1–1.0	+	Up to 50% degradation	9.0	0.4	Beachy *et al.*, 1985
β-Subunit (53 kDa)	β-Conglycinin	Soybean	Petunia	ND[e]	+	0.01	+	+	1.1	1.3	Bray *et al.*, 1987
			Tobacco	ND	+	<0.01	+	+	1.1	1.3	
19-kDa	Zein	Maize	Petunia	+	−	0	—	—	0.95	1.22	Ueng *et al.*, 1988
9-kDa + 9-amino acid insert	Napin	Rapeseed	Tobacco	+	+	ND	ND	ND	0.3	2.4	Radke *et al.*, 1988
15-kDa zein	Phaseolin/zein chimeric	French bean & maize	Tobacco	+	+	1.6	+	+	0.85	1.1	Hoffman *et al.*, 1987
23-kDa	Zein 4	Maize	Tobacco	ND	*[f]	0			0.93	5.0	Schernthaner *et al.*, 1988
β-Subunit (40 kDa)	Legumin B4	Field bean	Tobacco	ND	+	0.43–17.23ng /μg globulin	+	ND	2.7	0.4	Baumlein *et al.*, 1988
11S (58 kDa)	Legumin A	Pea	Petunia	+	+	0.2	ND	+	1.2	0.4	Ellis *et al.*, 1988
Lectin (31 kDa)	Lec1	Soybean	Tobacco	+	+	0.2	+	ND	Flanked by 3 genes		Okamuro *et al.*, 1986
PHA-L (35.5 kDa)	dlec2	French bean	Tobacco	ND	+	0.02	+	Some breakdown	1.2	1.6	Sturm *et al.*, 1988

PHA-L (35.5 kDa)	dlec2	French bean	Tobacco	+	+	0.05	+	Some break-down	1.1	1.4	Voelker *et al.*, 1989
	Pdlec2			+	+	0.001	+		0.80	1.2	
2S (13 kDa)	AT2S1	*Arabidopsis*	Tobacco	+	+	0.1	+	+	1.8	1.5	De Clercq *et al.*, 1990
2S (12 kDa)	Phaseolin/Brazil nut	French bean & Brazil nut	Tobacco	+	+	3.0–8.0	+	+	0.80	1.2	Altenbach *et al.*, 1989
2S (13 kDa)	Plec1/2S chimeric	Soybean & Brazil nut	Rapeseed	+	+	0.02–0.06	+	+	NA[i]		Guerche *et al.*, 1990
β-Subunit (48 kDa + 15 amino acid in-sert)	Phaseolin	French bean	Tobacco	+	+	0.002	+	Some break-down	0.86	1.2	Hoffman *et al.*, 1988
2S (13 kDa)[g]	AT2S1	*Arabidopsis*	*Arabidopsis*	ND	+	0.1	+	+	NA		Vandekerckhove *et al.*, 1989
			Rapeseed	ND	+	<0.1	+	+	NA		
2S (13 kDa)[h]	AT2S1	*Arabidopsis*	*Arabidopsis*	+	+	1.0–2.0	+	+	1.8	1.5	De Clercq *et al.*, 1990
			Rapeseed	+	+	1.0–2.0	+	+	1.8	1.5	
			Tobacco	+	+	1.0–2.0	+	+	1.8	1.5	
α-Subunit	CaMV 35S/ & CaMV 19S/ β-conglycinin chimeric	Soybean	Petunia	–	–	35S 10–50X >19S	+	Breakdown products	NA		Lawton *et al.*, 1987
									0.37	NA	

[a]Seed specific.
[b]Protein product, as % of total protein in transgenic seeds.
[c]In most cases, signal peptide removal, proteolytic cleavage, glycosylation, and assembly included.
[d]Detection of expected size of protein product.
[e]Not determined.
[f]mRNA detected in seeds.
[g]With six amino acid substitutions.
[h]Sequence modifications including 5, 8, and 12 additional methionine residues in the protein products.
[i]Not available.

The absence of α-zein proteins in petunia and tobacco seeds may be due in part to the low level of the mRNA synthesized, but protein stability is also partially responsible. The α-zein protein is stable in endosperm, leaves, and roots of transgenic tobacco plants (Schernthaner *et al.*, 1988), but it is unstable in cotyledon tissue (Schernthaner *et al.*, 1988; Ohtani *et al.*, 1990). Interestingly, the β-zein protein accumulates in tobacco seeds to 1.5% of the total seed protein when the coding sequence is controlled by a dicot storage protein (phaseolin) promoter (Hoffman *et al.*, 1987).

14.1.9.2 Promoter Analysis

Transfer and expression of genes in transgenic plants provides a powerful system for functional analysis of promoter sequences. By generating deletions, insertions, and modifications in the promoter sequences and using synthetic DNA sequences and reporter genes, elements responsible for gene regulation can be identified and studied. Table 14.2 summarizes analyses of promoters of seed storage protein genes.

The 5′ flanking sequences responsible for enhancing tissue-specific and developmental expression of both globulin and prolamine storage protein genes have been partially characterized. These include a 170-bp (−78 to −257) upstream DNA element in the soybean β-conglycinin's α′-subunit gene (Chen *et al.*, 1988), a 166-bp 5′ region of the LMW wheat glutenin gene (Colot *et al.*, 1987), a 549-bp 5′ flanking region of the barley B1 hordein gene (Marris *et al.*, 1988) and a 933-bp 5′ sequence of the 23-kDa maize zein gene (Schernthaner *et al.*, 1988). In a few instances, shorter DNA sequence elements were identified that appear to play a critical role in determining tissue-specific gene expression. For example, in the LMW wheat glutenin gene, there is the

```
TGTAAAGG
  A    AT
  C
```

element preceding the coding region by 300 bp (Colot *et al.*, 1987). In the soybean β-conglycinin gene, there are the

```
AACCCA
  G
  C
```

repeats that regulate the level of expression (Chen *et al.*, 1986). The A/T-rich motifs (−650 to −640: TTAATTTTAAG; −657 to −667: AATATTTTAAT) in β-phaseolin gene (Bustos *et al.*, 1989) have been suggested to have a regulatory function.

TABLE 14.2
Transgenic Analysis of Seed Storage Protein Promoter Sequences

Protein	Gene	Plant source	Host plant	Gene construct	Expression and important DNA regions				Refs.
					Developmental	Spatial	Level	DNA Element	
α′-Subunit	β-Conglycinin	Soybean	Petunia	5′deletion +14 to −8500	+	+	(+1 to −257) level 20> (+1 to −159)	A AGCCCA C	Chen *et al*, 1986
HMW LMW	Glutenin	Wheat	Tobacco	5′deletion HMW_{pro}/CAT LMW_{pro}/CAT	+ +	−438 to −5 −326 to −160	ND[b] ND	T GG TGAAAAAT C	Colot *et al.*, 1987
Prolamine	B1 Hordein	Barley	Tobacco	$B1_{pro}$/CAT	+	−512 to +37	ND	ND	Marris *et al.*, 1988
α′-Subunit	β-Conglycinin	Soybean	Tobacco	$CaMV_{pro}$ + NT insert[a]/CAT	+	−257 to −78	−257 to −78	ND	Chen *et al.*, 1988
23-kDa	Zein4	Maize	Tobacco	$Z4_{pro}$/CAT $Z4_{pro}$/CAT	+ +	−940 to −7 −940 to −7	ND ND	ND ND	Schernthaner *et al.*, 1988
β-Subunit	Phaseolin	French bean	Tobacco	Ph_{pro}/GUS $CaMV_{pro}$ + A/T-rich/GUS	+ −	+ −	ND −628 to −682	ND TTAATTTTAAG & AATATTTTAAT	Bustos *et al.*, 1989

[a]Insertion of β-glycinin 5′ −257 to −78 sequence element into $CaMV_{pro}$ at different positions and orientation.
[b]Not determined.

14.2 Molecular Genetic Approaches for Improving Seed Storage Protein Quality

14.2.1 MOLECULAR APPROACHES

Advances in gene cloning, DNA sequence manipulation, and transformation/regeneration of plants have allowed the use of molecular genetic approaches to improve the nutritional quality of seed storage proteins. Several strategies are now available for this undertaking and there are research efforts in each of these areas.

14.2.1.1 Protein Sequence Modification

This was one of the earliest ideas for using recombinant DNA technology to improve seed protein quality. Since most seeds contain one or a few types of storage proteins, it is an attractive idea to isolate a gene encoding a superabundant protein and modify its coding sequence to contain more codons for essential amino acids. This can be accomplished by insertion of essential amino acid-rich sequence(s) or by site-directed mutagenesis of existing codons to increase the essential amino acid content.

14.2.1.2 Synthetic Genes

With the development of DNA synthesis technology, it became possible to design and synthesize a gene encoding a protein rich in essential amino acids. This approach appears particularly tempting as a novel protein can be tailor-made to have desirable nutritional quality.

14.2.1.3 Overexpression of Homologous Genes

Seed proteins differ in amino acid composition. The overall amino acid composition of a seed is often determined by a few abundant species of storage proteins. However, some less abundant or minor proteins may contain elevated levels of a specific essential amino acid. Increasing the synthesis and accumulation of such a protein by overexpressing the gene encoding it could improve the concentration of a particular essential amino acid. An active promoter and/or an increase in the gene copy number represent two possible approaches to overexpress the gene product. Alternatively, one could suppress or inactivate a gene that encodes a major storage protein with a poor essential amino acid content and thereby elevate the levels of other seed proteins and improve the overall essential amino acid balance.

14.2.1.4 Transfer and Expression of Heterologous Genes

The development of plant transformation and regeneration methodology allows the transfer of genes between different plant species, even between plants, animals, and microbes. Thus, genes encoding essential amino acid-rich proteins can be identified and isolated from various sources and introduced into important seed crops. Proper expression of these heterologous genes at suitable levels could improve the concentration of essential amino acids.

14.2.2 FACTORS TO BE CONSIDERED

The preceding review on the biochemistry, molecular biology, and transgenic expression of seed storage protein genes reveals that these proteins possess unique properties, ordered structures, and specific mechanisms of synthesis, processing, and deposition. Genes encoding storage proteins contain sequence elements that regulate their developmental and spatial expression, as well as the rate of their transcription in transgenic plants. Thus, many factors may impact the genetic engineering of seed storage proteins.

14.2.2.1 Protein Solubility

Different types of seed storage proteins have characteristic solubilities. Thus, amino acid sequence modifications which involve substantial variation in polar and charged residues might cause solubility changes in the proteins that interfere with their association in the ER and assembly in the protein bodies. It is reassuring to find, however, that the sulfur-rich β-zein prolamine is accumulated stably in the seeds of transgenic tobacco and petunia (Hoffman *et al.*, 1987; Schernthaner *et al.*, 1988).

14.2.2.2 Protein Structure

Modification and insertion of amino acids in seed storage proteins could lead to changes in their secondary structure. For example, introduction of cysteine residues into a storage globulin could generate disulfide bonds, and modification of repeated peptides in prolamines might interfere with hydrogen bonding or other types of interactions between adjacent repeats. Changes in secondary structure could affect the interactions between subunit polypeptides and the assembly of oligomers which are important in the aggregation and deposition of storage protein in the protein bodies. It is also possible that changes in protein structure may expose regions sensitive to proteolytic enzymes.

14.2.2.3 Cleavage/Processing

Many seed storage proteins are synthesized as precursor polypeptides that undergo specific proteolytic cleavages to form subunit polypeptides. If cleavage sites are altered by sequence modification or insertion, processing of precursors might be prevented and assembly of oligomers might not occur. Similarly, for proteins that contain sugar residues, altering the glycosylation sites could result in changing the structure and properties of the storage proteins, although the absence of the sugar component does not appear to affect protein transport.

14.2.2.4 Transport and Deposition

Newly synthesized storage proteins are transported from the ER through the Golgi body to the protein bodies, or, alternatively, from the ER directly into the protein bodies. A functional signal peptide is necessary for an engineered seed protein to enter the ER. Precursor processing and assembly of oligomers occurs in the ER, thus sequence modification of a storage protein that interferes with its processing and assembly into oligomers might result in its failure to be transported into the protein bodies.

Although molecular signals that specify the pathway of transport of storage proteins from the ER into the protein body have not been elucidated, results from studies on phytohemagglutinin indicate that the targeting signals are contained in the polypeptide rather than the carbohydrate component (Voelker *et al.*, 1989). Thus, it is important to maintain the proper targeting sequences in the polypeptide of engineered seed proteins.

14.2.2.5 Protein Stability

The modified or heterologous proteins introduced into transgenic seeds must be stable or be subjected to as little proteolytic degradation as possible. This is one of the most critical factors that affect the consequences of changes in amino acid composition. Sequence modifications that interfere with processing, assembly, and transport might render the proteins susceptible to degradation. In addition, the sequence modification itself might create or expose regions in the polypeptides that are sensitive to proteolytic degradation.

14.2.2.6 Degradation during Germination

Seed storage proteins are degraded by enzymes during germination to supply carbon, nitrogen, and sulfur for the young seedling. It is important that the genetically engineered proteins be easily degraded so that they can serve this function. Since the proteolytic enzymes produced in germinating seeds do not appear to be sequence specific (Shutov and Vaintraub, 1987), transgenic seeds probably will be able to break down and mobilize the introduced heterologous or modified proteins. Most studies to date suggest that the introduced proteins are degraded in seeds during germination.

14.2.2.7 Multigene Family

Seed storage proteins are frequently encoded by multigene families, and the amount of seed storage protein in a mature seed represents the accumulation of the products of all the active gene members. The number of active members of a gene family could be large. Thus, an introduced gene must be expressed at high levels in order to make a significant difference in the overall amino acid composition of the seed.

14.2.2.8 Introns

Many seed protein genes contain introns. In one experiment it was demonstrated that β-phaseolin minigene, that is, a phaseolin gene missing its five introns, was expressed in transgenic tissues, indicating that the introns and their splicing are not necessary for generating stable phaseolin mRNA and that no alternative splice site was introduced by the removal of the introns (Chee *et al.*, 1986). Complementary DNAs encoding seed proteins have been used for chimeric gene construction and these appear not to affect tissue specificity and developmental expression of the promoters (Refs. in Table 14.1). However, in a transient expression study Callis *et al.* (1987) reported that an intronless maize alcohol dehydrogenase-1 (Adh1) gene produced between 50- to 100-fold less mRNA than that obtained from the normal gene containing 9 introns. Thus, when a cDNA is used as a coding sequence for transgenic expression of seed proteins, the level of its expression remains a factor for consideration and evaluation.

14.2.2.9 Regulation of Expression

The expression of seed storage protein genes is regulated at multiple levels, including transcription rate, time, and cell and tissue specificity. It is critical to maintain these parameters for modified or heterologous genes in transgenic plants. Studies using transgenic systems have shown that the 5′ flanking regions of seed protein genes contain regulatory signals that confer temporal, spatial, and rate of expression. These signals appear to function in transgenic plants with distant evolutionary relationships. While this is reassuring, some monocot promoters may not be properly recognized in dicot plants (Ueng *et al.*, 1988; Schernthaner *et al.*, 1988). Whether the reverse is also true remains to be elucidated.

14.2.3 ADVANTAGES AND DISADVANTAGES OF THE VARIOUS MOLECULAR APPROACHES

With reference to the multitude of factors that may affect the expression of a modified or heterologous seed protein gene in transgenic plants, we will now consider the advantages and disadvantages of the various molecular approaches for improvement of seed protein quality. The protein sequence modification approach has the advantage of using a major seed protein gene, which has been isolated from an important seed crop and is well characterized. However, a critical and demanding task is to be able to identify regions of the candidate gene that allow modification without causing adverse effects, as discussed earlier. The synthetic gene approach faces an even more challenging task in this regard, as it requires the creation of a novel seed protein that has proper structure, stability, and nutritional quality.

The other two approaches, namely the overexpression of homologous genes and the transfer of heterologous genes encoding essential amino acid-rich proteins, share the advantage of using naturally occurring genes. Thus, no sequence modifications of the coding sequence are required. The feasibility of using an essential amino acid-rich seed protein to enhance the content of these amino acids in seeds has been demonstrated (Altenbach *et al.*, 1989). However, to use these two approaches, a gene encoding a high-essential amino acid protein, preferably a seed storage protein, must first be identified and isolated. The search for candidate genes can be time consuming.

14.3 Current Research Approaches

14.3.1 PROTEIN SEQUENCE MODIFICATIONS

Maize seed is an important source of protein for human and animal consumption, but it contains low levels of lysine and tryptophan. This is because the major maize storage proteins, zeins, are totally lacking lysine and tryptophan. To study the feasibility of introducing lysine residues into zeins, Wallace *et al.* (1988) created lysine and tryptophan codons in various positions in a 19-kDa α-zein coding sequence (cDNA) by nucleotide substitutions and oligonucleotide insertions. A 450-bp open reading frame (ORF) from a simian virus 40 (SV40) coat protein was also inserted into the zein sequence as a control modification (Figure 14.1). Messenger mRNAs for the modified zeins were synthesized *in vitro* and injected into frog oocytes to study their synthesis and processing.

The results showed that the modifications, which included the addition of single or multiple lysine residues, or short lysine- and tryptophan-rich oligopeptides, did not affect translation, signal peptide cleavage, or stability of the zeins. All the modified zeins, with the exception of the one containing the SV40 coat protein insert, were also shown to aggregate in membrane-bound vesicles in the oocytes (Figure 14.2). This study showed that introduction of lysine and tryptophan residues (up to four) into α-zein proteins does not affect their stability and aggregation into the protein bodies.

Proteins of legume seeds and many other dicot crops are deficient in sulfur amino acids. Phaseolin is the major storage protein in French bean seeds. The β-

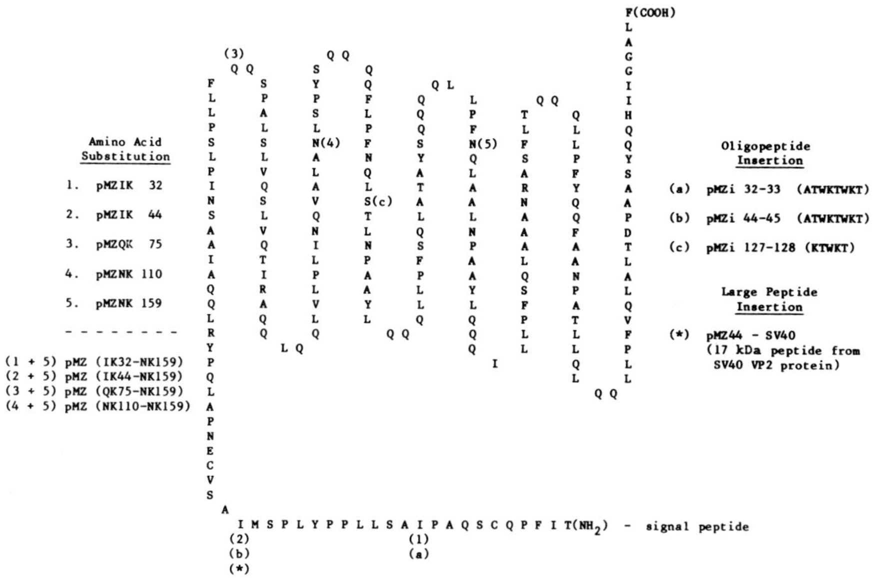

Figure 14.1 Modification of a maize α-zein. Shown is the amino acid sequence of a 19-kDa α-zein. The nine peptide repeats in the central region of the molecule are aligned and are separated by glutamine-rich turns. Also shown are the positions of modifications including amino acid substitutions (1–5), small peptide insertions (a–c), and a large peptide insertion (*, 17-kDa SV40 VP2 protein). For more details, see Wallace *et al.* (1988) with permission of the authors and publisher; copyright 1988 by the AAAS.

phaseolin gene encodes a 47-kDa subunit polypeptide containing 3 methionine residues (0.64 mol%) (Slightom *et al.*, 1983). Hoffman *et al.* (1988) inserted a 45-bp oligonucleotide containing six methionine codons into the third exon of a β-phaseolin genomic clone (Figure 14.3), resulting in a modified gene with 9 methionine codons. The high-methionine (hiMet) and the normal phaseolin genes were transferred into tobacco plants to study their expression.

The hiMet phaseolin gene was developmentally regulated in a tissue-specific manner (Figure 14.4); the hiMet phaseolin protein was glycosylated and assembled into trimers in the same manner as the normal protein. Although the levels of mRNA transcripts for both the hiMet and normal phaseolin genes were comparable, the quantity of hiMet protein was only about 0.2% of the normal phaseolin. Electron microscopic immunocytochemical studies suggest that the hiMet phaseolin was degraded either in the Golgi vesicles or the protein bodies.

Recently, Lawrence *et al.* (1990) reported the 3-dimensional structure of phaseolin. Based on this study, the site in the sequence where Hoffman *et al.* (1988) made the 45-bp nucleotide insertion is important for stabilizing the

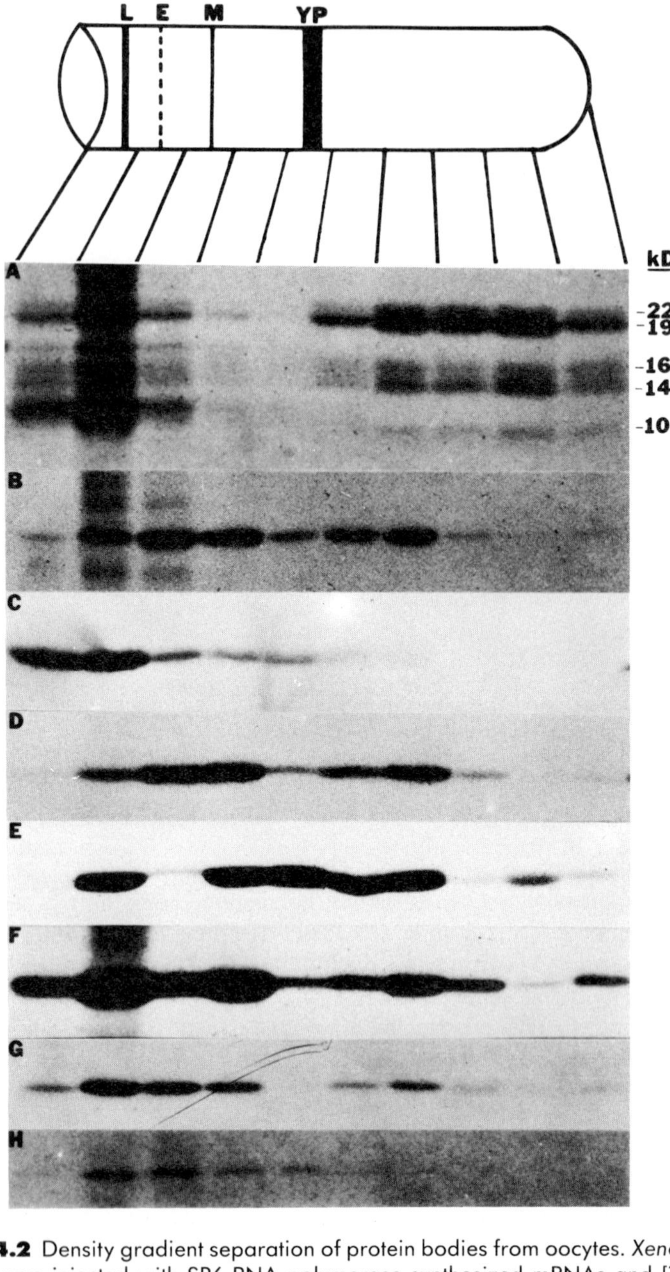

Figure 14.2 Density gradient separation of protein bodies from oocytes. *Xenopus laevis* oocytes were injected with SP6 RNA polymerase synthesized mRNAs and [^{3}H]leucine. After a 4-hr labeling period, the oocytes were homogenized and the organelles separated by centrifugation in a 10–15% metrizamide gradient. The distribution of radioactivity in the gradient was determined by gel electrophoresis and fluorography. Panels correspond to the following mRNAs (see Figure 14.1): A, native total zein mRNA; B,

```
              Xba I                                                                   Xba I
                ↓                                                                       ↓
   K   H   I        L   D   Q   M   R   M   M   D   Q   M   R   M   M   D   V         L   E   A   S
...AAG CAT ATT      CTA GAC CAG ATG AGA ATG ATG GAC CAG ATG AGG ATG ATG GAC GTT       CTA GAG GCC TCC...
...TTC GTA TAA GATC     TG GTC TAC TCT TAC TAC CTG GTC TAC TCC TAC TAC CTG CAA GATC       TC CGG AGG...

    Phaseolin                            45-bp Insertion                                    Phaseolin
```

Figure 14.3 Insertion of a methionine-rich sequence into a β-phaseolin gene. A 45-bp oligonucleotide duplex encoding six methionine residues was inserted into a unique Xba I site in the third exon of the French bean β-phaseolin gene. The modified hiMet phaseolin gene was transferred into tobacco. Adapted from Hoffman *et al.* (1988).

phaseolin trimers. This peptide insertion might have destabilized the trimers, rendering the hiMet phaseolin susceptible to proteolytic degradation.

In a related study, Vandekerckhove *et al.* (1989) replaced 6 nucleotides between the 6th and 7th cysteine codons in the *Arabidopsis* 2S albumin gene (AT2S1), such that a pentapeptide (YGGFL) with opiate activity, enkephalin, was generated. This peptide could be excised from the modified protein by cleavage of the flanking trypsin sites. The modified gene was transferred into rapeseed and *Arabidopsis* where it was expressed in a seed-specific manner. Up to 200 nmol of the neuropeptide was recovered per gram of transgenic seeds, indicating that the modified protein accounted for 3% of the total seed protein. This level of synthesis suggests that the modified protein is stable in the transgenic seeds.

The region between the 6th and 7th cysteine residues of the 2S albumins from different species has been found to vary in both length and sequence (Ampe *et al.*, 1986), suggesting that it may tolerate considerable structural change. This study demonstrated the feasibility of replacing 6 nucleotides in this variable region and suggests the further possibility of introducing essential amino acids, such as methionine or lysine, into this region. This possibility was demonstrated recently by De Clercq *et al.* (1990). In one of the experiments, a part of this variable region in the AT2S1 gene was deleted and replaced with a sequence that contained 11 additional methionine codons. This modified 2S gene was transferred into *Arabidopsis, Brassica napus,* and tobacco. The transgenic seeds of all three species were found to accumulate methionine-enriched 2S albumins at levels ranging from 1 to 2% of the total high salt-extractable seed protein.

←

unmodified α-zein mRNA; C, exogenous ^{3}H-labeled α-zein added to homogenate; D, isoleucine to lysine substitution at position 1; E, asparagine to lysine double substitution at positions 4 and 6; F, oligopeptide insertion at position 1; G, oligopeptide insertion at position 5; H, insertion of a 17-kDa fragment of the SV40 VP2 protein at position 2. The relative molecular masses of the zeins are indicated to the right of panel A. The zeins in panels B and G migrated at 19 kDa and the modified zein in panel H migrated at 35 kDa. Oocyte components: L, lipids; E, endomembranes; M, mitochondria; YP, yolk platelets. From Wallace *et al.* (1988) with permission of the authors and publisher; copyright 1988 by the AAAS.

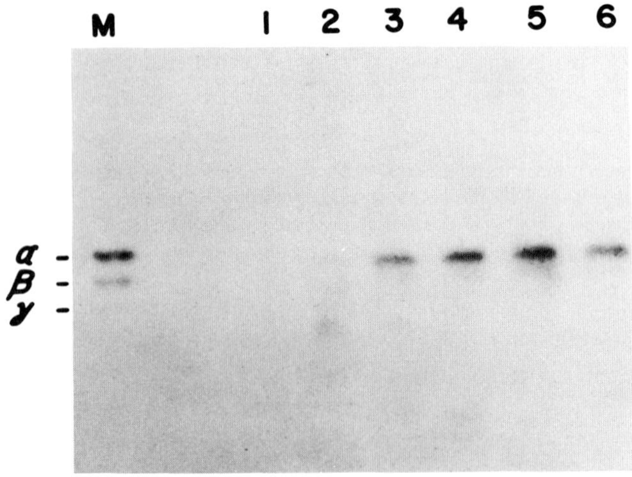

Figure 14.4 Developmental expression of the high-methionine phaseolin gene. An immunoblot assay was made for hiMet phaseolin accumulation in transgenic tobacco seeds at various times after pollination. Lanes: M, the three phaseolin subunit polypeptides, α (51 kDa), β (47 kDa), and γ (45 kDa); 1–6, protein extracts equivalent to 4 seeds at 8, 12, 14, 16, 17, and 21 days after pollination, respectively. From Hoffman *et al.* (1988) with permission of the authors and publisher.

Dickinson *et al.* (1990) recently reported the development of an *in vitro* assay to evaluate the effect of structural modifications on the assembly of glycinin, the major soybean storage protein. In this study, the modifications were generated by deletions and insertions in the cDNA sequence encoding the glycinin, and the modified sequences were then transcribed *in vitro* by using SP6 RNA polymerase. The mRNA transcripts were then translated in a rabbit reticulocyte lysate to produce radiolabeled glycinin polypeptides. The effect of amino acid modifications on the assembly of the glycinin subunits into oligomers was assayed as illustrated in Figure 14.5. Three types of assembly were assayed: self-assembly denotes the ability of the modified glycinins to assemble into 9S trimers by themselves; mixed-assembly to assemble into 9S trimers in the presence of unmodified glycinins; and reassembly to assemble into hexamers in the presence of dissociated glycinins. Using this system, it was found that modifications in the acidic polypeptide of glycinin did not affect the assembly, while alterations in the basic polypeptide usually eliminated assembly into trimers. In addition, extensive deletions or insertions in the hypervariable region of the C terminus in the acidic polypeptide were tolerated without affecting the subunit assembly. These results are consistent with the structural models of glycinin in which the basic polypeptide contributes more to the stability of the oligomers than the acidic polypeptide, and the hypervariable region of the acidic chain is

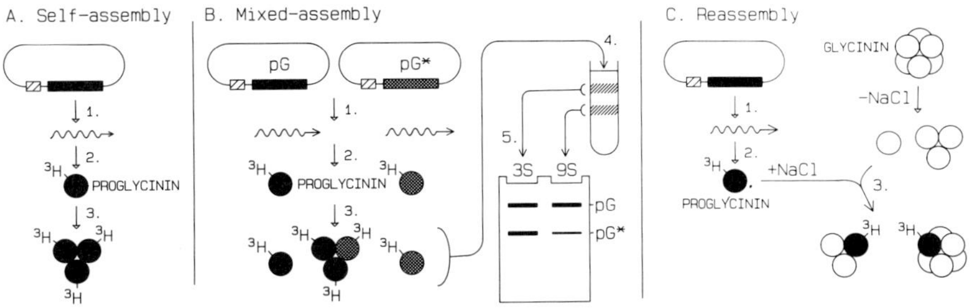

Figure 14.5 Proposed mechanism for self-assembly, mixed-assembly, and reassembly of soybean glycinin subunits. A. Self-assembly. Insertion/deletion modified glycinin cDNA sequences are transcribed *in vitro* using SP6 RNA polymerase (1) and the modified mRNAs are then translated in a rabbit reitculocyte lysate in the presence of [^{3}H]leucine (2). The ^{3}H-labeled proteins are allowed to assemble into oligomers (3). B. Mixed-assembly. Normal (pG) and modified (pG*) glycinin cDNAs are transcribed in the same reaction (1) and the mixture of mRNAs is translated as described (2). The mixed proteins are then allowed to assemble (3). C. Reassembly. Modified glycinin proteins are synthesized and labeled in the same reaction (1,2). Either monomers or trimers of the labeled protein are mixed with unlabeled dissociated glycinin and then allowed to reassociate into hexamers. After the assembly reaction, the proteins are separated into 3S monomers and 9S trimers by centrifugation in a sucrose gradient (4), and are then analyzed by polyacrylamide gel electrophoresis and fluorography (5). For the reassembly assay, step (4) is used to detect hexamer formation. From Dickinson *et al.* (1990) with permission of the authors and publisher.

located on the surface of the molecule and has little effect on the stability of the oligomers, as discussed in the protein structure section.

Dickinson *et al.* (1990) also inserted up to five methionine residues in the hypervariable region of the glycinin acidic polypeptide subunits (Table 14.3) and found that these modifications did not affect the assembly of the oligomers. This *in vitro* assembly assay may be of use in evaluating the effects of modifications on storage protein assembly before transferring the modified genes into plants.

14.3.2 SYNTHETIC GENES

There have been no reports on using a synthetic gene for seed protein improvement. However, in a study related to this approach, Yang *et al.* (1989) synthesized a 292-bp DNA encoding a polypeptide composed of 80% essential amino acids, the high essential amino acid encoding DNA (HEAAE-DNA) (Figure 14.6), and inserted it into the chloramphenicol acetyltransferase (CAT) coding sequence to generate a CAT-HEAAE fusion protein. This hybrid coding sequence was then placed under the transcriptional control of the nopaline synthase

TABLE 14.3
Assembly Assay Results from Insertion Modified Proglycinins[a]

Protein	Insertion[b]	Self-assembly	Mixed-assembly	Reassembly w/monomers	Reassembly w/trimers
G4		+	1.0	+	−
G4RM1	RM	+		+	
G4RM3	RMRMRM	+		+	
G4RM5	RMRMRMRMRM	+	0.9	+	−
G4GMT2	HGPMTEMNGP HGPMTGMNGPS	+	1.0	+	
G4GMR5	HGMRGMKHASFLSS	+	1.1	+	
G4GMR3	HASFLSSHASFLSS	+		+	
G4GMR11	HASFLSS	+		+	

[a]Insertion modifications were generated in the hypervariable region of the G4 proglycinin subunit protein by insertion of DNA duplexes into the glycinin cDNAs and the mutants produced were used in the assembly assay to evaluate the effects of the modifications (see Figure 14.5 for details). Note that some insertions will result in 1, 3, and 5 additional methionine residues in the proteins. From Dickinson *et al.* (1990) with permission of the authors and publisher.

[b]Insertion site in each case was immediately after position 330 for the RM repeats and immediately after position 304 for the other insertions (with numbering starting from the natural translation initiator methionine).

(NOS) promoter and the 3′ NOS untranslated sequence for polyadenylation. The chimeric gene was transferred into potato plants and its expression studied.

Messenger RNA and fusion protein of expected molecular size were detected in the tubers of the transgenic potato plants (Figure 14.7). The CAT-HEAAE protein accumulated 0.02 to 0.35% of the total tuber protein,which corresponds to about a 1.1% increase in the level of essential amino acids. This low level of protein synthesis is probably due to the instability of the HEAA-protein and the low activity of the NOS promoter.

14.3.3 OVEREXPRESSION OF HOMOLOGOUS GENES

There are no reports to date where this approach was used for seed protein improvement. However, there is information from relevant studies. For example, the amino acid composition of the major subunits of β-conglycinin (Thanh and Shibasaki, 1977) and the acidic and basic polypeptides of glycinin (Moreira *et al.*, 1979) reveal that some subunits contain higher levels of methionine than others. Thus, while the 48-kDa β-subunit of β-conglycinin has no methionine residues, several acidic chains of glycinin contain 4.5 Met residues/mol. Yamauchi *et al.* (1984) reported that a basic protein in the soybean variety

READING DIRECTION OF TOP STRAND (A)->->->

AATTCGGGGATCGTAAGAAATGGATGGATCGTCATCCATTTCTTCATCCATTTCTTAC
TTAAGCCCCTAGCATTCTTTACCTACCTAGCAGTAGGTAAAGAAGTAGGTAAAGAATG

GATCCATCCATTTCTTAAGAAATGGATGAAGAAATGGATGACGATCCATCCATTTCTT
CTAGGTAGGTAAAGAATTCTTTACCTACTTCTTTACCTACTGCTAGGTAGGTAAAGAA

CATCCATTTCTTCATCCATTTCTTACGATCAAGAAATGGATGAAGAAATGGATGAAGA
GTAGGTAAAGAAGTAGGTAAAGAATGCTAGTTCTTTACCTACTTCTTTACCTACTTCT

AATGGATGAAGAAATGGATGCATCCATTTCTTAAGAAATGGATGAAGAAATGGATGAA
TTACCTACTTCTTTACCTACGTAGGTAAAGAATTCTTTACCTACTTCTTTACCTACTT

GAAATGGATGACGATCGATCGTAAGAAATGGATGACGATCCATCCATTTCTTACGATC
CTTTACCTACTGCTAGCTAGCATTCTTTACCTACTGCTAGGTAGGTAAAGAATGCTAG

CCCGAATT
GGGCTTAA

<-<-<-READING DIRECTION OF BOTTOM STRAND(B)

SEQUENCE OF PROTEIN A

GlyAspArgLysLysTrpMetAspArgHisProPheLeuHisProPheLeuThrIleHisProPheLeu->
LysLysTrpMetLysLysTrpMetThrIleHisProPheLeuHisProPheLeuHisProPheLeuThr->
IleLysLysTrpMetLysLysTrpMetLysLysTrpMetLysLysTrpMetHisProPheLeuLysLys->
TrpMetLysLysTrpMetLysLysTrpMetThrIleAspArgLysLysTrpMetThrIleHisProPhe->
LeuThrIlePro

SEQUENCE OF PROTEIN B

GlyAspArgLysLysTrpMetAspArgHisProPheLeuThrIleAspArgHisProPheLeuHisPro->
PheLeuHisProPheLeuLysLysTrpMetHisProPheLeuHisProPheLeuHisProPheLeuHis->
ProPheLeuAspArgLysLysTrpMetLysLysTrpMetLysLysTrpMetAspArgHisProPheLeu->
HisProPheLeuLysLysTrpMetAspArgLysLysTrpMetLysLysTrpMetThrIleHisProPhe->
LeuThrIlePro

Figure 14.6 The nucleotide sequence of HEAAE-DNA and the derived protein sequences. The HEAAE-DNA encodes two proteins designated A and B. One or the other will be produced depending upon the orientation of the gene when it is inserted into the CAT sequence under the control of the nopaline synthase promoter. Protein A is composed of about 80% essential amino acids (those which are underlined). From Yang *et al.* (1989) with permission of the authors and publisher.

Raiden contained a significantly higher level of the sulfur amino acids (3.6%) than the major storage proteins (conglycinin, 0.6% and glycinin, 2.6%). This variability in amino acid composition suggests the potential of identifying essential amino acid-rich proteins in the seeds that can be improved by increasing their level of synthesis.

Recently, George and de Lumen (1991) identified a 10.8-kDa soybean protein which contained 12.1% Met and 2.5% Cys. This Met-rich protein gene may be used to improve the nutritional quality of soybean through this approach.

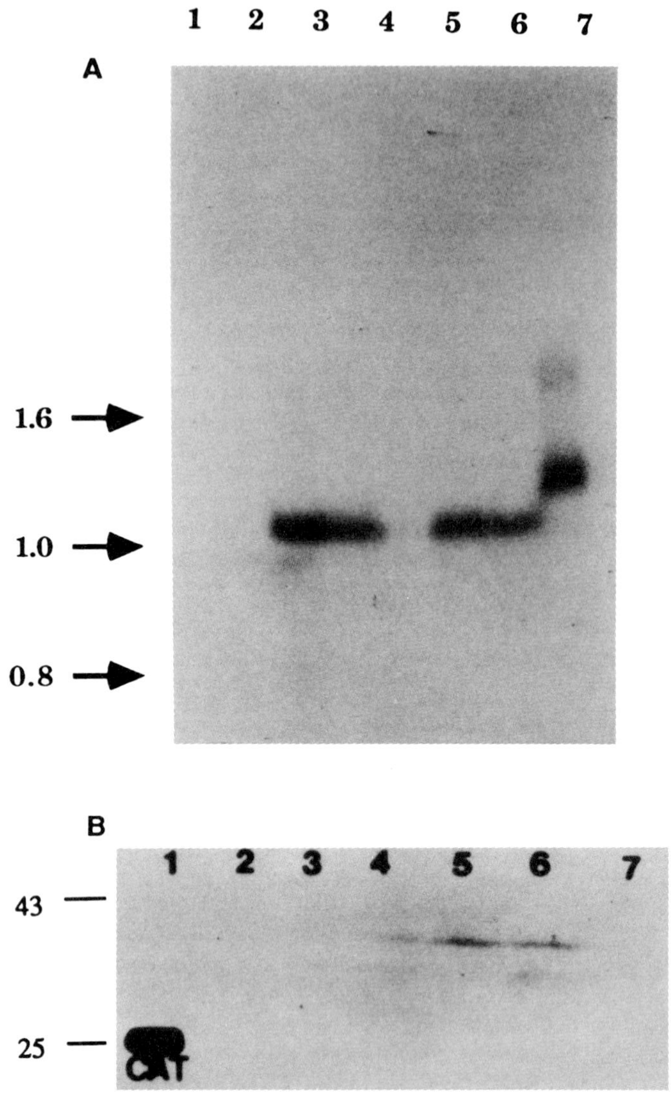

Figure 14.7 Analysis of expression of HEAAE-DNA in transgenic potato tubers. The HEAAE-CAT chimeric gene was transferred into two potato clones, K2 and K7, and its expression monitored. A. Northern blot assay of HEAAE-CAT mRNA. Poly(A) RNAs were isolated from control and transgenic potato tubers and analyzed by Northern blot using a HEAAE-DNA probe. Lanes: 1, negative control, K2 plant; 2–3, tubers of two independent transgenic K2 plants; 4, negative control, K7 plant; 5–6, tubers from two independent transgenic K7 plants; 7, positive DNA control, a 1.35-kb vector DNA fragment containing the HEAAE-CAT sequence. Expected HEAAE-CAT mRNA size is 1.1 kb. B. Immunoblot assay of HEAAE-CAT fusion protein. Total protein was extracted from control and transgenic tubers and the presence of HEAAE-CAT protein was assayed by Western blot analysis using CAT antibody. Lanes: 1, positive control, CAT protein; 2, negative control, K2 plant; 3–4, tubers from two independent transgenic K2 plants; 5–6, tubers of two independent transgenic K7 plants; 7, negative control, K7 plant. From Yang *et al.* (1989) with permission of the authors and publisher.

14.3.4 TRANSFER AND EXPRESSION OF HETEROLOGOUS GENES

In a study by Hoffman *et al.* (1987), the gene encoding a maize 15-kDa methionine-rich β-zein (7% Met and 4% Cys) was placed under the regulation of the French bean phaseolin gene 5′ flanking region and transferred into tobacco. The expression of this gene in the transgenic tobacco plants was tissue specific and developmentally regulated. The methionine-rich zein protein was deposited in both the endosperm and embryo tissues of tobacco seeds, and accumulated to 1.6% of the total tobacco seed protein.

The 2S albumin seed protein of the Brazil nut contains high levels of sulfur amino acids, 18% methionine and 8% cysteine (Ampe *et al.*, 1986; Sun *et al.*, 1987a). This gene has been used to enhance the methionine content of transgenic tobacco (Altenbach *et al.*, 1989) and rapeseed (Guerche *et al.*, 1990).

Altenbach *et al.* (1989) constructed a chimeric gene using a Brazil nut protein cDNA and the promoter of the French bean phaseolin gene. The chimeric gene was transferred into tobacco plants and mRNA for the methionine-rich protein was detected in the developing transgenic tobacco seeds. Subunit polypeptides with the expected size accumulated in the mature seeds, indicating that the tobacco seeds were able to properly process the Brazil nut precursor polypeptide (Figure 14.8). The methionine-rich protein was also found to disappear in germinating transgenic tobacco seeds. The level of synthesis of the methionine-rich protein varied from 3 to 8% of the total tobacco seed protein. Amino acid analysis revealed that the methionine content in transgenic seeds was enhanced by 10 to 30% over normal tobacco seeds (Table 14.4), demonstrating that significant increases in an essential amino acid of the seed protein are feasible with this approach. Concomitant increase in the cysteine content in the transgenic seeds, however, was not observed. Since the content of the total seed protein and the 2S protein fraction of transgenic tobacco remained the same in the normal seeds, it is possible that the synthesis of some cysteine-rich 2S tobacco seed proteins was suppressed as a result of the introduced protein.

Guerche *et al.* (1990) reported the expression of the Brazil nut methionine-rich 2S albumin gene in rapeseed. In this study, a soybean lectin promoter was used to direct the expression of the 2S albumin gene. The expression of this chimeric gene in transgenic rapeseed was found to be tissue specific and developmentally regulated. The protein product had a size consistent with correct processing of the precursor (Figure 14.9) and the protein expression level ranged from 0.02 to 0.06% of the total rapeseed protein. This level of protein accumulation is less than would be expected from the level of 2S albumin mRNA synthesized, suggesting that the low level of protein may be due either to inefficient translation of the chimeric mRNA or incorrect processing and targeting of the protein.

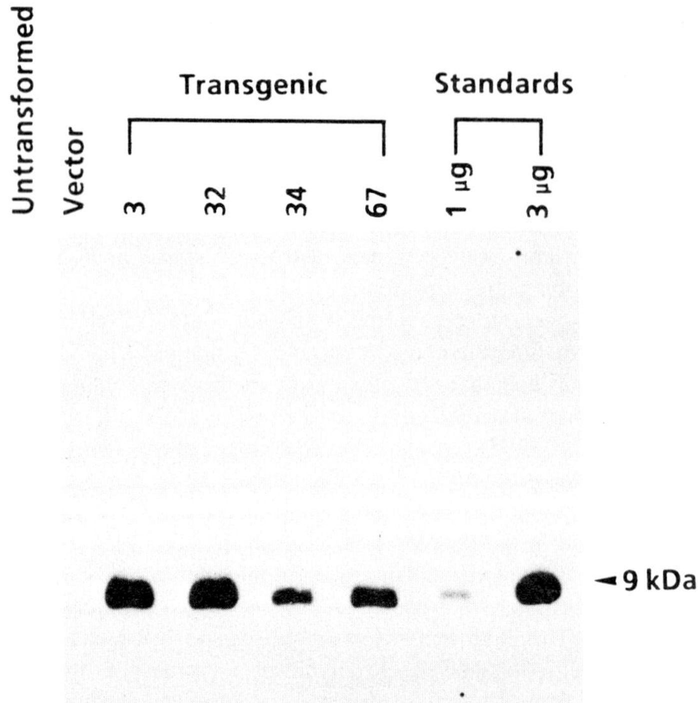

Figure 14.8 Expression of Brazil nut methionine-rich protein in transgenic tobacco seeds. Total protein was extracted from tobacco seeds and analyzed for the methionine-rich protein by an immunoblot assay using a monoclonal antibody. Lanes numbered 3, 32, 24, and 67 contained 30 μg of seed protein from plants transformed with the chimeric methionine-rich protein gene. Protein samples from control plants, either untransformed tobacco or tobacco transformed with only the vector plasmid are shown. The last two lanes are standards containing 1 and 3 μg purified Brazil nut methionine-rich protein. From Altenbach *et al.* (1989) with permission of the authors and publisher.

14.4 Future Prospects

Research on the biochemistry and molecular biology of the seed storage protein genes in the past ten years, as briefly reviewed in this chapter, has provided the basic knowledge and technology for the current efforts to improve seed proteins through molecular approaches. It has been repeatedly demonstrated that a seed protein gene can be transferred and expressed in transgenic plants in a developmentally regulated manner; the protein product is generally synthesized, transported, and properly deposited in the seeds. The feasibility of significant enhancement of essential amino acids in a transgenic seed by introduction of an

TABLE 14.4
Amino Acid Composition (mol %) of Tobacco Seed Proteins[a]

Amino acid	Tobacco	pARC12	3	32	34	67
Lys	2.37 (0.31)	2.03 (0.27)	2.38 (0.51)	2.06 (0.37)	2.41 (0.20)	2.43 (0.34)
His	2.13 (0.11)	2.01 (0.22)	2.06 (0.22)	2.12 (0.22)	1.99 (0.20)	2.02 (0.29)
Arg	9.73 (0.30)	10.00 (0.48)	9.60 (0.31)	9.82 (0.42)	9.68 (0.17)	9.66 (0.74)
Asx	8.44 (0.17)	8.41 (0.21)	8.09 (0.19)	8.47 (0.20)	8.30 (0.17)	8.55 (0.26)
Thr	4.08 (0.08)	4.04 (0.11)	3.83 (0.06)	3.96 (0.10)	3.98 (0.10)	3.90 (0.14)
Ser	5.30 (0.16)	5.19 (0.20)	5.18 (0.23)	5.30 (0.20)	5.64 (0.43)	5.35 (0.25)
Glx	19.65 (0.65)	19.83 (1.15)	19.56 (0.37)	19.86 (0.58)	19.56 (0.40)	19.77 (0.74)
Pro	4.14 (0.71)	4.58 (0.27)	4.42 (0.48)	4.63 (0.36)	4.38 (0.43)	4.07 (1.00)
Gly	8.64 (0.39)	9.20 (1.52)	10.60 (1.15)	8.93 (0.40)	10.17 (1.11)	9.54 (0.75)
Ala	6.05 (0.91)	5.67 (0.62)	6.25 (0.92)	5.54 (0.50)	5.35 (0.44)	5.64 (0.69)
½-Cys	2.23 (0.45)	1.99 (0.32)	2.09 (0.34)	2.12 (0.35)	2.27 (0.56)	2.21 (0.42)
Val	5.71 (0.35)	5.62 (0.19)	5.15 (0.35)	5.42 (0.26)	5.27 (0.37)	5.55 (0.32)
Met	3.60 (0.23)	3.54 (0.13)	4.47 (0.31)	4.74 (0.16)	3.95 (0.14)	4.31 (0.31)
Ile	4.19 (0.42)	4.11 (0.53)	3.87 (0.31)	3.91 (0.35)	4.02 (0.33)	4.22 (0.57)
Leu	7.07 (0.55)	7.04 (0.52)	6.40 (0.27)	6.69 (0.26)	6.68 (0.19)	6.80 (0.33)
Tyr	2.63 (0.26)	2.63 (0.24)	2.42 (0.21)	2.49 (0.19)	2.57 (0.22)	2.37 (0.22)
Phe	4.00 (0.60)	4.03 (0.36)	3.54 (0.41)	3.77 (0.22)	3.67 (0.27)	3.53 (0.47)

[a]Data presented is the average of five separate hydrolyses of two different preparations of total tobacco seed protein. The standard deviation is shown in parentheses below each value. Tobacco and PARC12 represent protein samples from control plants, either untransformed tobacco or tobacco transformed with only the vector plasmid. Columns numbered 3, 32, 34, and 67 represent seed proteins from transgenic tobacco expressing the methionine-rich protein chimeric gene. From Altenbach *et al.* (1989) with permission of the authors and publisher.

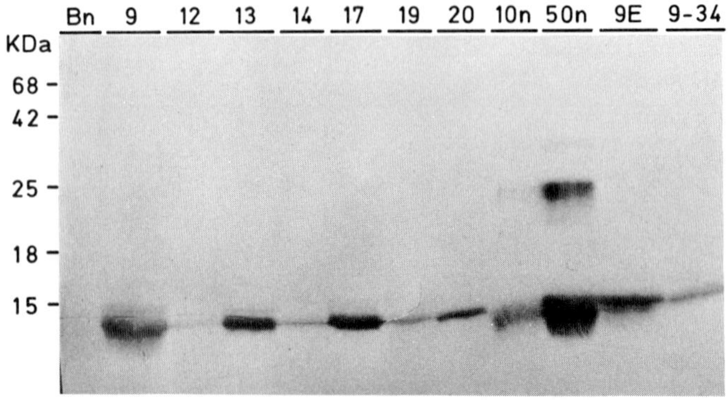

Figure 14.9 Western analysis for expression of a Brazil nut 2S albumin protein in the seeds or embryogenic line of transgenic *Brassica* plants. Total protein from *Brassica* seeds (100 μg) or embryogenic lines (50 μg) was separated by polyacrylamide gel electrophoresis, transferred to nitrocellulose, and then allowed to react with antibodies to total Brazil nut 2S albumins. Lanes: Bn, untransformed *Brassica* plant; 9, 12, 13, 14, 17, 19, and 20, individual transgenic plants; 9E, embryogenic line derived from plant 9; 10n and 50n, 10 ng and 50 ng of purified Brazil nut 2S protein; 9–34, 34-day-old seeds of plant 9. Unreduced Brazil nut 2S albumin has a molecular weight of 12,000. From Guerche *et al.* (1990) with permission of the authors and publisher.

essential amino acid-rich protein gene has also been demonstrated. However, to achieve the goal of developing major seed crops with improved nutritional quality, further progress in several research areas is critical.

First, we need to study and better understand how gene transcription is regulated. Previous experiments with transgenic plants have shown that the introduced seed protein genes are generally expressed at low levels. This amount of protein will not make a significant change in the percentage of essential amino acids in the seed. Research elucidating the structure and function of cis-acting regulatory elements and trans-acting factors may reveal mechanisms for enhancing transcription. Experiments to dissect and identify the elements controlling the level of expression are particularly relevant. In addition, studies on "position effects," gene dosage, posttranscriptional regulation including mRNA translational efficiency and stability, and protein targeting are also important. Other strategies to enhance the expression of an introduced seed storage protein are also worthy of study. For example, a compensatory increase in the synthesis of an introduced protein may be possible if a mechanism can be developed to decrease or suppress the expression and synthesis of the endogenous storage proteins.

Second, we need to know more about the structure of seed storage proteins, particularly in regard to the sequences that are important in maintaining the secondary structure, in the interactions between subunit polypeptides, and in the assembly of oligomers. This information will assist the identification of regions that have little effect on protein structure and which may be target sites for modification. Identification of variable regions in the coding sequences of seed storage protein genes is another valuable approach for this purpose. The use of the hypervariable region in the C terminus of the acidic polypeptide of the 11S soybean globulin (Dickinson *et al.*, 1990) and the variable region between the 6th and 7th cysteines of the *Arabidopsis* 2S albumin (Vandekerckhove *et al.*, 1989) as target sites for modifications are good examples of this approach.

Third, it is important to identify and isolate genes, preferably seed storage protein genes, which encode proteins containing high levels of a specific essential amino acid or a combination of these amino acids. Minor seed proteins are also potential candidates. We need to develop methods that can rapidly and reliably screen large numbers of seed samples for essential amino acid content. The gel electrophoresis method for methionine quantitation as developed by de Lumen and Kho (1987) is one example.

Fourth, and most critical, procedures for transformation and regeneration of important seed crops must be developed. Although continuous progress is being made, for example, the recent successful transformation and regeneration of fertile corn plants (Gordon-Kamm *et al.*, 1990), many important seed crops including legumes and other cereals remain resistant to these manipulations.

In the past five years, significant progress has been made in seed protein improvement based on molecular approaches. Further progress can be expected from continuing research in this area. Looking ahead, we need to consider issues concerning nutritional quality and food processing. If we take these issues into consideration during our current research and development, seeds with superior protein quality and usefulness can be engineered and produced.

References

Altenbach, S. B., Pearson, K. W., Leung, F. W., and Sun, S. S. M. (1987). *Plant Mol. Biol.* 8, 239–250.

Altenbach, S. B., Pearson, K. W., Meeker, G., Staraci, L. C., and Sun, S. S. M. (1989). *Plant Mol. Biol.* 13, 513–522.

Ampe, C., Van Damme, J., Castro, L. A. B., Sampaio, M. J. A. M., Van Montagu, M. and Vandekerchkove, J. (1986). *Eur. J. Biochem.* 159, 597–604.

Argos, P., Pedersen, K., Marks, M. D., and Larkins, B. A. (1982). *J. Biol. Chem.* 257, 9984–9990.

Argos, P., Narayana, S. V. L., and Nielsen, N. C. (1985). *EMBO J.* 4, 1111–1117.

Badenoch-Jones, J., Spencer, D., Higgins, T. J., and Millerd, A. (1981). *Planta* 153, 201–209.

Barton, K. A., Thompson, J. F., Madison, J. T., Rosenthal, R., Jarvis, N. P., and Beachy, R. N. (1982). *J. Biol. Chem.* 257, 6089–6095.

Baumlein, H., Wobus, U., Pustell, J., and Kafatos, F. (1986). *Nucl. Acids Res.* 14, 2707–2721.

Baumlein, H., Muller, A. J., Schlemann, J., Helbing, D., Manteuffel, R., and Wobus, U. (1988). *Biochem. Physiol. Pflanzen* 183, 205–210.

Beachy, R. N., Chen, Z. L., Horsch, R. B., Rogers, S. B., Hoffmann, N. J., and Fraley, R. T. (1985). *EMBO J.* 4, 3047–3053.
Bray, E. A., Naito, S., Pan, N.-S., Anderson, E., Dube, P., and Beachy, R. N. (1987). *Planta* 172, 364–370.
Bright, S. W. J., and Shewry, P. R. (1983). *CRC Critical Rev. in Plant Sci.* 1, 49–93.
Bustos, M. M., Guiltinan, M. J., Jorfano, J., Begum, D., Kalkan, F. A., and Hall, T. C. (1989). *The Plant Cell* 1, 839–853.
Callis, J., Fromm, M., and Walbot, V. (1987). *Genes and Devel.* 1, 1183–1200.
Cameron-Mills, V., and von Wettstein, D. (1980). *Carlsberg Res. Commun.* 45, 577–594.
Casey, R., and Domoney, C. (1987). *Plant Mol. Biol. Reporter* 5, 261–281.
Chappell, J., and Chrispeels, M. J. (1986). *Plant Physiol.* 81, 50–54.
Chee, P. P., Klassy, R. C., and Slightom, J. L. (1986). *Gene* 41, 47–57.
Chen, Z. L., Schuler, M. A., and Beachy, R. N. (1986). *Proc. Natl. Acad. Sci. U.S.A.* 83, 8560–8564.
Chen, Z. L., Oan, N. S., and Beachy, R. N. (1988). *EMBO J.* 7, 297–302.
Chrispeels, M. J., Higgins, T. J., and Spencer, D. (1982). *J. Biol. Chem.* 93, 306–313.
Colot, V., Robert, L. S., Kavanagh, T. A., Bevan, M. W., and Thompson, R. D. (1987). *EMBO J.* 6, 3559–3564.
De Clercq, A., Vandewiele, M., De Rycke, R., Van Damme, J., Van Montagu, N., Krebbers, E., and Vandekerckhove, J. (1990). *Plant Physiol.* 92, 899–907.
Dickinson, C. D., Evans, R. P., and Nielsen, N.C. (1988). *Nucl. Acids Res.* 16, 371.
Dickinson, C. D., Hussein, E. H. A., and Nielsen, N. C. (1989). *Plant Cell* 1, 459–469.
Dickinson, C. D., Scott, M. P., Hussein, E. H. A., Argos, P., and Nielsen, N. C. (1990). *Plant Cell* 2, 403–413.
Domoney, C., Ellis, T. H. N., and Davies, D. R. (1986). *Mol. Gen. Genet.* 202, 280–285.
Doyle, J. J., Schuler, M. A., Godette, W. D., Zenger, V., Beachy, R. N., and Slightom, J. L. (1986). *J. Biol. Chem.* 261, 9228–9238.
Ellis, T. H. N., Domoney, C., Castleton, J., Cleary, W., and Davies, D. R. (1986). *Mol. Gen. Genet.* 205, 164–169.
Ellis, J. R., Shirsat, A. H., Hepher, A., Yarwood, J. N., Gatehouse, J. A., Croy, R. R. D., and Boulter, D. (1988). *Plant Mol. Biol.* 10, 203–214.
Elliston, K. O., Imran, S., and Messing, J. (1988). *Plant Mol. Biol. Reporter* 6(2), 22–26.
FAO (1970). *FAO Nutritional Studies,* No. 24, Rome.
FAO (1973). In *FAO Nutritional Meet. Rep. Ser.* No. 52, *WHO Tech. Rep. Ser.*, No. 522, Rome.
Forde, B. G., Heyworth, A., and Kreis, M. (1985). *Nucl. Acids Res.* 13, 7323–7339.
Gallardo, D., Reina, M., Rigau,J., Boronat, A., and Palau, J. (1988). *Plant Sci.* 54, 211–218.
Gatehouse, J. A., Evans, I. M., Croy, R. R. D., and Boulter, D. (1986). *Phil. Trans. R. Soc. Lond.* B314, 367–384.
George, A. A., and de Lumen, B. O. (1991). *J. Agric. Food Chem.* 39, 224–227.
Goldberg, R. B. (1986). *Phil. Trans. R. Soc. Lond.* B314, 343–353.
Goldberg, R. B., Hoschek, G., and Vodkin, L. O. (1983). *Cell* 33, 465–475.
Goldberg, R. B., Barker, S. J., and Perez-Grau, L. (1989). *Cell* 56, 149–160.
Gordon-Kamm, W. J., Spencer, T. M., Mangano, M. L., Adams, T. R., Daines, R. J., Start, W. G., O'Brien, J. V., Chambers, S. A., Adams, W. R., Willets, N. G., Rice, T. B., Mackey, C. J., Krueger, R. W., Kausch, A. P., and Lemauz, P. G. (1990). *Plant Cell* 7, 603–618.
Greenwood, J. S., and Chrispeels, M. J. (1985). *Plant Physiol.* 79, 65–71.
Guerche, P., De Almeida, E. R. P., Schwarztein, M. A., *et al.* (1990). *Mol. Gen. Genet.* 221, 306–314.
Heidecker, G., and Messing, J. (1986). *Ann. Rev. Plant Physiol.* 37, 439–466.
Herman, E. M., Shannon, L. M., and Chrispeels, M. J. (1986). In *Molecular Biology of Seed*

Storage Proteins and Lectins. (Ed. by L. M. Shannon and M. J. Chrispeels), pp. 163–173. Rockville, Maryland: Amer. Soc. Plant Physiol.

Higgins, T. J. V. (1984). *Ann. Rev. Plant Physiol.* 35, 191–221.

Higgins, T. J. V., Chandler, P. M., Randall, P. J., Spencer, D., Beach, L. R., Blagrove, R. J., Kortt, A. A., and Inglis, A. S. (1986). *J. Biol. Chem.* 261, 11124–11130.

Hoffman, L. M., and Donaldson, D. D. (1985). *EMBO J.* 4, 883–889.

Hoffman, L. M., Donaldson, D. D., Bookland, R., Rashka, K., and Herman, E. M. (1987). *EMBO J.* 6, 3213–3221.

Hoffman, L. M., Donaldson, D. D., and Herman, E. M. (1988). *Plant Mol. Biol.* 11, 717–729.

Hurkman, W. J., and Beevers, L. (1982). *Plant Physiol.* 69, 1414–1417.

Jofuku, K. D., Okamura, J., and Goldberg, R. B. (1987). *Nature (London)* 328, 734–737.

Jones, J. D. G., Dunsmuir, P., and Bedbrook, J. (1985). *EMBO J.* 4, 2411–2418.

Kasarda, D. D., Okita, T. W., Bernardin, J. E., Baecker, P. A., immo, C. C., Lew, E. J.-L., Dietler, M. D., and Green, F. C. (1984). *Proc. Natl. Acad. Sci. U.S.A.* 81, 4712–4716.

Kim, W. T., Franceschi, V. R., Krishnan, H. B., and Okita, T. W. (1988). *Planta* 176, 173–182.

Kirihara, J. A., Tetri, J. B., and Messing, J. (1988a). *Gene,* 71, 359–370.

Kirihara, J. A., Hunsperger, J. P., Mahoney, W. C., and Messing, J. (1988b). *Mol. Gen. Genet.* 211, 477–484.

Kreis, M., Shewry, P. R., Forde, B. G., Rahman, S., and Miflin, B. J. (1983). *Cell* 34, 161–167.

Kreis, M., Shewry, P. R., Firde, B. G., and Miflin, B. J. (1985). *Oxford Surv. Plant Mol. Cell Biol.* 2, 253–317.

Kreis, M., Williamson, M. S., Forde, J., Schmutz, D., Clark, J., Buxton, B., Pywell, J., Marris, C., Henderson, J., Harris, N., Shewry, P. R., Forde, B. G., and Miflin, B. J. (1986). *Phil. Trans. R. Soc. Lond.* B314, 355–365.

Krishnan, H. B., Franceschi, V. R., and Okita, T. W. (1986). *Planta* 169, 471–480.

Kuhlemeier, C., Green, P. J., and Chua, N. H. (1987). *Ann. Rev. Plant Physiol.* 38, 221–257.

Larkins, B. A., and Hurkman, W. J. (1978). *Plant Physiol.* 62, 256–263.

Larkins, B. A., Pedersen, K., Marks, M. D., and Wilson, D. R. (1984). *Trends Biochem. Sci.* 9, 306–308.

Lawrence, M. C., Suzuki, E., Varghese, J. N., *et al.*, (1990). *EMBO J.* 9, 9–15.

Lawton, M. A., Tierney, M. A., Nakamura, I., Anderson, E., Komeda, Y., Dube, P., Hoffman, N., Fraley, R. T., and Beachy, R. N. (1987). *Plant Mol. Biol.* 9, 315–324.

Lilley, G. G., Caldwell, J. B., Kortt, A. A., Higgins, T. J., and Spencer, D. (1989). In *Proceedings of the World Congress on Vegetable Protein Utilization in Human Foods and Animal Feedstuffs*. (Ed. by T. H. Applewhite), pp. 497–502. Champaign, Illinois: American Oil Chemists' Soc.

Lycett, G. W., Delauney, A. J., Zhao, W., *et al.* (1984). *Plant Mol. Biol.* 3, 91–96.

Maier, U. G., Brown, J. W. S., Toloczyki, C., and Feix, G. (1987). *EMBO J.* 6, 17–22.

Maier, U. G., Brown, J. W. S., Schmitz, L. M., *et al.* (1988). *Mol. Gen. Genet.* 212, 241–245.

Marks, M. D., Pedersen, K., Wilson, D. R., DiFonzo, N., and Larkins, B. A. (1984). *Curr. Top. Plant Biochem. Physiol.* 3, 9–18.

Marris, C., Gallois, P., Copley, J., and Kreis, M. (1988). *Plant Mol. Biol.* 10, 359–366.

Matta, N. K., and Gatehouse, J. A. (1982). *Heredity* 48, 383–392.

Meinke, D. W., Chen, J., and Beachy, R. N. (1981). *Planta* 153, 130–139.

Moreira, M. A., Hermodson, M. A., Larkins, B. A., and Nielsen, N. C. (1979). *J. Biol. Chem.* 254, 9921.

Nelson, O. E. (1968). In *New Approaches to Breeding for Improved Plant Protein*. Proc. of a panel meeting organized by FAO/IAEA, Sweden.

Nelson, O. E. (1980). In *Advances in Cereal Science and Technology*. (Ed. by Y. Pomeranz), pp. 41–71. St. Paul, Minnesota: Amer. Assoc. Cereal Chem.

Nielsen, N. C., Dickinson, C. D., Cho, T. J., *et al.*, (1989). *The Plant Cell* 1, 313–328.
Ohtani, T., Galili, G., Wallace, J. C., Thompson, G. A., and Larkins, B. A. (1991). *Plant Mol. Biol.* 16, 117–128.
Okamuro, J. K., Jofuku, K. D., and Goldberg, R. N. (1986). *Proc. Natl. Acad. Sci. USA*. 83, 8240–8244.
Okita, T. W., Cheesbrough, V., and Reeves, C. D. (1985). *J. Biol. Chem.* 260, 8203–8213.
Payne, P. I. (1983). In *Seed Proteins.* (Ed. by J. Daussant, J. Mosse, and J. Vaughan), p. 223. London: Academic Press, Inc.
Pedersen, K., Argos, P., Naravana, S. V. L., and Larkins, B. A. (1986). *J. Biol. Chem.* 261, 6278–6284.
Phillips, R. L., and McClure, B. A. (1985). *Cereal Chem.* 62, 213–218.
Plietz, P., and Damaschun, G. (1986). *Stud. Biophys.* 3, 153–173.
Plietz, P., Damaschun, G., Muller, J. J., and Schlesier, B. (1983a). *FEBS Lett.* 162, 43–46.
Plietz, P., Damaschun, G., Zirrer, D., Gast, K., and Schlesier, B. (1983b). *Intl. J. Biol. Macromol.* 5, 356–360.
Plietz, P., Drescher, B., and Damaschun, G. (1988). *Biochem. Physiol. Pflanzen* 183, 199–203.
Radke, S. E., Andrews, B. M., Moloney, M. M., Crouch, M. L., Kridl, J. C., and Knauf, V. C. (1988). *Theor. Appl. Genet.* 75, 685–694.
Reeck, G. R., and Hedgcoth, C. (1985). *FEBS Lett.* 180, 291–294.
Reichelt, R., Schwenke, K.-D., Konig, T., Pahtz, W., and Wangermann, G. (1980). *Biochem. Physiol. Pflanzen* 175, 653–663.
Schernthaner, J. P., Matzke, M. A., and Matzke, A. J. M. (1988). *EMBO J.* 7, 1249–1255.
Sengupta-Gopalan, C., Reichert, N. A., Barker, R. F., Hall, T. C., and Kemp, J. D. (1985). *Proc. Natl. Acad. Sci. U.S.A.* 82, 3320–3324.
Shotwell, M. A., and Larkins, B. A. (1989). In *The Biochemistry of Plants, Comprehensive Treatise,* Vol. 15. (Ed. by A. Marcus), pp. 297–345. San Diego: Academic Press, Inc.
Shutov, A. D., and Vaintraub, I. A. (1987). *Phytochemistry* 26, 1557–1566.
Slightom, J. L., Sun, S. S. M., and Hall, T. C. (1983). *Proc. Natl. Acad. Sci. U.S.A.* 80, 3320–3324.
Soave, C., Rhigetti, R., Difonzo, N., and Salamini, F. (1981). *Genetics* 97, 363–377.
Soave, C., Reggiani, R., Difonzo, N., and Salamini, F. (1982). *Biochem. Genet.* 11, 1027–1037.
Sturm, A., Voelker, T. A., Herman, E. M., and Chrispeels, M. J. (1988). *Planta* 175, 170–183.
Sun, S. S. M., Slightom, J. L., and Hall, T. C. (1981). *Nature (London)* 289, 37–41.
Sun, S. S. M., Leung, F. W., and Tomic, J. C. (1987a). *J. Agri. Food Chem.* 35, 232–235.
Sun, S. S. M., Altenbach, S. B., and Leung, F. W. (1987b). *Eur. J. Biochem.* 159, 597–604.
Taylor, J. R. N., Schussler, L., and Liebenberg, N. v. d. W. (1985). *S. Afr. J. Bot.* 51, 35–40.
Thanh, V. H., and Shibasaki, K. (1977). *Biochim. Biophys. Acta.* 490, 370–384.
Ueng, P., Galli, G., Sapanara, V., *et al.*, (1988). *Plant Physiol.* 86, 1281–1285.
Valentini, G., Soave, C., and Ottaviano, E. (1979). *Heredity* 42, 33–40.
Vandekerckhove, J., Van Damme, J., Van Lijsebettens, M., *et al.*, (1989). *Bio/Technology* 7, 929–932.
Vodkin, L. O., Rhodes, P. R., and Goldberg, R. B. (1983). *Cell* 34, 1023–1031.
Voelker, T. A., Herman, E. M., and Chrispeels, M. J. (1989). *Plant Cell* 1, 95–104.
Wallace, J. C., Galli, G., Kawata, E. E., *et al.*, (1988). *Science* 240, 662–664.
Wilson, D. M., and Larkins, B. A. (1984). *J. Mol. Evol.* 29, 330–340.
Yamaguchi, M. (1980). In *World Vegetables.* (Ed. by M. Yamaguchi), p. 395. Westport, Connecticut: AVI.
Yamauchi, F., Sato, K., and Yamagishi, T. (1984). *Agric. Biol. Chem.* 48, 645–650.
Yang, M. S., Espinoza, N. O., Nagpala, P. G., *et al.*, (1989). *Plant Sci.* 64, 99–111.
Youle, R. J., and Huang, A. H. C. (1981). *Amer. J. Bot.* 68, 44–48.

Index

ISBN 0-12-428781-6

90065

9 780124 287815